Communications
in Computer and Information Science   25

Ana Fred   Joaquim Filipe   Hugo Gamboa (Eds.)

Biomedical Engineering Systems and Technologies

International Joint Conference, BIOSTEC 2008
Funchal, Madeira, Portugal, January 28-31, 2008
Revised Selected Papers

 Springer

Volume Editors

Ana Fred
Hugo Gamboa
Instituto Superior Técnico (IST)
and Instituto de Telecomunicações (IT)
Avenida Rovisco Pais 1, 1049-001 Lisbon, Portugal
E-mail: {afred, hugo.gamboa}@lx.it.pt

Joaquim Filipe
Institute for Systems and Technologies of Information,
Control and Communication (INSTICC)
and Instituto Politécnico de Setúbal (IPS)
Department of Systems and Informatics
Rua do Vale de Chaves, Estefanilha, 2910-761 Setúbal, Portugal
E-mail: jfilipe@insticc.org

Library of Congress Control Number: 2008940839

CR Subject Classification (1998): J.3, I.2, C.3, I.5.4

ISSN   1865-0929

ISBN-10   3-540-92218-0 Springer Berlin Heidelberg New York
ISBN-13   978-3-540-92218-6 Springer Berlin Heidelberg New York

springer.com

© Springer-Verlag Berlin Heidelberg 2008
Printed in Germany

Typesetting: Camera-ready by author, data conversion by Scientific Publishing Services, Chennai, India
Printed on acid-free paper   SPIN: 12587520   06/3180   5 4 3 2 1 0

Preface

This book contains the best papers of the First International Joint Conference on Biomedical Engineering Systems and Technologies (BIOSTEC 2008), organized by the Institute for Systems and Technologies of Information Control and Communication (INSTICC), technically co-sponsored by the IEEE Engineering in Medicine and Biology Society (EMB), ACM SIGART and the Workflow Management Coalition (WfMC), in cooperation with AAAI.

The purpose of the International Joint Conference on Biomedical Engineering Systems and Technologies is to bring together researchers and practitioners, including engineers, biologists, health professionals and informatics/computer scientists, interested in both theoretical advances and applications of information systems, artificial intelligence, signal processing, electronics and other engineering tools in knowledge areas related to biology and medicine.

BIOSTEC is composed of three co-located conferences; each specializes in one of the aforementioned main knowledge areas, namely:

- BIODEVICES (International Conference on Biomedical Electronics and Devices) focuses on aspects related to electronics and mechanical engineering, especially equipment and materials inspired from biological systems and/or addressing biological requirements. Monitoring devices, instrumentation sensors and systems, biorobotics, micro-nanotechnologies and biomaterials are some of the technologies addressed at this conference.

- BIOSIGNALS (International Conference on Bio-inspired Systems and Signal Processing) is a forum for those studying and using models and techniques inspired from or applied to biological systems. A diversity of signal types can be found in this area, including image, audio and other biological sources of information. The analysis and use of these signals is a multidisciplinary area including signal processing, pattern recognition and computational intelligence techniques, amongst others.

- HEALTHINF (International Conference on Health Informatics) promotes research and development in the application of information and communication technologies (ICT) to healthcare and medicine in general and to the specialized support to persons with special needs in particular. Databases, networking, graphical interfaces, intelligent decision support systems and specialized programming languages are just a few of the technologies currently used in medical informatics. Mobility and ubiquity in healthcare systems, standardization of technologies and procedures, certification, privacy are some of the issues that medical informatics professionals and the ICT industry in general need to address in order to further promote ICT in healthcare.

The joint conference, BIOSTEC, received 494 paper submissions from more than 40 countries in all continents. Of these, 65 papers were published and presented as full papers, i.e., completed work (8 pages/30' oral presentation), 189 papers reflecting work-in-progress or position papers were accepted for short presentation, and another 86 contributions were accepted for poster presentation. These numbers, leading to a "full-paper" acceptance ratio below 14% and a total oral paper presentation acceptance ratio below 52%, show the intention of preserving a high-quality forum for the next editions of this conference. This book includes revised and extended versions of a strict selection of the best papers presented at the conference.

The conference included a panel and six invited talks delivered by internationally distinguished speakers, namely: Sergio Cerutti, Kevin Warwick, F. H. Lopes da Silva, Vipul Kashyap, David Hall and Albert Cook.

We must thank the authors, whose research and development efforts are recorded here. We also thank the keynote speakers for their invaluable contribution and for taking the time to synthesize and prepare their talks. The contribution of all Program Chairs of the three component conferences was essential to the success of BIOSTEC 2008. Finally, special thanks to all the members of the INSTICC team, whose collaboration was fundamental for the success of this conference.

September 2008

Ana Fred
Joaquim Filipe
Hugo Gamboa

Organization

Conference Chair

Ana Fred   IST - Technical University of Lisbon, Portugal
Joaquim Filipe   Polytechnic Institute of Setúbal / INSTICC, Portugal
Hugo Gamboa   Institute of Telecommunications, Lisbon, Portugal

Program Co-chairs

Luis Azevedo   Instituto Superior Técnico, Portugal (HEALTHINF)
Pedro Encarnação   Universidade Católica Portuguesa, Portugal
       (BIOSIGNALS)
Teodiano Freire Bastos Filho   Universidade Federal do Espírito Santo, Brazil
       (BIODEVICES)
Hugo Gamboa   Instituto de Telecomunicações, Portugal
       (BIODEVICES)
Ana Rita Londral   ANDITEC, Portugal (HEALTHINF)
António Veloso   FMH, Universidade Técnica de Lisboa, Portugal
       (BIOSIGNALS)

Organizing Committee

Paulo Brito   INSTICC, Portugal
Marina Carvalho   INSTICC, Portugal
Bruno Encarnação   INSTICC, Portugal
Vitor Pedrosa   INSTICC, Portugal
Vera Rosário   INSTICC, Portugal
Mónica Saramago   INSTICC, Portugal

BIODEVICES Program Committee

Julio Abascal, Spain
Amiza Mat Amin, Malaysia
Rodrigo Varejão Andreão, Brazil
Ramon Pallàs Areny, Spain
Luis Azevedo, Portugal
Rafael Barea, Spain
Antonio Barrientos, Spain
Roberto Boeri, Argentina
Luciano Boquete, Spain
Susana Borromeo, Spain

Eduardo Caicedo Bravo, Colombia
Enrique A. Vargas Cabral, Paraguay
Sira Palazuelos Cagigas, Spain
Leopoldo Calderón, Spain
Alicia Casals, Spain
Gert Cauwenberghs, USA
Ramón Ceres, Spain
Luca Cernuzzi, Paraguay
Alberto Cliquet Jr., Brazil
Fernando Cruz, Portugal

Pedro Pablo Escobar, Argentina
Marcos Formica, Argentina
Juan Carlos Garcia Garcia, Spain
Gerd Hirzinger, Germany
Jongin Hong, UK
Giacomo Indiveri, Switzerland
Dinesh Kumar, Australia
Eric Laciar Leber, Argentina
José Luis Martinez, Spain
Manuel Mazo, Spain
Paulo Mendes, Portugal
José del R. Millán, Switzerland
Joseph Mizrahi, Israel
Raimes Moraes, Brazil
Pedro Noritomi, Brazil
Maciej J. Ogorzalek, Poland
Kazuhiro Oiwa, Japan
José Raimundo de Oliveira, Brazil
Evangelos Papadopoulos, Greece
Laura Papaleo, Italy

Francisco Novillo Parales, Ecuador
Qibing Pei, USA
José M. Quero, Spain
Antonio Quevedo, Brazil
Alejandro Ramirez-Serrano, Canada
Adriana María Rios Rincón, Colombia
Joaquin Roca-Dorda, Spain
Adson da Rocha, Brazil
Joel Rodrigues, Portugal
Carlos F. Rodriguez, Colombia
Mario Sarcinelli-Filho, Brazil
Fernando di Sciascio, Argentina
Jorge Vicente Lopes da Silva, Brazil
Amir M. Sodagar, USA
Juan Hernández Tamames, Spain
Mário Vaz, Portugal
António Veloso, Portugal
Peter Walker, UK
Jonathan Wolpaw, USA
Miguel Yapur, Ecuador

BIOSIGNALS Program Committee

Andrew Adamatzky, UK
Cedric Archambeau, UK
Magdy Bayoumi, USA
Peter Bentley, UK
Paolo Bonato, USA
Marleen de Bruijne, Denmark
Zehra Cataltepe, Turkey
Gert Cauwenberghs, USA
Mujdat Cetin, Turkey
Wael El-Deredy, UK
Eran Edirisinghe, UK
Eugene Fink, USA
Luc Florack, The Netherlands
David Fogel, USA
Alejandro Frangi, Spain
Sebastià Galmés, Spain
Aaron Golden, Ireland
Rodrigo Guido, Brazil
Bin He, USA
Roman Hovorka, UK
Helmut Hutten, Austria
Christopher James, UK

Lars Kaderali, Germany
Gunnar W. Klau, Germany
Alex Kochetov, Russian Federation
T. Laszlo Koczy, Hungary
Georgios Kontaxakis, Spain
Igor Kotenko, Russian Federation
Narayanan Krishnamurthi, USA
Arjan Kuijper, Austria
Andrew Laine, USA
Anna T. Lawniczak, Canada
Jason J. S. Lee, Taiwan,
    Republic of China
Kenji Leibnitz, Japan
Marco Loog, Denmark
David Lowe, UK
Mahdi Mahfouf, UK
Luigi Mancini, Italy
Elena Marchiori, The Netherlands
Fabio Martinelli, Italy
Martin Middendorf, Germany
Mariofanna Milanova, USA
Charles Mistretta, USA

Gabor Mocz, USA
Kayvan Najarian, USA
Tadashi Nakano, USA
Asoke K. Nandi, UK
Antti Niemistö, USA
Maciej J. Ogorzalek, Poland
Kazuhiro Oiwa, Japan
Jean-Chistophe Olivo-Marin, France
Ernesto Pereda, Spain
Leif Peterson, USA
Gert Pfurtscheller, Austria
Vitor Fernão Pires, Portugal
Chi-Sang Poon, USA
José Principe, USA
Chi-Sang Poon, USA
Nikolaus Rajewsky, Germany
Dick de Ridder, The Netherlands
Joel Rodrigues, Portugal
Marcos Rodrigues, UK
Virginie Ruiz, UK

Heather Ruskin, Ireland
William Zev Rymer, USA
Gerald Schaefer, UK
Dragutin Sevic, Serbia
Iryna Skrypnyk, Finland
Alan A. Stocker, USA
Jun Suzuki, USA
Andrzej Swierniak, Poland
Boleslaw Szymanski, USA
Asser Tantawi, USA
Lionel Tarassenko, UK
Gianluca Tempesti, UK
Anna Tonazzini, Italy
Duygu Tosun, USA
Bart Vanrumste, Belgium
Didier Wolf, France
Andrew Wood, Australia
Guang-Zhong Yang, UK
Eckart Zitzler, Switzerland

HEALTHINF Program Committee

Osman Abul, Turkey
Arnold Baca, Austria
Iddo Bante, The Netherlands
Jyh-Horng Chen, Taiwan,
    Republic of China
Kay Connelly, USA
Amar Das, USA
Adrie Dumay, The Netherlands
Eduardo B. Fernandez, USA
Alexandru Floares, Romania
Jose Fonseca, Portugal
Toshio Fukuda, Japan
M. Chris Gibbons, USA
David Greenhalgh, UK
Jin-Kao Hao, France
Tin Ho, USA
John H. Holmes, USA
Chun-Hsi Huang, UK
Benjamin Jung, Germany
Stavros Karkanis, Greece
Andreas Kerren, Sweden
Georgios Kontaxakis, Spain

Nigel Lovell, Australia
Andrew Mason, New Zealand
Boleslaw Mikolajczak, USA
Ahmed Morsy, Egypt
Laura Roa, Spain
Jean Roberts, UK
Joel Rodrigues, Portugal
Marcos Rodrigues, UK
George Sakellaropoulos, Greece
Meena Kishore Sakharkar, Singapore
Ovidio Salvetti, Italy
Nickolas Sapidis, Greece
Sepe Sehati, UK
Boris Shishkov, The Netherlands
Iryna Skrypnyk, Finland
John Stankovic, USA
Ron Summers, UK
Adrian Tkacz, Poland
Athanasios Vasilakos, Greece
Aristides Vagelatos, Greece
Taieb Znati, USA

BIODEVICES Auxiliary Reviewers

José M. R. Ascariz, Spain
Leandro Bueno, Spain
Natalia López Celan, Argentina
Andre Ferreira, Brazil
Vicente González, Paraguay
Muhammad Suzuri Hitam, Malaysia

Joaquin Roca-Gonzalez, Spain
Rober Marcone Rosi, Brazil
Evandro Ottoni Teatini Salles, Brazil
Hugo Humberto Plácido Silva, Portugal
Andrés Valdéz, Argentina

BIOSIGNALS Auxiliary Reviewers

Qi Duan, USA
Soo-Yeon Ji, USA
Yuri Orlov, Singapore

Ting Song, USA
Bruno N. Di Stefano, Canada

HEALTHINF Auxiliary Reviewers

Sara Colantonio, Italy
Teduh Dirgahayu, The Netherlands
Ana Sofia Fernandes, Portugal

Hailiang Mei, The Netherlands
Davide Moroni, Italy
John Sarivougioukas, Greece

Invited Speakers

Sérgio Cerutti
Kevin Warwick
Fernando Henrique Lopes da Silva
Vipul Kashyap

David Hall

Albert Cook

Polytechnic University of Milan, Italy
University of Reading, UK
University of Amsterdam, The Netherlands
Partners HealthCare System, Clinical
    Informatics R&D, USA
Research Triangle Institute in North Carolina,
    USA
University of Alberta, Faculty of
    Rehabilitation Medicine, Canada

Table of Contents

Part II: BIOSIGNALS

Part III: HEALTHINF

Invited Papers

Using the Web and ICT to Enable Persons with Disabilities

Albert M. Cook

Faculty of Rehabilitation Medicine
University of Alberta
Edmonton, Alberta
Canada
al.cook@ualberta.ca

Abstract. In order to lead full and productive lives, persons with disabilities need to have the same access to information and communication systems as the rest of the population. Advances in information and communication technologies (ICT) are occurring quickly, and the capability of technologies to meet the needs of persons with disabilities is growing daily. Future developments in assistive technologies (AT) and the successful application of these technologies to meet the needs of people who have disabilities are dependent on exploitation of these ICT advances. The goal for ICT universal design (or design for all) is to have an environment with enough embedded intelligence to be easily adaptable to the varying cognitive, physical and sensory skills of a wide range of individual's in order to meet their productivity, leisure and self care needs. If ICT advances are not adaptable enough to be accessible to persons with disabilities it will further increase the disparity between those individuals and the rest of the population leading to further isolation and economic disadvantage. On the other hand, availability of these technologies in a transparent way will contribute to full inclusion of individuals who have disabilities in the mainstream of society.

Keywords: Assistive technologies, information and computer technologies, persons with disabilities.

1 ICT and Persons with Disabilities Technology and Progress

Societal Progress requires change, much of which is accomplished through advances in technology. In his book, *A Short History of Progress*, Ronald Wright [1] points out that this characteristic has been true for millions of years as societies have advanced through greater utilization of technology.

Wright goes on to describe the problems that technology typically creates such as over consumption, environmental ruin, and separation of classes. These problems are amplified for people who have disabilities, and they lead to a gap in the access to work, self care and community participation for persons with disabilities compared to the general population. Since people with disabilities often depend on technologies for societal participation, the lack of availability of accessible technology or the obsolescence of accessible technologies isolates them further. This is an extension of the

A. Fred, J. Filipe, and H. Gamboa (Eds.): BIOSTEC 2008, CCIS 25, pp. 3–18, 2008.
© Springer-Verlag Berlin Heidelberg 2008

concept of the "digital divide" that separates people along socioeconomic lines based on their access to ICT. I refer to it as the "disability gap".

2 Advances in Information and Communication Technologies (ICT)

The 21st Century is characterized by a continuous move from a machine-based to a knowledge based economy [2]. In this shift, the basis of competence is changing to knowledge skills from machine skills. Information currently amounts to 75% of value added to products, and this will continually increase. Connectivity will be the key to business success. There is also a move from a regional or national scope of business influence to a global scope, in which virtual networks dictate organizational structures.

Key players in business development are becoming communication suppliers with the move from host-based to network based systems. Telephone, cable TV and internet service providers control commercial growth. Along with these changes networks will become more graphically-based moving increasingly from text-based systems. In order to lead full and productive lives, persons with disabilities need to have the same access to this new information and communication system as the rest of the population.

2.1 What Can We Expect from Technology in the Next 20 Years?

The cost of information technology is continually dropping for comparable or increased computing power and speed. There is also a greater understanding of the biological/physical interface for the control of computers. The human computer interface (HCI) is being developed to be more human-like, more user oriented and more intelligent-providing additional capabilities for searching, processing and evaluating information.

There are a number of changes that are likely to occur over the next few years [3]. There will be an increase I automated transactions between individuals and organizations enabling people to complete all transactions without face-to-face interactions. It is expected that we will achieve equalized access to the web and information between the developed and developing world. Embedded systems will dramatically increase with application such as "intelligence in the doorknob" that recognizes the owner and doesn't require key manipulation. We are likely to see much greater understanding of the biological to physical interface for the control of computers.

2.2 Changes in Mainstream Tech with AT Implications

There are many examples of emerging mainstream technologies with potential for assisting people with disabilities to access ICT systems. A few of these are described in this section.

Display-based assistive technologies present an array of choices for a user to select from [4]. This often referred to as scanning since the choices are highlighted sequentially and then chosen using some sort of gross movement. One of the problems associated with this approach is that there must be a physical display for making selections.

This often requires the overall system to be larger and more bulky or places a display between a user and a communication partner. A new development is a direct retinal display that creates image that overlays view of real object [5]. The retinal display is low powered because it is shined on retina directly. Scanning light into the eye allows the image to overlay an object such as a communication partner's face-enabling eye contact and small size. The scanning array could be the retinal image, since display scans across the retina power levels can be kept low kept low for safety. Another development is 3-D displays that create a more intuitive view of objects, events and activities [5]. This type of display may be helpful to individuals who have cognitive disabilities. It might also create new challenges for individuals with visual limitations.

Embedded automatic speech recognition is being developed for PDAs because of the need for keyboards with more and more functions and the limitations of very small keyboards [6]. This feature could be very useful to reduce individuals who have limited hand function or for those who cannot see the keyboard to make entries.

3 Changes in Rehabilitation Practice

Rehabilitation practice is moving from institutional to community-based services. Patients are discharged earlier form acute care into acute rehabilitation and then into the community for a major portion of their rehabilitation. Accompanying this change is a movement from a "medical model" of rehabilitation in which the emphasis is on recovery via surgery or medical care to a social model of disability in which the emphasis is on maximizing participation in society. These changes have been accompanied by an increasingly active role for consumers of assistive technologies in selection and application of these technologies. These changes have profound implications for rehabilitation engineering. Devices must be self sufficient since they have to operate in the community and cannot be dependent on expert staff to monitor and adjust them repeatedly. The increasing role for consumers means that persons with disabilities need to be included in all aspects of research and development form conception to final testing. We can no longer justify a role for persons with disabilities as only the final evaluator of our technologies.

4 Meeting the ICT Needs of Persons with Disabilities

Over the centuries, our ability to make tools is what distinguishes us as human, but our tools ultimately control us by making us dependent on them [1]. This dependence is less optional for people who have disabilities.

4.1 Impact of Technology Advances on People Who Have Disabilities

Technology advances increase the gap between people who have disabilities and those who don't [1]. All societies become hierarchical with an upward concentration of wealth (including aggregations of technology tools) that ensures that "there can never be enough to go around", and this disparity contributes to the "digital divide" and the "disability gap". As advances occur more quickly, the gap widens faster and the people who are poor and/or disabled loose out even more completely and faster.

This is a characteristic of cultural and societal "progress" over centuries-technology drives change, and creates both positive and negative outcomes in the process.

The prognosis is not good for people with disabilities unless there is considerable effort to keep them connected to ICT and thereby to commerce, employment and personal achievement. There two fundamental approaches to this problem (1) make mainstream technologies accessible to people who have disabilities, or (2) design special purpose technologies specifically for people with disabilities. The former approach is referred to as *universal design* or *design for all*. The second approach utilizes *assistive technologies*.

4.2 Implications for Assistive Technologies

Access to ICT for people with disabilities is a significant global problem, and it has major implications for assistive technologies. There is a constant challenge to keep ICT systems accessible to persons who have disabilities as mainstream advances occur and adaptations become potentially incompatible with the new systems. Communication technologies change rapidly, and each change may result in the need to re-design accessible interfaces. We are closer to goal of having assistive technology adaptations available when the mainstream consumer product ships, but there are still many problems with "workarounds" necessary to make mainstream operating system, productivity software and internet access accessible to people with disabilities.

Development and maintenance of access to ICT must be driven by the needs of people with disabilities. Developments which broaden the scope, applicability and usability of the human technology interface will be driven, at least in part by the needs of people who have disabilities. The Internet (e-mail and chat rooms) have the advantage of anonymity, and this can be a major benefit to individuals who have disabilities. Because the person's disability is not immediately visible, people who have disabilities report that they enjoy establishing relationships with people who experience them first as a person and then learn of their disability. For example, Blackstone,[7] describes some of the advantages of e-mail for individuals who have disabilities. Since the receiver of the message reads it at a later time composition can be at a slower speed. The person with a disability can communicate with another person without someone else being present, establishing a greater sense of privacy than situations in which an attendant is required. It is also possible to work form any location-avoiding some transportation problems.

4.3 Universal Design

Increasingly, commercial products are being designed to be usable by all people, to the greatest extent possible, without the need for adaptation or specialized design (NC State University, The Center for Universal Design, 1997).

4.3.1 General Principles of Universal Design
Features are built into products to make them more useful to persons who have disabilities (e.g., larger knobs; a variety of display options--visual, tactile, auditory; alternatives to reading text--icons, pictures) are built into the product. This is much less expensive than modifying a product after production to meet the needs of a person

with a disability. The North Carolina State University Center for Universal Design, in conjunction with advocates of universal design, have compiled a set of principles of universal design, shown in Box 1. This center also maintains a Web site on universal design (www.design.ncsu.edu/cud).

ONE: EQUITABLE USE
The design is useful and marketable to people with diverse abilities.
TWO: FLEXIBILITY IN USE
The design accommodates a wide range of individual prefer-ences and abilities.
THREE: SIMPLE AND INTUITIVE USE
Use of the design is easy to understand, regardless of the user's experience, knowledge, language skills, or current concentration level.
FOUR: PERCEPTIBLE INFORMATION
The design communicates necessary information effectively to the user, regardless of ambient conditions or the user's sensory abilities.
FIVE: TOLERANCE FOR ERROR
The design minimizes hazards and the adverse consequences of accidental or unintended actions.
SIX: LOW PHYSICAL EFFORT
The design can be used efficiently and comfortably and with a minimum of fatigue.
SEVEN: SIZE AND SPACE FOR APPROACH AND USE
Appropriate size and space is provided for approach, reach, manipulation, and use regardless of user's body size, posture, or mobility.

Box 1. Principles of Universal Design From North Carolina State University, The Center for. Universal Design, 1997

4.3.2 Universal Design for ICT

In universal design for ICT the barriers are technological rather than political and economic barriers that characterize architectural and commercial product design [8]. The goal of universal design for ICT is to have an environment with enough embed-ded intelligence to be easily adaptable. The features of future information services are that there will be no clearly predefined service and little distinction between interper-sonal communication and access to information. Services will need to be highly inter-active, inherently multimedia, sensory multimodal (i.e., access via auditory or visual means is equally possible). To achieve this cooperation between users or representa-tives of users is critical in a variety of contexts of use. The overall goal is to have access to information involving communities of users with a wide range of motor, sensory and cognitive skills.

In addition to Universal Design for ICT, access to capabilities of mainstream tech-nologies includes individualized assistive technologies that are easily – customized. This in return requires an increased understanding of the biological/physical interface for the control of assistive technologies and expanded availability of embedded systems networks.

4.4 A Working Definition of Assistive Technologies

The *International Classification of Functioning, Disability and* Health (ICF) is a system developed by the World Health Organization (WHO) [9] that is designed to describe and classify health and health related states. These two domains are described by body factors (body structures and functions) and individual and societal elements (activities and participation) [9]. The ICF recognizes two contextual factors that modify health and health related states: the environment and personal factors [9]. Environmental elements include assistive technologies in relation to activities of daily living, mobility, communication, religion and spirituality as well as in specific contexts such as education, employment and culture, recreation and sport [9]. Other environmental elements such as access to public and private buildings, and the natural and built outdoor environments, also have implications for assistive technologies.

A commonly used definition of assistive technology is from the Technical Assistance to the States Act in the United States (Public Law (PL) 100-407): *Any item, piece of equipment or product system whether acquired commercially off the shelf, modified, or customized that is used to increase, maintain or improve functional capabilities of individuals with disabilities.*

4.4.1 Hard and Soft Technologies
Odor [10] has distinguished between hard technologies and soft technologies. Hard technologies are readily available components that can be purchased and assembled into assistive technology systems. The main distinguishing feature of hard technologies is that they are tangible. On the other hand, soft technologies are the human areas of decision making, strategies, training, concept formation, and service delivery as described earlier in this chapter. Soft technologies are generally captured in one of three forms: (1) people, (2) written, and (3) computer. These aspects of technology, without which the hard technology cannot be successful, are much harder to obtain. Soft technologies are difficult to acquire because they are highly dependent on human knowledge rather than tangible objects. This knowledge is obtained slowly through formal training, experience, and textbooks such as this one. The development of effective strategies of use also has a major effect on assistive technology system success. Initially the formulation of these strategies may rely heavily on the knowledge, experience, and ingenuity of the assistive technology practitioner. With growing experience, the assistive technology user originates strategies that facilitate successful device use. There is a false belief that progress is solely driven by "hard" technological change The gap between the general public and persons with disabilities can only be closed by gains in both soft and hard technologies.

4.4.2 Mainstream Technologies to Specially Designed Technologies: A Range of Options
As illustrated in Figure 1, the needs of people with disabilities can be met in a number of ways. Off the shelf "standard" (i.e., mainstream technologies) commercially available devices (especially those designed using the principles of universal design) can

often be used by people with a variety of disabilities. For example, standard personal computers designed for the general population are often used by persons with disabilities. Sometimes these need to be modified however, to make them useable. Another type of commercially available device is one that is mass-produced but specifically designed for individuals with disabilities *(special commercially available devices)*. These devices often need to be modified to meet the needs of a specific individual. Our goal is to reduce the amount of modification necessary and to make mainstream technologies as accessible as possible. However, there will always be a portion of the disabled population that will require specifically designed assistive technologies.

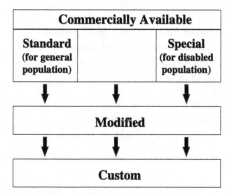

Fig. 1. This diagram shows the progression from commercially available devices for the general population and commercially available devices for special populations to modified devices and custom devices. From [4].

4.4.3 Mainstream Derivatives of Assistive Technology
There have been many developments originally intended for persons with disabilities that have found there way into the mainstream ICT market. A number of these are listed in Box 2.

4.5 The Human Technology Interface for ICT

4.5.1 General Concepts
It is estimated that as many as 40 million persons in the United States alone have physical, cognitive, or sensory disabilities [11]. The world-wide impact is significantly larger. If these people are to compete on an equal basis with non-disabled individuals, then it is extremely important that the internet be accessible to all. As the internet becomes more and more dependent on multimedia representations involving complex graphics, animation, and audible sources of information, the challenges for people who have disabilities increase. In order for access to the Internet to be useful to people with disabilities, the accessibility approach must be independent of individual devices. This means that users must be able to interact with a *user agent* (and the document it renders) using the input and output devices of their choice based on their specific needs. A **user agent** is defined as software to access Web content

Closed Captioning
Voice Recognition
On-screen Keyboards
Speech Synthesis
Digitized Speech
Computer Keyboard Equivalents
MouseKeys
StickyKeys
T9 Disambiguation
Word Prediction
Abbreviation Expansion
Single Latches on Laptops
On/Off Push button Toggle Switches
Call-out Control Descriptions
Screen Enlargement System
Color Schemes
Wearable Computers
Head Tracking Devices
Brainwave Recognition Units
Single Switch Hardware Interfaces

Box 2. Mainstream applications of assistive technologies

(www.w3.org/wai). This includes desktop graphical browsers, text and voice browsers, mobile phones, multimedia players, and software assistive technologies (e.g., screen readers, magnifiers) that are used with browsers. The person with a disability interacts with technology through the Human Technology Interface (HTI) [4].

The graphical user interface (GUI) has both positive and negative implications for persons with disabilities. The positive features are those that apply to non-disabled users (e.g., use of icons, recognition rather than recall memory, screen icons for the same task look the same, operations such as opening and closing files are always the same). The GUI is the standard user interface because of its ease of operation for novices and its consistency of operation for experts. The latter ensures that every application behaves in basically the same way. People with motor disabilities may not have the necessary physical (eye-hand coordination) and visual skills to navigate the GUI. Modification of the GUI to allow specialized access (see Figure 1) can also be more challenging for GUI-based operating systems.

As networks are expanded and more devices (e.g., cell phones, PDAs) have open architectures, it will be possible to download profiles, adaptations and special instructions that enable adaptable systems to be developed to meet the needs of people who have disabilities.

4.5.2 Access for Motor Impairment
There are a significant number of people who cannot effectively use standard keyboards, mouse controls or switches. It is likely that we will see a much greater understanding of the biological/physical interface for the control of computers in the future [3].

One approach that may offer promise is the brain computer interface (BCI). BCI systems may be grouped into a set of functional components including the input device, amplification, feature extraction, feature translation and user feedback [12]. Signals are mathematically analyzed to extract features useful for control [13]. Features or signals that have been used include slow cortical potentials, P300 evoked potential, sensorimotor rhythms recorded from the cortex and neuronal action potentials recorded within the cortex). A typical task for a user is to visualize different movements or sensations or images.

Another approach to cursor control is the use of a digital camera and image recognition software to track a particular body feature to control an on-screen mouse cursor [14]. The most easily tracked feature is the tip of the nose, but the eye (gross eye position not point of gaze), lip, chin and thumb have also been used. Non-disabled subjects used this approach and fund that the camera mouse was accurate but slower than a typical hand-controlled mouse. Using an on-screen keyboard the camera mouse was half as fast as a regular mouse in a typing task, but the accuracy obtained was equivalent on each system. More and more computers have built-in cameras, so the camera mouse requires only software to capture the body feature image and interpret its movement as mouse commands. This may lead to wider application of this technique.

There are many other approaches that are used to provide access to and control over technologies for people with severe motor disabilities [4].These range form keyboards of various type, to automatic speech recognition to mouse and mouse emulators systems to single and multiple switches.

A survey of Internet use was conducted with a large group of individuals with spinal cord injury from [15]. For this group the rate of internet access was higher (66%) than the general population (43%). Access to the internet differed by race, employment status, income, education and marital status. Primary uses by the SCI participants were social (e-mail, chat rooms) and information seeking (health-related information, on-line shopping), and they found that internet access contact reduced many of the barriers to full community participation that they normally face (e.g., transportation, telephone use, and need for personal attendants for outside trips). Internet access did not reduce inter-personal contact and isolate people with disabilities from social interaction.

4.5.3 Access for Cognitive Impairment

Cognitive disabilities include a wide range of skills and deficiencies. Learning disabilities typically involve significant difficulties in understanding or in using either spoken or written language, and these difficulties may be evident in problems with reading, writing, mathematical manipulation, listening, spelling or speaking [16]. These limitations make it increasingly difficult to access complicated Web sites that may include flashing pictures, complicated charts, and large amounts of audio and video data. While there are assistive technologies that are specifically designed to address these areas (discussed later in this chapter), many of the technological tools are useful for all students, and are part of instructional technology [17]. Even the so-called assistive technologies have features (e.g., multimedia, synthetic speech output, voice recognition input) that are useful to all learners.

For individuals with acquired cognitive disabilities due to injury (e.g., traumatic brain injury) or disease (e.g., stroke (CVA) or dementia) changing features such as font size, background/foreground color combinations, contrast, spacing between words, letters and paragraphs and using graphics can all improve access to screen-based information. Another technological concept for these individuals is a cognitive prosthesis, which is a custom-designed computer-based compensatory strategy that directly assists in performing daily activities[1]. It may also include additional technologies such as a cell phone, pager, digital camera or low tech approaches.

Persons with intellectual disabilities have difficulties with memory, language use and communication, abstract conceptualization, generalization and problem identification/problem solving. Characteristics of the HTI that are important for these individuals include simplicity of operation, capacity of the technology to support repetition, consistency in presentation, and inclusion of multiple modalities (e.g., speech, sounds and graphical representations) [18].

An example of technology designed for cognitive needs is the Planning and Execution Assistant and Trainer (PEAT). It is a PDA-based personal planning assistant designed to assist individuals with cognitive disorders due to brain injury, stroke, Alzheimer's disease, and similar conditions [19]. EAT employs artificial intelligence to automatically generate plans and also to revise those plans when unexpected events occur. PEAT uses a combination of manually entered schedules and a library of stored scripts describing activities of daily living (e.g., morning routine or shopping). Scripts can be used for both planning and for execution. Planning involves a simulation of the activity with key decision points presented and prompts (auditory and visual) supplied necessary to aid the individual through the planning process. The plan to be executed can be either the stored script or a modified script based on the simulation. The PEAT artificial intelligence software generates the best strategy to execute the required steps in the plan [20]. PEAT also automatically monitors performance, and corrects schedule problems when necessary.

Another mainstream development is the "Smart House" that includes intelligent appliances networked together and internet enabled as well as other features. There are a number of systems that are required for a Smart House. These components must also be "intelligent" meaning that they have the capability to be automatically controlled and to be networked. This includes intelligent components (e.g.,doors, windows, drapes, elevators), intelligent furniture (e.g., beds, cupboards, drawers) and intelligent appliances. Connecting these systems with each other and with the outside world are communication services (phone, cable). The Smart house network provides the infrastructure that links all systems in the house such that any device can be operated from anywhere inside or outside the house with control alone or in groups. External access to the network is via cell phone, PDA or computer. Since each Smart house system has a unique IP address any appliance can be accessed remotely.

The Smart House can carry out a number of actions to support its inhabitants. Examples include: getting the occupants up in the morning, presenting information (e.g. morning news, traffic and weather reports) as needed, help with planning and preparing meals and getting them off to work or school on time. During the day the Smart House can look after things and help get the chores done. Examples of things that the

---

[1] Institute for Cognitive Prosthetics, http://www.brain-rehab.com/definecp.htm

Smart house will enable remotely are using a mobile phone to shut down a coffee pot on your way to work, monitoring the babysitter at home, getting notification from the refrigerator that you're out of milk on the way home from work.

Some of the central systems in the Smart House are appliances. A good appliance design has obvious Functions that are intuitive, easy and safe to operate such that it can be operated by a wide range of users. In addition, an intelligent/connected appliance becomes part of a collaborative system that satisfies the goals of the Smart House. For example an intelligent trash can can keep track of what has been thrown out and automatically places an online order for delivery of refills. Intelligent appliances make it possible to carry out a number of functions. For example, the refrigerator can be checked remotely via its digital camera t see if milk is needed. A lawn watering system can be started from across the country, the homeowner can check to see if she forgot to turn off the iron or the owner can start preheating the oven so as soon as you get home, you can pop in tonight's dinner. Currently available appliances include the screen fridge and the Internet enabled microwave oven. The former allows monitoring of contents, automatic generation of shopping lists based on consumption of food items and relating a cookbook recipe to the contents of the refrigerator. In addition to all the standard cooking features included in a standard microwave oven, the internet enabled microwave enables a cook to search for recipes, and to download cooking-related information, such as cooking time and microwave power level, for automatic cooking.

While these appliances are luxuries for an able-bodied person they become major assists to independence for person with disabilities. For persons with cognitive disabilities additional functions such as assessment of the occupants current state and the state of various home utilities to aid with common activities of daily living, provides feedback should residents become disoriented or confused and report medical emergencies automatically; an orientation and direction finding device that senses the current location (via GPS) and gives directions to a desired location for individuals who cannot read maps because of visual or cognitive disabilities. One example of this approach is the Bath Smart House, a three-bedroom "show-home" which elderly people will be able to test, staying for a few days with their care providers [21]. On feature is a wall-mounted locator box that indicates items that might be mislaid. The user simply presses button & a warbling sound is emitted from the lost object. A telephone uses photographs of people instead of buttons on the front to facilitate recall. For night vigilance a pressure sensor is located under the legs of the bed enabling the Smart leg detects when bed is occupied. If the resident gets up in the dark, lights fade up in the bedroom and if resident goes into the toilet lights fade up. The lights fade off when the resident is back in bed. This can help to avoid falls.

4.5.4 Access for Auditory Impairment

Since web pages are a mixture of text, graphics, and sound, people who are deaf may be prevented from accessing some information unless alternative methods are available. The primary approach for thee individual is the use of the Microsoft Synchronized Accessible Media Interchange (SAMI), which allows authors of Web pages and multimedia software to add closed captioning for users who are deaf or hard of hearing. This approach is similar to the use of closed captioning for television viewers. The W3C WAI SMIL (www.w3.org/WAI) is designed to facilitate multimedia presentations in which an author can describe the behavior of a multimedia presentation, associate hyperlinks with media objects, and describe the layout of the presentation on a screen.

Trainable hearing aids adjust automatically to the environments in which they are used through access to embedded information networks. This allows automatic adaptation to changing noise levels and environments.

4.5.5 Access for Visual Impairment

Individuals with visual disabilities have less access to the internet, are on line less and are more likely to be on line from work than home than individuals without disabilities [22]. Severity of impairment and existence of multiple impairments each reduce the access and use further. Individuals under 65 have greater use and access than those over 65. This finding is important given the high prevalence of visual impairment in the population over 65. People who are employed are more likely to use computers and the internet, whether they are disabled or not.

A study of computer use by individuals who had visual impairments involved a group of individuals, half of whom reported no useable vision, and the other half had variable amounts of vision Gerber [23]. Half of the respondents had been blind since birth, 85% had some university education and 73% were employed. The leading reason why technology was important and helpful was access to employment and the creation of flexibility in finding work. For some individuals computer access allowed telecommuting and access to employment from home. The computer also allowed employed individuals to create a cultural identity by being successfully employed. The second major benefit of computer use identified was independent access to information including newspapers and magazines as well as web-based sources. As in the case of persons with spinal cord injuries, the social connections made through the internet, such as independently sending and receiving e-mail using adapted computers were identified as major benefits of internet use. Being shut out of advances because of lack of accessibility, especially as computers and software change, was a major fear for many of the participants. For example, if a new version of Windows™ is developed, it may not be compatible with the accessible screen reader or Braille display the person has been using. We discuss this issue further later in this chapter. As more and more appliances, entertainment products and productivity tools for work become electronically sophisticated with many new features, people with visual impairments worry that they will be left behind because of lack of access.

4.6 Visual Access to the Internet

Enhancements in web pages and internet design in general are becoming more and more dependent on multimedia representations involving complex graphics, animation, and audible sources of information. For individuals who have visual limitations, these changes increase the challenges for obtaining access. While the most obvious barriers are for those who are blind, people who have learning disabilities and dyslexia also find it increasingly difficult to access complicated Web sites that may include flashing pictures, complicated charts, and large amounts of audio and video data.

The W3C WAI user agent guidelines are based on several principles that are intended to improve the design of both types of user agents. The first is to ensure that the user interface is accessible. This means that the consumer using an adapted input system must have access to the functionality offered by the user agent through its user interface. Second, the user must have access to document content through the provision of control of the style (e.g., colors, fonts, speech rate, and speech volume) and format of a document. A third principle is that the user agent help orient the user to where he is in the document or series of documents. In addition to providing alternative representations of location in a document (e.g., how many links the document contains or the number of the current link), a well-designed navigation system that uses numerical position information allows the user to jump to a specific link. Finally, the guidelines call for the user agent to be designed following system standards and conventions. These are changing rapidly as development tools are improved.

Communication through standard interfaces is particularly important for graphical desktop user agents, which must make information available to assistive technologies. Technologies such as those produced by the W3C include built-in accessibility features that facilitate interoperability. The standards being developed by the W3C WAI provide guidance for the design of user agents that are consistent with these principles. The guidelines are available on the W3C WAI Web page (www.w3.org/wai).

4.6.1 Other ICT Access

Cellular telephones are becoming more powerful with capabilities approaching that of personal computers. This expanded capability will provide significant advantages for people with disabilities, especially those with low vision or blindness. describes Three changes will be particularly valuable to people who have disabilities: (1) standard cell phones will have sufficient processing power for almost all the requirement of persons with visual impairments, (2) software will be able to be downloaded into these phones easily, (3) wireless connection to a worldwide network will provide a wide range of information and services in a highly mobile way [24]. Because many of these features will be built into standard cell phones the cost will be low and reachable by persons with disabilities. A major advance will occur if the cell phone industry moves away from proprietary software to an open source format providing the basis for a greater diversity of software for tasks such as text-to-speech output, voice recognition and optical character recognition in a variety of languages. Many applications for people with disabilities will be able to be downloaded from the internet. With expanded availability of embedded systems, it will be possible for a user to store their customized programs on the network and download them as needed form any remote location.

Downloading a talking book program into a cell phone can provide access to digital libraries for persons who are blind. Outputs in speech or enlarged visual displays can be added as needed by the user. With a built-in camera and network access a blind person could obtain a verbal description of a scene by linking to on-line volunteers who provide descriptions of images. These applications will depend on the increasing application of universal design in information technology products [25]. These applications include ATMs, cell phones, vending machines and other systems that are encountered on a daily basis [25].

5   Infrastructure for Future Accessibility

The infrastructure for future accessibility consists of: (1) an expanded, smarter and more available "real" and "virtual" internet, (2) Home automation systems that are smarter and have greater interconnectivity, (3) universal design principles that are applied more widely, (4) alternative approaches for accessing information technologies, and (5) special-purpose assistive technologies.

The Infrastructure for future accessibility will depend on several factors. These include: Web-based virtual systems, home automation, universal design for ICT, alternatives for accessing information technologies and special-purpose assistive technologies. In addition there is n on going need for the development of soft technology tools.

If ICT advances are not adaptable enough to be accessible to persons with disabilities it will further increase the disparity between those individuals and the rest of the population leading to further isolation and economic disadvantage. On the other hand, availability of these technologies in a transparent way will contribute to full inclusion of individuals who have disabilities in the mainstream of society.

6   Conclusions

The move to the information age offers great promise for persons with disabilities. It also holds great threats for persons with disabilities. Constant vigilance is required to insure that information technologies remain accessible and responsive to the needs of persons with disabilities. The future for persons with disabilities will not be driven by advances in technology, but rather by how well we can take advantage of those advances for the accomplishment of the many tasks of living that require technological assistance.

7   Summary

Anticipated changes in technologies coupled with the focus on the social aspects of disability, provide a significant opportunity for major advances in the degree to which individuals with disabilities can participate in all aspects of life, including work, school, leisure and self care.

Technological advances will be particularly important as the percentage of the population that is elderly rises. Concepts from universal design will be important in ensuring that this segment of the population remains active and is able to participate in society. This new group of elderly individuals will also be more experienced with computers and other technologies than their predecessors and they may well demand greater performance and adaptability from both assistive technologies and mainstream ICT (e.g., telephones, internet communication). The percentage of individuals with long-term disabilities who join the over 65 age group will also increase. These individuals will have been long-term users of assistive technologies, and their experience will have major implications for developments to meet future needs.

While much of what I have described is conjecture, it is based on modest extrapolation from the current state of the art. There are some things that we know with a high degree of certainty. We know that computer systems will be faster, have more memory be smaller and be less expensive for the same or greater functionality. We also know that the communication channel bandwidth will continue to increase allowing much more information and much more sophisticated information processing. Finally, it is clear that people with disabilities will continue to assert their right to fully participate in society.

Technological advances also raise questions for people who have disabilities. The most important of these is whether accessibility will keep pace with technological developments. For example, will assistive technologies for input and output be compatible with the user agents and operating systems of tomorrow. A second major question is whether the needs of persons with disabilities will be a driving force in future technological developments. Will people who have disabilities have to adapt to the existing technologies based on characteristics for non-disabled people or will universal design become a greater reality? In the latter case, adaptations will become less important and accessibility will become the rule rather than the exception.

For people who have disabilities, there are significant implications of emerging information processing technologies. If not closely monitored, these could result in less rather than more access to the new information economy for persons with disabilities. Despite the wider use of universal design principles, there will still be a need for effective assistive technology design and application if individuals with disabilities are to realize the full potential of the new information age.

References

1. Wright, R.A.: A Short History of Progress. Anansi Publications, Toronto (2004)
2. Ungson, G.R., Trudel, J.D.: The emerging knowledge-based economy. IEEE Spectrum 36(5), 60–65 (1999)
3. Applewhite, A.: 40 years: the luminaries. IEEE Spectrum 41(11), 37–58 (2004)
4. Cook, A.M., Polgar, J.M.: Cook and Husey's Assistive Technologies: Principles and Practice, 3rd edn. Elsevier, St. Louis (2007)
5. Lewis, J.R.: In the eye of the beholder. IEEE Spectrum 41(5), 24–28 (2004)
6. Kumagai, J.: Talk to the machine. IEEE Spectrum 39(9), 60–64 (2004)
7. Blackstone, S.: The Internet: what's the big deal. Augment Commun. News 9(4), 1–5 (1996)
8. Emiliani, P.L.: Assistive technology (AT) versus Mainstream Technology (MST): The research perspective. Tech. Disab. 18, 19–29 (2006)
9. World Health Organization: International classification of functioning disability and health-ICF, Geneva, World Health Organization (2001)
10. Odor, P.: Hard and soft technology for education and communication for disabled people. In: Proc. Int. Comp. Conf., Perth, Western Australia (1984)
11. Lazzaro, J.L.: Helping the web help the disabled. IEEE Spectrum 36(3), 54–59 (1999)
12. Mason, S.G., Birch, G.E.: A general framework for brain-computer interface design. IEEE Trans. Neural Systems and Rehab. Engr. 11, 70–85 (2003)

13. Fabiani, G.E., Mcfarland, D.J., Wolpaw, J.R., Pfurtscheller, G.: Conversion of EEG activity into cursor movement by a brain-computer interface (BCI). IEEE Trans. Neural Systems and Rehab. Engr. 12, 331–338 (2004)
14. Betke, M., Gips, J., Fleming, P.: The camera mouse: Visual tracking of body features to provide computer access for people with severe disabilities. IEEE Trans. Neural Systems and Rehabilitation Engineering 10(1), 1–10 (2002)
15. Drainoni, M., Houlihan, B., Williams, S., Vedrani, M., Esch, D., Lee-Hood, E., Weiner, C.: Patterns of internet use by persons with spinal cord injuries and relationship to health-related quality of life. Arch. Phys. Med. Rehab. 85, 1872–1879 (2004)
16. Edyburn, D.L.: Assistive technology and students with mild disabilities: from consideration to outcome measurement. In: Edyburn, D., Higgins, K., Boone, R. (eds.) Handbook of Special Education Technology Research and Practice, pp. 239–270. Knowledge by Design, Whitefish Bay (2005)
17. Ashton, T.M.: Students with learning disabilities using assistive technology in the inclusive classroom. In: Edyburn, D., Higgins, K., Boone, R. (eds.) Handbook of Special Education Technology Research and Practice, pp. 229–238. Knowledge by Design, Whitefish Bay (2005)
18. Wehmeyer, M.L., Smith, S.J., Palmer, S.B., Davies, D.K., Stock, S.E.: Technology use and people with mental retardation. International Review of Research in Mental Retardation 29, 291–337 (2004)
19. Levinson, R.L.: The planning and execution assistant and trainer. J. head trauma rehabilitation 12(2), 769–775 (1997)
20. LoPresti, E.F., Mihailidis, A., Kirsch, N.: Assistive technology for cognitive rehabilitation: State of the art. Neuropsychological Rehabilitation 14(1), 5–39 (2004)
21. Mann, W.C.: Smart technology for aging, disability and independence. John Wiley and Sons, Ltd, Chichester (2005)
22. Gerber, E., Kirchner, C.: Who's surfing? Internet access and computer use by visually impaired youth and adults. J. Vis. Impair. Blindness 95(3), 176–181 (2001)
23. Gerber, E.: The benefits of and barriers to computer use for individuals who are visually impaired. J. Vis. Impair Blindness 97(5), 536–550 (2003)
24. Fruchterman, J.R.: In the palm of your hand: A vision of the future of technology for people with visual impairments. J. Vis. Impair Blindness 97(10), 585–591 (2003)
25. Tobias, J.: Information technology and universal design: An agenda for accessible technology. J. Vis. Impair Blindness 97(10), 592–601 (2003)

Hybrid Brains – Biology, Technology Merger

Kevin Warwick

University of Reading, U.K.

Abstract. In this paper an attempt has been made to take a look at how the use of implant and electrode technology can now be employed to create biological brains for robots, to enable human enhancement and to diminish the effects of certain neural illnesses. In all cases the end result is to increase the range of abilities of the recipients. An indication is given of a number of areas in which such technology has already had a profound effect, a key element being the need for a clear interface linking the human brain directly with a computer. An overview of some of the latest developments in the field of Brain to Computer Interfacing is also given in order to assess advantages and disadvantages. The emphasis is clearly placed on practical studies that have been and are being undertaken and reported on, as opposed to those speculated, simulated or proposed as future projects. Related areas are discussed briefly only in the context of their contribution to the studies being undertaken. The area of focus is notably the use of invasive implant technology, where a connection is made directly with the cerebral cortex and/or nervous system.

Tests and experimentation which do not involve human subjects are invariably carried out *a priori* to indicate the eventual possibilities before human subjects are themselves involved. Some of the more pertinent animal studies from this area are discussed including our own involving neural growth. The paper goes on to describe human experimentation, in which neural implants have linked the human nervous system bi-directionally with technology and the internet. A view is taken as to the prospects for the future for this implantable computing in terms of both therapy and enhancement.

Keywords: Brain-Computer Interface, Biological systems, Implant technology, Feedback control.

1 Introduction

Research is being carried out in which biological signals of some form are measured, are acted upon by some appropriate signal processing technique and are then employed either to control a device or as an input to some feedback mechanism [17,21]. In many cases neural signals are employed, for example Electroencephalogram (EEG) signals can be measured externally to the body, using externally adhered electrodes on the scalp [26] and can then employed as a control input. Most likely this is because the procedure is relatively simple from a research point of view and is not particularly taxing on the researchers involved. However, reliable interpretation of EEG data is extremely complex – partly due to both the compound nature of the multi-neuronal signals being measured and the difficulties in recording such highly attenuated.

A. Fred, J. Filipe, and H. Gamboa (Eds.): BIOSTEC 2008, CCIS 25, pp. 19–34, 2008.
© Springer-Verlag Berlin Heidelberg 2008

In the last few years interest has also grown in the use of real-time functional Magnetic Resonance Imaging (fMRI) for applications such as computer cursor control. This typically involves an individual activating their brain in different areas by reproducible thoughts [28] or by recreating events [27]. Alternatively fMRI and EEG technologies can be combined so that individuals can learn how to regulate Slow Cortical Potentials (SCPs) in order to activate external devices [12]. Once again the technology is external to the body. It is though relatively expensive and cumbersome.

It is worth noting that external monitoring of neural signals, by means of either EEG analysis or indeed fMRI, leaves much to be desired. Almost surely the measuring technique considerably restricts the user's mobility and, as is especially the case with fMRI, the situation far from presents a natural or comfortable setting. Such systems also tend to be relatively slow, partly because of the nature of recordings via the indirect connection, but also because it takes time for the individual themselves to actually initiate changes in the signal. As a result of this, distractions, both conscious and sub-conscious, can result in false indicators thus preventing the use of such techniques for safety critical, highly dynamic and, to be honest, most realistic practical applications. Despite this, the method can enable some individuals who otherwise have extremely limited communication abilities to operate some local technology in their environment, and, in any case, it can serve as a test bed for a more direct and useful connection.

The definition of what constitutes a Brain-Computer Interface (BCI) is extremely broad. A standard keyboard could be so regarded. It is clear however that various wearable computer techniques and virtual reality systems, e.g. glasses containing a miniature computer screen for a remote visual experience [15], are felt by some researchers to fit this category. Although it is acknowledged that certain body conditions, such as stress or alertness, can be monitored in this way, the focus of this paper is on bidirectional BCIs and is more concerned with a direct connection between a biological brain and technology, and ultimately a human and technology.

2 *In vivo* Studies

Non-human animal studies can be considered to be a pointer for what is potentially achievable with humans in the future. As an example, in one particular animal study the extracted brain of a lamprey, retained in a solution, was used to control the movement of a small wheeled robot to which it was attached [19]. The lamprey innately exhibits a response to light reflections on the surface of water by trying to align its body with respect to the light source. When connected into the robot body, this response was utilised by surrounding the robot with a ring of lights. As different lights were switched on and off, so the robot moved around its corral, trying to position itself appropriately.

Meanwhile in studies involving rats, a group of rats were taught to pull a lever in order to receive a suitable reward. Electrodes were then chronically implanted into the rats' brains such that the reward was proffered when each rat thought (one supposes) about pulling the lever, but before any actual physical movement occurred. Over a period of days, four of the six rats involved in the experiment learned that they did not in fact need to initiate any action in order to obtain a reward; merely thinking about it was sufficient [2].

In another series of experiments, implants consisting of microelectrode arrays have been positioned into the frontal and parietal lobes of the brains of two female rhesus macaque monkeys. Each monkey learned firstly how to control a remote robot arm through arm movements coupled with visual feedback, and it is reported that ultimately one of the monkeys was able to control the arm using only brain derived neural signals with no associated physical movement. Notably, control signals for the reaching and grasping movements of the robotic arm were derived from the same set of implanted electrodes [3, 16].

Such promising results from animal studies have given the drive towards human applications a new impetus.

3 Robot with a Biological Brain

Human concepts of a robot may involve a little wheeled device, perhaps a metallic head that looks roughly human-like or possibly a biped walking robot. Whatever the physical appearance our idea tends to be that the robot might be operated remotely by a human, or is being controlled by a simple programme, or even may be able to learn with a microprocessor/computer as its brain. We regard a robot as a machine.

In a present project neurons are being cultured in a laboratory in Reading University to grow on and interact with a flat multi-electrode array. The neural culture, a biological brain, can be electronically stimulated via the electrodes and its trained response can be witnessed.

The project now involves networking the biological brain to be part of a robot device. In the first instance this will be a small wheeled robot. The input (sensory) signals in this case will be only the signals obtained from the wheeled robot's ultrasonic sensors. The output from the biological brain will be used to drive the robot around. The goal of the project initially will be to train the brain to drive the robot forwards without bumping into any object. Secondly, a separate biological brain will be grown to be the thinking process within a robot head (called Morgui) which houses 5 separate sensory inputs.

What this means is that the brain of these robots will shortly be a biological brain, not a computer. All the brain will know is what it perceives from the robot body and all it will do will be to drive the robot body around or control the robot head respectively. The biological brain will, to all intents and purposes, be the brain of the robot. It will have no life, no existence outside its robotic embodiment.

Clearly this research alters our concept of what a robot is, particularly in terms of ethical and responsibility issues. If a role of animal research is to open up possibilities for future human trials, then in this case the research could well be opening a window on the ultimate possibility of human neurons being employed in a robot body. All the 'human' brain would know would be its life as a robot.

4 Human Application

At the present time the general class of Brain-Computer Interfaces (BCIs) for humans, of one form or another, have been specifically developed for a range of applications including military weapon and drive systems, personnel monitoring and for games consoles. However, by far the largest driving force for BCI research to date has been the requirement for new therapeutic devices such as neural prostheses.

The most ubiquitous sensory neural prosthesis in humans is by far the cochlea implant [7]. Here the destruction of inner ear hair cells and the related degeneration of auditory nerve fibres results in sensorineural hearing loss. As such, the prosthesis is designed to elicit patterns of neural activity via an array of electrodes implanted into the patient's cochlea, the result being to mimic the workings of a normal ear over a range of frequencies. It is claimed that some current devices restore up to approximately 80% of normal hearing, although for most recipients it is sufficient that they can communicate to a respectable degree without the need for any form of lip reading. The typically modest success of cochlea implantation is related to the ratio of stimulation channels to active sensor channels in a fully functioning ear. Recent devices consist of up to 32 channels, whilst the human ear utilises upwards of 30,000 fibres on the auditory nerve. There are now reportedly well over 10,000 of these prostheses in regular operation.

Studies investigating the integration of technology with the human central nervous system have varied from merely diagnostic to the amelioration of symptoms [29]. In the last few years some of the most widely reported research involving human subjects is that based on the development of an artificial retina [20]. Here, small electrode arrays have been successfully implanted into a functioning optic nerve. With direct stimulation of the nerve it has been possible for the otherwise blind recipient to perceive simple shapes and letters. The difficulties with restoring sight are though several orders of magnitude greater than those of the cochlea implant simply because the retina contains millions of photodetectors that need to be artificially replicated. An alternative is to bypass the optic nerve altogether and use cortical surface or intracortical stimulation to generate phosphenes [4].

Most invasive BCIs monitor multi-neuronal intracortical action potentials, requiring an interface which includes sufficient processing in order to relate recorded neural signals with movement intent. Problems incurred are the need to position electrodes as close as possible to the source of signals, the need for long term reliability and stability of interface in both a mechanical and a chemical sense, and adaptivity in signal processing to deal with technological and neuronal time dependence. However, in recent years a number of different collective assemblies of microelectrodes have been successfully employed both for recording and stimulating neural activity. Although themselves of small scale, nevertheless high density connectors/transmitters are required to shift the signals to/from significant signal processing and conditioning devices and also for onward/receptive signal transmission.

Some research has focussed on patients who have suffered a stroke resulting in paralysis. The most relevant to this paper is the use of a '3rd generation' brain implant which enables a physically incapable brainstem stroke victim to control the movement of a cursor on a computer screen [13,14]. Functional Magnetic Resonance Imaging (fMRI) of the subject's brain was initially carried out to localise where activity was most pronounced whilst the subject was thinking about various movements. A hollow glass electrode cone containing two gold wires and a neurotrophic compound (giving it the title 'Neurotrophic Electrode') was then implanted into the motor cortex, in the area of maximum activity. The neurotrophic compound encouraged nerve tissue to grow into the glass cone such that when the patient thought about moving his hand, the subsequent activity was detected by the electrode, then amplified and transmitted by a radio link to a computer where the signals were translated into control signals to bring about movement of the cursor. With two electrodes in place, the subject successfully learnt to move the cursor around by thinking about different movements. Eventually the patient reached a level of control where no abstraction was needed – to move the cursor he simply thought about moving the cursor. Notably, during the period that the implant was in place, no rejection of the implant was observed; indeed the neurons growing into the electrode allowed for stable long-term recordings.

Electronic neural stimulation has proved to be extremely successful in other areas, including applications such as the treatment of Parkinson's disease symptoms. With Parkinson's Disease diminished levels of the neurotransmitter dopamine cause over-activation in the ventral posterior nucleus and the subthalamic nucleus, resulting in slowness, stiffness, gait difficulties and hand tremors. By implanting electrodes into the subthalamic nucleus to provide a constant stimulation pulse, the over activity can be inhibited allowing the patient, to all external intents and purposes, to function normally [18].

5 Brain within a Brain

Ongoing research, funded by the UK Medical Research Council, is investigating how the onset of tremors can be accurately predicted such that merely a stimulation current burst is required rather than a constant pulsing [10]. This has implications for battery inter-recharge periods as well as limiting the extent of in-body intrusive signalling. The deep brain stimulator can be used to collect local field potential (LFP) signals generated by the neurons around the deep brain electrodes [10]. Determining the onset of events can be investigated by using fourier transforms to transfer the time based signal to a frequency based spectrogram to determine the change in frequency at the critical time period. However, in addition to that, the frequency changes in the period of time immediately prior to the tremor occurrence can give important information.

Fig.1 shows the results of an initial attempt to train an artificial neural network to indicate not only that a Parkinsonian tremor is present but also that one is very likely to occur in the near future. The aim of this research is that, once a reliable predictor has been obtained, the stimulating pulsing will only be enacted when a tremor is predicted, in order to stop the actual physical tremor occurring before it even starts in the first place.

The bottom trace in Fig.1 shows emg (muscular) signals, measured externally, associated with movement due to the tremors. It can be seen that the tremors in this incident actually start at around the 45 to 50 second point. The trace just above this indicates the corresponding electrical data measured as deep brain Local Field Potentials in the Sub-Thalamic Nucleus of the patient involved. It can be witnessed how, in this case, the electrical data takes on a different form (in terms of variance at least) at around the 45 to 50 second point. The four top plots meanwhile indicate the outputs from 4 differently structured artificial neural networks, based on multi-layer perceptrons with different numbers of neurons in the hidden (middle) layer.

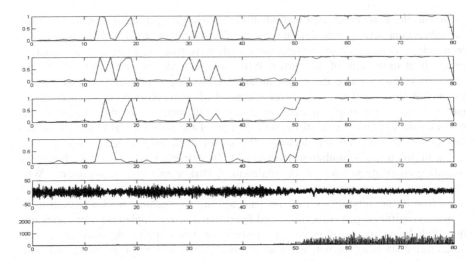

Fig. 1. Time plot of the onset of a Parkisonian tremor incident with corresponding artificial neural network indicators

It can be seen how, for each network, the output of the network goes high (logic 1) at the 45 to 50 second point, to indicate the presence of a Parkinsonian tremor. This is all well and good, what is important however is that the output of the networks also briefly goes high around the 30 second point and this can be seen as an indication of the fact that a tremor will shortly occur. Ongoing research is involved with selection of the type and number of inputs to the network, presently these being based on the energy spectrum in different frequency ranges. The networks are also being tested on considerable amounts of resting data, that is long periods of brain activity where no tremors at all actually occur in patients. Clearly the aim is that a network will not give false predictions of tremors.

In fact false positive predictions are not so much of a critical problem. The end result with a false positive is that a stimulation may occur when it is not strictly necessary. In any event no actual tremor would occur, which is indeed a good outcome, however unnecessary energy would have been used – in fact if numerous false predictions occurred the intelligent stimulator would tend toward the present 'blind' stimulator. Effectively the occasional false positive prediction is perhaps not a problem, unless it became a regular occurrence. The good news is that results show that the network can be readily tuned to avoid false positives anyway.

6 General Implant Studies

Some of the most impressive human research to date has been carried out using the microelectrode array, shown in Figure 2. The individual electrodes are only 1.5mm long and taper to a tip diameter of less than 90 microns. Although a number of trials not using humans as a test subject have occurred [1], human tests are at present limited to two studies. In the second of these the array has been employed in a recording only role [5,6,8], most notably recently as part of the 'Braingate' system. Essentially activity from a few neurons monitored by the array electrodes is decoded into a signal to direct cursor movement. This has enabled an individual to position a cursor on a computer screen, using neural signals for control combined with visual feedback. The first use of the microelectrode array (Figure 2) will be discussed in the following section as this has considerably broader implications which extend the capabilities of the human recipient.

A key selection point at the present time are what type of implant to employ, as several different possibilities exist, ranging from single electrode devices to multielectrode needles which contain electrode points at different depths to multielectrode arrays which either contain a number of electrodes which penetrate to the same depth (as in Figure 2) or are positioned in a banked/sloped arrangement. A further key area of consideration is the exact positioning of a BCI. In particular certain areas of the brain are, apparently, only really useful for monitoring purposes whilst others are more useful for stimulation.

Actually deriving a reliable command signal from a collection of captured neural signals is not necessarily a simple task, partly due to the complexity of signals recorded and partly due to time constraints in dealing with the data. In some cases however it can be relatively easy to look for and obtain a system response to certain anticipated neural signals – especially when an individual has trained extensively with the system. In fact neural signal shape, magnitude and waveform with respect to time are considerably different to the other signals that it is possible to measure in this situation.

If a greater understanding is required of neural signals recorded, before significant progress can be made, then this will almost surely present a major problem. This is especially true if a number of simultaneous channels are being employed, each requiring a rate of digitization of (most likely) greater than 20KHz in the presence of unwanted noise. For real time use this data will also need to be processed within a few milliseconds (100 milliseconds at most). Further, although many studies have looked into the extraction of command signals (indicating intent) from measured values, it is clear that the range of neural activity is considerable. Even in the motor area not only are motor signals present but so too are sensory, cognitive, perceptual along with other signals, the exact purpose of which is not clear – merely classifying them as noise is not really sufficient and indeed can be problematic when they are repeated and apparently linked in some way to activity.

It is worth stressing here that the human brain and spinal cord are linking structures, the functioning of which can be changed through electronic stimulation such as that provided via an electrode arrangement. This type of technology therefore offers a variety of therapeutic possibilities. In particular the use of implanted systems when applied to spinal cord injured patients, in whom nerve function is disordered, was described in [22] as having the following potential benefits (among others):

1. Re-education of the brain and spinal cord through repeated stimulation patterns
2. Prevention of spinal deformity
3. Treatment of intractable neurogenic and other pain
4. Assisting bladder emptying
5. Improving bowel function

6. Treatment of spasticity
7. Improvement of respiratory function – assisting coughing and breathing
8. Reduction of cardiovascular maleffects
9. Prevention of pressure sores – possibly providing sensory feedback from denervated areas
10. Improvement and restoration of sexual function
11. Improved mobility
12. Improved capability in daily living, especially through improved hand, upper limb and truncal control

Sensate prosthetics is another growing application area of neural interface technology, whereby a measure of sensation is restored using signals from small tactile transducers distributed within an artificial limb [7]. The transducer output can be employed to stimulate the sensory axons remaining in the residual limb which are naturally associated with a sensation. This more closely replicates stimuli in the original sensory modality, rather than forming a type of feedback using neural pathways not normally associated with the information being fed back. As a result it is supposed that the user can employ lower level reflexes that exist within the central nervous system, making control of the prosthesis more subconscious.

One final noteworthy therapeutic procedure is Functional Electrical Stimulation (FES), although it is debatable if it can be truly referred to as a BCI, however it aims to bring about muscular excitation, thereby enabling the controlled movement of limbs. FES has been shown to be successful for artificial hand grasping and release and for standing and walking in quadriplegic and paraplegic individuals as well as restoring some basic body functions such as bladder and bowel control [11]. It must be noted though that controlling and coordinating concerted muscle movements for complex and generic tasks such as picking up an arbitrary object is proving to be a difficult, if not insurmountable, challenge.

In the cases described in which human subjects are involved, the aim on each occasion is to either restore functions since the individual has a physical problem of some kind or it is to give a new ability to an individual who has very limited motor abilities. In this latter case whilst the procedure can be regarded as having a therapeutic purpose, it is quite possible to provide an individual with an ability that they have in fact never experienced before. On the one hand it may be that whilst the individual in question has never previously experienced such an ability, some or most other humans have – in this case it could be considered that the therapy is bringing the individual more in line with the "norm" of human abilities.

It is though also potentially possible to give extra capabilities to a human, to enable them to achieve a broader range of skills – to go beyond the "norm". Apart from the, potentially insurmountable, problem of universally deciding on what constitutes the "norm", extending the concept of therapy to include endowing an individual with abilities that allow them to do things that a perfectly able human cannot do raises enormous ethical issues. Indeed it could be considered that a cochlea implant with a wider frequency response range does just that for an individual or rather an individual who can control the curser on a computer screen directly from neural signals falls into this category. But the possibilities of enhancement are enormous. In the next section we consider how far things could be taken, by referring to relevant experimental results.

7 Human Enhancement

The interface through which a user interacts with technology provides a distinct layer of separation between what the user wants the machine to do, and what it actually does. This separation imposes a considerable cognitive load upon the user that is directly proportional to the level of difficulty experienced. The main issue it appears is interfacing the human motor and sensory channels with the technology. One solution is to avoid this sensorimotor bottleneck altogether by interfacing directly with the human nervous system. It is certainly worthwhile considering what may potentially be gained from such an invasive undertaking.

Advantages of machine intelligence are for example rapid and highly accurate mathematical abilities in terms of 'number crunching', a high speed, almost infinite, internet knowledge base, and accurate long term memory. Additionally, it is widely acknowledged that humans have only five senses that we know of, whereas machines offer a view of the world which includes infra-red, ultraviolet and ultrasonic. Humans are also limited in that they can only visualise and understand the world around them in terms of a limited dimensional perception, whereas computers are quite capable of dealing with hundreds of dimensions. Also, the human means of communication, essentially transferring an electro-chemical signal from one brain to another via an intermediate, often mechanical medium, is extremely poor, particularly in terms of speed, power and precision. It is clear that connecting a human brain, by means of an implant, with a computer network could in the long term open up the distinct advantages of machine intelligence, communication and sensing abilities to the implanted individual.

As a step towards this more broader concept of human-machine symbiosis, in the first study of its kind, the microelectrode array (as shown in Figure 2) was implanted into the median nerve fibres of a healthy human individual (myself) in order to test *bidirectional* functionality in a series of experiments. A stimulation current direct onto the nervous system allowed information to be sent to the user, while control signals were decoded from neural activity in the region of the electrodes [9, 23]. In this way a number of experimental trials were successfully concluded [24, 25]: In particular:

1. Extra sensory (ultrasonic) input was successfully implemented and made use of.
2. Extended control of a robotic hand across the internet was achieved, with feedback from the robotic fingertips being sent back as neural stimulation to give a sense of force being applied to an object (this was achieved between New York (USA) and Reading(UK))
3. A primitive form of telegraphic communication directly between the nervous systems of two humans was performed.
4. A wheelchair was successfully driven around by means of neural signals.
5. The colour of jewellery was changed as a result of neural signals – as indeed was the behaviour of a collection of small robots.

Fig. 2. A 100 electrode, 4X4mm Microelectrode Array, shown on a UK 1 pence piece for scale

In each of the above cases it could be regarded that the trial proved useful for purely therapeutic reasons, e.g. the ultrasonic sense could be useful for an individual who is blind or the telegraphic communication could be very useful for those with certain forms of Motor Neurone Disease. However each trial can also be seen as a potential form of augmentation or enhancement for an individual. The question then arises as to how far should things be taken? Clearly enhancement by means of BCIs opens up all sorts of new technological and intellectual opportunities, however it also throws up a raft of different ethical considerations that need to be addressed directly.

8   On Stimulation

After extensive experimentation it was found that injecting currents below 80μA onto the median nerve fibers had little perceivable effect. Between 80μA and 100μA all the functional electrodes were able to produce a recognizable stimulation, with an applied voltage of 40 to 50 volts, dependant on the series electrode impedance. Increasing the current above 100μA had no apparent additional effect; the stimulation switching mechanisms in the median nerve fascicle exhibited a non-linear thresholding characteristic.

During this experimental phase, it was pseudo randomly decided whether a stimulation pulse was applied or not. The volunteer (myself), wearing a blindfold, was unaware of whether a pulse had been applied or not, other than by means of its effect in terms of neural stimulation. The user's accuracy in distinguishing between an actual pulse and no pulse at a range of amplitudes is shown in Figure 3.

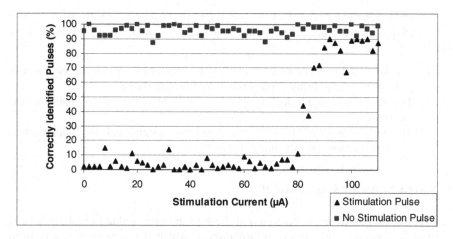

Fig. 3. Effect of stimulation amplitude on the number of correctly identified pulses and absence of pulses (over 100 trials)

In all subsequent successful trials, the current was applied as a bi-phasic signal with pulse duration of 200 µsec and an inter-phase delay of 100 µsec. A typical stimulation waveform of constant current being applied to one of the MEA's implanted electrodes is shown in Fig 4.

It was, in this way, possible to create alternative sensations via this new input route to the nervous system. Of the 5 enhancement features mentioned in the previous section, this one will be described, as an example, in further detail. Background information on the other enhancements can be found in a number of references, e.g. [9,23.24,29].

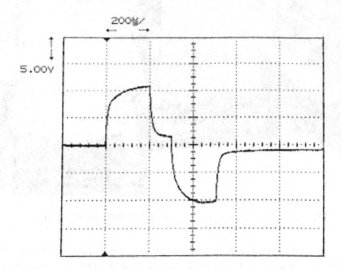

Fig. 4. Voltage profile during one bi-phasic stimulation pulse cycle with a constant current of 80µA

It must be reported that it took 6 weeks for my brain to repetitively recognize the stimulating signals accurately. This time period can be due to a number of contributing factors:

(a) The team had to learn which signals (what amplitude, frequency etc.) would be best in order to bring about a recognizable stimulation.
(b) The recipient's brain had to learn to recognize the new signals it was receiving.
(c) The bond between the recipient's nervous system and the implant was physically changing (becoming stronger).

9   Extra Sensory Experiment

An experiment was set up to determine if the human brain is able to understand and successfully operate with sensory information to which it had not previously been exposed. Whilst it is quite possible to feed in such sensory information via a normal human sensory route, e.g. electromagnetic radar or infra-red signals are converted to visual, what we were interested in was feeding such signals directly onto the human nervous system, thereby bi-passing the normal human sensory input.

Fig. 5. Experimentation and testing of the ultrasonic baseball cap

Ultrasonic sensors were fitted to the rim of a baseball cap (see Figure 5) and the output from these sensors, in the form of a proportional count, was employed to bring about a direct stimulation of the nervous system. Hence when no objects were in the vicinity of the sensors, no stimulation occurred, and as an object moved close by so the rate of stimulation pulses being applied increased in a linear fashion up to a pre-selected maximum rate. No increase in stimulation occurred when an object moved closer than 10cm to the sensors.

The ultrasonic sensors were open type piezoelectric ceramic transducers with coni-cal metal resonators and operated at 40 KHz. These were used in a pair, one for transmit and one for receive, to give maximum sensitivity for small and distant ob-jects. The most useful range for the experimentation was found to be 2 – 3m, this be-ing also dependent on the size of object. A simple microcontroller was programmed to perform the echo ranging on the pair of transducers, and provide the range to the first detectable object only. This was translated into a stimulation pulse train, which operated on a single pin of the electrode array. Pins on the array had been tested for their suitability for stimulation by the earlier experimentation in which the recipient identified the presence or absence of stimulation pulse trains at various amplitudes and repetition frequencies.

It was found that very little learning was required for the new ultrasonic sense to be used effectively and successfully – merely a matter of 5/6 minutes. This said it must be remembered that it had already taken several weeks for the recipient's brain to successfully, accurately recognize the current signals being injected.

As a result, in a witnessed experiment, the recipient, whilst wearing a blindfold, was able to move around successfully within a cluttered laboratory environment, al-beit at a slower than normal walking pace. The sensory input was "felt" as a new form of sensory input (not as touch or movement) in the sense that the brain made a direct link between the signals being witnessed and the fact that these corresponded in a linear fashion to a nearby object.

10 Conclusions

External input-output interfaces with human and animal brains have been studied for many years. These are sometimes referred to as Brain-Computer Interfaces (BCIs) even though the interface may be external to the (human) body and its sensorimotor mechanism. In this paper an attempt has been made to put such systems in perspec-tive. Emphasis has been placed on such interfaces that can be obtained by means of implanted devices through invasive surgery and actual direct neural connections. In particular a number of trials in this area have clearly shown the possibilities of moni-toring, stimulating and enhancing brain functioning.

Although there is no distinct dividing line it is quite possible as far as humans are concerned to investigate BCIs in terms of those employed for direct therapeutic means and those which can have an enhanced role to play. It is clear that the interac-tion of electronic signals with the human brain can cause the brain to operate in a dis-tinctly different manner. Such is the situation with the stimulator implants that are successfully used to counteract, purely electronically, the tremor effects associated with Parkinson's disease. Such technology can though potentially be employed to modify the normal functioning of the human brain and nervous system in a number of different ways.

The same stimulator, with slightly different positioning, has been shown to elicit feelings of sadness or happiness in the recipient. Given the nature of the intelligent stimulator described here it would appear to be possible to monitor, in real time, a human brain with a computer brain, and for the computer brain to predict when the human is going to feel sad – quite some time before they actually feel sad. In theory a signal could then be injected at that time to make them feel happy, or at least to stop them actually ever feeling sad in the first place. Maybe this could be regarded as an electronic anti-depressant. There are of course questions about recreational use here – but this would need a deep brain implant which might well prove to be rather too onerous for most people.

Perhaps understandably, invasive BCIs are presently far less well investigated in University experiments than their external BCI counterparts. A number of animal trials have though been carried out and the more pertinent have been indicated here along with the relevant human trials and practice. In particular the focus of attention has been given to the embodiment of grown neural tissue within a technological body. Whilst only 1,000 or so neurons are involved this presents an interesting research area in a number of ways. But once the number of such neurons used increases 1,000 or 1,000,000-fold, it also raises enormous philosophical and ethical issues. For example is the robot 'thinking' and what rights should it have?

The potential for BCI applications for individuals who are paralysed is enormous, where cerebral functioning to generate command signals is functional despite the motor neural pathways being in some way impaired – such as in Lou Gehrig's disease. The major role is then either one of relaying a signal of intention to the appropriate actuator muscles or to reinterpret the neural signals to operate technology thereby acting as an enabler. In these situations no other medical 'cure' is available, something which presents a huge driver for an invasive implant solution for the millions of individuals who are so affected. Clearly though, bidirectional signalling is important, not only to monitor and enact an individual's intent but also to provide feedback on that individual's resultant interaction with the real world. For grasping, walking and even as a defensive safety stimulant, feedback is vital. This paper has therefore focussed on such studies.

Where invasive interfaces are employed in human trails, a purely therapeutic scenario often exists. In a small number of instances, such as use of the microelectrode array as an interface, an individual has been given different abilities, something which opens up the possibilities of human enhancement. These latter cases however raise more topical ethical questions with regard to the need and use of a BCI. What might be seen as a new means of communication for an individual with an extreme form of paralysis or a new sensory input for someone who is blind, opening up a new world for them, can also be seen as an unnecessary extra for another individual, even though it may provide novel commercial opportunities. What is therapy for one person may be regarded as an enhancement or upgrading for another.

Whilst there are still many technical problems to be overcome in the development of BCIs, significant recent experimental results have indicated that a sufficient technological infrastructure now exists for further major advances to be made. Although a more detailed understanding of the underlying neural processes will be needed in the years ahead, it is not felt that this will present a major hold up over the next few years, rather it will provide an avenue of research in which many new results will shortly appear through trials and experimentation, possibly initially through animal studies although it must be recognised that it is only through human studies that a full analysis can be made and all encompassing conclusions can be drawn. Nevertheless the topic opens up various ethical questions that need to be addressed and as such, research in this area should, I believe, only proceed in light of a pervasive ethical consensus.

Acknowledgements. The Author would like to acknowledge the considerable assistance and input of the Consultant Neurosurgeons Mr. Peter Teddy, Mr. Amjad Shad, Mr. Ali Jamous and Mr. Tipu Aziz and researchers Iain Goodhew, Mark Gasson, Ben Whalley and Ben Hutt. Ethical approval for the author's research was obtained from the Ethics and Research Committee at the University of Reading, UK and with regard to the neurosurgery aspect, the Oxfordshire National Health Trust Board overseeing the Radcliffe Infirmary, Oxford, UK.

References

1. Branner, A., Normann, R.: A multielectrode array for intrafascicular recording and stimulation in the sciatic nerve of a cat. Brain Research Bulletin 51, 293–306 (2000)
2. Chapin, J.K.: Using multi-neuron population recordings for neural prosthetics. Nature Neuroscience 7, 452–454 (2004)
3. Carmena, J., Lebedev, M., Crist, R., O'Doherty, J., Santucci, D., Dimitrov, D., Patil, P., Henriquez, C., Nicolelis, M.: Learning to control a brain-machine interface for reaching and grasping by primates. Plos Biology 1(2), article number e2 (2003)
4. Dobelle, W.: Artificial vision for the blind by connecting a television camera to the visual cortex. ASAIO J. 46, 3–9 (2000)
5. Donoghue, J.: Connecting cortex to machines: recent advances in brain interfaces. Nature Neuroscience Supplement 5, 1085–1088 (2002)
6. Donoghue, J., Nurmikko, A., Friehs, G., Black, M.: Advances in Clinical Neurophysiology, Supplements to Clinical Neurophysiology. In: Development of a neuromotor prosthesis for humans, ch. 63, vol. 57, pp. 588–602 (2004)
7. Finn, W., LoPresti, P. (eds.): Handbook of Neuroprosthetic methods. CRC Press, Boca Raton (2003)
8. Friehs, G., Zerris, V., Ojakangas, C., Fellows, M., Donoghue, J.: Brain-machine and brain-computer interfaces. Stroke 35(11), 2702–2705 (2004)
9. Gasson, M., Hutt, B., Goodhew, I., Kyberd, P., Warwick, K.: Invasive neural prosthesis for neural signal detection and nerve stimulation. Proc. International Journal of Adaptive Control and Signal Processing 19(5), 365–375 (2005)
10. Gasson, M., Wang, S., Aziz, T., Stein, J., Warwick, K.: Towards a demand driven deep brain stimulator for the treatment of movement disorders. In: Proc. 3rd IEE International Seminar on Medical Applications of Signal Processing, 16/1–16/4 (2005)
11. Grill, W., Kirsch, R.: Neuroprosthetic applications of electrical stimulation. Assistive Technology 12(1), 6–16 (2000)
12. Hinterberger, T., Veit, R., Wilhelm, B., Weiscopf, N., Vatine, J., Birbaumer, N.: Neuronal mechanisms underlying control of a brain-computer interface. European Journal of Neuroscience 21(11), 3169–3181 (2005)
13. Kennedy, P., Bakay, R., Moore, M., Adams, K., Goldwaith, J.: Direct control of a computer from the human central nervous system. IEEE Transactions on Rehabilitation Engineering 8, 198–202 (2000)

14. Kennedy, P., Andreasen, D., Ehirim, P., King, B., Kirby, T., Mao, H., Moore, M.: Using human extra-cortical local field potentials to control a switch. Journal of Neural Engineering 1(2), 72–77 (2004)
15. Mann, S.: Wearable Computing: A first step towards personal imaging. Computer 30(2), 25–32 (1997)
16. Nicolelis, M., Dimitrov, D., Carmena, J., Crist, R., Lehew, G., Kralik, J., Wise, S.: Chronic, multisite, multielectrode recordings in macaque monkeys. Proc. National Academy of the USA 100(19), 11041–11046 (2003)
17. Penny, W., Roberts, S., Curran, E., Stokes, M.: EEG-based communication: A pattern recognition approach. IEEE Transactions on Rehabilitation Engineering 8(2), 214–215 (2000)
18. Pinter, M., Murg, M., Alesch, F., Freundl, B., Helscher, R., Binder, H.: Does deep brain stimulation of the nucleus ventralis intermedius affect postural control and locomotion in Parkinson's disease? Movement Disorders 14(6), 958–963 (1999)
19. Reger, B., Fleming, K., Sanguineti, V., Simon Alford, S., Mussa-Ivaldi, F.: Connecting Brains to Robots: an artificial body for studying computational properties of neural tissues. Artificial life 6(4), 307–324 (2000)
20. Rizzo, J., Wyatt, J., Humayun, M., DeJuan, E., Liu, W., Chow, A., Eckmiller, R., Zrenner, E., Yagi, T., Abrams, G.: Retinal Prosthesis: An encouraging first decade with major challenges ahead. Opthalmology 108(1) (2001)
21. Roitberg, B.: Noninvasive brain-computer interface. Surgical Neurology 63(3), 195 (2005)
22. Warwick, K.: I Cyborg. University of Illinois Press (2004)
23. Warwick, K., Gasson, M., Hutt, B., Goodhew, I., Kyberd, P., Andrews, B., Teddy, P., Shad, A.: The application of implant technology for cybernetic systems. Archives of Neurology 60(10), 1369–1373 (2003)
24. Warwick, K., Gasson, M., Hutt, B., Goodhew, I., Kyberd, P., Schulzrinne, H., Wu, X.: Thought Communication and Control: A First Step Using Radiotelegraphy. IEE Proceedings on Communications 151(3), 185–189 (2004)
25. Warwick, K., Gasson, M., Hutt, B., Goodhew, I.: An Attempt to Extend Human Sensory Capabilities by means of Implant Technology. In: Proc. IEEE Int. Conference on Systems, Man and Cybernetics, Hawaii (2005)
26. Wolpaw, J., McFarland, D., Neat, G., Forheris, C.: An EEG based brain-computer interface for cursor control. Electroencephalogr. Clin. Neurophysiol. 78, 252–259 (1990)
27. Pan, S., Warwick, K., Gasson, M., Burgess, J., Wang, S., Aziz, T., Stein, J.: Prediction of parkinson's disease tremor onset with artificial neural networks. In: Proc. IASTED Conference BioMed 2007, Innsbruck, Austria, pp. 341–345 (2007)
28. Warwick, K.: The promise and threat of modern cybernetics. Southern Medical Journal 100(1), 112–115 (2007)
29. Warwick, K., Gasson, M.N.: Practical Interface Experiments with Implant Technology. In: Sebe, N., Lew, M., Huang, T.S. (eds.) ECCV/HCI 2004. LNCS, vol. 3058, pp. 7–16. Springer, Heidelberg (2004); Brain-computer interface based on event-related potentials during imitated natural reading. International Journal of Psychology 39(5-6) (suppl.S.), p.138
30. Yoo, S., Fairneny, T., Chen, N., Choo, S., Panych, L., Park, H., Lee, S., Jolesz, F.: Brain-computer interface using fMRI: spatial navigation by thoughts. Neuroreport 15(10), 1591–1595 (2004)
31. Yu, N., Chen, J., Ju, M.: Closed-Loop Control of Quadriceps/Hamstring activation for FES-Induced Standing-Up Movement of Paraplegics. Journal of Musculoskeletal Research 5(3), 173–184 (2001)

From the Bench to the Bedside: The Role of Semantic Web and Translational Medicine for Enabling the Next Generation Healthcare Enterprise

Vipul Kashyap

Clinical Informatics R&D, Partners Healthcare System
93 Worcester St, Wellesley, MA 02481, U.S.A.
vkashyap1@partners.org

Abstract. The success of new innovations and technologies are very often disruptive in nature. At the same time, they enable novel next generation infrastructures and solutions. These solutions introduce great efficiencies in the form of efficient processes and the ability to create, organize, share and manage knowledge effectively; and the same time provide crucial enablers for proposing and realizing new visions. In this paper, we propose a new vision of the *next generation healthcare enterprise* and discuss how Translational Medicine, which aims to improve communication between the basic and clinical sciences, is a key requirement for achieving this vision. This will lead therapeutic insights may be derived from new scientific ideas - and vice versa. Translation research goes from bench to bedside, where theories emerging from preclinical experimentation are tested on disease-affected human subjects, and from bedside to bench, where information obtained from preliminary human experimentation can be used to refine our understanding of the biological principles underpinning the heterogeneity of human disease and polymorphism(s). Informatics and semantic technologies in particular, has a big role to play in making this a reality. We identify critical requirements, viz., data integration, clinical decision support and knowledge maintenance and provenance; and illustrate semantics-based solutions *wrt* example scenarios and use cases.

Keywords: Semantic Web technologies, Translational Medicine, Data Integration, Clinical Decision Support, Knowledge Maintenance and Provenance, Resource Description Framework (RDF), Web Ontology Language (OWL), Electronic Medical Record, Clinical Trials, Eligibility Criteria, Molecular Diagnostic Tests, Genetic Variants, Hypertrophic Cardiomyopathy, Family History, Business Object Models, Ontologies, Business Rules.

1 Introduction

The success of new innovations and technologies are very often disruptive in nature. At the same time, they enable novel next generation infrastructures and solutions. These solutions often give rise to creation of new markets and/or introduce great efficiencies. For example, the standardization and deployment of IP networks resulted in

A. Fred, J. Filipe, and H. Gamboa (Eds.): BIOSTEC 2008, CCIS 25, pp. 35–56, 2008.

introducing novel applications that were not possible in older telecom networks. The Web itself has revolutionized the way people look for information and corporations do business. Web based solutions have dramatically driven down operational costs both within and across enterprises. The Semantic Web is being proposed as the next generation infrastructure, which builds on the current web and attempts to give information on the web a well defined meaning [1].

On the other hand, the healthcare and life sciences sector is playing host to a battery of innovations triggered by the sequencing of the Human Genome coupled with a more proactive approach to Medicine. There is increased emphasis of prevention and wellness of the individual as opposed to disease prevention; and significant activity has focused on Translational Research, which seeks to accelerate "translation" of research insights from biomedical research into clinical practice and vice versa.

In this paper, we explore the convergence of these areas of activities and thought processes discussed above. We present semantics-based solution approaches that have the potential of accelerating the creation and sharing of knowledge across the healthcare and life sciences spectrum. We discuss next the vision of the next generation healthcare enterprise and discuss how Translational Medicine is a key requirement for that vision.

1.1 The Next Generation Healthcare Enterprise: A Vision Statement

There is a great need to get away from the short term goals of treating current diseases and conditions and focus on longer term strategies of enhancing the well-being and quality of life of an individual. There is no reason why the interaction of a patient with the healthcare enterprise should only happen when the patient is suffering from a clinical condition or disease; and that the services offered to patients should be limited to the resolution of the disease or clinical condition. In fact, it is a well known fact that adopting the approach of disease prevention will result in reducing the load on current healthcare infrastructure. From this perspective, the vision of the next generation healthcare enterprise may be articulated as follows:

The Next Generation Healthcare Enterprise provides services across the healthcare and life sciences (HCLS) spectrum targeted towards delivering optimum wellness, therapy and care for the patient. These holistic services cut across biomedical research, clinical research and practice and ***create a need for*** *the accelerated adoption of genomic and clinical research into clinical practice.*

There are certain significant implication of this vision statement we would like to emphasize. A key consequence of this vision is that not only should the healthcare enterprise the current needs of a patient, but also anticipate future needs and implements interventions that can potentially prevent diseases and other adverse clinical events. A key component of being able to do this is to anticipate ahead of time the potential diseases a patient could have. This could potentially be done by sequencing the genome of the patient and assessing the disposition of a patient towards diseases and adverse clinical events. Given that genomics, proteomics and metabolomics are areas which are targeted toward developing new therapies and diagnostic techniques, which need to be tested via clinical research studies, there clearly creates a need for knowledge sharing, communication and collaboration across the healthcare and life sciences spectrum, which is indeed the underlying goal of Translational Medicine!

1.2 Translational Medicine

Translational Medicine, aims to improve the communication between basic and clinical science so that more therapeutic insights may be derived from new scientific ideas - and vice versa. Translation research [2] goes from bench to bedside, where theories emerging from preclinical experimentation are tested on disease-affected human subjects, and from bedside to bench, where information obtained from preliminary human experimentation can be used to refine our understanding of the biological principles underpinning the heterogeneity of human disease and polymorphism(s). From a more prosaic perspective, insights available from clinical practice can also enable better identification of prospective participants and The products of translational research, such as molecular diagnostic tests are likely to be the first enablers of personalized medicine (see an interesting characterization of activity in the healthcare and life sciences area in [3]). We will refer to this activity as Translational Medicine in the context of this chapter.

1.3 Paper Outline

The organization of this chapter is as follows. We begin in Section 2 with two use case scenarios, which demonstrate the need for knowledge and information sharing across genomic research, clinical research and clinical practice. This is followed in Section 3, with an analysis of various stakeholders in the fields of healthcare and life sciences, along with their respective needs and requirements. In Section 4, we discuss a service oriented architecture for Translational Medicine followed by identification of key functional requirements that need to be supported, viz. *data integration, clinical decision support* and *knowledge maintenance and provenance*. In Section 5, we present a solution approach for data integration based on semantic web specifications such as the Resource Description Framework (RDF) [4] and the Web Ontology Language (OWL) [5]. In Section 6, a solution approach for clinical decision support that uses the OWL specification and a Business Rules Engine is presented. Section 7 presents an implementation of Knowledge Maintenance and Provenance using the OWL specification. This is followed by a discussion of the Conclusions and Future work in Section 8.

2 Use Case Scenarios

We discuss two use case scenarios, one which illustrates the use of molecular diagnostic tests based on genomic research in the context of clinical practice, and the other which illustrates the use of data collected in the context of clinical practice for clinical research.

2.1 Translation of Genomic Research into Clinical Practice

We anticipate that one of the earliest manifestations of translational research will be the adoption of therapies and tests created from genomics research into clinical practice. Consider a patient who presents with shortness of breath and fatigue in a doctor's clinic. Subsequent examination of the patient reveals the following information:

- Abnormal heart sounds which could be represented in a structured physical exam.
- Discussion of the family history of the patient reveals that his father had a sudden death at the age of 40, but his two brothers were normal.
- Based on the finding of abnormal heart sounds, the doctor may decide (or an information system may recommend) to order an ultrasound for the patient.
- The finding of the ultrasound may reveal cardiomyopathy, based on which the doctor can decide (or the information system may recommend) to order molecular diagnostic tests to screen genes such as MYH7, MYBPC3, TNN2, etc. for genetic variations:
- If the patient tests positive for pathogenic variants in any of the above genes, the doctor may want to recommend that first and second degree relatives of the patient consider testing; and select treatment based on the above-mentioned data. He can stratify for treatment by clinical presentation, imaging and non-invasive physiological measures in the genomic era, for e.g., non-invasive serum proteomics.
- The introduction of genetic tests introduces further stratification of the patient population for treatment. Whenever a patient is determined to be at a high risk for sudden death, they are put under therapeutic protocols based on drugs such as Amiadorone or Implantable Cardioverter Defibrillator (ICD). It may be noted that the therapy is determined purely on the basis of phenotypic conditions which in the case of some patients may not have held to be true. In this case whereby molecular diagnostic tests may indicate a risk for cardiomyopathy, phenotypic monitoring protocol may be indicated.

2.2 Translation of Clinical Practice into Clinical Research

The identification, recruitment, and enrollment of eligible subjects for clinical trials is extremely resource intensive and an obstacle for the conduct of clinical research. Currently, methods for identifying potential research subjects in clinical settings involve scanning (often manually) patient lists by single fields of data collected in the EMR - diagnosis or drug administration, for example. There is a great potential for reusing the data collected during the normal course of patient – physician interaction in the course of clinical practice to support different types of clinical research functions, e.g., patient recruitment, adverse event detection and resolution, following a patient through a clinical trial. We now discuss a use case scenario related to patient recruitment below:

A clinical trials manager manages several clinical research protocols and is interested in obtaining target enrollment for each study as efficiently as possible. She has limited staff to evaluate patients for studies, and only wants to schedule or meet with patients who are likely to be eligible.

John Doe is a Patient at Hospital. He is admitted to the Hospital for some adverse clinical condition. Upon admission to the hospital, an entry in the EMR system is created for John. As part of the admissions process (or a fixed amt of time afterward), selected diagnosis and treatment information are entered into his medical record.

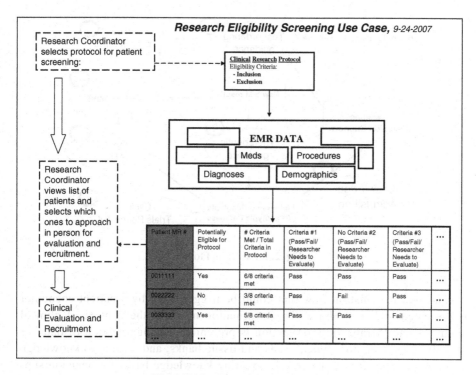

Fig. 1. Patient Recruitment Use Case

The eligibility criteria for various clinical trials that are planned for launch are used to search against all patients' (including John Doe's) EMR data. The results of these queries on Joe Doe's data (along with similar results from all other patients in the hospital) will appear in a report. The report will enable clinical trials investigators to identify Joe Doe, along with any other patients, who are potentially eligible for one or more clinical trials.

The clinical trials manager opens an application each morning to find a list of patients that are potentially eligible for one or more of open clinical trials. This list specifies which of all the clinical trials protocols available at the hospital that the patient is potentially eligible for, and specifies the # of criteria the patient matches versus the total # of eligibility criteria for each study. This use case is illustrated in Fig. 1 above.

3 Information Needs and Requirements

In this section, we present an information flow (illustrated in Fig. 2. below), in the context of which we identify various stakeholders and their information needs and requirements.

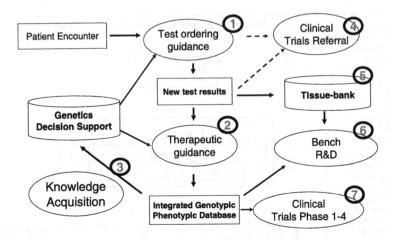

Fig. 2. Information Flow

A canonical information flow that could be triggered off by a patient encounter is presented in Fig. 2 above. The information needs in the context of clinical diagnosis and therapeutic intervention is presented. The aggregation of data for identifying patients for clinical trials and tissue banks; and leading to knowledge acquisition especially in the context creating knowledge bases for decision support, mappings between genotypic and phenotypic traits; is also presented. An enumeration of the information requirements is presented below in Table 1 below.

Table 1. Information Needs and Requirements

Step Number	Information Requirement	Application	Stakeholder(s)
1	Description of Genetic Tests, Patient Information, Decision Support KB	Decision Support, Electronic Medical Record	Clinician, Clinical Trials Investigator, Patient
2	Test Results, Decision Support KB	Decision Support, Database with Genotypic-Phenotypic associations	Clinician, Patient, Healthcare Institution
3	Database with Genotypic-Phenotypic associations	Knowledge Acquisition, Decision Support, Clinical Guidelines Design	Knowledge Engineer, Clinical Trials Investigator, Clinician
4	Test Orders, Test Results	Clinical Trials Management Software	Clinical Trials Investigator

Table 1. (*Continued*)

Step Number	Information Requirement	Application	Stakeholder(s)
5	Tissue and Specimen Information, Test Results	LIMS	Clinician, Life Science Researcher
6	Tissue and Specimen Information, Test Results, Database with Genotypic – Phenotypic associations	Lead Generation, Target Discovery and Validation, Clinical Guidelines Design	Life Science Researcher, Clinical Trials Investigator
7	Database with Genotypic – Phenotypic associations	Clinical Trials Design	Clinical Trials Designer

4 Service Oriented Architectures for Translational Medicine

In the previous section, we presented an analysis of information requirements for translational medicine. Each requirement identified in terms of information items, has multiple stakeholders, and is associated with different contexts, such as:

- Different domains such as genomics, proteomics or clinical information
- Research (as required for drug discovery) as opposed to operations (for clinical practice)
- Different applications such as the electronic medical record (EMR), and laboratory information systems (LIMS)
- Different services such as decision support, data integration and knowledge-provenance-related services

In this section, we build upon this analysis and present a conceptual architecture required to support a cycle of learning from innovation and its translation into the clinical care environment. The components of the conceptual architecture illustrated in Fig. 3 are as follows:

Portals. This is the user interface layer and exposes various personalized portal views to various applications supported by the architecture. Different stakeholders such clinical researchers, lab personnel, clinical trials designers, clinical care providers, hospital administrators and knowledge engineers can access information through this layer of the architecture.

Applications. Various stakeholders will access a wide variety of applications through their respective portals. The two main applications, viz. the Electronic Health Record (EHR) system and Laboratory Information Management Systems (LIMS) are illustrated in the architecture. We anticipate the emergence of novel applications that integrate applications and data across the healthcare and life science domains , which are collectively identified as "translational medicine" applications. Conceptual Architecture for Translational Medicine

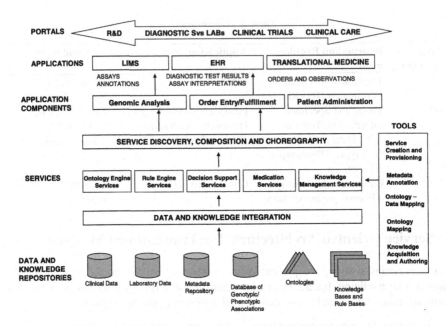

Fig. 3. Service Oriented Architecture for Translational Medicine

Service Discovery, Composition and Choreography. Newly emerging applications are likely to be created via composition of pre existing services and applications. This component of the architecture is responsible for managing service composition and choreography aspects. Tools that support annotations of services and help define and create new services are used to create service descriptions that can be consumed by this component to enable service discovery, composition and choreography.

Services. The services that need to be implemented for enabling Translational Medicine applications can be characterized as (a) business or clinical services and (b) infrastructural or technological services. Examples of clinical services are clinical decision support services and medication services, which address some aspect of functionality needed to support clinical systems requirements. Some examples of technological services are ontology and rule engine services, which provide specific informatics services such as classification or inferencing and may be invoked by clinical services to implement their functionality. Another example is a knowledge management service which implements functionality for creation and maintenance of various ontologies and knowledge bases. Tools that support knowledge authoring are used to create knowledge that can be consumed by these services.

Data and Knowledge Integration. This component of the architecture enables integration of genotypic and phenotypic patient data and reference information data. This integrated data could be used for enabling clinical care transactions, knowledge acquisition of clinical guidelines and decision support rules, and for hypothesis discovery for identifying promising drug targets. Examples of knowledge integration would be merging of ontologies and knowledge bases to be used for clinical decision support. Tools that support creation of mappings and annotations are used to create mappings across knowledge and data sources that are consumed by the data and knowledge integration component.

Data and Knowledge Repositories. These refer to the various data, metadata and knowledge repositories that exist in healthcare and life sciences organizations. Some examples are databases containing clinical information and results of laboratory tests for patients. Metadata related to various knowledge objects (e.g., creation data, author, category of knowledge) are stored in a metadata repository. Knowledge such as genotypic-phenotypic associations can be stored in a database, whereas ontologies and rule bases can be stored in specialized repositories managed by ontology engines and rule bases. Tools that support metadata annotation can be used to create metadata annotations that can be stored in the metadata repository.

Functional Requirements
The architecture discussed in the earlier section helps us identify crucial functional requirements that need to be supported for enabling translational medicine as follows:

Service Discovery, Composition and Choreography. The ability to rapidly provision new services is crucial for enabling new emerging applications in the area of translational medicine. This involves the ability to define and develop new services from pre existing services on the one hand and the ability to provision and deploy them in an execution environment on the other. This may involve composition of infrastructural and business services. For instance, one may want to develop a new service that composes an ontology engine and a rule engine service for creating new semantics-based decision support services. From a clinical perspective, one may want to compose clinical protocol monitoring services and notification services (which monitor the state of a patient and alert physicians if necessary) with appropriate clinical decision support services and medication dosing services to offer sophisticated decision support.

Data and Knowledge Integration. This represents the ability to integrate data across different types of clinical and biological data repositories. In the context of the use case scenarios discussed in Section 3, there is a need for integration and correlation of clinical and phenotypic data about a patient obtained from the EMR with molecular diagnostic test results obtained from the LIMS. Furthermore, the integrated information product will need to be used in different contexts in different ways as identified in the information requirements enumerated in Table 1. Effective data integration would require effective knowledge integration where clinically oriented ontologies such as SNOMED [6] and ICD-10 [7] may need to be integrated with biologically oriented ontologies such as BioPAX [8] and Gene Ontology [9].

Decision Support. Based on the clinical use case discussed in Section 3, there is a need for providing guidance to a clinician for ordering the right molecular diagnostic tests in the context of phenotypic observations about a patient and for ordering appropriate therapies in response to molecular diagnostic test results. The decision support functionality spans both the clinical and biological domains and depends on effective integration of knowledge and data across data repositories containing clinical and biological data and knowledge.

Knowledge Maintenance and Provenance. All the functional requirements identified above (service composition, data integration and decision support) critically depend on domain-specific knowledge that could be represented as ontologies, rule bases, semantic mappings (between data and ontological concepts), and bridge ontology mappings (between concepts in different ontologies). The healthcare and life sciences domains are experiencing a rapid rate of new knowledge discovery and change. A knowledge change "event" has the potential of introducing inconsistencies and changes in the current knowledge bases that inform semantic data integration and decision support functions. There is a critical need to keep knowledge bases current with the latest knowledge discovery and changes in the healthcare and life sciences domains.

We now present some solution approaches based on semantic web technologies for data integration, decision support and knowledge maintenance and provenance.

5   Data and Information Integration

We now describe with the help of an example, our solution approach for data integration based on semantic web specifications such as RDF [4] and OWL [5], to bridge clinical data obtained from an EMR, clinical trials data obtained from a Clinical Trials Management System (CTMS) and genomic data obtained from a LIMS. The data integration approach consists of the following steps:

1. Creation of a domain ontology identifying key concepts across the clinical and genomic domains. In some cases one need to incorporate multiple ontologies that reflect different standards in use.
2. Design and creation of wrappers that exposes the data in a given data repository in a RDF view.
3. Specification of mapping rules that provide linkages across data retrieved from different data repositories. In the case of multiple ontologies, the mapping rules could also identify inter-ontology linkages.
4. A user interface for: (a) specifications of data linkage mappings; and (b) Visualization of the integrated information.

We begin with a discussion of semantic data integration architecture underlying the implementation, followed by a discussion of the various steps presented above.

5.1   Semantic Data Integration Architecture

The semantic data integration architecture is a federation of data repositories as illustrated in Fig. 4 below and has the following components.

Data Repositories. Data repositories that participate in the federation offer access to all or some portion of the data. In the translational medicine context, these repositories could contain clinical data stored in the EMR or CTMS; or genomic data stored in the LIMS. Data remains in their native repositories in a native format is and not moved to a centralized location, as would be the case in data warehouse based approach.

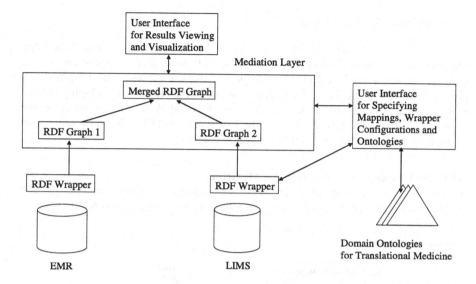

Fig. 4. Semantic Data Integration Architecture

Domain Ontologies. Ontologies contain a collection of concepts and relationships that characterize the knowledge in the clinical and genomic domains. They provide a common reference point that supports the semantic integration and interoperation of data.

RDF Wrappers. Wrappers are data repository specific software modules that map internal database tables or other data structures to concepts and relationships in the domain ontologies. Data in a repository is now exposed as RDF Graphs for use by the other components in the system.

Mediation Layer. The mediation layer takes as input mapping rules that may be specified between various RDF graphs and computes the merged RDF graphs based on those mapping rules.

User Interfaces. User interfaces support: (a) Visualization of integration results; (b) design and creation of domain ontologies; (c) configuration of RDF wrappers; and (d) specification of mapping rules to merge RDF graphs.

The main advantage of the approach is that one or more data sources can be added in an incremental manner. According to the current state of the practice, data integration is implemented via one-off programs or scripts where the semantics of the data is hard coded. Adding more data sources typically involves rewriting these programs and scripts. In our approach, the semantics are made explicit in RDF graphs and the integration is implemented via declarative specification of mappings and rules. These can be configured to incorporate new data sources via appropriate configurations of mappings, rules and RDF wrappers, leading to a cost and time-effective solution. Furthermore, different ontologies can be incorporated by specifying inter-ontology mappings as opposed to the current approach of performing expensive Extract-Transform-Load (ETL) processes.

5.2   Ontologies for Translational Medicine

A key first step in semantic data integration is the definition of a domain ontology spanning across multiple domains; or creation of inter-ontology mappings across multiple ontologies that reflect different perspectives, e.g., research and practice a given (clinical) domain. There are multiple collaborative efforts in developing ontologies in this area being undertaken in the context of the W3C Interest group on the Semantic Web for the Healthcare and the Life Sciences [10]. We now present an example of some ontologies and inter-ontology mappings.

5.2.1   An Ontology for Translational Medicine

We present a portion of an ontology for translational medicine that spans clinical and genomic domains (Fig. 5) and contains the following key concepts and relationships.

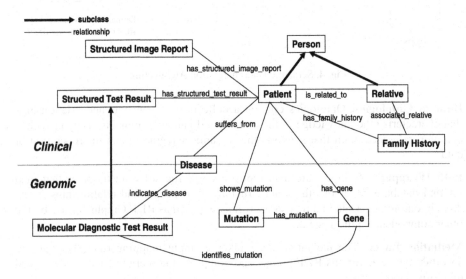

Fig. 5. A Domain Ontology for Translational Medicine

The core concept is **Patient**, which characterizes patient state information, such as value of various patient state parameters, the results of diagnostic tests and his/her family and lineage information. Patient is modeled as a subclass of **Person**. The **is_related** relationship and the **Relative** concept, model the relationships and relatives of a Person respectively. The **Family History** concept and the **has_family_history** relationship models information of family members that may have had a given disease. The **Structured Test Result** concept models results of laboratory tests, while radiological reports and observations are modeled by the **Structured Image Report** concept. The **Molecular Diagnostic Test Result** represents the results of a molecular diagnostic test result and is modeled as a sublass of the Structured Test Result class. Molecular diagnostics identify mutations (represented using the **identifies_mutation** relationship) and indicates diseases (represented using the **indicates_disease** relationship) in a patient. The **Gene** concept and the **has_genome** relationship is used to model genes and the genomic constitution of a patient. Genetic variants or mutation of a given gene are represented using the **Mutation** concept. The **Disease** concept characterizes the disease states which can be diagnosed about a patient, and is related to the patient concept via the **suffers_from** relationship.

5.2.2 Mappings across Multiple Clinical Ontologies

We now present examples of two clinical ontologies, which model clinical information from different perspectives. The DCM ontology is based on clinical data models for the clinical practice context, called Detailed Clinical Models [11] and is derived from the HL7/RIM [12] a key information modeling standard proposed by HL7. The SDTM ontology is based on clinical data models for submission of clinical trials data to the FDA and is based on the SDTM model [13] proposed by CDISC. We illustrate the differences in these ontologies based on OWL representations of a systolic blood pressure measurement in the DCM and SDTM ontologies, respectively.

```
SystolicBloodPressureMeasurement equivalentClass
ClinicalElement
that key value "SnomedCodeForSystolicBP"
and magnitude only float[>= 0.0, <= 220.0]
and units value mmHg
```

In the DCM ontology, **SystolicBloodPressure** is modeled as a subclass of a generic **ClinicalElement** class, which is related to a particular snomed class (identified as "SnomedCodeForSystolicBP") whose magnitude is identified as a range of values and the units are described as mmHg.

```
SystolicBloodPressureMeasurement equivalentClass
VSTEST that VSTESTCD value "SYSBP"
and units value mmHg
```

In the SDTM ontology, **SystolicBloodPressure** is modeled as a subclass of a generic **VSTEST** class that represents vital signs and observations. The subclass is identified by a controlled value called "SYSBP" which identifies all the vital signs observations that capture systolic blood pressure.

It may be the observed that the same information is modeled using different perspectives in the SDTM and DCM ontologies respectively. An important advantage of using semantic web technologies is the ability to represent mappings across multiple clinical ontologies in a declarative manner. In the examples below, DCM ontology elements have the subscript "dcm:" whereas the SDTM ontology elements have the subscript "sdtm:"

Class Mappings. The mappings between the two versions of systolic blood pressure can be specified as below.

```
dcm:SystolicBloodPressure equivalentClass
sdtm:VSTEST that VSTESTCD value sdtm:SYSBP
```

Property Mappings. The mappings between the various properties that are used in the definition of these classes can be specified as below.

```
dcm:key equivalentProperty sdtm:VSTESTCD
dcm:units equivalentProperty sdtm:VSORRESU
dcm:magnitude equivalentProperty sdtm:VSSORRES
```

Mappings of Controlled Terms. It may be noted that the specification of class mapping contain controlled terms such as "sdtm:SYSBP". These controlled terms need to be mapped to concepts in standardized vocabularies such as Snomed (prefix "snomed:") and NCI Thesaurus (prefix "ncit:") which are then mapped to each other.

```
sdtm:SYSBP sameAs ncit:NCITCodeForSystolicBP
snomed:SnomedCodeForSystolicBP
sameAs ncit:NCITCodeForSystolicBP
```

Mappings of Units. Finally units used different standards need to be mapped to each other.

```
dcm:mmHg sameAs sdtm:VSSORRES.mmHg
```

5.3 Use of RDF to Represent Clinical and Genomic Data

As discussed in Section 5.1, RDF wrappers perform the function of transforming information as stored in internal data structures in LIMS, EMR and CT systems into RDF-based graph representations. The RDF graphs illustrated in Fig. 6 below, represent clinical data related to a patient with a family history of *Sudden Death*. Nodes (boxes) corresponding Patient ID and Person ID are connected by an edge labeled *related_to* modeling the relationship between a patient and his father. The name of the patient ("Mr. X") is modeled as another node, and is linked to the patient node via an edge labeled *name*. Properties of the relationship between the patient ID and person ID nodes are represented by reification[1] (represented as a big box) of the edge labeled *related_to* and attaching labeled edges for properties such as the *type* of relationship (*paternal*) and the *degree* of the relationship (*1*).

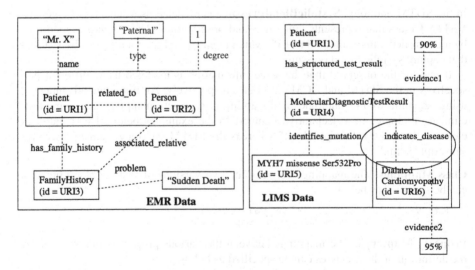

Fig. 6. RDF Representation of Clinical and Genomic Data

---

[1] Reification is a process by which one can view statements (edges) in RDF as nodes and assign properties and values to them.

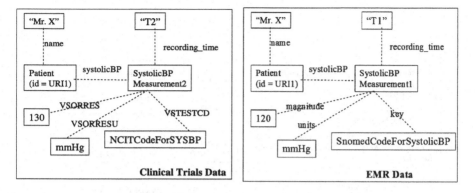

Fig. 7. Multiple Representations of Clinical Data

Genomic data related to a patient evaluated for a given mutation (*MYH7 missense Ser532Pro*) is illustrated. Nodes (boxes) corresponding to Patient ID and Molecular Diagnostic Test Result ID are connected by an edge labeled *has_structured_test_result* modeling the relationship between a patient and his molecular diagnostic test result. Nodes are created for the genetic mutation *MYH7 missense Ser532Pro* and the disease *Dialated Cardiomyopathy*. The relationship of the test result to the genetic mutation and disease is modeled using the labeled edges *identifies_mutation* and *indicates_disease* respectively. The degree of evidence for the dialated cardiomyopathy is represented by reification (represented as boxes and ovals) of the *indicates_disease* relationship and attaching labeled edges *evidence1* and *evidence2* to reified edge. Multiple confidence values expressed by different experts can be represented by reifying the edge multiple times.

Consider the RDF graphs in Fig. 7 that illustrate multiple RDF representations of similar clinical data. On the left hand side is a representation of a systolic blood pressure measurement represented using the DCM ontology discussed in the previous section. The edge labels, *magnitude*, *units* and *key* are properties from the DCM ontology. On the right hand side is a representation using the SDTM ontology with the corresponding labels, *VSORRES*, *VSORRESU* and *VSTESTCD*.

5.4 The Integration Process

The data integration process is an interactive one and involves a human end user, who in our case may be a *clinical* or *genomic researcher*. RDF graphs from different data sources are displayed. The end user previews them and specifies a set of rules for linking nodes across different RDF models. Some examples of simple rules that could be merging of nodes that have same IDs or URIs, introduction of new edges based on pre-specified declarative rules specified by subject matter experts and informaticians. New edges that are inferred (e.g., suffers_from) may be added back to the system based on the results of the integration. Sophisticated data mining that determines the confidence and support for these new relationships might be invoked. This integration process results in generation of merged RDF graphs as in Fig. 8 and Fig. 9 below.

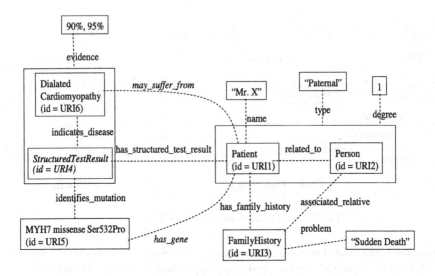

Fig. 8. Merged RDF graph that integrates genomic and clinical data

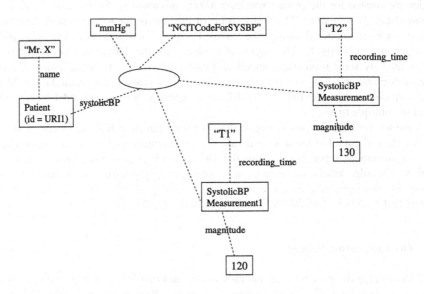

Fig. 9. Merged RDF graph that integrates EMR and clinical trials data

6 Clinical Decision Support

Various forms of clinical decision support functionality have been implemented in clinical systems used by healthcare delivery organizations. Examples of typical applications are: (a) Computerized physician order entry (CPOE) applications, which allow physicians to enter orders (e.g., medications, laboratory and radiology tests) into the computer system interactively providing real-time decision support as part of the ordering process; (b) automatic alerting, which automatically identifies serious clinical conditions and notifies the patient's attending physician while suggesting potential treatments for patient's condition; (c) Adverse drug-events monitor (ADE), which reviews patients' medication profiles for pairs of interacting drugs (physiological changes, reflected in abnormal laboratory results, that may occur as a result of an adverse drug-drug interaction may be accounted for); and (d) outpatient reminders and results manager; an application that helps clinicians review and act upon test results in a timely manner.

In order to maintain the currency and consistency of decision support knowledge across all clinical information systems and applications, we are implementing a rules based approach for representing and executing decision support knowledge [16]. At Partners Healthcare System, we are in piloting a clinical decision support service implemented using a commercial industry strength business rules engine [14]. Ontology-based approaches, which seek to use OWL-based ontology engines in conjunction with Business Rules Engines, have also been used for implementing Clinical Decisions Support [15]. Similar approaches are also being investigated for automating the process for eligibility criteria for screening patients for admission to a clinical trial. We illustrate our approach with the help of this example decision rule:

```
IF the patient's LDL Test Result is greater than 120
AND the patient has a contraindication to Fibric Acid
THEN Prescribe the Zetia Lipid Management Protocol
```

6.1 Business Rules Management Servers

The Business Object Model corresponding to the decision rule discussed above can be created as follows.

```
Class Patient: Person
method get_name(): string;
method has_genetic_test_result(): StructuredTestResult;
method has_liver_panel_result(): LiverPanelResult;
method has_ldl_result(): real;
method has_contraindication(): set of string;
method has_mutation(): string;
method has_therapy(): set of string;
method set_therapy(string): void;
method has_allergy(): set of string;

Class StructuredTestResult
method get_patient(): Patient;
method indicates_disease(): Disease;
method identifies_mutation(): set of string;
method evidence_of_mutation(string): real;

Class LiverPanelResult
method get_patient(): Patient;
method get_ALP(): real;
method get_ALT(): real;
method get_AST(): real;
method get_Total_Bilirubin(): real;
method get_Creatinine(): real;
```

The model describes patient state information by providing a class and set of methods that make patient state information, e.g., results of various tests, therapies, allergies and contraindications, available to the rules engine. The model also contains classes which represent classes corresponding to complex tests such as a liver panel result and methods that retrieve information specific to those tests, e.g. methods for retrieving creatinine clearance and total bilirubin. The methods defined in the object model are executed by the rules engine which results in invocation of services on the clinical data repository for retrieval of patient data.

The Business Object Model provides the vocabulary for specifying various clinical decision support rules. The rule-based specification of the example decision rule is given below.

```
IF the_patient.has_ldl_result() > 120
AND ((the_patient.has_liver_panel_result().get_ALP() ≥ <Normal>
    AND the_patient.has_liver_panel_result().get_ALT() ≥ <Normal>
    AND the_patient.has_liver_panel_result().get_AST() ≥ <Normal>
    AND the_patient.has_liver_panel_result().get_Total_Bilirubin()
                                                        ≥ <Normal>
    AND the_patient.has_liver_panel_result().get_Creatinine()
                                                      ≥ <Normal>)
    OR "Fibric Acid Allergy" ∈ the_patient.has_allergy())
THEN the_patient.set_therapy("Zetia Lipid Management Protocol")
```

The above rule represents the various conditions that need to be specified (the IF part) so that the system can recommend a particular therapy for a patient (the THEN part). The following conditions are represented on the IF part of the rule:

1. The first condition is a simple check on the value of the LDL test result for a patient.
2. The second condition is a complex combination of conditions that check whether a patient has contraindication to Fibric Acid. This is done by checking whether the patient has an abnormal liver panel or an allergy to Fibric Acid.

6.2 Ontology-Based Approaches

The rule-based specification discussed in the previous section can be modularized by separating the definition of a *"Patient with a contraindication to Fibric acid"*, from the decisions that are recommended once a patient is identified as belonging to that category. The definitions of various patient states and classes can be represented as axioms in an ontology that could be executed by an OWL ontology inference engine. At execution time, the business rules engine can invoke a service that interacts with the ontology engine to infer whether a particular patient belongs to a given class of patients, in this case, whether a patient has a contraindication to Fibric Acid. The ontology of patient states and classes can be represented as follows.

```
Class Patient
ObjectProperty hasLiverPanelResult
ObjectProperty hasAllergy

Class LiverPanelResult
ObjectProperty hasALP
ObjectProperty hasALT
ObjectProperty hasAST
```

```
ObjectProperty hasTotalBilirubin
ObjectProperty hasCreatinine

Class Allergy
Class FibricAcidAllergy
FibricAcidAllergy subclass Allergy

Class AbnormalALPResult
Class AbnormalALTResult
Class AbnormalASTResult
Class AbnormalTotalBilirubinResult
Class AbnormalCreatinineResult

Class AbnormalLiverPanelResult equivalentClass
LiverPanelResult
that hasALP only AbnormalALPResult
and hasALT only AbnormalALTResult
and hasAST only AbnormalASTResult
and hasTotalBilirubin only AbnormalTotalBilirubinResult
and hasCreatinine only AbnormalCreatinineResult

Class PatientContraindicatedtoFibricAcid equivalentClass
Patient
that hasAllergy some FibricAcidAllergy
and hasLiverPanel only AbnormalLiverPanel
```

The representation of an axiom specifying the definition of a patient with Fibric Acid contraindication enables the knowledge engineer to simplify the rule base significantly. The classification of a patient as being contraindicated to Fibric Acid is now performed by the Ontology Engine.

```
IF the_patient.has_contraindiction() contains
                           "Fibric Acid Contraindication"
THEN the_patient.set_therapy("Zetia Lipid Management Protocol")
```

7 Knowledge Maintenance and Provenance

In the previous section, we discussed an ontology-based approach for modularization, due to which changes in decision support logic can be isolated. The location of these changes can be identified as belonging of to a given ontology, and the rules impacted by the change can be easily determined. This is a key challenge in the healthcare and life sciences as knowledge continuously changes in this domain.

There is a close relationship between knowledge change and provenance. Issues related to when, and by whom was the change effected are issues related to knowledge provenance and provides useful information for maintaining knowledge. The issue of representing the rationale behind the knowledge change involves both knowledge change and provenance. On the one hand, the rationale behind the change could be that a knowledge engineer changed it, which is an aspect of provenance. On the other hand, if the change in knowledge is due to the propagation of a change in either a knowledge component or related knowledge, it is an aspect of knowledge change propagation as invoked in the context of knowledge provenance. We address the important issue of knowledge change propagation in this section. Consider the definition in natural language of fibric acid contraindication:

> A patient is contraindicated for Fibric Acid if he or she has an allergy to Fibric acid or has an abnormal liver panel.

Suppose there is a new (hypothetical) biomarker for fibric acid contraindication for which a new molecular diagnostic test is introduced in the market. This leads to a redefinition of a fibric acid contraindication as follows.

> The patient is contraindicated for Fibric Acid if he has an allergy to Fibric Acid or has elevated Liver Panel or <u>has a genetic mutation</u>

Let's also assume that there is a change in clinically normal range of values for the lab test AST which is the part of the liver panel lab test. This leads to a knowledge change and propagation across various knowledge objects that are sub-components and associated with the fibric acid contraindication concept. The definition of "Fibric Acid Contraindication" changes which is triggered by changes at various levels of granularity. This is discussed below and a diagrammatic representation of the OWL specification and the changes are illustrated in Fig. 10.

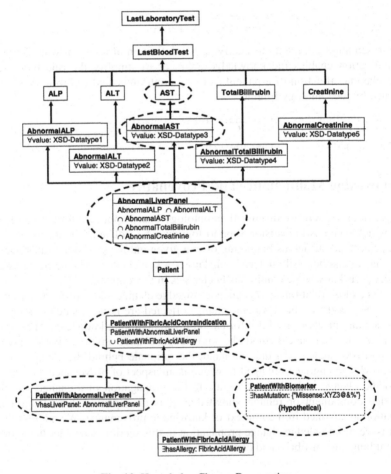

Fig. 10. Knowledge Change Propagation

A potential sequence of change propagation steps are enumerated below:

1. The clinically normal range of values for the AST lab test changes.
2. This leads to a change in the abnormal value ranges for the AST lab test.
3. This leads to a change in the definition of an abnormal Liver Panel.
4. This leads to a change in what it means to be a patient with an abnormal liver panel.
5. The definition of Fibric Acid contraindication changes due to the following changes.

   • The change in the definition of a patient with an abnormal liver panel as enumerated in steps 1-4 above.
   • Introduction of a new condition, viz., a patient having a mutation: "Missense: XYZ3@&%" (hypothetical). This is a new condition which could lead to a change what it means to be a patient with a contraindication to Fibric Acid.

OWL specifications of the knowledge illustrated in Fig. 10 above can be imported into an OWL-reasoner. A change in the definition is implemented as a difference between two concepts and the OWL-reasoner is invoked to determine consistency and identify the other changes. Towards this end the OWL-reasoner performs both datatype reasoning (e.g., checking for ranges of values) and OWL-based inferencing.

8 Conclusions

We have presented in this paper, use cases for translational medicine and for reuse of healthcare and clinical data, which cuts across various spheres of activity in the healthcare and life sciences, viz., various biomedical research areas, clinical trials and clinical practice. A set of crucial functional requirements, i.e., data integration, clinical decision support and knowledge maintenance and propagation, were identified to enable realization of the use case scenario. Solutions based on semantic web specifications and semantic tools and technologies were presented.

There is a growing realization that Healthcare and Life Sciences is a knowledge intensive field and the ability to capture and leverage semantics via inference or query processing is crucial for enabling translational medicine. Given the wide canvas and the relatively frequent knowledge changes that occur in this area, we need to support incremental and cost-effective approaches to support "as needed" data integration. Scalable and modular approaches for knowledge-based decision support that enable better maintenance for knowledge in the face of change is required. Automated semantics-based knowledge update and propagation is key to keeping the knowledge updated and current. Personalized/Translational Medicine cannot be implemented in a scalable, efficient and extensible manner without Semantic Web technologies.

Acknowledgements. We would like to acknowledge Tonya Hongsermeier at Partners Healthcare for introducing us to one of the use cases and for interesting discussions around semantics. The members of the W3C Task Force on Clinical Observations Interoperability are also acknowledge for their invaluable input and feedback to some of the ideas presented in this paper.

References

1. Berners-Lee, T., Hendler, J., Lassila, O.: The Semantic Web. Scientific American 284(5), 34–43 (2001)
2. Journal of Translational Medicine,
   http://www.translational-medicine.com/info/about
3. Kashyap, V., Neumann, E., Hongsermeier, T.: Tutorial on Semantics in the Healthcare and LifeSciences. In: The 15th International World Wide Conference (WWW 2006), Edingburgh, UK (May 2006), http://lists.w3.org/Archives/Public/www-archive/2006Jun/att-0010/Semantics_for_HCLS.pdf
4. Resource Description Framework, http://www.w3.org/RDF
5. OWL Web Ontology Language Review, http://www.w3.org/TR/owl-features
6. Snomed International, http://www.ihtsdo.org/our-standards/snomed-ct
7. International Classification of Diseases,
   http://www.who.int/classifications/icd/en/
8. BioPAX, http://www.biopax.org
9. The Gene Ontology, http://www.geneontology.org
10. W3C Semantic Web Healthcare and Life Sciences Interest Group,
    http://www.w3.org/2001/sw/hcls
11. Detailed Clinical Models, http://www.detailedclinicalmodels.org
12. HL7 Reference Information Model,
    http://hl7projects.hl7.nscee.edu/projects/design-repos
13. CDISC Standards – Study Data Tabulation Model/Study Data Submission Model,
    http://www.cdisc.org/models/sds/v3.1/index.html
14. Goldberg, H., Vashevko, M., Postilnik, A., Smith, K., Plaks, N., Blumenfeld, B.: Evaluation of a Commercial Rules Engine as a basis for a Clinical Decision Support Service. In: Proceedings of the Annual Symposium on Biomedical and Health Informatics, AMIA 2006 (2006)
15. Kashyap, V., Morales, A., Hongsermeier, T.: Implementing Clinical Decision Support: Achieving Scalability and Maintainability by combining Business Rules with Ontologies. In: Proceedings of the Annual Symposium on Biomedical and Health Informatics, AMIA 2006 (2006)
16. Greenes, R.A., Sordo, M.A., Zaccagnini, D., Meyer, M., Kuperman, G.: Design of a Standards-Based External Rules Engine for Decision Support in a Variety of Application Contexts: Report of a Feasibility Study at Partners HealthCare System. In: Proceedings of MEDINFO – AMIA (2004)

Part I

BIODEVICES

Active Annuloplasty System
for Mitral Valve Insufficiency

Andrés Díaz Lantada[1], Pilar Lafont[1], Ignacio Rada[2], Antonio Jiménez[2],
José Luis Hernández[2], Héctor Lorenzo-Yustos[1], and Julio Muñoz-García[1]

[1] Grupo de Investigación en Ingeniería de Máquinas, Universidad Politécnica de Madrid,
c/ José Gutiérrez Abascal, nº 2, 28006, Madrid, Spain (+34) 913364217
adiaz@etsii.upm.es, plafont@etsii.upm.es
[2] Hospital Central de la Defensa, Glorieta del Ejército, s.n., 28047, Madrid, Spain

Abstract. Active materials are capable of responding in a controlled way to different external physical or chemical stimuli by changing some of their properties. These materials can be used to design and develop sensors, actuators and multifunctional systems with a large number of applications for developing medical devices.

Shape-memory polymers are active materials with thermo-mechanical coupling (changes in temperature induce shape changes) and a capacity to recover from high levels of distortion, (much greater than that shown by shape-memory alloys), which combined with a lower density and cost has favoured the appearance of numerous applications. In many cases, these materials are of medical standard, which increases the chances of ultimately obtaining biocompatible devices.

This paper presents the procedure for designing, manufacturing, and programming the "shape-memory" effect and in vitro trials for an active annuloplasty ring for the treatment of mitral valve insufficiency, developed by using shape-memory polymers.

Keywords: Shape-Memory Polymers (SMP), Mitral valve Insufficiency, Annuloplasty ring, Laser Stereolithography, Silicone mould vacuum casting, Biomaterials.

1 Introduction to Mitral Valve Insufficiency

1.1 Mitral Valve Insufficiency

The mitral valve is made up of two components whose mission is to channel the blood from the auricle to the left ventricle. Firstly, there is the so-called mitral valve complex comprising the mitral ring, the valves of the mitral valve, and the commissures joining both valves. Apart from the mitral valve complex itself, this valve has the so-called "tensor" complex, which in turn comprises the tendinous chords which continue with the papillary muscles attached to the left ventricle.

A failure of any of these elements leads to functional changes in the mitral apparatus, such as mitral insufficiency, explained below, and hemodynamic repercussions.

Mitral insufficiency is defined as the systolic regurgitation of blood from the ventricle to the left auricle, due to incompetence in mitral valve closing. This can arise

A. Fred, J. Filipe, and H. Gamboa (Eds.): BIOSTEC 2008, CCIS 25, pp. 59–72, 2008.

for three main reasons: a) primary disease of the mitral valve; b) an anatomic or functional alteration in the papillary chords and muscles, and c) a disorder in the correct function of the auricle and the left ventricle [1].

Valve reconstruction is currently the preferred treatment for mitral insufficiency provided this is possible. With the aid of preoperative transesophagic echocardiography lesions can be located and their extent seen so a surgeon can evaluate if valve repair is possible and design an exact plan for any operation required. Nowadays, the object of this surgery is not simply limited to eliminating mitral insufficiency but in many cases to reconstructing the geometry of the entire mitral valve apparatus to ensure a durable repair. Surgically restoring the geometry to normal consists in: a) augmenting or reducing the abnormal vellums; b) replacing broken or short tendinous chords using "Goretex" type sutures, and c) annuloplasty [2], [3].

1.2 Treating Mitral Insufficiency with Annuloplasty

Carpentier's description of a rigid prosthetic ring to allow a selective reduction of the entire mitral ring opened the way to modern mitral repair [4]. Annuloplasty consists in inserting the said ring-shaped device via the jugular vein into the coronary sinus and after applying traction, retraction or heat, it reduces its perimeter, thereby reducing the mitral ring and improving the contact between the valve vellums, which in turn leads to a reduction in the patient's degree of mitral insufficiency.

Since then, a series of implants have been developed that can be basically classified as rigid or flexible and total or partial. Rigid monoplane implants have been displaced due to the large number of experimental and clinical works showing that the perimeter of the mitral ring constantly changes in size and shape during the heart cycle. The recent findings showing that these changes are produced in a three-dimensional way with a paraboloid hyperbolic shaped ring has given rise to new rigid three-dimensional prosthesis. However, these designs fail to take account of the continuous changes in this structure. Duran proposes replacing the most conventional devices for other flexible or semi-rigid designs that reproduce the three-dimensional shape, such as the one marketed by Medtronic Inc. [5], [6].

1.3 Desirable Improvements: Postoperative and Progressive Procedures

However, inserting a device to close the mitral valve means making additional demands on the heart that may lead to postoperative problems. It would be ideal to insert a ring with the same shape as the patient's mitral ring, and when they have recovered from the operation, progressively act on this ring (in several stages) and remotely. This seeks to maintain a balanced situation and not excessively overload the patient's heart during the operation [7], [8], [9], [10].

In this way, the progressive closing of the patient's mitral ring can be controlled and by using non-invasive inspection technologies the extent of improvement in the patient's mitral insufficiency can be evaluated after each stage of acting on the ring [11].

2 Solving Mechanical Operation Using Shape-Memory Polymer Based Devices

Shape-memory polymers (SMP) are materials that give a mechanical response to temperature changes. When these materials are heated above their "activation" temperature, there is a radical change from rigid polymer to an elastic state that will allow deformations of up to 300%. If the material is cooled after manipulation it retains the shape imposed; it "freezes", the said structure returning to a rigid but "unbalanced" state. When the material is heated above its activation temperature, it recovers its initial undeformed state.

The cycle can be repeated numerous times without degrading the polymer and most suppliers can formulae different materials with activation temperatures of between −30 °C and 260 °C, depending on the application required.

They are therefore active materials that present thermomechanical coupling and a high capacity for recovery from deformation, (much greater than shown by shape-memory alloys), which combined with a lower density and cost has favoured the appearance of numerous applications. Their properties permit applications for manufacturing sensing devices– actuators, especially for the aeronautics, automobile and medical industry [2], [12], [13], [14].

The main problems associated with the use of shape-memory polymers are the lack of structured processes for developing devices based on these materials. The design process and the relevant transformation processes for these devices need to be more thoroughly investigated.

The main advantages of shape-memory polymers are:

- They are new materials with the ability to change their geometry from an initial deformed shape to a second predetermined shape during the manufacturing process.
- They are more economical than shape-memory alloys.
- Different additives can be used to change their properties "a la carte", to better adapt them to the end application.
- The levels of deformation that can be attained are much greater than those obtainable from using shape-memory alloys.
- They can also be more easily processed and allow the use of "Rapid Prototyping Technologies", which speeds up the production of devices.
- More complex geometries and actuators can be obtained than with developments based on shape-memory alloys.

However, due to their recent appearance, in many cases their mechanical and thermomechanical properties are still not completely typified, which gives rise to doubts concerning how devices based on these materials will react. One of the basic aims of current research is to increase knowledge of the properties of these polymers by seeking to typify them as clearly as possible. Depending on the results of this research, performance models can be developed to facilitate the design of prototypes and actuators with these materials [15].

When compared to other medical devices, both surgical and implantable ones, they have additional advantages to those mentioned above:

- They are frequently medical grade materials that can be more easily adapted to biocompatible applications.
- The combined use of preoperative inspection technologies and CAD-CAE-CAM technology (design, calculation and computer aided manufacture) means that prostheses and customised devices can be made to measure for patients.
- Different additives can be used to change their properties "a la carte", to better adapt them to the end application.
- Their activation temperature and properties can be programmed, thanks to the amount of copolymers that can be "designed" and the use of additives.
- These polymers can be biodegradable and used in drug release devices.
- By being able to combine them with other materials and use reinforcing fibres means their performance strength can be enhanced.

Among the medical devices developed that endorse the advantages of using these polymers, the most notable are self-expanding stents, thrombectomy devices, intelligent sutures, drug release devices and active catheters [16], [17], [18], [19].

Regarding annuloplasty rings, non-commercial devices have been found based on shape-memory polymers. The Sorin Group's Memo 3D manages to reduce its shape by using a shape-memory alloy (Nitinol type, similar to those used in the manufacture of self-expanding stents). However, the change of shape is produced during the operation itself on making contact with human body temperature, which means that no postoperative measures are possible [20], [21], [22].

However, the capacity of shape-memory polymers for recovering their shape when faced with forces of up to around 7 MPa means that a 3 mm thick annuloplasty ring in this material, similar to devices currently in use, will stand a circumferential force of between 4 to 12 N that is stood by the patient's mitral ring [9], [23].

In accordance with the above, what is proposed is a ring made of shape-memory polymer and electrical resistances or heaters distributed in its interior to activate the "Shape-memory effect", accompanied by the required shape change.

Firstly, the ring adapts to the end size required (that needed to eliminate the mitral insufficiency) and with the resistances already in place. The ring is then uniformly heated to a temperature higher than the transition temperature (situated for the end product between 41 °C and 43 °C) and is forced to take on the expanded transitory shape (to do this cone-shaped tools can be used with a cross section similar to that of the mitral ring), letting it cool down to room temperature. The device also consists of a battery to power the resistances and heat them.

The rise in temperature of the resistances causes a local rise in temperature, which, if suitably controlled leads to a change in phase of the SMP and therefore a reduction in size.

Using an associated electronic control enables the resistances to be operated in pairs and at different times, in order to carryout the progressive or "step by step" operation required on the ring. Figure 1 shows a preliminary design.

Fig. 1. Preliminary active annuloplasty ring design. SMP with internal resistances for heating.

A patent for this device has been granted under the title of "Active annuloplasty system for the progressive treatment of valve insufficiencies and other cardiovascular pathologies" with Document Number P200603149 after being evaluated by the Spanish Patents and Trade Marks Office.

An alternative to heating by contact (either using resistances or a heating filament), is based on the insertion of magnetic or metallic nanoparticles into the shape-memory polymer itself and then heating it by induction [24]. These nanoparticles can be distributed in the core of the polymer during the process of mixing and casting its components. This option allows certain problems associated with the consumption and need for batteries to be eliminated, but "step by step" operation is made difficult. A diagram of this possibility is shown in Figure 2.

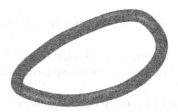

Fig. 2. Preliminary active annuloplasty ring design. SMP with magnetic or metallic nanoparticles for heating by induction.

The following sections present the design alternatives and the prototypes obtained, as well as the first "in vitro" trials performed, the results, and future recommendations for optimising the results. The development has been carried out in collaboration between researchers from Universidad Politécnica de Madrid (UPM) and doctors from the Hospital Central de la Defensa, Madrid.

3   Designing and Manufacturing Prototypes

3.1   CAD-CAE-CAM Technologies in the Design of Medical Devices

Computer aided design and calculation technologies, (CAD – Computer Aided Design and CAE – Computer Aided Engineering), have become an essential tool for developing medical device products. They enable alternative shapes and designs to be obtained quickly, as well as making it easier to evaluate their advantages by being able to analyse stress, deformations, ergonomics or dynamic response. They are also highly valuable for comparing and selecting the different materials that can be used. In addition, when combined with preoperative inspection techniques, they serve to design implantable devices made to measure for the patient and simulate their implantation [25], [26].

Figure 3 shows alternative ring designs for annuloplasty made by using the "Solid Edge v.18" computer design package. With the help of these programmes it is very simple to change the parameters of a design, which enables a shape to be adapted to the size of a particular patient's mitral ring or change the thickness of rings depending on how long the device is required to last.

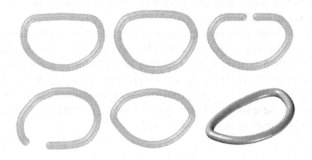

Fig. 3. Alternative designs for annuloplasty rings produced with CAD technologies

Bellow is explained how prototypes are manufactured from the designs shown and the advantages of using rapid prototyping and rapid tooling technologies.

3.2 The Importance of Rapid Prototyping Technologies

These new technologies enable physical parts to be obtained in a short time (days) directly from the computer aided designs. To a large extent they help reduce and optimise design iterations, thereby speeding up production start-up, which means they are highly valued by Industry [25], [26], [27].

The Product Development Laboratory of Universidad Politécnica de Madrid (UPM) has advanced technologies for developing rapid prototypes that combine computer aided design tools (CAD), layer model technology manufacture (LMT) and the reproduction of these models using silicone moulds. The two technologies used as a support for developing the device are described below; to be precise laser stereolithography and vacuum casting in silicone moulds.

Laser Stereolithography
Rapid prototyping technologies first appeared in 1987 with the American Company 3D Systems's laser stereolithography, currently the most widespread technology. It is based on being able to activate a polymerisation reaction in a liquid state epoxy resin by means of laser beam projection with a power and frequency suited to the type of

resin. The laser gradually "draws" on the surface of the liquid resin, following a path marked out by the CAD 3D file that contains the part geometry. When the liquid state monomers are exposed to ultraviolet radiation they polymerise into a solid state. The operation is repeated until the end part is obtained in epoxy resin by successively superimposing polymerised layers.

Figure 4 shows the physical models obtained in epoxy resin by laser stereolithography using an SLA-350 machine available at the Product Development Laboratory of Universidad Politécnica de Madrid, from the designs shown in Figure 3. Together with the annuloplasty designs, also shown is a 3 mm thick, 30 mm outer diameter toroidal ring to give the image an idea of scale.

Fig. 4. Physical models obtained by laser stereolithography from files containing the 3D part geometry

Rapid Tooling Technologies by Rapid Shape Copying
Rapid Tooling technologies enable instruments and mould parts to be quickly and cheaply obtained, by reproducing the geometry of physical models. Moreover, the parts manufactured in these moulds are obtained in the end materials, with the properties required for the end application. In order to manufacture the annuloplasty rings used in the first in vitro trials, soft silicone moulds obtained by rapid shape copy were used.

The parts obtained by stereolithographic procedure are particularly suitable for checking sizes and appearance. They can also be used as models for obtaining silicone moulds, which are subsequently used to obtain polyurethane resin replicas, more resistant and suited to working trials, and which also possess shape-memory properties. With the vacuum casting process different types of dual component resins can be used, with changeable properties, and the prototypes obtained reproduce the mould cavities with great precision (roughnesses around 50 μm) [28].

The chosen material is a polyurethane resin from MCP Iberia company with reference 3174 which is supplied in dual component form, which means it can be cast (after mixing the two components) in silicone moulds that reproduce the prototype shape required. For the induction heating alternative, magnetic particles can be added to the polymer core when mixing the two components supplied. The properties of the material are shown in Table 1.

Table 1. Properties of the shape-memory polymer for obtaining prototypes. Trade name: MCP Iberia 3174 polyurethane resin.

Property	Value
Shape change temperature	72 – 76 °C
Resistance to traction	51 MPa
Resistance to compression	68 MPa
Traction elasticity modulus	1750 MPa
Thermal conductivity	0,2 W / (m·K) a 25 °C

It must be pointed out that shape change temperature of the polyurethane resin used is not suited to the in vitro end trials, nor fits the initial specifications which required a range of 41 to 43 °C to activate the shape-memory effect. However, this polyurethane resin has been used because it is easy to cast and can be shaped in silicone moulds that enable prototypes to be made very quickly (less than 5 days from the computer files to the prototype in the end material).

Fig. 5. Silicone Moulds obtained from stereolithographic models

Fig. 6. Different polyurethane resin prototypes obtained under vacuum casting in silicone moulds. Both open and closed rings were made to analyse alternative performances.

Other mould manufacture technologies are currently being used for casting polymers with alternative memory shapes which do not attack the silicone moulds and whose transition temperature can be set from 4 to 6 °C above that of the human body, in accordance with what was explained concerning the conception of the device. Contact has been made with companies like CRG Industries that develop shape-memory polymers, the transition temperature being able to be set according to the client's requirements.

Figure 5 shows silicone moulds obtained from the physical epoxy resin models obtained in Figure 4 which enable prototypes to be obtained from the material with shape-memory. Enhanced design models have led to the construction of new silicone moulds and the obtaining of prototype annuloplasty rings, both solid ones and with side slits for housing the heating resistances. These are shown in Figure 6.

4 Programming the Shape-Memory Effect

When the annuloplasty rings have been shaped to ensure the mitral valve closes properly, they need to undergo heat deformation at 80 °C in the case of polyurethane resin, (higher temperature than that needed to activate the shape-memory effect) to increase their cross section until it coincides with the patient's mitral valve that is in a state of insufficiency.

So, a temporary shape is obtained and the ring can be implanted without submitting the patient's heart to an additional overload due to a sudden reduction in the passage section of the mitral valve. Subsequently, after the operation the recovery effect of the original shape is activated, which produces a gradual, controllable closure of the valve and a controlled recovery of mitral regurgitation.

To perform this deformation process called "shape-memory effect programming", tools were used that were obtained by laser stereolithography in the form of a cone base with a similar cross section to that of the mitral ring. Figures 7, 8 and 9 show different tools and the deformations caused to different ring prototypes thanks to the use of a counter-shape that acts as a press on the tool and the prototype.

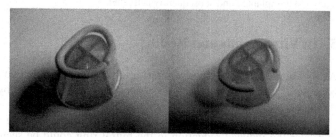

Fig. 7. Design and prototype of base for programming shape-memory effect. Deformation induced to the ring at 80 °C to obtain a temporary shape.

The tool shown in Figure 8 has interchangeable actuator heads for programming the rings in accordance with the different designs shown in Figure 3.

Fig. 8. Alternative device with interchangeable heads (depending on the type of ring to be implanted). Deformation induced at 80 °C to obtain a temporary shape.

Figure 9 shows an annuloplasty ring with the temporary shape already applied and prepared for implant and the first in vitro trials. With the aid of a cone base a 15% increase in cross section was produced (maximum inner diameter ring size pass from 26 to 28 mm), which will be used to evaluate the subsequent shape-memory effect recovery in in vitro trials. Heating resistances have also been included in the circumferential groove for activating the shape-memory effect.

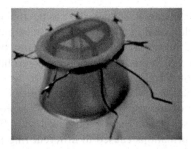

Fig. 9. Active ring with heating resistances and with the temporary shape already applied. It is ready to be implanted and subsequently activated.

5 In Vitro Trials and Results

As an animal model for performing the first in vitro trials two pig hearts were used because of their similarity to human ones, as is demonstrated by their being used for biological valve replacement operations.

Before the implant 7 staples were also placed around the ring valve of each heart so that later the induced reduction in the mitral ring could be seen by X-ray. However, the difficulty in taking X-rays in the same position after the operation as before it, led to this process being rejected as a tool for quantitative evaluation. In the end, the videos and photographs taken during the activation of the memory effect turned out to be better for evaluating the results.

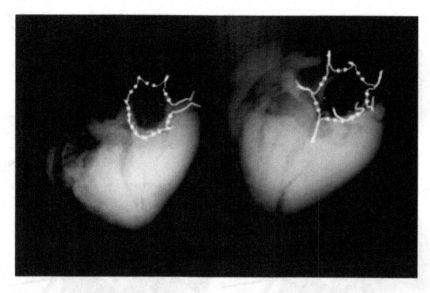

Fig. 10. Active annuloplasty ring prototypes implanted in two pig hearts

Figure 11 shows the process for activating the shape-memory process in the ring and reducing the associated mitral ring cross section. The 4,7 Ω resistances were each supplied with power in series by a transformer at 12 V, which obtained an intensity of 364 mA, similar to what can be supplied by implantable commercial devices. The images were taken at 30 second intervals and show a 10.7% reduction in cross section in an operating period of 150 seconds. This shows a 71%, recovery compared to effort since the increase in cross section induced was 15%.

By interrupting heating the temperature decreases and the recovery process is halted, which means the required effect can be obtained step by step. By recommencing the heating process the recovery will continue, although in these first in vitro trials heating was done continuously in order to evaluate the maximum recovery that could be obtained and the duration of the entire process. Temperature was measured continuously using a thermocouple.

Future prototypes will have a power source similar to that used in other implantable devices such as pacemakers or drug supply pumps, and in addition with a protective textile coating (which will also act as heat insulation). On the other hand, the power needed to activate the memory effect in the end device at a temperature ranging from 41 to 43 °C will be considerably lower than in the first in vitro trials that were performed with polyurethane resin prototypes, as explained before. This lower voltage (around 10 times less) for the end prototypes will lead to lower consumption and smaller size components, making it easier to obtain a biocompatible device.

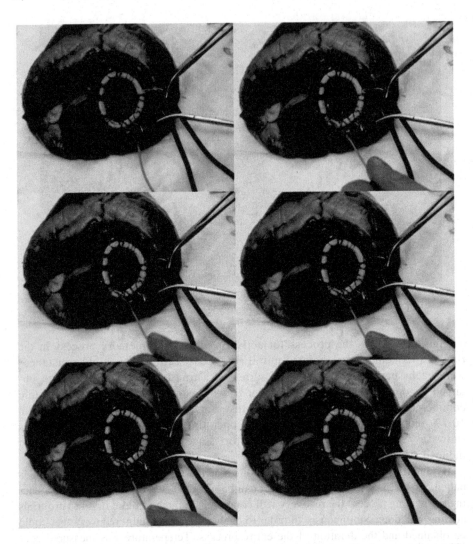

Fig. 11. Activating the shape-memory effect using heating resistances. The images, taken at 30 second intervals, are ordered from left to right and top to bottom and show the progressive closure of the mitral ring.

Despite it being desirable for cross section reduction to reach 15% to 20%, it is very important to point out the material's capacity for recovery by overcoming the forces imposed by the mitral ring of the hearts used.

The conclusions of the study carried out, the possible improvements to the annuloplasty system and the operating process that can be performed in the short term, are set out below.

6 Future Improvements and Conclusions

For the postoperative and progressive treatment of mitral insufficiency the use of an annuloplasty device made of shape-memory polymer has been proposed, with electrical resistances distributed inside it to activate the "shape-memory" effect so that the required change in shape can be induced to reduce mitral regurgitation.

This provides an alternative to current devices, that do not permit any change of shape after implantation, and therefore any errors committed during the operation cannot be corrected.

The design, manufacturing, "shape-memory" effect programming and in vitro trials of an annuloplasty ring for treating mitral insufficiency, developed by using shape-memory polymers, have been presented. This was done in collaboration with researchers from Universidad Politécnica de Madrid (UPM) and doctors from the Hospital Central de la Defensa Madrid.

Using computer aided manufacture and design technologies has enabled different designs and prototypes to be produced in parallel, as well as rapid improvements to obtain the devices that were used in the in vitro trials.

Future actions regarding improvements in the shape-memory programming process lead to optimising the reduction in mitral ring cross section up to the required 15% to 20%. Using alternative shape-memory polymers with a lower activation temperature will also result in more suitable devices, since they will require a smaller size heating system and will be easier to manufacture.

However, it is very important to point out the material's capacity for recovery by overcoming the forces imposed by the mitral ring of the hearts used, which shows the feasibility of developing an active annuloplasty system based on the use of shape-memory polymers.

References

1. Díaz Rubio, M., Espinós, D.: Tratado de Medicina Interna. Editorial Médica Panamericana (1994)
2. Gómez Durán, C.: Estado Actual de la Cirugía Mitral Reconstructiva. Rev. Esp. Cardiol. 57, 39–46 (2004)
3. Hernández, J.M., et al.: Manual de Cardiología Intervencionista. Sociedad Española de Cardiología, Sección de Hemodinámica y Cardiología Intervencionista (2005)
4. Carpentier, A.: Cardiac Valve Surgery - The French Correction J. Thorac. Cardiovasc. Surg. 86(3), 323–337 (1983)
5. Duran, C.M.G.: Duran Flexible Annuloplasty Repair of the Mitral and Tricuspid Valves: Indications, Patient Selection, and Surgical Techniques Using the Duran Flexible Annuloplasty Ring. Medtronic Inc (1992)
6. Okada, Y., et al.: Comparison of the Carpentier and Duran Prosthetic Rings Used in Mitral Reconstruction. Ann. Thorac. Surg. 59, 658–662 (1995)
7. Flameng, W., et al.: Recurrence of Mitral Valve Regurgitation after Mitral Valve Repair in Degenerative Valve Disease. Circulation 107, 1609–1613 (2003)
8. Gillinov, A.M., Cosgrove, D.M., et al.: Mitral Valve Repair. Cardiac Surgery in the Adult, pp. 933–950. McGraw-Hill, New York (2003)

9. Kaye, D., et al.: Feasibility and Short-Term Efficacy of Percutaneous Mitral Annular Reduction for the Therapy of Heart Failure-Induced Mitral Regurgitation. Circulation 108, 1795–1797 (2003)
10. St. Goar, F., et al.: Endovascular Edge-to-Edge Mitral Valve Repair: Short-Term Results in a Porcine Model. Circulation 108, 1990–1993 (2003)
11. Yamaura, Y., et al.: Three-dimensional Echocardiographic Evaluation of Configuration and Dynamics of the Mitral Annulus in Patients Fitted with an Annuloplaty Ring. J. Heart Valve Dis. 6, 43–47 (1997)
12. Lendlein, A., Kelch, S.: Shape-Memory Polymers. Angew. Chemie. Chem. Int. 41, 2034–2057 (2002)
13. Lendlein, A., Kelch, S., Kratz, K.: Shape-Memory Polymers. Encyclopedia of Materials: Science and Technology (2005)
14. Tonmeister, P.A.: Shape-Memory Polymers Reshape Product Design. Plastics Engineering 14, 10–11 (2005)
15. Volk, B., et al.: Characterization of Shape-Memory Polymers. NASA Langley Research Centre. Texas A&M University (2005)
16. Lendlein, A., Langer, R.: Biodegradable elastic shape-memory polymers for potential biomedical applications. Science 296, 1673–1676 (2002)
17. Lendlein, A., et al.: Light-induced Shape-Memory Polymers. Nature 434, 879–882 (2005)
18. Wilson, T., et al.: Shape-Memory Polymer Therapeutic Devices for Stroke. In: Proc. SPIE, vol. 6007, pp. 157–164 (2005)
19. Small, W., et al.: Laser-activated Shape-Memory Polymer Intravascular Thrombectomy Device. Optics Express 13, 8204–8213 (2005)
20. Nusskern, H.: Thermische Stellelemente in der Gerätetechnik. Feinwerktechnik, Mikrotechnik, Messtechnik, 9, Carl Hanser Verlag (1995)
21. Pelton, A., Stöckel, D.: Medical uses of Nitinol. Materials Science Forum 327–328, 63–70 (2000)
22. Tautzenberger, P., et al.: Vergleich der Eigenschaften von Thermobimetallen und Memory-Elementen. Metall 41, 26–32 (1987)
23. Shandas, R., Mitchell, M., et al.: A general method for estimating deformation and forces imposed in vivo on bioprosthetic heart valves with flexible annuli: in vitro and animal validation studies. J. Heart Valve Dis. 10, 495–504 (2001)
24. Mohr, R., et al.: Initiation of shape-memory effect by inductive heating of magnetic nanoparticles in thermoplastic polymers. PNAS 103, 3540–3545 (2006)
25. Kucklick, T.R.: The Medical Device R&D Handbook. CRC Taylor & Francis (2005)
26. Schwarz, M.: New Materials Processes, and Methods Technology. CRC Taylor & Francis (2005)
27. Freitag, D., Wohlers, T.: Rapid Prototyping: State of the Art. Manufacturing Technology Information Analysis Centre (2003)
28. Lafont, P., et al.: Rapid Tooling: Moldes Rápidos a Partir de Estereolitografía. Rev. Plast. Mod. 79, 150–156 (2000)

Towards a Probabilistic Manipulator Robot's Workspace Governed by a BCI

Fernando A. Auat Cheeín[1], Fernando di Sciascio[1], Teodiano Freire Bastos Filho[2], and Ricardo Carelli[1]

[1] Institute of Automatic, National University of San Juan
Av. San Martin 1109-Oeste, San Juan, Argentina
{fauat,fernando,rcarelli}@inaut.unsj.edu.ar
[2] Electrical Engineering Deparment, Federal University of Espirito Santo
Av. Fernando Ferrari 514, Vitoria, Brazil
tfbastos@ele.ufes.br

Abstract. In this paper, probabilistic-based workspace scan modes of a robot manipulator are presented. The scan modes are governed by a Brain Computer Interface (BCI) based on Event Related Potentials (Synchronization and Desynchronization events). The user is capable to select a specific position at the robot's workspace, which should be reached by the manipulator. The robot workspace is divided into cells. Each cell has a probability value associated to it. Once the robot reaches a cell, its probability value is updated. The mode the scans are made is determined by the probability of all cells at the workspace. The updating process is governed by a recursive Bayes algorithm. A performance comparison between a sequential scan mode and the ones proposed here is presented. Mathematical derivations and experimental results are also shown in this paper.

Keywords: Brain-Computer Interface, Robot Manipulator, Probabilistic Scan Mode.

1 Introduction

Brain Computer Interfaces have got a great impulse during the last few years. The main reasons for this growing are the availability of powerful low-cost computers, advances in Neurosciences and the great number of people devoted to provide better life conditions to those with disabilities. These interfaces are very important as an augmentative communication and as a control channel to people with disorders like amyotrophic lateral sclerosis (ALS), brain stroke, cerebral palsy, and spinal cord injury ([1] and [2]).

The main point of a BCI is that the operator is capable to generate commands using his/her EEG (electroencephalographic) signals in order to accomplish some specific actions ([2], [3], [4] and [5]). Thus, an operator using a BCI can control, for example, a manipulator, a mobile robot or a wheelchair (amongst other devices). The EEG frequency bands have enough information to build an alphabet of commands in order

A. Fred, J. Filipe, and H. Gamboa (Eds.): BIOSTEC 2008, CCIS 25, pp. 73–84, 2008.
© Springer-Verlag Berlin Heidelberg 2008

to control/command some kind of electronic device [6]. In this paper a BCI, which is controlled through alpha waves from the human brain, is used. Although the EEG signal acquisition/conditioning, which is part of this BCI, was developed in other work of the authors ([7]), one of the objectives of this paper is to illustrate its versatility, mainly in terms of the simple algorithms used.

Event related potentials (ERP) in alpha frequency band are used here. Such potentials are ERD (Event Related Desynchronization) and ERS (Event Related Synchronization), well described in the following sections. This BCI has a Finite State Machine (FSM) which was tested in a group of 25 people.

The main contributions of this paper are the scan mode algorithms proposed to allow the user to command a manipulator (Bosch SR-800), based on a probabilistic scan of the robot's workspace. The workspace is divided into cells. Each cell contains three values: its position (x, y) at the robot's workspace plane and a probability value. This value indicates the accessibility of that element. Once a particular cell is accessed, its probability is updated based on *Bayes'* rule.

This paper is organized as follows: a brief description of the sequential scan mode of the manipulator's workspace is presented in section 2. The probabilistic scan modes proposed are shown in Section 3. Section 4 shows the results for a *Montecarlo* experimentation, where the probabilistic evolution of the whole workspace and of a specific cell is presented. Section 5 shows the conclusions of this work.

2   Sequential Scan Mode

As a brief introduction, the sequential scan mode of the robot workspace developed in [7] is presented here.

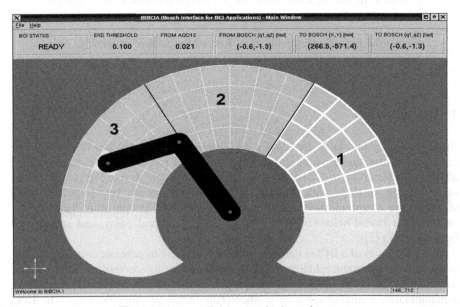

Fig. 1. Main zone division at robot's workspace

The workspace is previously divided into three main zones as it can be seen in Fig. 1. The system iteratively scans from *zone 1* to *zone 3* until one of them is selected by the user (using EEG signals). Once it is so, the selected zone is scanned row by row until one is selected. Once a row is selected, the system scans cell by cell (switching columns) iteratively inside the selected row. After a cell is selected by the user, the robot reaches the position given by that cell.

3 Probabilistic Scan Modes

The two probabilistic scan modes shown in this paper are based on *Bayes* rule for updating probability values of the cells at the manipulator's workspace. The scan modes are shown in the following sections.

3.1 First Approach of a Probabilistic Scan Mode

The first approach of a probabilistic scan mode works as follows:

1. The workspace's resolution is set to 72 cells and can be easily changed, decreasing or increasing this number. The workspace behaves as a *pmd* (probabilistic mass distribution).
2. Each cell has its own initial probability. This value can be previously determined by some heuristic method (for example: if the BCI operator is right-handed, then cells to the right of the workspace will have higher accessing probability than the ones to the left). However, it is also possible to set all cells to a probability near zero, in order to increase or decrease them depending on the times they are accessed by the user. In this work, the first case was adopted.
3. Let a and b be the higher and lower probabilities cells respectively. Then, the workspace is divided into three zones according to these values. Table 1 shows how division is made. Let $P(C_i \mid G)$ be the probability of cell C_i given a group G to which it belongs.
4. Every zone at the workspace is divided in three sub-zones under the same philosophy presented before. Each one of these sub-zones contains a set of probabilistic weighted cells.
5. The scan mode proceeds as follows:

   I. First, the zone with the highest probability value at the workspace is highlighted. If that zone is not selected by the operator, the second highest probabilistic zone is highlighted. If it is not selected, the highlight passes to the third and last zone. The scan keeps this routine until a zone is selected.

   II. When a zone is selected, the highlight shows first the sub-zone with the highest probability inside the zone previously selected. The scan, in this case, is exactly the same used in the last step.

   III. When a sub-zone is selected, then the scan highlights first the cell with the highest probability of occupancy. If it is not selected, the scan passes to the next cell value. This routine keeps going on until a cell is selected. Once a position is selected, the probability value of the cell, sub-zone, zone and complete workspace is updated. The update of the probabilities values is made by the *Bayes'* rule.

Table 1. Workspace's Zones Definitions

a	highest probability cell value
b	lowest probability cell value
$\left\{ c_i : b + \dfrac{2}{3}(a-b) < P(C_i \mid G) \le a \right\}$	zone 1: the set of all cells which probabilities are the highest of the workspace
$\left\{ c_i : (b + \dfrac{(a-b)}{3}) < P(C_i \mid G) \le (b + \dfrac{2}{3}(a-b)) \right\}$	zone 2: the set of all cells which probabilities are of middle range
$\left\{ c_i : b \le P(C_i \mid G) \le (b + \dfrac{(a-b)}{3}) \right\}$	zone 3: the set of all cells with the lower probability of the workspace.

As it can be seen, the number of cells that belong to a sub-zone or a zone is variable. Then, the organization of the zones at robot's workspace is dynamic. This allows improving the scan mode in order to access in a priority way to the most frequently used cells.

The probability update of each cell at the workspace is based on the recursive *Bayes'* rule. Once a cell is reached by the user, its probability value changes according to eq. (1).

Let C be any cell at robot's workspace and G a set to which that cell belongs. Thus, the updating algorithm is given by,

$$P_k(C \mid G) = \frac{P_k(G \mid C)P_{k-1}(C \mid G)}{P_k(G \mid C)P_{k-1}(C \mid G) + P_k(G \mid \overline{C})P_{k-1}(\overline{C} \mid G)}. \tag{1}$$

Though (1) is mainly used in very simple applications [7], it fits as an updating rule for the purpose of this work.

Equation (1) can be re-written in (2), where a scale factor was used.

$$P_k(C \mid G) = \eta P_k(G \mid C) P_{k-1}(C \mid G). \tag{2}$$

According to the Total Probability Theorem [8], η is the scale factor, which represents the total probability of $P(G)$. In (1), $P_{k-1}(C \mid G)$ is the prior probability of a cell given the primary set to which it belongs at time $k-1$. $P_k(G \mid C)$ is the transition probability which represents the probability that a given cell C belong to a set G. Finally, $P_k(C \mid G)$ is the posterior probability -at instant k- of the cell used given the zone to which it belongs.

In order to make sense to the use of the recursive *Bayes* algorithm, an initial probability value must be given to all cells at the workspace.

Figure 2 shows the evolution of a cell's probability when it is accessed successively by the user.

The cell used in Fig. 2, for example, has an initial value of 0.05 but it is increased each time the cell is accessed by the user. As was expected, the maximum value a cell can reach is one. When this situation occurs, the whole workspace is scaled. This scaling does not change the scan mode because the relative probability information

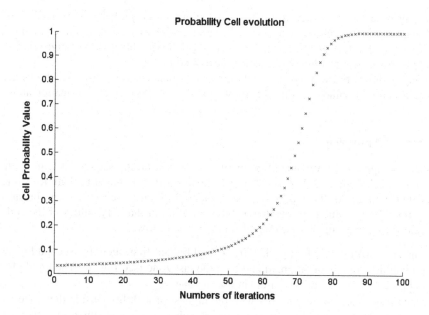

Fig. 2. Evolution of Cell's probability when successively accessed

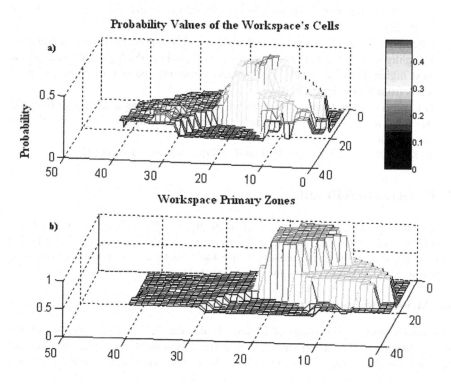

Fig. 3. Probabilistic distribution of a workspace for a right-handed user

remains without changes, i.e., if a cell p has the maximum probability over all cells, after scaling, p will continue being the cell with the highest weight. A more extended development of this algorithm can be seen at [9]. Once the updating algorithm is complete, the scan algorithm is released as described in Section 3.

Figure 3 shows the workspace's *pmd* for a right-handed user. Fig. 3.a shows the cells probability's value and Fig. 3.b shows the different zones of the manipulator's workspace.

3.2 Second Approach

This second approach investigated in this work is based on the sequential scan mode algorithm. Each zone or sub-zone -as those shown in Fig. 1- has a probability value associated with it. As the workspace is considered as *pmd* then each zone or sub-zone's probability value is calculated as the sum of all probability values of the cells that belong to that group. The scan mode proceeds as follows:

1. The zone with the highest probability is highlighted first; then, the second higher probability zone is highlighted and then the last zone (see Fig. 1). The highlighting process repeats until the user chooses a zone.
2. Once a zone is chosen, the row with the highest probability -inside that zone- is highlighted. A row of a zone is known as sub-zone. If this sub-zone is not selected by the user after a period of time, the highlight passes to the next higher probability value row. This process is repeated iteratively until a row is selected by the user.
3. Once a sub-zone is chosen, the cell with the highest probability of that sub-zone is highlighted. If it is not chosen after a period of time, the highlight passes to the next higher probability cell. The process continues and if no cell is chosen, it starts from the beginning cell.
4. If a cell is chosen, then its probability is updated according to the *Bayes* rule (Eq. 3). Then, workspace *pmd*, sub-zone's probabilities and all zone's probabilities are also updated.
   The sampling time used in all scan modes is the same one used in [7].

4 Experimental Results

This section is entirely dedicated to compare the three scan types: sequential and probabilistic ones. For this purpose, a *Montecarlo* experiment was designed [10]. This experiment shows the performance of the three methods by measuring the time needed to reach different cells at the robot's workspace.

4.1 Montecarlo Experiment

The robot's workspace consists of 72 cells. It also can be considered as a 4×18 matrix. According to this, a cell's position is defined by a number of row and a number of column at that matrix. The number of a row and a column can be considered as a random variable. To generate a random position of a cell destination, the following algorithm was implemented.

i. An uniform random source generates two random variables: x and y.
ii. The random variable x is mapped into the rows of the 4×18 matrix workspace.
iii. The random variable y is mapped into the columns of the 4×18 matrix workspace.
iv. When a position is generated, both scan types begin. The time needed to reach the cell is recorded.
v. After the system reaches the position proposed, a next process point generation is settled -the algorithm returns to point i-.

4.2 Mapping Functions

Let f_x be a mapping function such as:

$$f_x : A \rightarrow B$$
$$x \rightarrow m$$

where,

$$\begin{cases} A = \{x : x \in [0,1) \subset \Re\} \\ B = \{m : m \in \{1,2,3,4\} \subset \aleph\} \end{cases} \quad (3)$$

and let f_y be another mapping function such as:

$$f_y : A \rightarrow C$$
$$y \rightarrow n$$

where,

$$\begin{cases} A = \{y : y \in [0,1) \subset \Re\} \\ B = \{n : n \in \{1,2,3,...,18\} \subset \aleph\} \end{cases} \quad (4)$$

Equations (3) and (4) show the domain and range of the mapping functions. Finally, the mapping is made according to the following statements.

i. Let δ be the sum of all weights at robot's workspace, that is, $\delta = \sum_{i \in B} \sum_{j \in C} P_{ij}$, where

P_{ij} is the probability value of a cell located at the $i-$row and $j-$column.

ii. Let $x \in A$ be an outcome of the uniform random source for f_x.

- If $0 \le x < \dfrac{\sum_{i=1, j \in C} P_{ij}}{\delta}$ then $f_x(x) = i = 1$. This means that the value of $x \in A$ should

be lower than the sum of all cell's values in row one -over δ - to $f_x(x)$ be equal to one.

- If $\dfrac{\sum\limits_{i=1,j\in C} P_{ij}}{\delta} \le x < \dfrac{\sum\limits_{i=2,j\in C} P_{ij}}{\delta}$ then $f_x(x)=i=2$. This means that $x \in A$ should be

  greater or equal to the sum of all cell's values in row one and lower than the sum of all cell's values in row 2.

- The same process continues up to the last row, whose expression is: if

  $\dfrac{\sum\limits_{i=3,j\in C} P_{ij}}{\delta} \le x < \dfrac{\sum\limits_{i=4,j\in C} P_{ij}}{\delta}$ then $f_x(x)=i=4$.

- Each time a cell is selected, the mapping functions vary. It is so because they are dependent with the probability value of the cells.

- For the mapping over the columns, the procedure is the same, however in this case, the sum is made over the set B (four rows).

Concluding, the mapping presented here is dynamic because it is updated each time a cell varies its probability value. For the case implemented in this work (a right-handed user) the initial mapping functions are represented in Figs. 4.a and 4.b. In Fig. 4.b is also possible to see that column 10 has higher probability than column 1. It is also important to see that, if all cells at robot's workspace have the same probability weight, then the mapping functions would be uniform. Thus, each row or column would have the same probability to be generated.

4.3 Montecarlo Simulation Results

The objective of *Montecarlo* experiments was to test the performance of both scanning methods: probabilistic and sequential ones. The performance is measured in function of the time needed to access a given position. This position is generated by the uniform random source. After 500 trials the mean time needed to access a random position by the first approach of the probabilistic scan was of 20.4 seconds. For the second approach of the probabilistic scan the mean time needed was of 16.8 and for the sequential scan was of 19.8 seconds. The three results are in the same order but the probabilistic second approach of the scan mode requires less time.

Consider now only the right side of the workspace, which is, according to Fig. 3, the most accessed side. The mean time of access for all points belonging to the workspace right side is of 8.4 seconds under the first approach of the probabilistic scan instead of 11.3 seconds corresponding to the second approach of the probabilistic scan mode. Under sequential scan, the mean time is of 14.8 seconds. The probabilistic scan mode first approach is 43% faster than the sequential scan for cells over the right side of the workspace while the second approach is 23.7% faster.

Figure 5 shows how a low probability valued cell in the probability scan first approach evolves after successive callings. The cell passes through the different zones of cells according to its actual probability value. After 240 iterations -or callings-, the cell has passed through three zones and its performance has also been improved as long as its weight. In Fig. 5, one can see that at the beginning, 32 seconds were needed to access that cell.

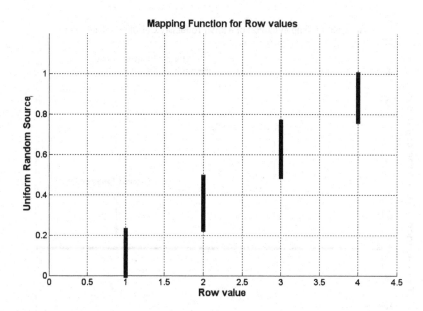

Fig. 4. (a). Mapping function for the four values of rows

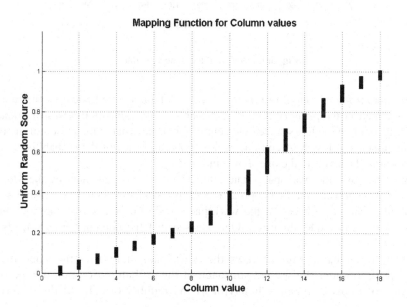

Fig. 4. (b). Mapping function for the 18 values of columns

After 240 iterations, only 14 seconds were needed. This time is smaller than the one needed on the sequential scan mode which is of 18 seconds. Figure 5 also shows when the cell changes zones. Thus, if its probability increases, the cell passes from,

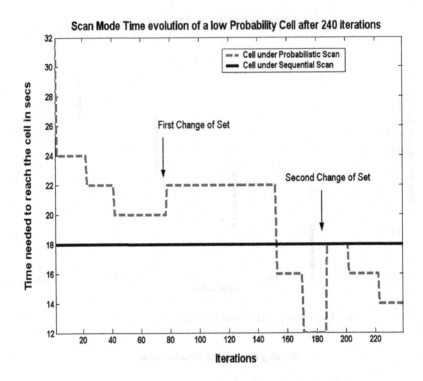

Fig. 5. Evolution of a cell access time

for example, *primary zone 2* to *primary zone 1*. Although a cell could be the first in being scanned in the *primary zone 2*, if it increases its value and passes to *primary zone 1*, it could be the last scanned element in this zone. That is the reason of the two time increments in Fig. 5. A cell under the second approach of the probabilistic scan shows similar behavior to the one shown in Fig. 5.

Figure 6 shows the workspace state after 500 iterations generated by the *Montecarlo* experiment using the first approach of the probabilistic scan. Fig. 6.a shows the probability state of each cell at the workspace while Fig. 6.b shows the new three zones of the scan mode algorithm. One can see that the non-connectivity tends to disappear.

On the other hand, Fig. 7 shows the workspace state after the same iterations of Fig. 6 under the second approach of the probabilistic scan, though this scan do not imply a dynamic behavior of the number of cells of the different zones.

As it can be seen from Figs. 6 and 7, probabilistic distribution of the workspace depends on the type of scan mode used. Both probabilistic scan modes presented in this work show a better performance respect to the sequential scan mode.

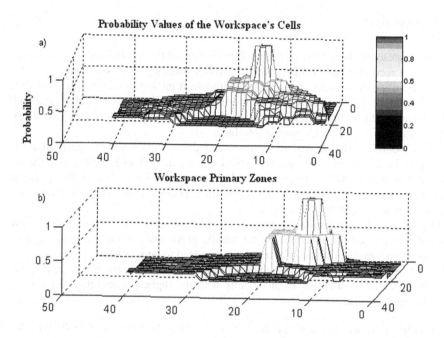

Fig. 6. Workspace state after 500 iterations

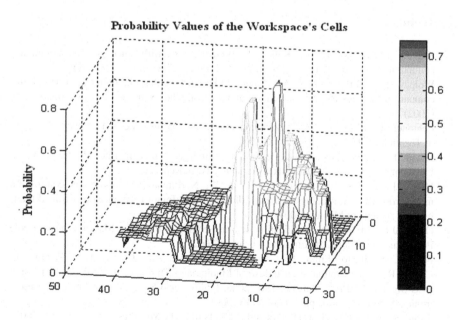

Fig. 7. Workspace state after 500 iterations under the second approach of the probabilistic scan mode

5 Conclusions

The work presented here showed the implementation of two probabilistic scan modes, based on a recursive *Bayes* algorithm, of a robot manipulator's workspace. A comparison between these methods and a sequential scan mode showed that the probabilistic scan improves the access time of the most frequently accessed cells. Although this system could be implemented in several Human-Machine Interfaces, it was primary designed for a Brain-Computer Interface.

Experimental results show that the time needed to access a specific position at the workspace is decreased each time the position is reached. This is so because the recursive Bayes algorithm implemented updates the probability value of that position once it is reached. A decrement of the access time means that the user of the Interface needs less effort to reach the objective.

In this work, a right-handed workspace distribution case was presented. This case showed that all cells to the right of the middle point -half of the main workspace-have the higher probability and the lower time needed to be accessed.

Finally, it is possible to say that the system learns the user's workspace configuration. It pays special attention to those cells with the highest probability minimizing the time needed to access them.

Acknowledgements. The authors thank CAPES (Brazil), SPU and CONICET (Argentina) and FAPES (Brazil), for their financial support to this research.

References

1. Kubler, A., Kotchoubey, B., Kaiser, J., Wolpaw, J.R., Birbaumer, N.: Brain-computer communication: unlocking the locked. Psychol. Bull. 127(3), 358–375 (2001)
2. Wolpaw, J.R., Birbaumer, N., McFarland, D.J., Pfurtscheller, G., Vaughan, T.M.: Brain-computer interfaces for communication and control. Clin. Neurophysiol. 113(6), 767–791 (2002)
3. Lehtonen, L.: EEG-based brain computer interfaces. Master's thesis, Helsinky University of Technology, Helsinky, Finlandia (2003)
4. Felzel, T.: On the possibility of developing a brain-computer interface (BCI). Technical University of Darmstadt, Darmstadt, Germany, Tech. Rep. (2001)
5. Millán, J., Renkens, F., Mouriño, J., Gerstne, W.: Non-invasive brain-actuated control of a mobile robot. In: Proceedings of the 18th International Joint Conference on Artificial Intelligence, Acapulco, Mexico (2004)
6. Ochoa, J.B.: EEG brain classification or brain computer interface, Master's thesis, Ecole Polytechnique Federale de Lusanne, Lusanne (2002)
7. Ferreira, A., Bastos Filho, T.F., Sarcinelli Filho, M., Auat Cheeín, F.A., Postigo, J., Carelli, R.: Teleoperation of an Industrial Manipulator Through a TCP/IP Channel Using EEG Signals. In: Proceedings of ISIE 2006, International Symposium on Industrial Electronics, Montreal, vol. 1, pp. 3066–3071 (2006)
8. Thrun, S., Burgard, W., Fox, D.: Probabilistic Robotics. The MIT Press, Massachusetts (2005)
9. Papoulis, A.: Probabilidad, variables alatorias y procesos estocásticos. Eunibar, Barcelona (1980)
10. Ljung, L.: System Identification. Prentice Hall, New Jersey (1987)

New Soft Tissue Implants Using Organic Elastomers

David N. Ku

GWW School of Mechanical Engineering
Georgia Institute of Technology
Atlanta, GA 30332-0405
USA

Abstract. Typical biomaterials are stiff, difficult to manufacture, and not initially developed for medical implants. A new biomaterial is proposed that is similar to human soft tissue. The biomaterial provides mechanical properties similar to soft tissue in its mechanical and physical properties. Characterization is performed for modulus of elasticity, ultimate strength and wear resistance. The material further exhibits excellent biocompatibility with little toxicity and low inflammation. The material can be molded into a variety of anatomic shapes for use as a cartilage replacement, heart valve, and reconstructive implant for trauma victims. The biomaterial may be suitable for several biodevices of the future aimed at soft-tissue replacements.

Keywords: Biomaterial, mechanics, elasticity, strength, wear, biocompatibility, medical devices.

1 Introduction

Most of the existing biomaterial technology is limited to materials such as silicones, Teflon®, polyethylene, metals and polyurethanes that do not exhibit the mechanical and physical properties of natural tissue. These materials are stiff, difficult to manufacture, and not initially developed for medical implants. Artificial tissue substitutes have not been found to withstand the rigors of repetitive motion and associated forces of normal life. Cadaver tissue is limited in supply and due to the risk of infection is coming under increased scrutiny by FDA and is not accepted in Europe.

As an example, one of the most successful medical implants is the artificial knee replacement for the treatment of arthritis. Arthritis and joint pain as a result of injury are major medical problems facing the US and patients worldwide. Worldwide, approximately 190 million people suffer from osteoarthritis. This condition affects both men and women, primarily over 40 years of age. The spread of arthritis is also fueled by the rise of sports injuries. Activities enjoyed by many can translate into injury or joint damage that may set up a process of deterioration that can have devastating effects decades later. The number of patients that have arthritis is staggering and growth is expected with baby boomers entering the prime "arthritis years" with prolonged life expectancies. Growth of the world's elderly is expected to increase three times faster than that of the overall population.

A. Fred, J. Filipe, and H. Gamboa (Eds.): BIOSTEC 2008, CCIS 25, pp. 85–95, 2008.

Current standard treatment is to surgically implant a total knee replacement (TKR) that is made of metals such as cobalt chromium or titanium. These devices are highly rigid, providing no shock absorption. Further, the metal integrates poorly with bone and the HMWPE caps often create microparticulate debris with strong inflammation. The invasive surgery is not indicated for those under age 60 and usually reserved for end-stage patients. A revision procedure is technically demanding, and amputation may be required.

An alternative biodevice may be a soft tissue replacement. Arthritis stems from damage to cartilage, the soft tissue between the bones. Damaged cartilage leads to grinding of bone on bone and eventual pain and limited joint function. Biodevices that replace the soft tissue would restore diarthroidal joint function much better and protect further damage by a more natural stress distribution.

A similar problem exists for heart valves. Prosthetic heart valves are made from metal and pyrolitic carbon which do not function like native heart valves. The use of hard materials creates high shear zones for hemolysis and platelet activation. Tissue valves are subject to calcification, again a problem of hard tissues not acting like natural soft tissue. An alternative would be to design a biodevice with soft tissue flexibility and endothelial cell covering to provide a wide-open central flow and low-thrombogenic surface.

Yet another example for the need for soft tissue replacements is a replacement part for reconstructive surgery after traumatic accidents or cancer resection. For many parts of the body, a replacement shape needs to have smooth contours as well as soft tissue compliance to yield a natural shape. The base biomaterial should not have a chemical composition that is non-organic, such as silicone, which can induce a hyperimmunogenic response.

Soft tissue replacements should start with a biomaterial that has compliance ranges similar to human soft tissue, be strong and wear resistant, manufactured to personal shapes, and have long-term biocompatibility. Cellular in-growth or preloading of cells can then be performed on this established scaffold. These features are demonstrated in a new biomaterial described in this paper.

2 Methods

2.1 Biomaterials

Soft tissue-like devices can be made from polymers such as poly vinyl alcohol as thermoset materials. As an example, a PVA cryogel can be made according to the full descriptions in US Patent U.S. Patent Numbers 5,981,826 and 6,231,605. The cryogels are made in a two stage process. In the first stage a mixture of poly (vinyl alcohol) and water is placed in a mold, and repeatedly frozen and thawed, in cycles, until a suitable cryogel is obtained.

Poly(vinyl alcohol) having an average molecular weight of from about 85,000 to 186,000, degree of polymerization from 2700 to 3500, and saponified in excess of 99% is preferred for creating soft tissue-like mechanical properties. High molecular weight poly (vinyl alcohol) in crystal form is available from the Aldrich Chemical Company. The PVA is then solubilized in aqueous solvent. Isotonic saline (0.9% by weight NaCl, 99.1% water) or an isotonic buffered saline may be substituted for water to prevent osmotic imbalances between the material and surrounding tissues if the cryogel is to be used as a soft tissue replacement.

Once prepared, the mixture can be poured into pre-sterilized molds. The shape and size of the mold may be selected to obtain a cryogel of any desired size and shape. Vascular grafts, for example, can be produced by pouring the poly (vinyl alcohol)/water mixture into an annular mold. The size and dimensions of the mold can be selected based upon the location for the graft in the body, which can be matched to physiological conditions using normal tables incorporating limb girth, activity level, and history of ischemia.

A new biomaterial, commercially available as Salubria® from SaluMedica, LLC, Atlanta, GA is similar to human tissue in its mechanical and physical properties. The base organic polymer is known to be highly biocompatible and hydrophilic (water loving). The hydrogel composition contains water in similar proportions to human tissue. Unlike previous hydrogels, Salubria is wear resistant and strong, withstanding millions of loading cycles; yet it is compliant enough to match normal biological tissue. The material can be molded into exact anatomic configurations and sterilized without significant deterioration.

2.2 Mechanical Characterization

2.2.1 Tensile Testing
Tensile test specimens were cut from sheets of Salubria. They were tested in accordance with ASTM 412 (die size D) in tension to failure using an Instron Model 5543 electro-mechanical load frame pulling at a rate of 20 inches per minute.

2.2.2 Stress-Strain Constitutive Relationship
The stress is a function of the load and the cross-sectional area. However, the cross-sectional area is difficult to measure for soft-tissues. The stretch ratio relates the final and initial area with an assumption of material incompressibility. Therefore, the ultimate stress calculation is a function of the load at the breaking point of the sample, the stretch ratio and the initial cross-sectional area. The initial cross-sectional area is the product of the initial width of the sample, w_o, and the initial thickness, t_o.

Stretch Ratio: $\lambda = C / C_o$ (1)

Initial Cross-Section Area: $A_o = w_o * t_0$ (2)

Final Stress: $\sigma_{ult} = \dfrac{F_{ult} * \lambda}{A_o}$ (3)

In order to get an estimation of the pressure in an intact tube the following simplified assumptions were used. It was assumed that a tubular specimen will burst when the circumferential wall stress is equal to the ultimate stress σ_{ult}. However, when an artery is under physiologic load conditions it is in a state of plane strain and undergoes finite two dimensional stretches.

The stretch ratios are:

$$\lambda_\theta = \frac{r}{R} \qquad\qquad \lambda_z = \frac{l}{L}$$

Rewritten to solve for the pressure, P:

$$P = \frac{T\sigma_\theta}{\lambda_\theta^2 \lambda_z R}$$

This equation can be used to convert pressure and stress using the stress and the stretch ratios for tubular specimens.

2.2.3 Unconfined Compression, Creep, and Creep recovery

Cylindrical unconfined compression samples were cast in a custom mold. Samples were tested in unconfined compression on an Instron Model 5543 electro-mechanical load frame and on a DMTA IV dynamic mechanical analyzer. Rates of 1% strain per second and 20 inches per minute were tested.

2.2.4 Ultimate Strength

Ring specimens were pulled in tension until they failed. Ring specimens of Salubria were preconditioned twenty five cycles. Then the specimens were distracted at 0.1mm/s, 1m/s, 10 mm/s, and 100mm/s using a MTS 858 Mini Bionix Test System. Comparisons are made to normal coronary arteries using identical protocols. The load at failure was recorded as the ultimate load, and the ultimate stress was calculated. Failure of the ring specimen was defined as a complete tear of the ring through the entire wall. The stress was derived based on the assumption of incompressibility and was defined as the ratio of load and cross-sectional area. The stretch ratio was defined as the ratio of the final and initial circumference. The final stress at failure represented the ultimate strength for the tension tests. To determine the final stress, an equation was derived based on the assumption of incompressibility [3] which means that the initial volume Vo and final volume V are equal. In the present experiments the stretch ratio is defined as the ratio of the final and initial circumference, Equation (1). The ultimate stress, σ_{ult} defined by the load at the breaking point of the sample divided by the final cross-sectional area, was calculated using Equation (3).

2.2.5 Fatigue Resistance

Ring specimens were cycled at different cycles, and then pulled in tension until failure. The frequency of the cyclic tests was set at 2 Hz because this value is close to physiologic frequency of heart beats (~1.2 Hz) and strain rates effects testing showed that there are no significant difference to do cyclic test under 1HZ to 5HZ. The purpose of the cyclic tests was to experimentally determine how the fatigue affects the ultimate strengths of porcine common carotid arteries.

2.2.6 Cyclic Compression

A cyclic compression study was performed to assess the response of Salubria biomaterial cylinders to repetitive compressive loading at physiologic stress of approximately

1.3 MPa. The specimen was loaded for 1 million loading cycles at a rate of approximately 1.5 Hz. Dimensional integrity was measured using an optical comparator and mechanical modulus of elasticity was determined at 20% strain.

2.2.7 Wear Testing

An accelerated wear tester was built to test the wear rate of Salubria biomaterial. Polished stainless steel rollers with a diameter of 1 5/8 inches (chosen to approximate the average radius of the femoral condyles) are rotated so that they slide and roll across the test sample, creating a peak shear load of approximately 90N (0.2 MPa). Separate testing has shown the coefficient of friction for polished stainless steel against Salubria biomaterial to be equivalent to that of a porcine femoral condyle with the cartilage surface abraded away to subchondral bone, or roughly 4 times that of a porcine femoral condyle with an intact cartilage surface. The rotating cylinders exert a normal load of approximately 200N on the sample. The wear tester was operated at a rate that subjects the sample to 1000 wear cycles per hour where one cycle is defined as one roller to sample contact. Data was collected with the sample lubricated and hydrated with water (a worst case scenario since synovial fluid should provide some surface lubrication). Wear rate was measured by weight loss of the sample over a number of cycles. Salubria biomaterial was tested for >10,000,000 cycles against polished stainless steel rollers.

2.3 Biocompatibility

Biomaterials for use in humans must pass a full complement of biocompatibility tests as specified by ISO and the USFDA. The material was tested for ability to produce cytotoxicity, intracutaneous irritation, sensitization by Kligman maximization, Ames mutagenicity, chromosomal aberration, and chronic toxicity.

2.3.1 Rabbit Osteochondral Defect Model

In addition to the standard biocompatibility testing, the ability of Salubria biomaterial to withstand load or cause local inflammatory responses in a widely used rabbit osteochondral defect model was assessed (e.g., Hanff et al., 1990). A cylindrical plug (3.3 mm diameter, 3 mm depth) was implanted in the right knee of each of sixteen New Zealand white rabbits. An unfilled drill hole was made in the left knee of each animal to serve as an operative control.

After three months of implantation in the patellofemoral groove, eight rabbits were sacrificed for histologic analysis of the implant site and surrounding synovium. The remaining eight animals were sacrificed and the implant assessed for any change in physical properties. In addition, the distant organ sites that are known targets for PVA injected intravenously were assessed histologically for any sign of toxicity due to implantation. Tissues assessed included liver, spleen, kidney, and lymph node.

2.3.2 Particulate Inflammation

Salubria biomaterial was tested for particulate toxicity or inflammatory reaction in the joint. The study design was based on a study conducted by Oka et al. (2000) on a similar PVA-based biomaterial in comparison to UHMWPE. Particulate sizes for the study varied from approximately 1 micron to 1000 microns. The total volume of particulate injected over the 2 divided doses was designed to represent full-thickness wear of 10 x 10 mm diameter cartilage implants.

2.3.3 Ovine Knee Inflammation Model

In vivo testing was performed using a meniscus shaped device made of Salubria and implanted into the sheep knee joint. Devices were removed at 2 week, 3 week, 2 month, 4 month and 1 yr. intervals. Animals were fully load-bearing on the day of operation and after. Full range of motion with no disability was observed. Gross examination of surrounding tissues and histology of target end organs (liver, kidney, spleen, and lymph nodes) were evaluated for acute or chronic toxicity.

3 Results

3.1 Tensile Testing

Plots of stress versus strain in tension (figure 1) show a non-linear response. Due to the non-linearity of the loading curve, tangent modulus values at a defined percent strain are used to characterize the material stiffness. Tangent modulus ranges from 1.2-1.6 MPa. Ultimate tensile strength is 6-10 MPa. The stress-strain curve exhibits a non-linear elastic behavior similar to natural soft tissue.

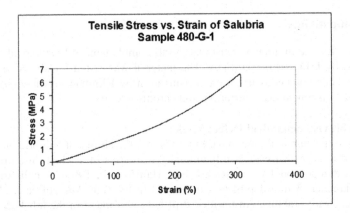

Fig. 1. Representative Tension Curve for Salubria soft tissue organic elastomer

3.2 Unconfined Compression

Figure 2 is a representative curve of stress versus strain in compression. Compression loading curves show a non-linearity suggesting that Salubria is a viscoelastic material similar to cartilage. Tangent modulus values in compression range from 0.1 to 7 MPa. Plastic compressive failure occurs at or above 65% strain.

3.3 Creep and Creep Recovery

Creep and creep recovery experiments were performed to assess the performance of Salubria biomaterial under long-term static loading at physiologic loads of up to 480

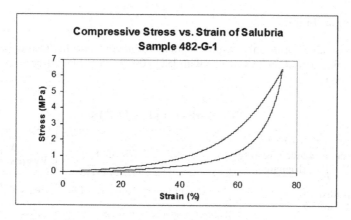

Fig. 2. Representative Compression Curve

N-force. High loads were applied for 24 hours creating deformation of 50% of the initial height. Initial loading demonstrates a biphasic visco-elastic behavior. After 24 hours of recovery in saline, sample height returned to within 5% of the original height. The compressive modulus of the material before the test and after creep recovery was unchanged.

3.4 Ultimate Strength

Sixty-four specimens were pulled at four different stain rates. Ultimate stress increased as a weak function of increasing strain rates. The ultimate stress at 100 mm/s was 4.54 MPa, greater than the 3.26 MPa at 0.1 mm/s. The differences between 0.1mm/s and 100 mm/s was highly significant with $p<0.001$. The differences between 0.1 mm/s and 10 mm/s gave $p=0.013$; and 1 mm/s to 100 mm/s was $p=0.018$. The difference between 1 mm/s and 10 mm/s was not statistically significant. Strain rates between 1 and 100 mm/s correspond to a cyclic frequency of 0.5 Hz to 5 Hz for fatigue testing.

3.5 Cyclic Compression

Repetitive compressive loading at physiologic stress of approximately 1.3 MPa was imposed. After 1 million loading cycles at a rate of approximately 1.5 Hz, there was minimal change (<5%) in sample dimensions and no change in modulus of elasticity at 20% strain.

3.6 Wear Testing

Results for one formulation of Salubria biomaterial tested for >2,000,000 cycles against polished stainless steel rollers demonstrated minimal wear with no statistical differences in thickness after testing.

3.7 Biocompatibility

The following table outlines the results of standard biocompatibility testing performed on Salubria biomaterial, in accordance with ISO 10993-1 and FDA Blue Book Memorandum #G95-1.

BIOCOMPATIBILITY TESTING

ISO 10993-1 Recommended Testing Requirement	Test Performed	Test Results
Cytotoxicity, ISO 10993-5	ISO MEM Elution L929 cells, GLP.	Pass
	Direct Contact Neurotoxicity	Pass
Sensitization and Irritation, ISO 10993-10	Kligman Maximization Method	Pass
	Primary Vaginal Test: Repeat Exposure	Pass
Sub-acute and Sub-Chronic Toxicity, ISO 10993-6	Sub-acute and sub-chronic toxicity	Pass
Genotoxicity, ISO 10993-3	Ames Mutagenicity: Dimethyl Sulfoxide Extract, 0.9% Sodium Chloride Extract	Pass
	Chromosomal Abberation	Pass
Implantation, ISO 10993-6 and 10993-11	Subacute or site Specific Implantation with chronic Toxicity	Pass
	Biocompatibility study in Rabbits following Intra-articular injections.	Pass
	Rabbit Trochlear Osteochondral Defect	Pass

3.8 Rabbit Defect Testing

After three months of implantation in the patellofemoral groove, eight rabbits were sacrificed for histologic analysis of the implant site and surrounding synovium. Tissues assessed included liver, spleen, kidney, and lymph node. The Salubria biomaterial was well-tolerated with subchondral bone formation surrounding the implant, no fibrous tissue layer or inflammatory response, no implant failures or evidence of wear debris formation, no osteolysis, and no toxic effects on the implant site or distant organ tissues.

At time of explantation, the samples were essentially unchanged (see Fig. 3). Based on histologic examination in comparison to the operative control, there was no evidence of inflammatory reaction in either the surrounding cartilage/bone (see Fig. 4) or in the synovium. In fact, a layer of normal hyaline cartilage partially covered the implant surface. The cartilage surface of the patella also showed no changes in the area articulating against the Salubria implant. There was no sign of toxicity on histologic examination of the distant organ sites.

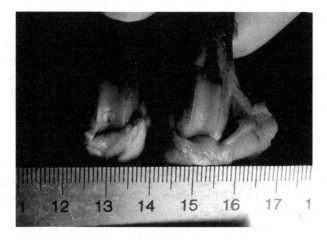

Fig. 3. The left-hand knee shows a Salubria implant in the patellofemoral joint of a rabbit knee after 3 months implantation. The right-hand knee is an operative control.

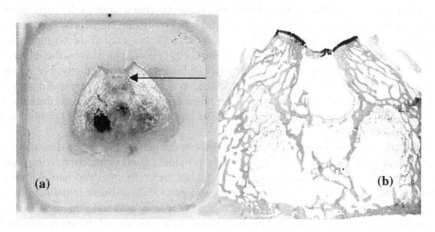

Fig. 4. (a) This digital scan of a paraffin tissue block containing a Salubria implant demonstrates that the implant remains in place over 3 months of implantation in the rabbit patellofemoral groove. (b) Hematoxylin and eosin stain of a section from the tissue block in (a) showing the implant site – the implant has been removed during the staining process. There is no evidence of inflammatory reaction; the surrounding cartilage and bone are normal in histologic appearance.

On excision for mechanical testing, the sample edges are firmly adherent to the surrounding bone. The indentation force (i.e., the force required to cause a certain amount of sample deformation) of the implant is unchanged from a non-implanted control. Comparison material characterization testing showed that the implanted samples were not different from non-implanted controls.

3.9 Particulate Inflammation Study

Salubria biomaterial was tested for particulate toxicity or inflammatory reaction in the joint. Salubria particulates were biologically well-tolerated. The biomaterial particulate was deposited on the superficial synovium with minimal inflammatory reaction. There was no evidence of migration from the joint space or toxicity in the knee or at distant organ sites. There was no evidence of third body wear or osteolysis.

3.10 Ovine Knee Implant Inflammation Study

For the goat study, the native articular cartilage surfaces were protected in the test group compared to extensive damage from the control meniscectomy group. No local inflammation was noted on MRI or histology. No distant organ inflammation was seen in these large animals, confirming the observations in the rabbits.

4 Discussion

The biomaterial described here exhibits the requisite characteristics for soft tissue replacements. For knee cartilage, the material has non-linear viscoelastic properties similar to native tissue. The strength and fatigue properties exceed the requirements for a fully loaded knee articular joint (Stammen, 2001). For heart valves, the material must be moldable to complex anatomies and exhibit low thrombogenicity. For reconstructive anatomic parts, the biomaterial should be easily molded to custom shapes and have low inflammation potential. The biomaterial presented here exhibits these properties and opens the potential for soft tissue replacements that more closely match the anatomic and physiologic requirements.

Although ring specimens and dumbbell shape specimens are both one-dimensional tests, ring specimens were used because they provide a good connection. Ring samples can relieve the experimental error that comes from the inappropriate clamping dumbbell specimens which can cause the specimens to slide or break in the neighborhood of the clamp. There may be damage from preparing uniaxial dumbbell shaped strips. Dumbbell strips would also be difficult to obtain because of the small diameter of the tubular samples.

The biocompatibility testing for Salubria reflects previous biocompatibility testing on other PVA-based biomaterials. PVA hydrogels in the literature are non-carcinogenic with rates of tumorigenicity similar to the well-accepted medical-grade materials, silicone and polyethylene. Nakamura (2001) reports on a 2-year carcinogenicity study conducted on a PVA-based biomaterial subcutaneously implanted in rats. Pre-clinical investigation of other PVA-based hydrogels and Salubria biomaterial demonstrates that these materials are biocompatible in the joint space (Oka et al., 1990). The rabbit is the most commonly published cartilage repair model with study lengths varying from 3 months to 1 year, with little difference in results at 3 months from those at 1 year. These studies indicate that 3 months is sufficient to assess biocompatibility and early treatment failure in the rabbit model. These results are further confirmed by clinical results on SaluCartilage[TM].

Based on this study, Salubria soft tissue biomaterial has been shown to be biocompatible with long-term implantation. There is no evidence of inflammatory response or local or distant toxicity. Furthermore, the biomaterial has stable, durable physical properties over the period of implantation in joints and would be suitable for use as structure deceives such as a cartilage replacement. The biomaterial presented here opens the potential for soft tissue replacements that more closely match the anatomic and physiologic requirements for human biodevices.

Acknowledgements. The biomaterials were supplied by SaluMedica, LLC and some of the studies were performed at SaluMedica, LLC.

References

1. Hanff, G., Sollerman, C., Abrahamsson, S.O., Lundborg, G.: Repair of osteochondral defects in the rabbit knee with Gore-TEXTM (expanded polytetrafluoroethylene). Scandinavian Journal of Plastic and Reconstructive Hand Surgery 24, 217–223 (1990)
2. Nakamura, T., Ueda, H., Tsuda, T., Li, Y.-H., Kiyotani, T., Inoue, M., Matsumoto, K., Sekine, T., Yu, L., Hyon, S.-H., Shimizu, Y.: Long-term implantation test and tumorigenicity of polyvinyl alcohol hydrogel plates. Journal of Biomedical Materials Research 56(2), 289–296 (2001)
3. Oka, M., et al.: Development of an artificial articular cartilage. Clin. Mater 6(4), 361–381 (1990)
4. Stammen, J.A., Williams, S., Ku, D.N., Guldberg, R.E.: Mechanical properties of a novel PVA hydrogel in shear and unconfined compression. Biomaterials 22(8), 799–806 (2001)

Computer Aids for Visual Neuroprosthetic Devices

Samuel Romero[1,*], Christian Morillas[2], Juan Pedro Cobos[2], Francisco Pelayo[2],
Alberto Prieto[2], and Eduardo Fernández[3]

[1] Department of Computer Science, University of Jaén, Campus Las Lagunillas s/n, Jaén
E-23071, Spain
sromero@ujaen.es
[2] Department of Computer Architecture and Technology, University of Granada
C/ Periodista Daniel Saucedo Aranda s/n, Granada, E-18071, Spain
{cmorillas,juanp}@atc.ugr.es, {fpelayo,aprieto}@ugr.es
[3] Bioengineering Institute, University Miguel Hernández, Avenida de la Universidad s/n
Elx, E-03202, Spain
e.fernandez@umh.es

Abstract. Given that vision is the sense providing most information to the human being, any affection related to it significantly reduces the quality of life of the patients. Computing is showing to be a promising approach to provide therapies for patients suffering from low vision or even blindness. We describe some specific tools based on software and hardware solutions oriented to the development of this kind of aids. We present a system performing bioinspired image processing on a portable equipment, including a neuromorphic coding module that generates spike event patterns corresponding to information obtained from the processed images. These events drive a computer-controlled platform for neural stimulation of the visual cortex, intended for eliciting the perception of patterns of phosphenes in the visual field of blind subjects.

Keywords: Bioinspired real-time image processing, visual neuroprostheses, vision aids, neuromorphic encoding, artificial vision, phosphene, electrical neurostimulation, blindness.

1 Introduction

Vision provides us with the majority of the information in relation to our environment. About 40% of our sensorial input comes from our eyes. The impact of a damage in the visual system of a subject, with a partial or total loss of vision, severely affects the performance of daily tasks, social interactions, labour, etc. This implies a significant reduction in the quality of life of subjects affected of any impairment in their visual system.

World Health Organization estimates that roughly 37 million persons are completely blind, and low vision affects up to 124 million [1]. The number of affected

---

* Corresponding author.

A. Fred, J. Filipe, and H. Gamboa (Eds.): BIOSTEC 2008, CCIS 25, pp. 96–108, 2008.

patients is expected to increase due to the ageing of population in developed countries, and to a variety of pathologies and accidents affecting one or more of the components of the complex visual system.

Major causes of blindness are age-related macular degeneration (AMD), diabetic retinopathy, cataracts, glaucoma, optic nerve damage, and ocular. For some of these affections, optical, surgical or pharmacological solutions are available. However, in a high number of cases, these impairments are non-curable, and no treatment is offered to these patients.

Given this, a strong research is being conducted in the development of therapies for visual affections, with a variety of techniques, as retinal cell transplantation, the use of growth factors, or gene therapy, mainly applied to retinitis pigmentosa (RP), or optical aids for low vision.

In the last few decades, the development of computer engineering and its application to biomedical engineering has opened a new alternative for therapy design in visual impairments.

In the case of full blindness, there are a number of research groups pursuing the development of visual neuroprostheses, based on electrically stimulating neural tissue in the visual pathway, so that visual perceptions are elicited, and a rudimentary, but functional form of sight can be provided. Depending on the point of the visual pathway on which the neurostimulation interface is placed, we can classify visual neuroprostheses as retinal [2], optic nerve [3], or cortical implants ([4, 5, 6]).

In some of these approaches, especially for cortical neuroprostheses, the processing carried out by the first stages of the visual pathways should be somehow replaced or emulated, so we need some sort of retina-like image pre-processing, and a neuromorphic encoding similar to the one found at the ganglion cell level (which is the output of the retina, and goes, after some other stages, to the visual cortex).

Whatever is the selected interface for visual neurostimulation, the employment of this kind of devices implies a high degree of complexity, and all of them require computer engineering support to deal with the control and information processing required to make these devices work.

In any visual neuroprosthesis, a high number of channels for stimulation are desirable. The more neurostimulation points, the more phosphenes are expected to appear in the visual field of the patient, composing richer patterns of percepts. This implies handling a series of parameters that need to be selected and tuned for every patient, and for every channel, which might range from only 16 electrodes to several thousands in a near future.

The computer system we describe in this paper is a software/hardware platform conceived for research with visual neuroprostheses. The purpose of the research platform is to provide automated and patient-driven procedures for prosthesis parameter tuning and psychophysical testing. The computer-controlled neurostimulator serves as an abstraction layer to hide the complexity of handling such an intricate implant. The platform is part of a set of tools designed to cover different needs in the development of a full visual prosthesis, such as an artificial retina model, or an automated synthesizer for embedded circuits to obtain a portable, low power consumption controller for the stimulator.

In the next section, a general processing architecture is proposed, which takes camera images as input and carries out a series of bioinspired operations, leading to different results that can be employed in a range of visual rehabilitation applications. This section also describes in detail the retina-like processing model selected for our systems.

Section 3 depicts the neuromorphic encoding model, so images filtered by the retinal processor are transformed into a stream of electrode addresses. This way, the amplitude-to-frequency transformation better matches the way the brain represents visual information at the output of the retina.

In Section 4 the neurostimulation research platform is explained, starting by describing the main components of the CORTIVIS prosthesis, and next exposing the system architecture of the software and hardware components. We include a description of the current threshold finding procedures, as well as the set of psychophysical test implemented in this platform. We pay special attention to the phosphene mapping and re-mapping procedures, given the importance of being able to elicit recognizable patterns of percepts.

Finally, we briefly mention some other additional tools we also have developed, in order to complete a suite of systems devoted to visual rehabilitation.

2 Retina-Like Image Pre-processing

The CORTIVIS (Cortical Visual Neuroprosthesis for the Blind) consortium, of which the authors are members, implements a bioinspired retinal processing model as part of a system designed to transform the visual world in front of a blind individual into multiple electrical signals that could be used to stimulate, in real time, the neurons at his/her visual cortex [7, 8], as we will describe later.

Even though the main objective of this project is the design of a complete system for neurostimulation, a part of the system is useful to develop non-invasive aids for visualization, or sensorial transduction to translate visual information into sound patterns.

Fig. 1 shows the reference architecture illustrating all the capabilities developed so far. The input video signal is processed by three parallel modules for the extraction or enhancing of image features, according to different processing modes.

The first module performs temporal filtering, as natural retinae also detect temporal changes in the visual field. This temporal enhancement is implemented by computing the differences between two or more consecutive frames, with varying strength in the periphery of the visual field (foveated distribution) as in biological retinae [9].

The spatial processing module performs an intensity and colour-contrast filtering over different combinations of the three colour planes of the frame. This spatial processing resembles the function of bipolar cells in the retina, computing a difference of Gaussians, although new filters in the form of any Matlab [10] expression over the colour or intensity channels can be added. The stereo processing module emphasizes closer objects by obtaining disparity maps at different resolutions, starting from image pairs captured by a couple of head mounted cameras (optionally assisted by a sonar rangefinder).

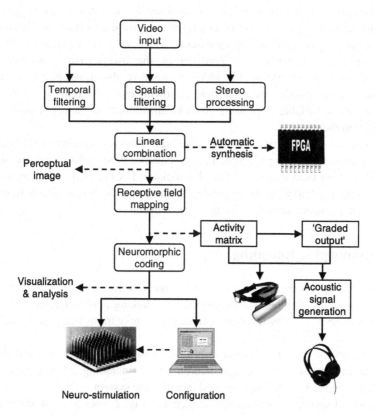

Fig. 1. Architecture of the bioinspired image processing system for the development of computer-based visual aids. After obtaining a linear combination of spatial, temporal and depth-related filtering, different outputs for a variety of applications are possible. The choices include automatic synthesis for programmable devices; sending the information to HMD for functional validation, to headphones for acoustic signalling, or delivering neurostimulation to the neural tissue to evoke visual perceptions in the blind subjects.

The next stage puts together the results computed by each of the previous modules. This way, we can integrate the most relevant features into a single compact representation, selecting the contribution of the most convenient features.

Depending on the application for which we employ the processing mode, the resolution for the output will be different; but, in general, a reduction of the resolution is required. For example, in a neuroprosthesis, this resolution corresponds to the number of electrodes in the implant. If we apply this scheme for a sensorial transduction system for the translation of visual information into audible patterns, the resolution of the output will be limited to the amount of different sounds that the patient is able to distinguish (or the number of spatial positions he/she is able to locate) without interference with his/her normal perception capabilities.

This process of resolution reduction corresponds to the concept of receptive field, this is, the area of the image (set of pixels) that contribute to the calculation of the value resulting in the reduced representation, which we call "activity matrix". The default configuration performs a partition of the image into rectangular non-overlapping areas of a uniform size, although we have also developed a tool for the definition of more complex receptive fields, allowing even different sizes and shapes, which also can be variable, depending on its localization, from the centre of the visual field to its periphery.

In the case of a neuroprosthesis, the next stage is the re-encoding of this information into a neuromorphic representation, as a sequence of stimulation events (spikes), which will be later used to drive a clinical stimulator. For this purpose we apply to the activity matrix the leaky integrate-and-fire spiking neuron model, which transforms intensity in the matrix (white level) into trains of events.

3 Neuromorphic Encoding

Features extracted by the image processing stage can be used in a complete neural stimulation system, after being transformed by a spiking neuron model. This model translates numerical activity levels (amplitude) into spike trains (frequency) which the stimulation device can handle.

Different neuron models can be found in the literature [11], and we decided to implement an integrate-and-fire spiking neuron model, including a leakage factor, because of the simplicity to be implemented in a discrete system.

The selected spiking neuron model, depicted in Fig. 2, uses a set of accumulators to sum up activity levels resulting of the current frame processing. Periodically, the integrated value is compared to a previously defined threshold. Whenever the threshold is reached, the accumulator is initialized to a resting level (resting potential) and a spike event is raised. The leakage factor avoids unexpected events due to ambient noise or residual activity from previously processed frames.

Each spike event generated is delivered to the stimulation device which has to form the electric waveform to be applied to the neural tissue.

All the events generated during a stimulation session can be stored for analysis. Fig. 3 shows a graphic representation of all the events produced by a white horizontal bar moving from bottom to top on a black background, considering a 10 by 10 channels stimulation device.

In order to test the effectiveness of this information coding method, we have developed a procedure for restoring activity matrix values from the temporal events sequence (i.e. an inverse spike to activity conversion). Restored activity matrix was applied again to the neuromorphic pulse coding stage, producing very precise results with small differences. Results obtained from the restoring stage are illustrated in Fig. 4.

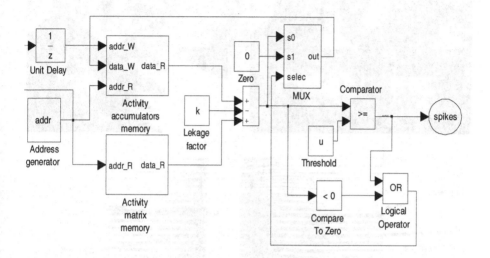

Fig. 2. Block diagram of the neuromorphic coding subsystem for a sequential implementation

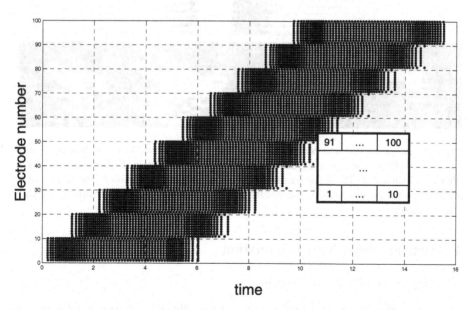

Fig. 3. Spike event trains produced by a horizontal bar pattern (with an equivalent height of 4.42 matrix rows) moving from bottom to top of the image (at an equivalent speed of 1.1 matrix rows/sec), and diagram of stimulation channels numbering. The visual processing includes a temporal contrast (ON and OFF).

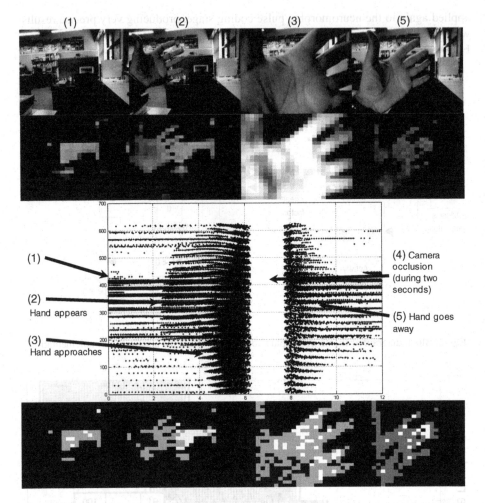

Fig. 4. Image processing and coding example. The first row shows four snapshots of a video sequence, with the corresponding activity matrix just below, obtained with a certain spatial filtering combination. The graph in the middle shows the representation of the spike events produced by the image sequence. Finally, the bottom row represents the reconstruction of each activity matrix.

4   Cortical Visual Neurostimulation

4.1   The Cortivis Prosthetic System

The model selected for the CORTIVIS prosthesis, illustrated in Fig. 5, takes one or two cameras, as input, which feed a bio-inspired retinal encoder (as described before), along with a leaky integrate-and-fire spiking neuron model, to determine the time instants each electrode (identified by its address) will be activated. This stream of electrode addresses is sent through a wireless link to the implanted section of the prosthesis. The RF link also provides energy for the implanted stimulator. This neurostimulator is finally connected to an array of microfabricated electrodes, which are inserted into the visual area of the brain cortex.

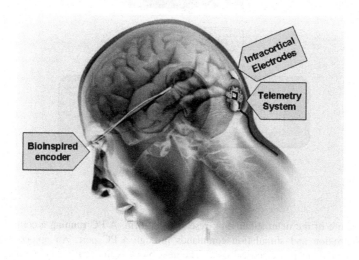

Fig. 5. Scheme of the visual prosthesis proposed by CORTIVIS. A camera acquires images, which are processed through a bioinspired encoder. The encoder sends stimulation commands wirelessly to the intracranial telemetry system. Finally, the array of microelectrodes stimulates the visual cortex of the subject.

4.2 Clinical Research Platform

System Architecture

Fig. 6 shows the building blocks of the clinical research station for the CORTIVIS prosthesis. A PC running the control software is connected to an electronic neurostimulator, through one of the PC's ports. An opto-coupling stage protects the patient against electrical risks, as required in the design of biomedical instruments.

The second stage of the platform is an electronic equipment which receives and decodes commands from the PC. Whenever a configuration word is received, it stores the waveform parameters (pulse width, number of pulses in a train and inter-pulse interval, etc.) for the corresponding channel in a configuration memory. If a stimulation command is sent from the PC, the equipment selects the corresponding output channel through a demultiplexor, and drives a Digital-to-Analog converter so that a biphasic waveform is sent to the output, according to the stored parameters for the corresponding electrode.

The output of the neurostimulator drives the intra-cranial implant. In our case, we have selected the Utah Electrode Array [12], which is a microfabricated array of 10x10 microelectrodes, although any other electrode array could be used with little modification.

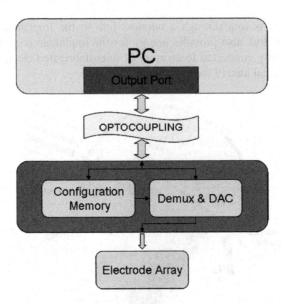

Fig. 6. Structure of the neurostimulation research platform. A PC running a control software sends configuration and stimulation commands through a PC port. An optocoupling stage protects the patient against electrical risks. The next block is the neurostimulation electronics. A configuration memory stores the waveform parameters for every channel, and a demultiplexing and digital-to-analog conversion block creates the corresponding waveform, and sends it through the corresponding electrode in the array.

Operation Modes

The experimental set-up can be employed in two different configurations, known as "stimulation" and "simulation/training" modes.

The neurostimulation mode corresponds to the set described before, that is, a PC controlling a neurostimulator, which delivers pulses to an implanted array of electrodes. The purpose of this set is to allow researcher tuning the set of parameters required to elicit phosphenes in the visual field of the patient, and then, run a series of psychophysical tests, in order to characterize the evoked perceptions.

Nevertheless, a second configuration is available for debugging and training purposes (with sighted volunteers). In this mode, the neurostimulator and the array of electrodes are replaced by a second PC with head-mounted displays (see Fig. 7). The first PC plays the same role as in the previous configuration. The commands sent through the communication port are received by the second PC, which implements simulation rules including random values for current threshold, and phosphene location in the visual field. The simulator in the second PC offers a representation of a set of phosphenes in a head mounted display, according to the information obtained in similar experiences with human visual intra-cortical microstimulation. This way, all the procedures can be essayed and refined before going to the surgery room.

Fig. 7. Usage example of the simulation/training mode of the experimental platform. In this case, a mapping session is undergoing having the subject locating the phosphenes projected onto the HMD in the visual field, by pointing on a tactile screen.

Current Threshold Finding
The first task after the patient has been safely implanted is finding the lowest current amplitude required to evoke a phosphene, for every channel of the implant.

Every procedure is patient-driven, so the response of the patient triggers the next step of the process, avoiding verbal interaction, so the process is faster. The basic algorithm selects every electrode, and sends pairs of configuration and stimulation commands to the stimulator with growing current amplitude, until the patient signals the event of a phosphene in his/her visual field, by clicking a mouse button. Then, the process is repeated for the next electrode.

In order to accelerate the process, we use a binary search scheme, and we set the starting point for the binary search for a channel to the threshold found for the previous channel. Applying this procedure, a set of 100 electrodes can be configured in less than 5 minutes (for a step of 1 second between consecutive stimulations). Fig. 8 shows pulses obtained with our platform.

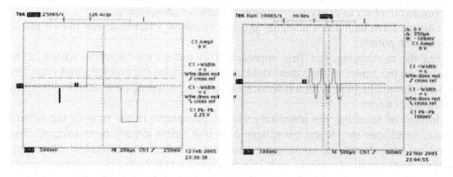

Fig. 8. On the left, example of biphasic pulse obtained with the experimental neurostimulation platform (vertical scale: 500 mV/div; horizontal scale: 200 microsec/div). On the right, example of pulse train obtained with the experimental neurostimulation platform (vertical scale: 100 mV/div; horizontal scale: 500 microsec/div).

Psychophysical Tests

After the threshold current has been found for every electrode, a set of tests is performed, so we can determine the psychophysical properties of the evoked percepts.

This way, a detailed set of perceptual tests has to be run, which again requires making this process as easy and agile as possible. Following the same philosophy as for the threshold finding procedure, a patient-driven automated scheme is employed.

The platform includes the following set of psychophysical essays:

- Brightness sensitivity: a change in certain parameters of the waveform (mainly amplitude) will modify the perceived brightness of the evoked phosphene. A pair of phosphenes is elicited, and the brightness of one of them changes until the patient finds no change.
- Spatial resolution: a pair of phosphenes produced by distant electrodes is evoked consecutively with closer and closer electrodes until the patient cannot differentiate them.
- Phosphene cluster count: a set of 1, 2 or 3 phosphenes from adjacent electrodes is elicited. The patient gives feedback on the number of phosphenes perceived.
- Motion mapping and orientation selectivity: a straight line of electrodes (row, column or diagonal) in the matrix consecutively get activated. The patient indicates the general direction of apparent motion of the phosphene.
- Simple pattern discrimination: a simple pattern (similar to Snellen symbols) and its "mirrored" pattern are consecutively activated in the electrode array. The subject tells if they seem to be different or similar.

4.3 Phosphene Mapping and Re-mapping

Experiments both with human and non-human subjects have shown that the correspondence between the spatial location of the stimulation point in the cortex and the position of the evoked phosphene in the visual field can present strong deformations [13].

In any case, a mapping between the location of the activated electrode in the array and the position of its corresponding phosphene in the visual field should be determined for every channel of the implant. Similarly, the reverse operation (re-mapping), indicating which electrodes should be activated to get a specific pattern of phosphenes is required in order to evoke recognizable patterns of phosphenes.

Our platform includes a mapping process based on a tactile screen placed just in front of the patient.

For the re-mapping, our first approach is to project the objective pattern to be evoked on the centre of the visual field, and then, select, for every desired point, the closest phosphene to it. With a reverse look up at the mapping table, its corresponding electrode is found.

Instead of selecting the absolutely closest phosphene in the map to the desired point, we choose the closest phosphene only if it hasn't already been selected. This way, we get patterns incorporating a maximum number of phosphenes, rather than having more precise locations with less percepts. Additionally, this selection procedure enhances the response whenever the distribution of the map is highly uneven.

An example of mapping and re-mapping is provided in Fig. 9, showing a remarkable enhancement in the representation of phosphene patterns.

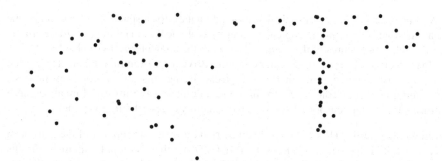

Fig. 9. On the left, phosphene pattern that would be elicited after direct selection of the top row and central column of an electrode array. Although the distribution of electrodes forms a "T" shape, the evoked pattern is unrecognizable, so a re-mapping is required. This set of phosphenes corresponds to a randomly generated mapping (25x25). On the right, after our re-mapping algorithm is applied to the figure on the left, a different set of electrodes is recruited, evoking a better recognizable pattern of phosphenes, much more similar to the desired "T" shape.

5 Additional Tools

Along with the system described in this paper, two other complementary tools also included as part of the architecture shown in Fig. 1 have been developed. The first one is an automatic synthesizer of digital circuit descriptions, so that the retina model defined in software can automatically be transformed into a configuration file for a programmable circuit (FPGA, Field Programmable Gate Array). This way, we have a hardware implementation of the retinal processing in a single chip, in order to obtain portability, low-power consumption, and real-time operation. This tool is further described in [14].

The second one hardware platform, named VIS2SOUND has been built. This system analyses images coming from two cameras mounted on eyeglasses following the retinal bioinspired model, and provides binaural signals through headphones, so that the patient is able to locate the spatial position and distance of the main object in the scene. This tool is described in [15].

6 Conclusions

Persons with visual impairments, such as low vision or blindness, will benefit in the future from therapies that are created around computer-based technologies. Although most of the systems under research are not yet available for clinical application, the first results seem to be promising.

However, the selection of visual information to be transmitted through a limited-bandwidth device (such as vision to sound encoders or implanted electrodes) will be a key aspect in the success of these therapeutic devices.

We have proposed a set of tools that look into nature for inspiration, so we mimic the visual information pre-processing carried out by biological retinae, in order to determine the most relevant features of the scene to be sent to the computerized visual aid.

This selection of information can be used to drive a set of implanted electrodes that will elicit visual perceptions in blind subjects. Being this also a complex process, computers play a key role in controlling and automating the use of clinical research platforms for visual neuroprostheses, as the system we describe in this paper.

Acknowledgements. This work has been carried out with the support of the European project CORTIVIS (ref. QLK6-CT-2001-00279), the National Spanish Grants DEPROVI (ref. DPI 2004-07032), IMSERSO-150/06, and by the Junta de Andalucía Project: P06-TIC-02007.

References

1. World Health Organization: Prevention of avoidable blindness and visual impairment (2005), http://www.who.int/gb/ebwha/pdf_files/EB117/B117_35-en.pdf
2. Humayun, M.S.: Visual perception in a blind subject with a chronic microelectronic retinal prosthesis. Vision Res. 43(24), 2573–2581 (2003)
3. Veraart, C.: Visual sensations produced by optic nerve stimulation using an implanted self-sizing spiral cuff electrode. Brain Res. 813, 181–186 (1998)
4. Dobelle, W.H.: Artificial Vision for the Blind by Connecting a Television Camera to the Visual Cortex. Asaio J. 46, 3–9 (2000)
5. Troyk, P., et al.: A Model for Intracortical Visual Prosthesis Research. Artif. Organs. 11, 1005–1015 (2003)
6. Fernández, E., Pelayo, F., Romero, S., Bongard, M., Marin, C., Alfaro, A., Merabet, L.: Development of a cortical visual neuroprosthesis for the blind: The relevance of neuro-plasticity. J. of Neural Eng. 12, R1–R12 (2005)
7. Cortivis website, http://cortivis.umh.es
8. Romero, S., Morillas, C., Martínez, A., Pelayo, F., Fernández, E.: A Research Platform for Visual Neuroprostheses. In: Computational Intelligence Symposium, SICO 2005, pp. 357–362 (2005)
9. Morillas, C., Romero, S., Martínez, A., Pelayo, F., Reyneri, L., Bongard, M., Fernández, E.: A Neuroengineering suite of Computational Tools for Visual Prostheses. Neurocomputing 70(16-18), 2817–2827 (2007)
10. The Mathworks website, http://www.mathworks.com
11. Gerstner, W., Kistler, W.: Spiking Neuron Models. Cambridge University Press, Cambridge (2002)
12. Normann, R.: A neural interface for a cortical vision prosthesis. Vision Res. 39, 2577–2587 (1999)
13. Normann, R.A., Warren, D.J., Ammermuller, J., Fernandez, E., Guillory, S.: High-resolution spatio-temporal mapping of visual pathways using multi-electrode arrays. Vision Res. 41, 1261–1275 (2001)
14. Martínez, A., Reyneri, L.M., Pelayo, F.J., Romero, S., Morillas, C.A., Pino, B.: Automatic generation of bio-inspired retina-like processing hardware. In: Cabestany, J., Prieto, A.G., Sandoval, F. (eds.) IWANN 2005. LNCS, vol. 3512, pp. 527–533. Springer, Heidelberg (2005)
15. Morillas, C., Pelayo, F., Cobos, J.P., Prieto, A., Romero, S.: Bio-inspired Image Processing For Vision Aids. In: 1st International Conference on Bio-inspired Systems and Signal Processing, vol. 2, pp. 63–69 (2008)

Tissue-Viability Monitoring Using an Oxygen-Tension Sensor

Dafina Tanase[1], Niels Komen[2], Arie Draaijer[3], Gert-Jan Kleinrensink[2],
Johannes Jeekel[2], Johan F. Lange[2], and Paddy J. French[1]

[1] Delft University of Technology, Mekelweg 4, 2628 CD Delft, The Netherlands
[2] Erasmus Medical Centre, Dr. Molewaterplein 40, 3015 GD Rotterdam, The Netherlands
[3] TNO Quality of Life, Utrechtseweg 48, 3704 HE Zeist, The Netherlands
d.tanase@tudelft.nl

Abstract. Many patients still die every year as a result of anastomotic leakage after surgery. An objective aid to monitor the anastomotic site pre- and postoperatively and detect leakage at an early stage is needed. We propose a miniature measurement system to detect adequate tissue oxygenation pre- and postoperatively (continuously for 7 days) on the colon. The complete sensor chip should include an oxygen-tension sensor (pO_2), a carbon dioxide tension sensor (pCO_2) and a temperature sensor. The work presented here focuses on the measurements done with the oxygen-tension and temperature sensors. In-vitro measurements have been initially performed to test the sensor system and in-vivo tests were carried out on the kidney and the intestines of male wistar rats. The results obtained so far have shown the suitability of this technique for clinical application, therefore sensor-system miniaturisation is presently underway.

Keywords: Oxygen-tension, sensor, tissue viability.

1 Introduction

An anastomosis is the surgical connection of two tubular segments to restore continuity (Fig. 1). Leakage of a colorectal anastomosis is a complication in which intestinal content leaks into the abdominal cavity due to a "defect" in the anastomosis. This defect can be caused by a reduced oxygen supply and it can lead to cell death and necrosis of the anastomosis. As a result, leakage can occur and as a consequence, peritonitis may develop and can lead further to sepsis, multiple-organ failure and ultimately death. Therefore, anastomotic leakage of a colorectal anastomosis is considered a potentially lethal complication.

The reported incidence varies between 3 % and 19 % [1]–[3], [7], with a mortality rate that can be as high as 32 % [4]. To date, no peroperative methods to avoid or predict anastomotic leakage, or any validated, objective parameters for detection of anastomotic leakage in an early postoperative phase, exist. Current diagnostic methods include observation of clinical signs and symptoms (fever and pain), while confirmation is obtained by imaging. These methods are faced with several disadvantages. When anastomotic leakage has progressed to a state of clinical

A. Fred, J. Filipe, and H. Gamboa (Eds.): BIOSTEC 2008, CCIS 25, pp. 109–122, 2008.

manifestation, the patient is already ill and treatment needs to be initiated. Imaging modalities, more specifically abdominal CT-scans and/with contrast enemas, are normally used to confirm a clinical diagnosis of anastomotic leakage, meaning the patient is already ill [5].

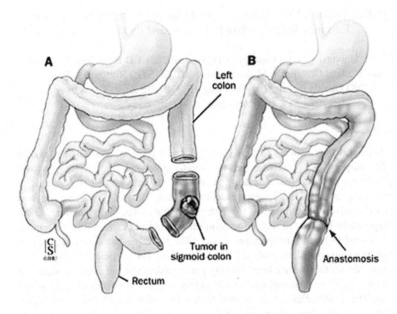

Fig. 1. Schematic representation of an anastomosis; the segment of the colon containing the tumor is removed and the two colon ends are joined into an anastomosis

At present, clinically relevant anastomotic leakage is usually diagnosed approximately 6 to 8 days after surgery [1], [6]. Some studies report an even longer interval (12 days) between operation and diagnosis of anastomotic leakage [7]. The long intervals between the construction of the anastomosis and the diagnosis of anastomotic leakage are detrimental for the prognosis, increasing mortality rates [8]. Therefore, a biomarker reflecting the viability of the anastomosis, could be a fast and objective diagnostic tool in addition to current methods, allowing diagnosis of anastomotic leakage before its clinical presentation.

In this respect, the main goal of this research is to develop a miniature, wireless sensor system to monitor tissue viability pre- and postoperative, continuously for 7 days. The complete sensor chip should include an oxygen-tension sensor (pO_2), and a carbon-dioxide tension sensor (pCO_2). As these parameters depend on temperature, a temperature sensor will be considered as well (Fig. 2). The present work focuses on the use of the oxygen-tension and temperature sensors in-vitro and in-vivo. Furthermore, a number of aspects regarding sensor-system miniaturisation will be presented.

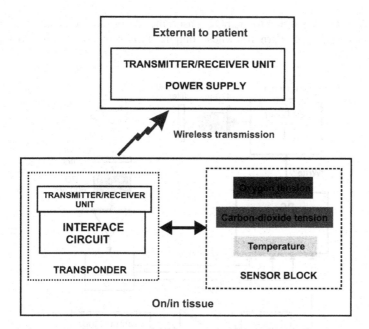

Fig. 2. Schematic of the wireless sensor system for tissue viability monitoring

The wireless sensor system is made up of two parts: one is external to the patient and consists of the transmitter-receiver unit and the power supply, while the second one is the implant. It consists of the sensor block and the transponder. Data from the implant is sent via the transponder to the external unit where it is stored and further processed.

2 Measurement Method

The measurements presented in this paper were performed with a pO2 sensor and a temperature sensor (Fig. 3), which were fabricated by TNO Quality of Life, The Netherlands [9]. The pO2 sensor consists of a coating at the tip of an optical fibre and works on the principle of dynamic quenching by oxygen of fluorescent particles immobilized in a gas permeable polymer. In this case, the fluorescent particle is ruthenium bathophenantroline, which enters an excited state caused by the LED excitation with a wavelength of 470 nm. The excited state of ruthenium is deactivated by the collision process with oxygen, the particles emitting light with a wavelength of 600 nm. The emitted signals are detected by a photodiode (PD) and converted to a digital signal using an on-board analogue-to-digital (ADC) converter. The sequence of data sent to the computer is BG=BackGround, I1 = Intensity 1 and I2 = Intensity 2 (Fig. 4). The BG measurement is taken prior to the LED pulse and serves as reference signal, while I1 and I2 are the two measurements taken at 0.2μs and 1.2μs, after the light pulse. The oxygen concentration is determined by measuring the fluorescence lifetime, as specified by the Stern-Volmer equation. The repetition rate of the light pulses is 20 kHz. In addition to the pO2 sensor, the sensor block contains the temperature sensor (NTC type, Farnell), whose output, T, is sent also via an ADC to the computer.

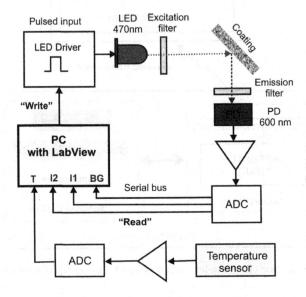

Fig. 3. Block diagram of the sensor system

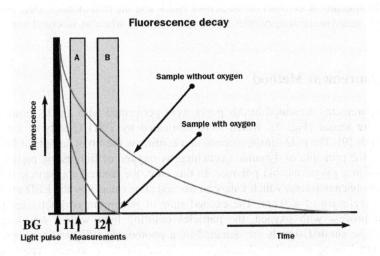

Fig. 4. Fluorescence decay and sequence of data sent to the computer (BG=BackGround measurement, I1 = Intensity 1, I2 = intensity 2)

The data from the two sensors is transmitted via the serial port and further processed using the software *LabView,* National Instruments. Fig. 5 shows a photograph of the sensor block with the two sensors and a magnified view of the fibre tip with the oxygen-sensitive coating.

Fig. 5. Photograph of the sensor block with the two sensors (top) and a magnified view of the fibre tip showing the oxygen-sensitive coating (bottom)

2.1 The pO$_2$-Sensitive Coating

The coating consists of the fluorescent ruthenium particles that are embedded in a polymer which is made porous by the use of the glass powder CPG (Controlled Porosity Glass). CPG can be added in different percentages to the polymer, so that coatings with different grades of porosity and different pore dimensions can be achieved. Ruthenium particles adhere to these pores forming a compact oxygen-sensitive structure, through which the oxygen molecules can diffuse.

The thickness of the coating is an important parameter with respect to the response time of the sensor. Fig. 6 shows the response time in water of two coatings with different thicknesses and the same amount of CPG particles (10%). It can be seen that the thinner coating reacts much faster than the thicker one, which is an important advantage for our medical application – the thinner the coating is, the faster the reaction time and the smaller the geometrical dimensions of the implanted sensor.

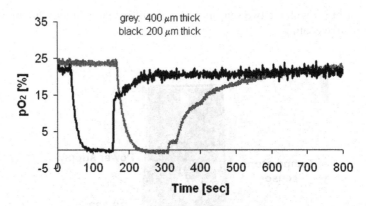

Fig. 6. Response time in water of two coatings with different thicknesses, but the same amount of CPG particles (10%)

3 Measurements and Results

The first set of measurements was performed in gas and water to test the reaction time of the sensor and its stability over 7 days. Furthermore, the stability of the sensor in two solutions with different pH-values, vinegar (pH=3) and sodium bicarbonate (pH=9) was investigated. Although placed in pH solutions with extreme values, the coating preserved well and the sensor was reliable over time and not pH dependent.

Apart from these preliminary measurements, the sensor was tested in blood samples. The measurement setup for the in-vitro testing is shown in Fig. 7. A blood sample was taken in a vacutainer and the optical fibre was immersed in the blood. During the measurements, the vacutainer was placed in a glass with water and the temperature was kept constant (37°C) using a hot plate. The measurement results are shown in Fig. 7. At the beginning, the sensor measured the oxygen pressure in air which corresponds to 160 mmHg. At time 1800 s, the sensor was placed in blood, indicating 35 mmHg. At time 6000 s, the blood separation effect became visible. The tip of the optical fibre was left in plasma, while the RBCs (red blood cells) sedimented at the bottom of the vacutainer. After blood mixture at time 8500 s, a slight rise in the blood pressure was measured due to the oxygen present in the head space. Later on, the blood separation effect became again visible (at time 15000 s).

The series of measurements continued with the animal study. The measurement setup is shown in Fig. 8. It consists of the pO_2 and temperature sensor block and a notebook for reading and processing the data from the sensors. The investigations were performed at the Erasmus Medical Centre in Rotterdam, using male wistar rats, 12 weeks old. Before starting the surgical procedure, the rats were anesthesized with a nosemask (Isoflurane 2%, FiO_2 = 60 %), shaved and placed on a warm plate.

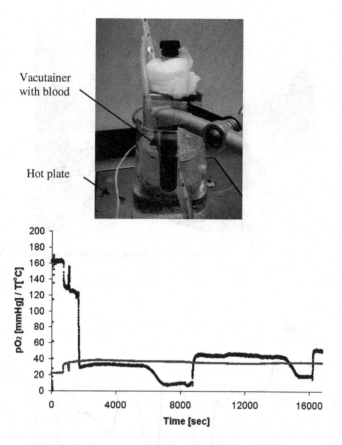

Vacutainer
with blood

Hot plate

Fig. 7. Measurement setup for in-vitro testing (top) and measurement results obtained in a blood sample kept at constant temperature, 37°C (bottom)

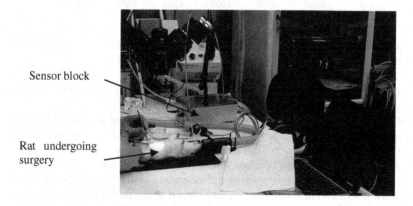

Sensor block

Rat undergoing
surgery

Fig. 8. Measurement setup for in-vivo testing at Erasmus Medical Centre, Rotterdam

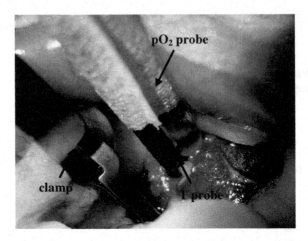

Fig. 9. Close-up of a rat kidney on which the sensor probes are placed together with the clamp that obstructs the renal artery

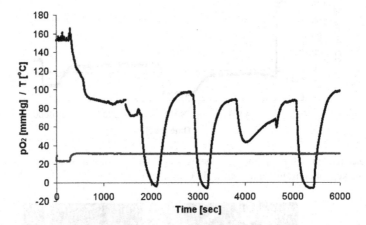

Fig. 10. Graphical representation showing the pO_2 changes due to three cycles of total ischemia and reperfusion. The temperature was constant during the experiment ($37^{\circ}C$).

Access to the internal anatomical structures of the animals was gained by laparotomy, a surgical incision into the abdominal wall. The incision length was 4 cm.

The first test was conducted by placement of the sensors on the rat kidney as shown in Fig. 9. During the tests, the renal artery was clamped to induce ischemia and afterwards released, for reperfusion. The cycle was repeated three times. The results are presented in Fig. 10, where the constant line at $37^{\circ}C$ indicates the temperature.

The next test was performed by exposing the ascending colon. The oxygen-tension and temperature sensors were fixed together and placed at pre-defined locations along the ascending colon, laterally (with respect to the mesocolon) and antimesenterial (opposed to the mesocolon) as shown in Fig. 11. The table and graph with the measurement results, for different sensor locations around the anastomosis are presented in Fig. 12.

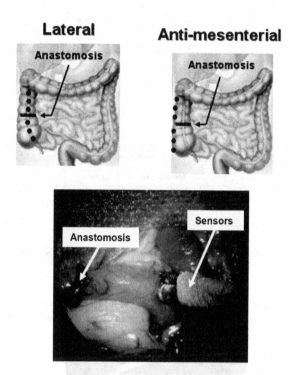

Fig. 11. Placement sites of the sensors along the colon (top) and a photograph showing the anastomosis with the sensors 1 cm away (bottom)

The lowest oxygen-tension values are obtained on the anastomosis (part 3 and part 8 in Fig. 12). This is an expected effect since transection and construction of the anastomosis compromises the circulation and consequently reduces perfusion and oxygenation at the anastomotic site. The farther from the anastomosis the pO_2 was measured, the better the oxygenation was and the higher the oxygen-tension values. The spikes on the graph are artefacts visible only at the moments when the sensors were moved from one tissue location to another, because then, for short periods of time, the fibre was in air. The temperature changes corresponding to tissue and air are visible on the temperature graph.

Another series of measurements were performed with the sensors on the small intestine (Fig. 13). In this case, the blood supply to the central part of the small intestine was obstructed by two strings that were fastened for ischemia and released for reperfusion. The measurement results during the ischemia-reperfusion experiment are shown in Fig. 14.

LATERAL				ANTIMESENTERY			
#	Location [cm]	Start time [sec]	pO₂ [mmHg]	#	Location [cm]	Start time [sec]	pO₂ [mmHg]
1	1	100	152->126	6	1	4900	122->209
2	0.5	1600	126->112	7	0.5	5600	209->125
3	On anastomosis	2050	112->43	8	On anastomosis	6400	125->38
4	0.5	2900	43->85	9	0.5	7500	38->100
5	1	4000	85->122	10	1	8200	100->100

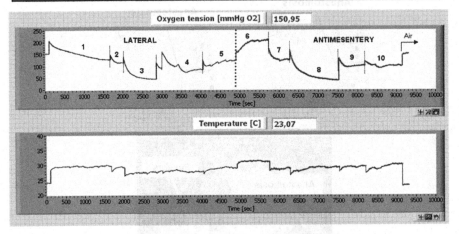

Fig. 12. The results of the tests performed on the colon, to determine the distribution of the oxygen radially and longitudinally with respect to the anastomosis, at different locations

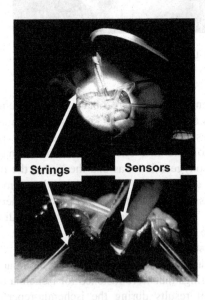

Fig. 13. An overall view and a close-up of the small intestine showing the strings (used to obstruct the blood flow) and the sensors

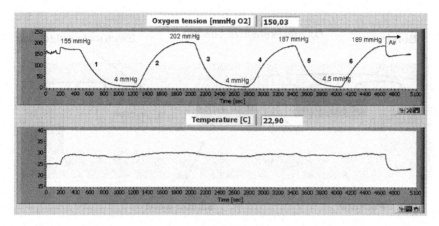

Fig. 14. The graphical representation of the ischemia-reperfusion experiment on the small intestine

At the beginning of the test, the sensors were placed on the small intestine and by fastening the strings the intestine was made ischemic (part 1). The values readily decreased to 4 mmHg, indicating total ischemia. Once the strings were released (part 2), an overshoot was noted, showing a maximum at 202 mmHg. Two other cycles were repeated to test the correctness of the measurement. Also in this case, the results of the tests met our expectations.

4 Sensor-System Miniaturisation

So far, the obtained results have shown the suitability of this technique for the clinical studies and therefore, a further step has been taken to miniaturise the existent sensor system (Fig. 15). The device consists of a light source (LED), a photodiode used as light detector and another photodiode used as temperature sensor (Fig. 16). The LED has been manually assembled in a cavity that was etched in silicon, while the light and temperature detectors were fabricated in the standard IC fabrication process at DIMES, directly on the wafer. At the moment, the sensor system is wire-bonded in a standard 16 DIL package to allow technical testing. The oxygen-sensitive coating, consisting of ruthenium prepowder, quartz and the polymer sylgard 182, has been manually applied on top of the chip and cured for 45 minutes at 100°C. A series of new tests with different coating compositions and thicknesses are currently underway at TNO Zeist and TU Delft, to find the best solution in terms of coating thickness and sensor response time.

Apart from these measurements, work is being performed on the read-out electronics and the wireless communication. Once the technical tests are finished, the total system will be placed on a flexible substrate for clinical use. On short-term, a new series of animal studies are planned on rats at Erasmus Medical Centre, with the complete system.

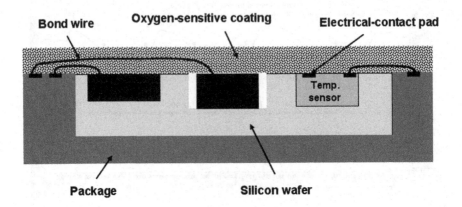

Fig. 15. Schematic of the miniature sensor system consisting of a light source (LED), a temperature sensor (photodiode) and a light detector (photodiode)

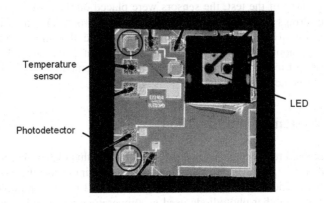

Fig. 16. Photograph of the miniature sensor system (chip size 1.5mm x 1.5mm)

5 Conclusions and Future Work

The article has presented a number of aspects regarding the use of an oxygen-tension sensor for tissue-viability monitoring. The optical method presented in section 2 has proven to be suitable for the detection of the oxygen tension in different environments (gas, water, blood), but most important in tissue.

The measurements on the blood samples (Fig. 7) have confirmed the blood separation effect which is normal sedimentation of red blood cells in a sample that is not stirred. The effect of blood clogging has posed some problems in the beginning, when coatings with rather large pores were used. This drawback has been avoided in the subsequent tests by the use of coatings with pores in the nanometer range. In this way, the red blood cells with a diameter of approximately 8 μm could not diffuse the coating. There are still some adhesions of blood cells on the surface of the coating, but so far measurements were carried out without any significant influences. More tests will eventually be necessary, on tissue, for long periods of time, to further investigate this effect.

The ischemia-reperfusion test on the rat kidney (Fig. 10) has proven the correctness of the measurements. Moreover, oxygen tension was measured for 3-4 hours on the rat kidney without sensor recalibration.

The series of measurements on the large intestine (Fig. 12) has shown a significant decrease (approximately 40 mmHg) in pO_2 on the anastomosis as compared to the other measurement sites on the colon. It was also shown that on two points (lateral and antimesentery) of the anastomosis, the values for the pO_2 were almost the same, indicating limited heterogeneity in pO_2 distribution as opposed to pO_2 distributions found in other organs.

Finally, the last measurements with the sensor on the small intestine (Fig. 14) have shown expected reactions of the sensor to the local changes on the intestine. When the intestine became ischemic, the pO_2 decreased and when the obstruction was released, the pO_2 increased significantly, with an overshoot.

The fabrication of the sensor system and the preliminary results were presented in section 4. Aspects regarding the coating and the integration of the sensor system with the read-out electronics into a wireless system are presently being investigated. Technical tests are carried out at the moment, while new measurements on animals are planned on a short-term basis.

Acknowledgements. Herewith we thank our colleagues Eduardo Margallo-Balbas and Johan Kaptein from Delft University of Technology for the fruitful discussions regarding sensor miniaturisation and for granting us permission to use some of their devices for our preliminary tests. We also thank Wim van der Vlist and Ruud Klerks from DIMES for their help with assembly and packaging of the devices.

References

1. Kanellos, I., Vasiliadis, K., Angelopoulos, S., et al.: Anastomotic leakage following anterior resection for rectal cancer. Tech. Coloproctol. 8 (suppl. 1), s79–s81 (2004)
2. Guenaga, K.F., Matos, D., Castro, A.A., Atallah, A.N., Wille-Jorgensen, P.: Mechanical bowel preparation for elective colorectal surgery. Cochrane Database Syst. Rev. 2, CD001544 (2003)
3. Peeters, K.C., Tollenaar, R.A., Marijnen, C.A., et al.: Risk factors for anastomotic failure after total mesorectal excision of rectal cancer. British Journal of Surgery 92(2), 211–216 (2005)
4. Choi, H.-K., Lau, W.-L., Ho, J.W.C.: Leakage after resection and intraperitoneal anastomosis for colorectal malignancy: analysis of risk factors. Dis. Colon. Rectum 49, 1719–1725 (2006)
5. Eckmann, C., Kujath, P., Schiedeck, T.H., Shekarriz, H., Bruch, H.P.: Anastomotic leakage following low anterior resection: results of a standardized diagnostic and therapeutic approach. Int. J. Colorectal Dis. 19(2), 128–133 (2004)
6. Alves, A., Panis, Y., Pocard, M., Regimbeau, J.M., Valleur, P.: Management of anastomotic leakage after nondiverted large bowel resection. J. Am. Coll. Surg. 189(6), 554–559 (1999)

7. Hyman, N., Manchester, T.L., Osler, T., Burns, B., Cataldo, P.A.: Anastomotic leaks after intestinal anastomosis: it's later than you think. Ann. Surg. 245(2), 254–258 (2007)
8. Macarthur, D.C., Nixon, S.J., Aitken, R.J.: Avoidable deaths still occur after large bowel surgery. British J. Surg. 85(1), 80–83 (1998)
9. Draaijer, A., Konig, J.W., Gans, O., Jetten, J., Douwma, A.C.: A novel optical method to determine oxygen in beer bottles. In: EMC Congress, France (1999)

Towards a Morse Code-Based Non-invasive Thought-to-Speech Converter

Nicoletta Nicolaou and Julius Georgiou

Dept. of Electrical and Computer Engineering, University of Cyprus, Kallipoleos 75
1678 Nicosia, Cyprus
{nicoletta.n,julio}@ucy.ac.cy

Abstract. This paper presents our investigations towards a non-invasive custom-built thought-to-speech converter that decodes mental tasks into morse
code, text and then speech. The proposed system is aimed primarily at people
who have lost their ability to communicate via conventional means. The investigations presented here are part of our greater search for an appropriate set of
features, classifiers and mental tasks that would maximise classification accuracy in such a system. Here Autoregressive (AR) coefficients and Power Spectral Density (PSD) features have been classified using a Support Vector
Machine (SVM). The classification accuracy was higher with AR features compared to PSD. In addition, the use of an SVM to classify the AR coefficients increased the classification rate by up to 16.3% compared to that reported in
different work, where other classifiers were used. It was also observed that the
combination of mental tasks for which highest classification was obtained varied from subject to subject; hence the mental tasks to be used should be carefully chosen to match each subject.

Keywords: Brain-Computer Interface, Electroencephalogram, Morse Code,
Thought Communication, Speech Impairment.

1 Introduction

The development of techniques that offer alternative ways of communication by
bypassing conventional means is an important and welcome advancement for
improving quality of life. This is especially desirable in cases where the conventional
means of communication, such as speech, is impaired. We envisage the development
of a simple and wearable system that communicates by converting thoughts into
speech via morse code and a text-to-speech converter.

In this paper we present preliminary investigations towards the development of
such a system. The investigations form part of our search for features, classifiers and
mental tasks that are appropriate for utilisation in our system. In particular, we
compare the classification accuracy obtained between combinations of mental task
pairs when (i) autoregressive (AR) coefficients and Power Spectral Density (PSD)
values are utilised as features; and (ii) Support Vector Machine (SVM), Linear
Discriminant Analysis (LDA) and Neural Network (NN) are utilised as classifiers.

A. Fred, J. Filipe, and H. Gamboa (Eds.): BIOSTEC 2008, CCIS 25, pp. 123–135, 2008.

Our investigations suggest that the combination of AR coefficients and SVM is more appropriate for our application, as an increase in classification accuracy ranging from 8.2-16.3% has been observed compared to classification of the same features using LDA and NN.

The paper is organised as follows. Section 2 provides a background into communication via thoughts and how morse code has been utilised for this purpose so far. This is followed by section 3 where a description of the system envisaged, the objectives that motivated these preliminary investigations and a description of the methods utilised are provided. The findings are presented in section 4 followed by a discussion towards how these could be interpreted and understood as part of the proposed system. The main conclusions and plans for future work emerging from these investigations are outlined in section 5.

2 Background

A number of conditions, such as amyotrophic lateral sclerosis, strokes and speech impairment, affect the ability to communicate with the environment through speech. However, the problem becomes more severe when limb or muscle control is also affected, since other means of communication e.g. typing, are eliminated. For this reason other biological signals that are not directly related to speech production have been utilized for communication purposes. Systems employing the electrical signals from muscles (EMG) [1] or eye movements (EOG) [2] have been developed. Even though communication can be achieved via the use of other biological signals, patients who are locked-in may have lost all muscle control and are unable to use such systems. In addition, such systems cause fatigue to the user. Thus, an alternative method is to utilise brain activity as an input signal to a device for communication purposes (brain-computer interface, BCI). A BCI is "a communication system that does not depend on the brain's normal output pathways of peripheral nerves and muscles" [3]. Since the first recording of the brain electrical activity from humans by Hans Berger in 1929, its use for investigating brain function and clinical evaluation of neurological disorders went hand in hand with speculations that there were other applications the brain activity could be used for. Ideas that this activity could be used to decipher thoughts, which could in turn be used for direct human-to-human and human-to-machine communication, were very popular, particularly in science fiction scenarios. However, the attention of the scientific community was turned in this direction due to the development of the required technology that allowed sophisticated online analysis of multichannel electroencephalogram (EEG) data, teamed with increased knowledge of the characteristics of the signals that make up the EEG, and increased recognition of the benefits that use of the EEG could offer to people whose traditional means of communication with the environment were severely impaired.

Interest of the scientific community in the development of BCI systems began early in the 1970's. The US Department of Defence was mainly responsible for initiating this interest, with the goal to look into the development of technologies that allowed a closer collaboration between humans and computers and that would improve human performance when engaged in tasks associated with high mental load – and in particular military tasks, [3]. The main goals of this research were not achieved and interest then turned towards the use of biological signals as a control signal for systems such as weaponry. Successful research by Vidal employed visual-evoked potentials to control the movement of a cursor through a 2-D maze [4]. From the beginning of the 90's until nowadays the number of research groups around the world involved in BCI research increased greatly as the means and the technology to obtain and analyse brain activity became available. Nowadays this technology is primarily developed for assistive purposes. A large number of disabilities, such as amyotrophic lateral sclerosis (ALS) and brainstem strokes render people unable to use conventional means, such as voluntary muscle control, to communicate with the environment, even though the brain function remains intact.

A BCI system consists of three main parts (see also figure 1):

(1) Input signal: this is the brain activity, which can be acquired either invasively or non-invasively. The most common form of the brain activity utilized in a BCI is the EEG, as it offers good temporal resolution – an essential feature for rapid communication -, relative simplicity and low cost of equipment necessary for EEG recordings. This, however, does not come without a cost, as the EEG signals are a mixture of activity originating from various brain sources, mixed with other activity of non-cerebral origin (artefacts). Different types of brain activity are utilised in BCIs, e.g. execution of different mental tasks (motor imagery, mental arithmetic, etc [5], [6]), self-regulation of amplitude of μ and β rhythms (e.g. [7]), and event-related potentials (e.g. [8], [9]).

(2) Feature Extraction and Translation: the digitized and amplified EEG signals are then processed with the goal of Signal-To-Noise ratio maximization. A main concern is the removal of artifact contamination, for which a number of methods have been proposed, e.g. Independent Component Analysis [10, 11]. For an online BCI system artifact removal must be performed automatically, which is currently one of the main problems that BCI technology is facing, and for which a limited number of solutions have been proposed [12, 13]. The clean EEG recordings are then processed using feature extraction methods to obtain a set of features which can be used as a means of identification of the different mental tasks. These are then classified (usually via supervised classifiers such as LDA and NNs) and translated into a set of commands that allow control of a device.

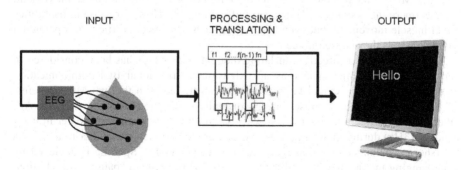

INPUT   PROCESSING &   OUTPUT
               TRANSLATION

Fig. 1. The main parts of which a BCI system is composed. The brain activity recorded by an electroencephalogram is processed and classified into a command for controlling a device.

(3) Output: the device to be controlled, such as a computer screen, a prosthetic device or a wheelchair. Some BCI protocols require that the user and/or the system be trained beforehand while others do not require any training.

Currently BCIs are suitable for basic environmental control through binary communication, such as word spelling (e.g. [14]). Even though it is possible to control demanding applications (e.g. [15]), currently this is limited by the trade-off between speed of communication and classification accuracy. Thus, the most common application still remains 1-dimensional cursor movement on a computer screen, which offers the ability to communicate with the environment when teamed with a "virtual keyboard". Communication can be achieved by mentally controlling the cursor movement on the screen for choosing letters on the "virtual keyboard" [16] or to highlight the desired character from a scrolling list [14]. Different mental tasks are associated with left/right and/or up/down cursor movement, thus allowing the subject to pick characters and spell words. Despite the simplicity of these applications, current BCI systems are faced with some unresolved problems: (i) the maximum subject-independent speed of communication reported is 25 bits/min [17], although higher speeds have been reported for individual cases, e.g. maximum of 7.6 letter/min [18] and 35 bits/min [19]. If we consider a character with 8 bit resolution then 25 bits/min is equivalent to 3.13 chars/min, which is not acceptable for normal speech communication; And (ii) current systems are bulky and non portable and, thus, are not practical for real-world applications. It is envisaged that the development of custom-built hardware as part of the proposed system will provide a solution to both these issues. In addition, these can be aided if the "virtual keyboard" is substituted by a simplified set of characters whose choice is directly associated with particular mental tasks, thus eliminating the intermediate step of cursor movement.

Such a potential simplification could be achieved via the use of Morse Code (MC), which has already been utilised for communication for disabled people. In MC transmission of information is based on short and long elements of sound (dots and dashes) and was originally created for telegraph communication. The elegance of MC lays in its simplicity and the high speech reception and transmission rates. A skilled MC operator can receive MC in excess of 40 words per minute [20]. The world record for understanding MC was set in 1939 and still stands at 75 words per minute [21]. Utilisation of MC for the disabled is commonly based on some form of muscle movement, such as operating a switch [22] or a sip-puff straw [23]. However, certain disabilities affect muscle movement, but even if not, then such systems are difficult to operate on a daily basis as they cause fatigue.

The use of MC for directly translating thoughts into words has been considered in very few BCI systems, mainly as an extension to traditional BCI communication methods. In [24] the "virtual keyboard" was substituted with the two MC elements, ".", and "-", and the user chose through mentally controlling cursor movement. Another MC-BCI system is described in [25] based on the attenuation of power in the μ band (8-13Hz) during motor imagery, whose duration corresponds either to a "." or a "-" (shorter or longer motor imagery duration respectively). Spelling is achieved by interchanging motor imagery with baseline task (representing a "pause"). In addition, [26] showed how communication in a BCI system could conceptually be achieved via a tri-state MC scheme and utilising a fuzzy ARTMAP as classifier. In such a system a ".", a "-" or a "space" would be represented by 3 mental tasks and the continuous

EEG would be sampled every, e.g., 0.5s, for decision making. It is also stated that the conversion of a mental task into one of the 3 MC elements would take 6ms of computation time; however this heavily depends on a number of operating system factors.

The concept behind the latter two systems is closer to the concept of the proposed system, as the intermediate step of cursor movement is eliminated. The use of MC is advantageous as it simplifies the dictionary to 3 symbols, the choice of which will be achieved through 2 mental tasks. This reduces the system complexity and improves communication speed. Hence, we envisage the development of a portable, embedded, custom and wearable MC-based BCI system that could be used either as an assistive or as an enhancing communication aid.

3 Performance Optimisation

The proposed system is shown in figure 2 and consists of 4 parts: (1) EEG signals are recorded from a patient performing two mental tasks, each corresponding to either a "." and "-" (depending on the task duration) or a "pause". The patient is mentally spelling letters and words in MC; (2) windows of specified duration of the recordings are processed and classified as ".", "-" or "pause"; (3) MC is then converted into text, which is in turn converted to speech via a text-to-speech converter (4). At this stage our priority is to maximise correct interpretation of EEG data. Computational efficiency is not a key consideration as we will be designing custom hardware tailored to the chosen processing methods. Therefore, it is imperative to firstly converge on a particular combination of signal processing methods that could be used reliably in the proposed system. The preliminary investigations presented in this paper are associated with part 2 of the proposed system and are part of our greater search for the optimal combination of features and classifiers.

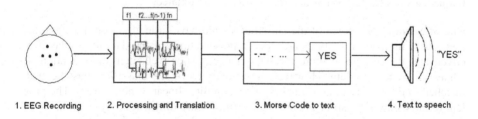

1. EEG Recording   2. Processing and Translation   3. Morse Code to text   4. Text to speech

Fig. 2. The proposed MC-BCI system

3.1 Methods

Feature Extraction. AR models are commonly utilised in EEG analysis [27]. More specifically, the estimated AR coefficients have been shown to capture well the differences between various mental tasks, and as a result are frequently used as features in mental task classification and BCIs (e.g. [5]). Eq. 1,

$$x_t = \sum_{\tau=1}^{p} a_\tau x_{t-\tau} + \varepsilon_t \tag{1}$$

represents an AR(p) model where p is the model order, x_t is the time series to be modelled (here it is the EEG data), a_τ, (τ=1,...,p) are the estimated coefficients of the pth-order AR model and ε_t is zero-mean random noise (commonly Gaussian with unit variance). In EEG analysis an AR(p) is fitted to each channel of EEG data and the pth dimensional vector of estimated coefficients represents the different mental tasks, as a variation of the coefficients depending on the mental task is observed. The AR model order used in EEG analysis ranges from 5 up to 13 [28]. For the specific dataset used here an order of 6 was chosen as suggested in [29]. Estimation of the coefficients is possile via a number of ways – here we used the method of Least Squares.

The second set of features utilised is PSD values obtained via parametric spectral analysis. In particular an AR(p) model (here p=6) is first fitted on the data and the power spectrum is subsequently obtained from the estimated coefficients via:

$$S(f) = \frac{\hat{\sigma}_p^2}{N \sum_{k=0}^{p} a_k e^{-j2\pi fk/N}} \tag{2}$$

where a_k, (k=1,...,p) are the estimated coefficients, f is a vector of chosen frequencies, $\hat{\sigma}_p^2$ is the estimated noise variance and N is the number of samples. The advantage of parametric methods for spectrum estimation is the ability to specify a set of frequencies of interest over which the spectrum is estimated.

Classification. The choice of the classifier should have little effect on the classification rate if the chosen features are good representations of the data to be classified. Given that the features capture the data characteristics well, then classification becomes an easier problem. However, the properties of the classifier must be well-matched to the feature dimensionality or separability (linear or non-linear). The problem of choosing a classifier is enhanced if the feature dimensionality is high, as this does not allow the visualisation of the features and, consequently, whether they are linearly separable or not.

SVMs offer a solution to this issue, as both linear and non-linear classification can be obtained simply by changing the "kernel" function utilized [30]. Due to the fairly new development of SVMs they are not commonly utilised in BCI systems (see [31] for an example). Thus, their performance for mental task classification has not been widely assessed and their application in such systems can be considered novel.

SVMs belong to the family of kernel based classifiers. The main concept of SVMs is to implicitly map the data into the feature space where a hyperplane (decision boundary) separating the classes may exist. This implicit mapping is achieved via the use of Kernels, which are functions that return the scalar product in the feature space by performing calculations in the data space. The simplest case is a linear SVM trained to classify linearly separable data. After re-normalisation, the training data, $\{x_i, y_i\}$ for $i=1, \ldots, m$ and $y_i \in \{-1,1\}$, must satisfy the constraints

$$x_i w + b \geq +1 \quad \text{for } y_i = +1 \tag{3}$$

$$x_i w + b \leq -1 \quad \text{for } y_i = -1 \tag{4}$$

where w is a vector containing the hyperplane parameters and b is an offset. The points for which the equalities in the above equations hold have the smallest distance to the decision boundary and they are called the support vectors. The distance between the two parallel hyperplanes on which the support vectors for the respective classes lie is called the *margin*. Thus, the SVM finds a decision boundary that maximises the margin. Finding the decision boundary then becomes a constrained optimization problem amounting to minimisation of $\|w\|^2$ subject to the constraints in (3) and (4) and is solved using Lagrange optimisation framework. The general solution is given by

$$f(x) = \sum_i \alpha_i y_i \langle x_i, x \rangle \tag{5}$$

In the case of non-linear classification, Kernels (functions of varying shapes, e.g. polynomial or Radial Basis Function) are used to map the data into a higher dimensional feature space in which a linear separating hyperplane could be found. The general solution is then of the form:

$$f(x) = \sum_i \alpha_i y_i K \langle x_i, x \rangle \tag{6}$$

Depending on the choice of the Kernel function SVMs can provide both linear and non-linear classification, hence a direct comparison between the two can be made without having to resort to utilisation of different classifiers.

Data. At this stage we utilise EEG data that is available online. The dataset chosen is well-known and has been used in various BCI applications. It contains EEG signals recorded by Keirn and Aunon during 5 mental tasks and is available from (http://www.cs.colostate.edu/~anderson). Each mental task lasted 10s and subjects participated in recordings over 5 trials and a number of sessions (subjects 2 and 7 participated in 1 session, subject 5 in 3 and subjects 1, 3, 4 and 6 in 2). The data was recorded with a sampling rate of 250Hz from 6 EEG electrodes placed at locations

C3, C4, P3, P4, O1 and O2, plus 1 EOG electrode (more details on the recording protocol can be found in [29]). The 5 mental tasks contained in the dataset are:

(1) Baseline: subjects are relaxed and should be thinking of nothing particular.

(2) Multiplication: subjects are asked to perform non-trivial mental multiplication problems; it is highly likely that a solution was not arrived at by the end of the allocated recording time.

(3) Rotation: a 3-dimensional geometric figure is shown on the screen for 30s, after which the subjects are asked to mentally rotate the figure about an axis.

(4) Letter composition: subjects are asked to mentally compose a letter, continuing its composition from where it was left off at the end of each trial. And

(5) Counting: subjects are asked to count sequentially by imagining the numbers being written on a blackboard and rubbed off before the next number is written. In each trial counting resumes from where it was left off in the previous trial.

Figure 3 shows a single randomly chosen 10s trial for each of the 5 mental tasks for subject 1 and electrode C3, for illustration purposes.

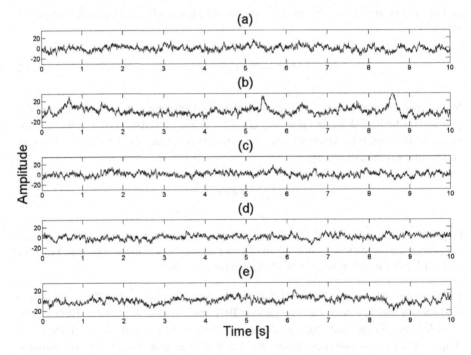

Fig. 3. Continuous 10s EEG recordings from subject 1 and electrode C3 during baseline (a), multiplication (b), letter composing (c), mental rotation (d) and counting (e). No major differences between the 5 tasks can be easily identified via visual inspection, which highlights the importance of appropriate feature extraction methods.

Choice of this specific dataset was twofold. Firstly, the dataset contains recordings from mental tasks that are traditionally associated with BCI systems. Secondly, it allows the investigation of a large combination of mental task pairs as it contains recordings from 5 different tasks – this will allow us to identify whether the choice of tasks is subject-dependent and whether the investigation of other non-traditional tasks should be encouraged. In addition, the specific dataset is a benchmark for BCI research and thus our results can be directly compared to results for the same dataset reported in the literature.

4   Results

To allow a direct comparison of the results with those presented in [32], we used data from 2 sessions and 4 subjects (subjects 1, 3, 5 and 6). The data was split in non-overlapping segments of 0.5s duration, resulting in 200 segments per task per subject, over 2 sessions. The SVM classification rate was averaged over 10 trials, where in each trial a randomly chosen set of 100 segments was used for training, with the remaining segments used for testing. Averaging the classification rate over a number of trials, where different datasets are used in each trial, mimises any potential biasing of the results from properties that may be specific to some of the data. All 10 pair combinations of the 5 mental tasks were classified and the pair of tasks with the maximum average classification rate for each subject was identified. The average classification rate was estimated as $(TP_1+TP_2)/2$, where TP_i (true positive) is the number of segments classified correctly for mental task i. The feature vectors describing each 0.5s segment are 36-dimensional in the AR(6) case and 300-dimensional in the PSD values case (6 AR coefficients and 50 PSD values per electrode; the final feature vectors consisted of the concatenated AR coefficients and PSD values for all electrodes respectively).

The classification results for the AR(6) features are presented in table 1. It can be seen that the choice of classifier had a positive effect on the classification accuracy. The use of an SVM increased the accuracy by up to nearly 13% compared to that obtained for the same features using LDA and by up to 16.3% using an NN (see table 2 for details), as presented in [32]. In theory, the choice of classifier has a smaller effect on the classification rate if the features utilised represent the data well. Nonetheless, the use of an SVM with RBF Kernel increases the classification rate by a large margin and, hence these results indicate that the use of an SVM is more appropriate for these features. In addition, the pair of tasks which provided the highest average classification was different than the equivalent pair from [ibid]. However, it was also observed that the task pair which gave highest average classification varied with each subject, in agreement with [ibid]. Hence a particular task pair for which optimal operation can be obtained should be identified for each subject. In addition, performance could be improved if the tasks utilised had a more intuitive connection with the way of thinking associated with MC.

Table 1. Maximum average classification rate (%) for AR(6) features with SVM. Results presented are averaged over 10 trials.

Subject	Class. Rate (%)	Tasks	Kernel
1	88.4	Letter vs multiplication	RBF
3	87.9	Letter vs counting	RBF
5	83.9	Roration vs counting	RBF
6	92.4	Counting vs multiplication	Linear

Table 2. Maximum average classification rate (%) for AR(6) features. Column 2 presents our results, while columns 3 and 4 give the best rates presented in (Huan and Palaniappan, 2004) for LDA and NN.

Subject	SVM	LDA	NN
1	88.4	80.2	78.9
3	87.9	73.6	73.9
5	83.9	71.4	67.6
6	92.4	84.3	77.6

Table 3. Maximum average classification rate (%) for power spectrum values with SVM. Results presented are averaged over 10 trials.

Subject	Class. Rate (%)	Tasks	Kernel
1	58.0	Letter vs multiplication	RBF
3	56.6	Letter vs counting	RBF
5	68.0	Roration vs counting	RBF
6	60.2	Counting vs multiplication	Polynomial

The classification rates for the PSD features are presented in table 3. The rates obtained are much lower than the ones reported in [26]. This could be attributed to three reasons. Firstly, in this work classification between pairs of tasks was obtained as opposed to between 3 tasks as in [ibid] hence a direct comparison cannot be made. Secondly, the PSD features are already of high dimension (300-dimensional). Considering the principles behind SVM operation, i.e. the transformation of the data into a feature space of even higher dimensions, then an SVM may not be the appropriate choice for classification when the feature vector is already of high dimension. Thirdly, the classification rates presented in [ibid] were averaged for a single training set whose ordering of the training patterns was randomly varied 10 times, hence the high classification rate reported may have been a side-effect of the particular choice of training set. This is in opposition to the classification rates reported here, which have been obtained for different sets of training and test data and, thus, are less biased by data specifics. In addition, another issue with utilisation of PSD values as features is the partial spectrum overlarp of certain artefacts (such as eye movements) with EEG activity, which can potentially adversely affect the classification rate.

The feature vectors were created by concatenating the estimated AR coefficients from all 6 electrodes. However, the wearability and portability of an MC-based BCI is facilitated by employing a small number of electrodes –ideally two, or even a single, electrode(s). It may be possible to obtain higher classification rates by utilising a single electrode that is more relevant to the specific mental task rather than using a combination of electrodes, all of which are not as relevant to the task. This is also advantageous as it decreases the feature dimensionality.

5 Conclusions

This paper presents the results of initial investigations in the search for appropriate features and classifiers towards the development of a thought-to-speech converter. The results indicate that the use of an SVM for the classification of AR coefficients is more appropriate than LDA and NN and will be utilised in the development of the proposed system.

The proposed system is promising as it offers the ability to communicate more efficiently via direct conversion of thoughts into speech. In order to ensure optimal operation other aspects of the system must also be investigated. Firstly, a more extensive set of features and classifiers will be examined such that the optimal combination in terms of maximising accuracy is determined – computational efficiency is not a consideration as the system will be customised and capable of parallel processing. Secondly, these investigations suggest that different combinations of mental tasks seem to be more appropriate for different subjects. We are going to look into finding a combination of tasks that are more intuitive and more closely related to the concept of MC, as this could improve classification accuracy and facilitate easier operation.

References

1. Shenoy, P., Miller, K.J., Crawford, B., Rao, R.P.N.: Online Electromyographic Control of a Robotic Prosthesis. IEEE Trans. on Biomed. Eng. 55(3), 1128–1135 (2008)
2. Hori, J., Sakano, K., Miyakawa, M., Saitoh, Y.: Eye movement communication control system based on EOG and voluntary eye blink. In: Miesenberger, K., Klaus, J., Zagler, W.L., Karshmer, A.I. (eds.) ICCHP 2006. LNCS, vol. 4061, pp. 950–953. Springer, Heidelberg (2006)
3. Wolpaw, J.R., et al.: Brain-Computer Interface Technology: a review of the First International meeting. IEEE Trans. on Rehab. Eng. 8(2), 164–173 (2000)
4. Vidal, J.J.: Real-time detection of brain events in EEG. Proc. of the IEEE 65, 633–664 (1977)
5. Guger, C., Schlögl, A., Neuper, C., Walterspacher, D., Strein, T., Pfurtscheller, G.: Rapid Prototyping of an EEG-based Brain-Computer Interface (BCI). IEEE Trans. on Neural Systems and Rehab. Eng. 9(1), 49–58 (2000)
6. Curran, E., Sykacek, P., Roberts, S.J., Penny, W., Stokes, M., Jonsrude, I., Owen, A.: Cognitive tasks for driving a Brain Computer Interfacing System: a pilot study. IEEE Trans. in Rehab. Eng. 12(1), 48–55 (2004)
7. Wolpaw, J.R., McFarland, D.J.: Control of a two-dimensional movement signal by a non-invasive brain-computer interface in humans. PNAS 101(51), 17849–17854 (2004)

8. Allison, B.Z., McFarland, D.J., Schalk, G., Zheng, S.D., Moore Jackson, M., Wolpaw, J.R.: Towards an independent brain-computer interfact using steady state visual evoked potentials. Clinical Neurophysiology 119, 399–408 (2008)

9. Farwell, L.A., Donchin, E.: Talking off the top of your head: toward a mental prosthesis utilizing event-related brain potentials. Electroenceph. Clin. Neurophysiology 70, 510–523 (1988)

10. Hyvärinen, A., Karhunen, J., Oja, E.: Independent Component Analysis. John Wiley & Sons, Chichester (2001)

11. Jung, T.-P., Makeig, S., Humphries, C., Lee, T.-W., McKeown, M.J., Iragui, V., Sejnowski, T.J.: Removing electroencephalographic artefacts by blind source separation. Psychophysiology 37, 163–178 (2000)

12. Nicolaou, N., Nasuto, S.J.: Automatic artefact removal from event-related potentials via clustering. J. of VLSI Signal Processing 48(1-2), 173–183 (2007)

13. Joyce, C.A., Gorodnitsky, I.F., Kutas, M.: Automatic Removal of Eye Movement and Blink Artefacts from EEG Data Using Blind Component Separation. Psychophysiology 41, 313–325 (2003)

14. Scherer, R., Muller, G.R., Neuper, C., Graimann, B., Pfurtscheller, G.: An Asynchronously controlled EEG-based virtual keyboard: improvement of the spelling rate. IEEE Trans. on Biomed. Eng. 51(6), 979–984 (2004)

15. del, R., Millán, J., Renkens, F., Mourino, J., Gerstner, W.: Noninvasive brain-actuated control of a mobile robot by human EEG. IEEE Trans. on Biomedical Engineering 51(6), 1026–1033 (2004)

16. Wolpaw, J.R., Birbaumer, N., McFarland, D.J., Pfurtscheller, G., Vaughan, T.M.: Brain-Computer Interfaces for communication and control. Clinical Neurophysiology 113, 767–791 (2002)

17. Vaughan, T.M., et al.: Brain-computer Interface Technology: a review of the second international meeting (Guest Editorial). IEEE Trans. on Neural Systems and Rehab. Eng. 11(2), 94–109 (2003)

18. Müller, K.-R., et al.: Machine learning for real-time single-trial EEG-analysis: From brain-computer interfacing to mental state monitoring. J. of Neuroscience Methods 167, 82–90 (2008)

19. Blankertz, B., Dornhege, G., Krauledat, M., Müller, K.-R., Curio, G.: The non-invasive Berlin Brain-Computer Interface: Fast acquisition of effective performance in untrained subjects. NeuroImage 37, 539–550 (2007)

20. Coe, L.: Telegraph: A History of Morse's invention and its predecessors in the United States. McFarland & Company (2003)

21. French, T.: McElroy, World's Champion Radio Telegrapher. Artifax Books (1993)

22. Park, H.-J., Kwon, S.-H., Kim, H.-C., Park, K.-S.: Adaptive EMG-driven communication for the disability. In: 1st Joint BMES/EMBS Conf. Serving Humanity, Advancing Technology, Atlanta, USA, October 13-19, p. 656 (1999)

23. Levine, S.P., Gauger, J.R.D., Bowers, L.D., Khan, K.J.: A comparison of Mouthstick and Morse code text inputs. Augmentative and Alternative Communication 2(2), 51–55 (1986)

24. Palaniappan, R.: Brain computer interface design using band powers extracted during mental tasks. In: 2nd International IEEE EMBS Conf. on Neural Eng., Arlington, Virginia, March 16-19, pp. 321–324 (2005)

25. Altschuler, E.L., Dowla, F.U.: Encephalolexianalyzer. United States Patent Number 5, 840, 040, November 24 (1998)

26. Palaniappan, R., Paramesan, R., Nishida, S., Saiwaki, N.: A new brain-computer interface design using Fuzzy ARTMAP. IEEE Trans. on Neural Systems and Rehab. Eng. 10(3), 140–148 (2002)
27. Wright, J.J., Kydd, R.R., Sergejew, A.A.: Autoregression Models of EEG. Biological Cybernetics 62, 201–210 (1990)
28. Lopes da Silva, F.: EEG analysis: Theory and Practice. In: Electroencephalography: Basic Principles, Clinical Applications and Related Fields, Ch. 6, pp. 1135–1163 (1998)
29. Keirn, Z.A., Aunon, J.I.: A new mode of communication between man and his surroundings. IEEE Trans. Biomed. Eng. 37, 1209–1214 (1990)
30. Burges, C.J.C.: A tutorial on Support Vector Machines for Pattern Recognition. In: Fayyad, U. (ed.) Data Mining and Knowledge Discovery, pp. 121–167. Kluwer Academic Publishers, Boston (1998)
31. Gysels, E., Celka, P.: Phase synchronisation for the recognition of mental tasks in a brain-computer interface. IEEE Trans. on Neural Systems and Rehab. Eng. 12(4), 406–415 (2004)
32. Huan, N.-J., Palaniappan, R.: Neural network classification of autoregressive features from electroencephalogram signals for brain-computer interface design. J. of Neural Eng. 1, 142–150 (2004)

Neurophysiologic and Cardiac Signals Simulator Based on Microconverter

Maurício C. Tavares[1,3], Carlos M. Richter[3], Tiago R. Oliveira[1], and Raimes Moraes[2]

[1] PDI/Contronic Sistemas Automaticos Ltda., Rua Rudi Bonow 275, Pelotas, 96070-310, Brazil
{mauricio.tavares,tiago.rockembach}@contronic.com.br
[2] Electric Engineering Department, Federal University of Santa Catarina, Campus Universitario
Trindade, Florianópolis, 88040-900, Brazil
raimes@eel.ufsc.br
[3] Biomedical Engineering Laboratory, Catholic University of Pelotas, Rua Felix da Cunha 412
Pelotas, 96080-000, Brazil
{mtavares,richter}@ucpel.tche.br

Abstract. This paper describes a microconverter-based electronic device developed to simulate auditory evoked potentials of short, middle and long latencies, ECG and electronystagmographic signals. The simulator reproduces real physiologic signals from sampled AEP and ECG waveforms instead of generating them based on equations. The main simulator part is the ADuC841 microconverter (single-cycle 20 MHz 8052 core, FLASH memory and two 12-bit DACs). Called SimPac I, the equipment is portable and easy to operate, being a worthy tool for calibration of AEP, ECG, ENG and VENG systems during manufacture and maintenance. It can also be used in the development and testing of DSP algorithms intended to filter and/or average the above mentioned signals. As result, examples of several waveforms generated by the SimPac I are shown.

Keywords: Neurophysiologic signals simulator, ECG simulator, ENG simulator, electro-medical equipments maintenance.

1 Introduction

The performance of medical equipments must be periodically assessed since patients may be harmed if these equipments do not work properly. However, performance analyzers for these equipments are expensive, preventing its wide use in third world countries. Therefore, the availability of low cost simulators is a real need in those countries [1].

1.1 Neurophysiologic and Cardiac Signals Characteristics

Auditory evoked potentials (AEP) play a fundamental role in the audiology practice. The analysis of electric potentials generated in response to acoustic stimulation resulted in many important applications in the oto-neurology field [3]. AEPs are

A. Fred, J. Filipe, and H. Gamboa (Eds.): BIOSTEC 2008, CCIS 25, pp. 136–147, 2008.

classified according to its latency. Potentials of short latency occur in up to 10 ms after the auditory stimulation. Middle latency potentials occur between 10 ms and 100 ms after the stimulus, and long latency potentials are registered after 100 ms from the stimulus. Short latency AEPs are known by the acronyms BAEP – Brainstem Auditory Evoked Potentials, BERA - Brainstem Electric Response Audiometry or ABR - Auditory Brainstem Response [4]. The electrocochleography (EcochG), used for cochlear evaluation, is also considered a short latency AEP. ABR is used for the assessment of the brainstem integrity and also for objective audiometry. Middle latency evoked potentials are identified by the acronyms MLR or MLAEP (Middle Latency Auditory Evoked Potential). MLAEP is indicated to evaluate dysfunctions that could commit the hearing pathways located between the brainstem and the primary cortex.

P300 and MMN (Mismatch Negativity) are the most used long latency AEPs in the clinical practice. P300 has a positive peal around the 300 ms latency and its use allows obtaining, in only one test, information on: the activity of the transition thalamus-auditory cortex, the auditory cortex, the hippocampus, the hearing attention and the cognition. P300 is elicited through a "rare paradigm" in which a few "rare" stimuli happen randomly in a series of "frequent" stimuli. A 20-30% rare on 80-70% frequent proportion is usual in the clinical practice [5]. Their difference may be in the intensity or in the frequency. The MMN test also uses rare and frequent stimuli, but the outcome reflects the central processing capacity [6].

ECG is certainly the most known bioelectric signal generated by the human body. That signal is sampled and analyzed for clinical diagnosis, surgical accompaniment and rehabilitation. There are several types of ECG simulators that are able to reproduce changes in amplitude, heart beat frequency and many types of arrhythmias [7]. They are applied to the development of new monitors and in the preventive and corrective maintenance.

Perfect body balance is very important for the organism orientation in the environment. That balance widely depends on the vestibular system which acts in cooperation with the visual system to maintain the vision focus during head movements. The vestibulo-ocular reflex (VOR) is the cerebral mechanism that makes it possible. The VOR assessment is carried out by the electronystagmography test (ENG) or by its variant, the vector-electronystagmography (VENG) [8]. The evaluation of ENG/VENG is based on the analysis of the nystagmus, that is, the reflex ocular movements which occur when the labyrinth receives caloric or rotational stimuli.

The developed patient simulator is called SimPac I having the purpose of serving as a tool for development, validation, adjustment and maintenance of AEP, ECG and ENG/VENG equipments.

2 Materials and Methods

SimPac I generates, in two channels, all short, middle and long latency AEP waveforms mentioned in the introduction. The morphology of those signals has been shown elsewhere [2]. Real signals of AEP were scanned to create an array of 500 values for each one. Their magnitudes were adjusted to match the dynamic range of the 12-bit D/A converters contained in the microconverter. Figure 1 shows the block diagram of SimPac I.

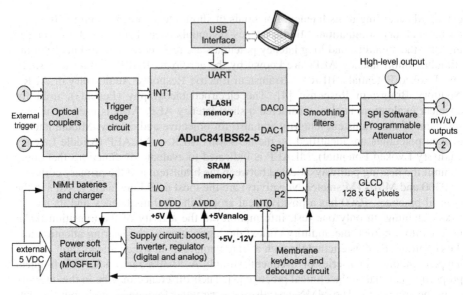

Fig. 1. Block diagram of the SimPac I hardware

2.1 Hardware

The microconverter (μC) ADuC841 [9] is the main hardware part. It is an optimized single-cycle 20 MHz 8052 core that contains four different memory blocks: 62 kbytes for code (Flash), 4 kbytes for data (Flash), 256 bytes of general-purpose RAM and 2 kbytes of internal XRAM. The simulator firmware was implemented in the internal memory. The ADuC841 is in charge of generating the waveforms from tables stored into the program memory, using the DDS technique (Digital Direct Synthesis) [10]. Power is supplied by four NiMH 1.2 V batteries or by a 5 V external source.

The equipment has an on/off circuit (Fig. 2) commanded by software in order to prevent its undue powering off while data is being written into the FLASH memory. That could happen if a conventional mechanical switch was utilized. The turn on/off circuit works together with an available membrane key. When the equipment is off, a short key strike activates the power on circuit through Q4 that, by its turn, commutates the MOSFET Q2 [11]. In that way, the battery power feeds the voltage regulator circuit that supplies the microconverter. An I/O pin activates the Q3 base that causes the supply retention. To shut down the equipment, the user should hold the same key pressed for 5 s. When that happens, the microconverter cut off the command signal applied to the Q3 base and the equipment is turned off as soon as the user releases the key. A goodbye message is also displayed on the GLCD.

The microconverter requires a +5 V supply and the GLCD, +5 and -12 V. The analog circuits are supplied by symmetrical voltage: ±5 V. Those voltages are generated by a circuit which combines boosts and inverters. The complete power circuit uses two LM2621 [12], one ADM8660 [13] and one ICL7662 [14]. The battery charger circuit was based on MAX712 [15]. The operator interacts with the equipment by means of a membrane keyboard. His actions are based on a menu shown on a graphic liquid crystal display (GLCD) with 8 kpixels. The GLCD is driven by the μC through the ports P0 and P2.

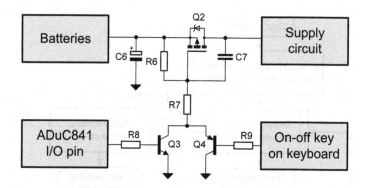

Fig. 2. Power on/off is carried out under software control

The simulator generates the AEP waveform after receiving an external pulse, since the averaging task demands synchronism between auditory stimuli and electric signal acquisition. The external pulse, a TTL level transition, has to be applied to an external trigger connector. The level transition may be either low-to-high or high-to-low depending on the operator choice that has to be programmed through the menu. In order to generate a cyclic waveform for tests, a key named "manual trigger" was made available. For any trigger selection, manual or external, the pulse goes through a Schmitt trigger circuit to eliminate bouncing. The manual trigger requests the interrupt INT0, while the external trigger activates the INT1.

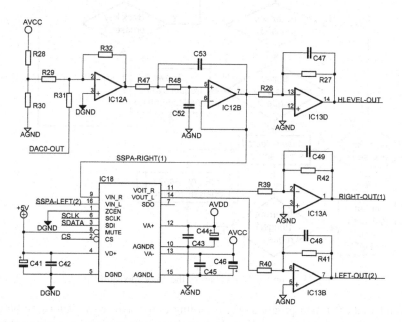

Fig. 3. Output signal conditioning circuit comprises a SPI Software Programmable Attenuator (SSPA) that has a 120 dB range. Filter and attenuator shown were implemented for both channels.

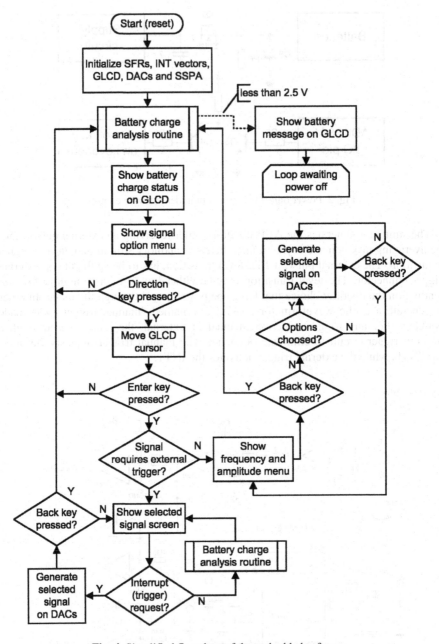

Fig. 4. Simplified flowchart of the embedded software

An internal µC timer, programmed according to the selected waveform to be generated, sets the D/A conversion. Each D/A output is processed by a conditioning circuit which filters (low-pass reconstruction) and lowers the signal amplitude to levels compatible with the ones sampled on the human body. This circuit is shown in

Figure 3. The amplitude adjustment is made by an attenuator that has a maximum range of 120 dB and 0.5 dB resolution [16], being programmed through the μC SPI port.

2.2 Embedded Software

The SimPac I embedded software was developed in C language with the μC/51 V1.20.04 compiler [17]. The software flowchart is shown in Figure 4. The compiled code was written into the FLASH program memory using WSD 6.7 [18].

When the equipment is turned on, the software exhibits a greeting message on the GLCD which is briefly followed by the main menu.

The channel 0 of the A/D converter samples the voltage supplied by the battery pack. The measured voltage is shown by a 5 level bar graph on the GLCD as commonly seen in cell phones. When the voltage is below 2.5V, message is displayed warning the operator.

After selecting the signal to be simulated on the menu, a sub-menu is shown to ask for the complementary parameters like frequency, amplitude, heart rate, ENG angular velocity and direction. When the external trigger demands the generation of the programmed waveform, the software sets the timer to send the samples to DAC at the programmed rate. The interrupt INT1 generates the signal by DDS. EcochG, ABR, MLR and LLR are generated in the same way. At each INT1 call, a finite loop transfers the samples stored into the tables to the D/A converters. The routine that generates the P300 and MMN signals alternatively reproduces common and rare signals (stored into different tables) to simulate the physiologic response to stimulation. At each INT1 interrupt, the software establishes the signal that will be generated. The sequence of generated waveforms is established by an oddball table also stored in the FLASH.

Table 1. Characteristics of the generated waveforms

Signal	Amplitude	Period/frequency	D/A Rate	Trigger
EcochG	0.5 μV typical	5 ms	100 kSPS	External
ABR	1 μV typical	10 ms	50 kSPS	External
MLR	3 μV typical	50 ms	10 kSPS	External
LLR	5 μV typical	500 ms	1 kSPS	External
P300	5 μV typical	500 ms	1 kSPS	External
MMN	5 μV typical	500 ms	1 kSPS	External
ECG	0.5, 1 and 2 mV	30, 60, 80, 120, 160, 200, 240 and 300 BPM	500 SPS	key
EEG	100 μV	-	500 SPS	key
ENG	0.1, 0.5, 1 and 2 mV	1 to 125 degrees/s	500 SPS	key
Sine	0.5, 1, 2 and 10 mV	0.01, 0.05, 0.1, 0,5, 10, 50, 60 and 100 Hz	500 SPS	key
Square	0.5, 1, 2 and 10 mV	0.01, 0.05, 0.1, 0.5, 20, 40, 50, 60 and 100 Hz	500 SPS	key

To generate ECG signals, the start up is given by the manual trigger button. The timer defines the D/A conversion rate. The different heart beat frequencies are simulated based on the period between two P-QRS-T complexes and the T duration. For ENG signals, the used algorithm is similar to the ECG one; however the angular velocity changes are simulated by modifying the sweeping step of the tables recorded into the FLASH memory. The waveform parameters used on the SimPac I software are shown in Table 1. The different sets of periods and D/A conversion rates are obtained by programming the reference timers before starting the generation of each signal.

3 Results

To work with the ADuC841 SMD package (LQFP), a PCB was designed to make easier the handling of its pins (Fig. 5). The remaining prototype circuit was assembled on a universal pre-drilled board. Software development and preliminary tests were

Fig. 5. The first prototype is based on a board designed to allow easier handling of the LQFP package. This board contains the ADuC841 (*left*) and the serial interface to access the internal FLASH (*right*). All other components were added to this board by means of a pre-drilled board.

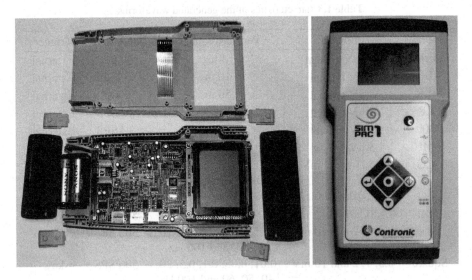

Fig. 6. Photo of the developed simulator

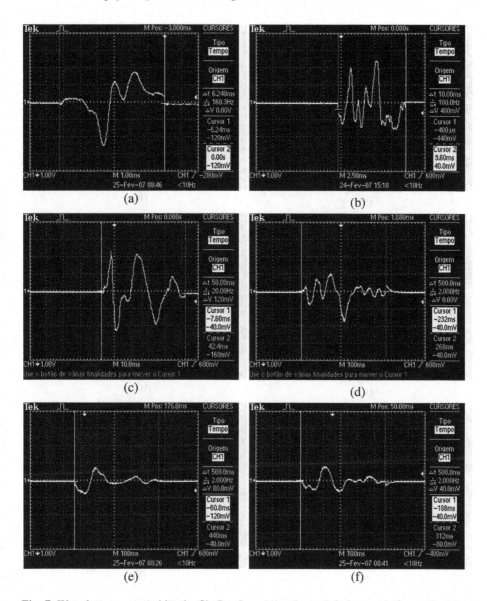

Fig. 7. Waveforms generated by the SimPac I as shown by a digital scope before attenuation (SSPA): EcochG (*a*); ABR (*b*); MLR (*c*); LLR (*d*); MMN frequent (*e*); MMN rare (*f*)

carried out using this experimental hardware. Later on, another PCB was developed to comprise all used parts, including an improved GLCD interface that accepts modules of three manufacturers (Intech ITM-12864K0, ITM-12864K0-002, Hantronix HDM64GS12-L30S) without the need of any board adjustment. This second assembly is shown in Figure 6. The board was housed into a plastic case that has six membrane keys. One of them is used to switch the power on/off. The other five are used to set the generated waveform pattern among the several options displayed on the GLCD.

144   M.C. Tavares et al.

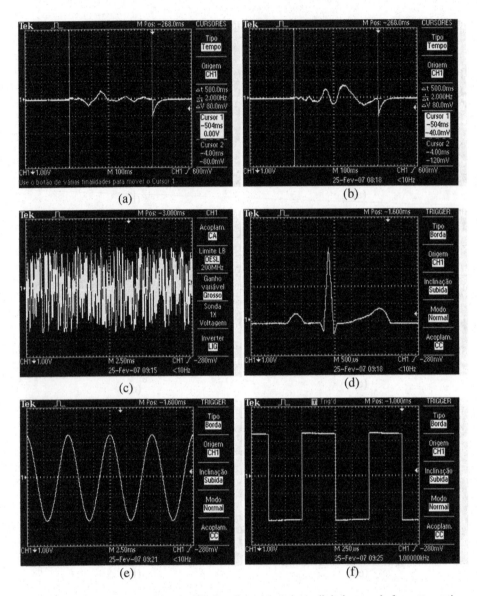

Fig. 8. Waveforms generated by the SimPac I as shown by a digital scope before attenuation (SSPA): P300 frequent (*a*); P300 rare (*b*); EEG (*c*); ECG (*d*); sine (*e*) and square (*f*)

The connectors for the input/output synchronism signal, DC power input (to recharge the batteries) and USB (for PC communication) are located on the right side of the plastic case.

Using a commercial AEP equipment [20], the developed SimPac I was exhaustively tested in bench by selecting all available parameters for synchronism, amplitude and period. Examples of the generated waveforms are presented in Figures 7 and 8. Figure 9 shows waveforms sampled by the commercial system. It is possible to

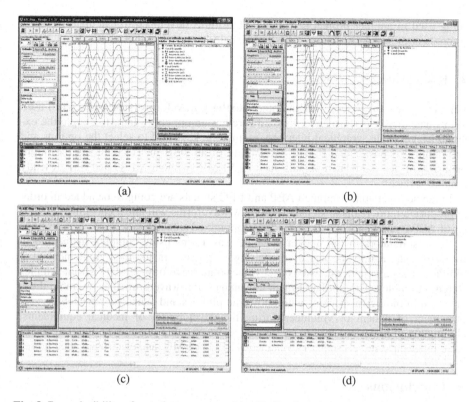

Fig. 9. Reproducibility of waveforms generated by SimPac I as shown by a commercial system that generated the external synchronism signal. Each waveform was obtained after averaging a stimulus set of: ABR (*a*); MLR with VEMP (*b*); LLR (*c*) and P300 frequent/rare (*d*).

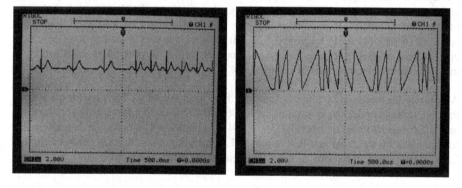

Fig. 10. Simulated ECG signal at 60 BPM and 120 BPM (*left*). Simulated ENG at several angular velocities, clockwise and counter clockwise (*right*).

note the perfect reproduction of the AEP signals generated by the SimPac I using the external trigger of the commercial equipment. Figure 10 shows ECG and ENG waveforms generated by SimPac I that were sampled from the DAC outputs using a digital scope (Rigol DS5102MA). The ENG waveforms generated by SimPac I are already being used in another project to validate algorithms developed to calculate the slow component of angular velocity of positional, caloric and rotary nystagmus.

4 Discussion

SimPac I generates ECG, ENG and other auxiliary waveforms (sine and square) used to evaluate circuit performance. Its hardware has an optical isolator to couple the external trigger input, graphical interface, membrane keyboard to select the options and soft-controlled power on/off. The availability of an USB interface allows the future inclusion of new simulated signals such as VEMP (without MLR) and SSR (Steady State Response).

In the next version, combined AEP and EEG signals are going to be added in order to produce a more realistic signal. These new waveforms will allow checking the ability of AEP equipments in extracting data from mixed signals.

Some simulators in the market generate waveforms using complex mathematical formulas that require the use of digital signal processors for their implementation. SimPac I uses a simple microconverter to generate waveforms by DDS from samples stored into tables. The SimPac I main advantages are: waveforms that resemble those observed in biological systems with known amplitudes and latencies; repeatability; replacement of patient or volunteer during the development of medical equipments; elimination of undesired factors such as the electrode-skin impedance, interference of other bioelectric signals such as spontaneous EEG or EMG, and electromagnetic noise.

5 Conclusions

This work has described an electronic device to generate ECG, ENG, AEPs and auxiliary waveforms. The waveforms generated by SimPac I were sampled by a commercial system and a digital oscilloscope. The tests carried out have shown the reliability and accuracy of programmed parameters: synchronism, amplitude, duration and repetition. This simulator is a worthy tool for the development of new equipment and software to process AEP, ECG and ENG. It can also be very useful to tune commercial systems produced to acquire these signals as well as in preventive and corrective maintenance of electro-medical equipments.

Acknowledgements. The authors thank to Brazilian Agency CNPq by the partial financial support under the grant number 310611/2005-9 (DTI engineering scholarship) and the research grant number 507363/2004-3. Thanks to Leonardo de Jesus Furtado for his work on the SimPac I layout and to Luciano Fagundes Kawski for his valuable help in developing the tools for the AEP signals sampling and conditioning.

References

1. Moraes, R.: PROQUALI – Family of Equipments for Medical-Assistance Quality Evaluation (in Portuguese). CNPq R&D Project Report 507363/2004-3 (2007)
2. Freitas, G.M., Oliveira, T.R., Moraes, R., Tavares, M.C.: Simulador de Potenciais Evocados Auditivos de Curta, Média e Longa Latência Baseado em Microconversor. In: Anais do CBEB 2006, vol. 1, pp. 1224–1227 (2006)

3. Ferraro, J.A., Durrant, J.D.: Potenciais Evocados auditivos: Visao Geral e Principios Basicos. In: Katz, J. (ed.) Tratado de Audiologia Clinica, 4th edn., Manole, Sao Paulo, pp. 315–336 (1999)
4. Chiappa, K.H.: Evoked Potentials in Clinical Medicine, 3rd edn. Lippincott-Raven Publishers, Philadelphia (1997)
5. Kraus, N., McGee, T.: Potenciais Auditivos Evocados de Longa Latencia. In: Katz, J. (ed.) Tratado de Audiologia Clinica, 4th edn., Manole, Sao Paulo, pp. 403–420 (1999)
6. Caovilla, H.H.: Audiologia Clínica. Atheneu, São Paulo (2000)
7. Prutchi, D., Norris, M.: Design and Development of Medical Electronic Instrumentation. Wiley, New Jersey (2005)
8. Castagno, L.A., Tavares, M.C., Richter, C.M., et al.: Sistema Computadorizado de Eletronistagmografia e Vectonistagmografia "UCPel/Castagno" (Versão 3.0). In: Anais do IV CBIS, pp. 26–31 (1994)
9. Analog Devices: MicroConverter 12-bit ADCs and DACs with Embedded High Speed 62-kB Flash MCU. Analog Devices Inc., Norwood (2003)
10. Grover, D., Deller, J.R.: Digital Signal Processing and the Microcontroller. Motorola University Press/Prentice Hall PTR, New Jersey (1999)
11. IRLML6402 HEXFET Power MOSFET,
   http://www.irf.com/product-info/datasheets/data/irlml6402.pdf
12. LM2621 Low Input Voltage, Step-Up DC-DC Converter,
   http://cache.national.com/ds/LM/LM2621.pdf
13. CMOS Switched-Capacitor Voltage Converters ADM660/ADM8660,
   http://www.analog.com/UploadedFiles/Data_Sheets/ADM660_8660.pdf
14. CMOS Voltage Converters ICL7662/Si7661,
   http://datasheets.maxim-ic.com/en/ds/ICL7662-Si7661.pdf
15. MAX712, MAX713 NiCd/NiMH Battery Fast-Charge Controllers,
   http://www.maxim-ic.com/quick_view2.cfm/qv_pk/1666
16. PGA2311 Stereo Audio Volume Control,
   http://focus.ti.com/lit/ds/symlink/pga2311.pdf
17. Wickenhäuser: uC/51 V1.20.04 User´s Manual. Wickenhäuser Elektrotechnik, Karlsruhe (2005)
18. Windows Serial Downloader V6.7, ftp://ftp.analog.com/pub/
   www/technology/dataConverters/microconverter/wsd_v6_7.exe
19. ARTEB Handheld Enclosures Product Drawings,
   http://www.rose-bopla.com/PDF_Files/Ch_04_Handhelds/
   ArtebDrawingsWeb.pdf
20. Contronic: Manual do Usuário - Módulo de Aquisição de Sinais Bioelétricos – MASBE – Rev. 4. Contronic, Pelotas (2007)
21. Contronic: Manual do Usuário - ATC Plus - Software para Audiometria de Tronco Cerebral – Build 2.1.X – Rev. 4. Contronic, Pelotas (2007)

Multipolar Electrode and Preamplifier Design for ENG-Signal Acquisition

Fabien Soulier[1,2], Lionel Gouyet[1,2], Guy Cathébras[1,2], Serge Bernard[1,3], David Guiraud[1,4], and Yves Bertrand[1,2]

[1] LIRMM, 161 rue Ada, 34392 Montpellier, France
Firstname.Lastname@lirmm.fr
[2] Université Montpellier II,
[3] Centre National de la Recherche Scientifique
[4] Institut National de Recherche en Informatique et Automatique

Abstract. Cuff electrodes have several advantages for *in situ* recording ENG signal. They are easy to implant and not very invasive for the patient. Nevertheless, they are subject to background parasitic noise, especially the EMG generated by the muscles. We show that the use of cuff electrodes with large numbers of poles can increase their sensitivity and their selectivity with respect to an efficient noise rejection. We investigate several configurations and compare the performances of a tripolar cuff electrode versus a multipolar one in numerical simulation.

One the other hand the use of cuff electrodes leads to the recording of the sum of the signals generated by all the axons within the nerve. This puts in evidence the need of signal separation techniques that require a great quantity of information. Again, we show that multipolar electrodes can solve this problem since poles can be switched one to another, provided that they are distributed along a regular tessellation.

Finally, we present the structure of an ASIC preamplifier processing a spatial filtering to obtain the Laplacian of the potential rejecting low-frequency noise.

1 Introduction

In a context of neural system pathologies such as spinal cord injury, Functional Electrical Stimulation (FES) techniques are the possible alternatives to restore lost sensory or motor abilities. These techniques consist in generating artificial contraction by electrical stimulation. In FES system a direct opened loop control doesn't allow efficient stimulation. In order to provide a loopback control we need sensory information (force, contact...) [1]. An attractive solution consists in using the natural sensors. The sensory information is propagated by associated afferent fibers. But unfortunately, in peripheral nerves the complete nerve activity due to the large number of axons makes the extraction of the studied signal particularly hard. Moreover the sensory signal seen through the nerve is a very low amplitude signal compared with the amplitude of parasitic signals. For instance, on a monopolar recording, electromyograms (EMG) created by muscle activity have amplitude about three orders of magnitude higher than the electroneurogram (ENG). In this context, the two main objectives to be able to exploit natural sensors are:

A. Fred, J. Filipe, and H. Gamboa (Eds.): BIOSTEC 2008, CCIS 25, pp. 148–159, 2008.

- to find a solution to separate the useful information from the complete ENG signal;
- to reject the parasitic external signals.

The classical solution consists in using multipolar electrodes, but from tripole [2] to nine pole electrode [3,4], the selectivity of the neural information is not efficient enough to be suitable in closed loop FES system. To achieve both a better sensitivity and efficient background noise rejection we propose a new configuration of the cuff electrode with a large number of poles regularly distributed onto the cuff. In this configuration, a group of poles can behave, with suitable low level analog signal processing, like a kind of a directive antenna. Moreover, the large number of poles will allow enough channels in order to apply source separation signal processing on the ENG. Of course, the directivity of the sensor relies on the quality of the subsequent low-level analog signal processing.

In this paper, we first show how to generalize the preprocessing operations on the recorded signal from tripolar to multipolar configuration using the Laplacian formalism. Then we discuss on the optimal pole placement around the nerve regarding tessellation methods. Both the electrode configuration and the associated preprocessing circuit result from this pole distribution and must be taken into account. We particularly focus on the hexagonal seven-pole electrode, presenting the associated seven input preamplifier and preliminary simulation results.

2   EMG Noise Rejection

Cuff electrodes have been the most used in the last ten years [5,6,7]. They are relatively easy to implant, they are not invasive for the nerve and implantation is very stable and thus allows chronic experiments.

ENG can be recorded as the potential difference created on the electrodes by the charges associated to the action potentials (AP) propagating along the nerve fibers. Fig. 1 shows a typical tripolar cuff electrode. When recording with this kind of electrode, a classic method to reject parasitic signals consists in calculating the average of the potential differences between the central pole and each of the outer poles [8,9]:

$$V_{\text{rec}} = \frac{(V_0 - V_1) + (V_0 - V_2)}{2} = V_0 - \frac{V_1 + V_2}{2} \tag{1}$$

The last expression shows that this operation consists in:

1. averaging the signal on the outer poles, *i.e.* applying a low-pass spatial filter.

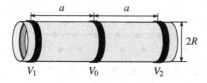

Fig. 1. Tripolar electrode cuff model

2. subtracting the result to the signal of the central pole, keeping only the high spatial frequencies.

Therefore, the recorded V_{rec} signal can be considered as spatial high-pass filtered.

More precisely, this filter is a second-order one considering that the expressions

$$\frac{1}{a}(V_2 - V_0) \quad \text{and} \quad \frac{1}{a}(V_0 - V_1) \tag{2}$$

evaluate the first derivative $\frac{dV}{dx}$. Thus the difference

$$\frac{1}{a}\left(\frac{(V_2 - V_0)}{a} - \frac{(V_0 - V_1)}{a}\right) = \frac{2}{a^2}\left(\frac{(V_1 + V_2)}{2} - V_0\right) \tag{3}$$

denotes the second derivative $\frac{d^2V}{dx^2}$ that is the one-dimensional *Laplacian* of the potential. We can identify in the last expression the equation (1) without the constant factor $-2a^{-2}$.

Laplacian filters can reject both homogeneous potentials and linearly varying ones like those created by far EMG sources. The purpose of this new design is to build two-dimensional Laplacian using more poles to obtain isotropic rejection.

3   Positioning the Poles

A tripolar cuff electrode [10] provides only one recording which is the superposition of all AP "seen" by the electrode at a given moment. The use of several poles on a cuff electrode (see Fig. 2) could allow us to record more signals, thus increase the quantity of neural data and facilitate the signal post-processing on the recording system.

In order to achieve optimal placement of poles, we must pay attention to three constraints:

1. The electrodes have to be placed all around the nerve, thus the poles have to be distributed onto the whole surface of the cuff.
2. The poles have to be equally spaced to simplify electronics in charge of analog signal preprocessing (weight coefficients in Laplacian preamplifier).
3. They have to be able to be substituted one to each other, so we take benefits of the maximum measurement locations, allowing powerful signal processing.

Since the cylindric shape of the cuff results from the wrapping of an initially plane device, these conditions imply to look for a regular tessellation of the plane as the positions of the poles or, more precisely, tessellations composed of regular polygons symmetrically tiling the plane. It is well known that there are exactly three type of regular tessellations [11]. They can be specified using the Schläfli symbols: $\{3,6\}$, $\{4,4\}$ and $\{6,3\}$.

Fig. 2. Multipolar electrode cuff model

{6,3}   {4,4}   {3,6}

Fig. 3. There are exactly three regular tessellations composed of regular polygons symmetrically tiling the plane

The first symbol in the Schläfli notation denotes the shape of the patch (triangle, square or hexagon). On the figure 3, each vertex corresponds to a pole. Each of them being surrounded by a number of equidistant poles given by the second Schläfli symbol, respectively 6, 4 and 3.

From the previous tessellations, one can build three kinds of electrodes by selecting one central pole and its closest neighbors. Namely, we can define a mesh of:

- triangular 4-pole electrodes,
- squared 5-pole electrodes,
- hexagonal 7-pole electrodes.

These candidates can be seen on the figure 4 and the resulting expressions for the Laplacian are:

$$V_{rec} = V_0 - \frac{1}{3}\sum_{i=1}^{3} V_i \qquad \text{for } \{6,3\} \qquad (4)$$

$$V_{rec} = V_0 - \frac{1}{4}\sum_{i=1}^{4} V_i \qquad \text{for } \{4,4\} \qquad (5)$$

$$V_{rec} = V_0 - \frac{1}{6}\sum_{i=1}^{6} V_i \qquad \text{for } \{3,6\} \qquad (6)$$

One can notice that the $\{4,4\}$ configuration correspond to the 2D Laplacian filter used in image processing [12].

V_1
V_0
V_2 V_3
{6,3}

{6,3}
V_1
V_0
V_2 - - V_6
V_4
{4,4}

V_2 V_1
V_6
V_0
V_3 V_5
V_4
{3,6}

Fig. 4. Three possible configurations of electrodes

4   Numerical Results

4.1   Action Potential Modeling

In order to evaluate the performances of multipolar electrodes, we need a model for the extracellular electric field created by an AP. Let us consider a $10\,\mu m$ diameter myelinated axon. Its Ranvier nodes are $1\,\mu m$ long, while their diameter is $6\,\mu m$ and their spacing is 1 mm. Let us call Ω the center of the Ranvier node. When the AP is present at this node, we can model it as a $6\,\mu m$ diameter circle, perpendicular to the axon axis, with a positive charge $+q$ at its center (Ω) and a negative charge $-q$ spread on the circle. The potential created at a point M of the space by this AP can be approximated by:

$$V(M) = \frac{qa^2}{8\pi\varepsilon_0\varepsilon_r r^3}\left(1 - \frac{3}{2}\sin^2\psi\right) \tag{7}$$

In this expression, a is the radius of the Ranvier node ($3\,\mu m$), r is the distance between Ω and M, while ψ is the angle between the axe of the axon and $\overrightarrow{\Omega M}$. This approximation, valid for $r \gg a$, is in good accordance with measurements. In particular, we can see that $V(M)$ is negative for $\psi = \pi/2$ [13, page 81]. Last, q can be easily estimated from the characteristics of the Ranvier node. For this study, we took $q \simeq 20\,\text{fC}$ and $\varepsilon_r \simeq 80$.

The model given by equation 7 was used to evaluate the sensitivity of the electrodes to AP occurring inside the nerve. Given the position of a single AP we can easily calculate the induced potential on each pole of the cuff, since they are very small. The figure 5 shows the voltage response of the three possible structures. The diameter of the cuff is $2R = 3\,\text{mm}$ while the spacing between poles is $d = R$. The AP is located in a plane at 20% of R from the center of the electrode. Last, the effect of the wrapping of the electrodes is taken into account to calculate the position of the poles.

Obviously, the 7-pole (hexagonal) electrode exhibits the most isotropic sensitivity. We choose therefore this particular configuration for the prototype (fig. 6) and focus on it in the next sections of the paper.

4.2   Tripolar and Heptapolar Electrodes Models

We have limited the next study to the comparison of the classical tripolar cuff with the heptapolar (hexagonal shape $\{3,6\}$) electrode. The induced potential on each pole of the heptapolar electrode is calculated like above. For the tripolar cuff, we need to average the potential on each ring. This lead to an elliptic integral we have solved using numerical methods.

In the following, we compare a tripolar cuff electrode, whose diameter is $2R = 3\,\text{mm}$ and ring spacing is $a = 4R$, with one patch of the hexagonal cuff. To get comparable results, this hexagonal cuff has the same diameter ($2R = 3\,\text{mm}$) and the spacing between poles is $d = R$.

For all the calculations, the coordinates were fixed as follow: the origin O is at the center of the cuff electrode. The Ox axis is the axis of the nerve (and, obviously, of the cuff). The Oy axis passes by the center of the considered patch (which is perpendicular to this axe). Last the Oz axe is placed to form a direct trihedron with Ox and Oy.

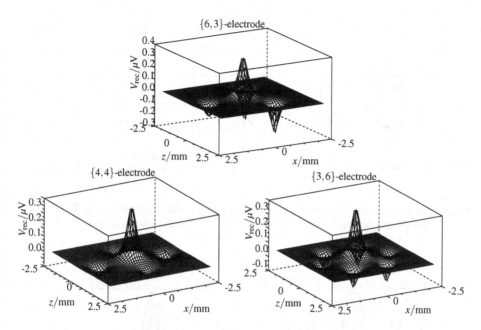

Fig. 5. XZ-sensitivity of multipolar cuff electrodes for an AP source located in a $y = 0.8R$ plane

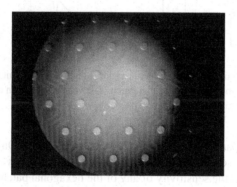

Fig. 6. Microphotograph of the $\{3,6\}$ multipolar cuff electrode with connection routing

4.3   Internal Sensitivity

Figure 7 shows the radial sensitivities of the tripolar cuff electrode and a wrapped hexagonal patch of the multipolar electrode. The vertical axis is the value of V_{rec} (in $dB\,\mu V$) calculated for an AP placed on the Oy axis, at abscissa yR. The graph shows clearly that while the sensitivity of the tripolar cuff is quasi constant on the section of the nerve, the sensitivity of the hexagonal patch is far higher (up to 30 dB) when considering an AP located between the center of the patch and the center of the cuff.

Figure 8 show the longitudinal sensitivities of the two considered electrodes. On the left hand-side of the figure, the AP is placed on the Ox axis, whereas on the other side,

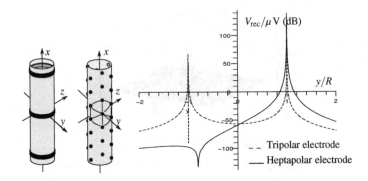

Fig. 7. Radial sensitivities of a tripolar cuff electrode, and a bent hexagonal patch. The vertical axis is in dB μV and the unit for the horizontal axis is the radius R of the electrode.

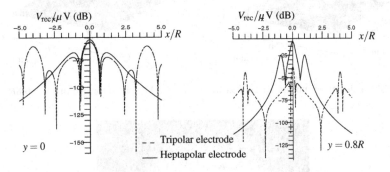

Fig. 8. Longitudinal sensitivities on the axis of the nerve and on an off-center (80 % of R) axis for a tripolar cuff electrode and a bent hexagonal patch. The vertical axis is in dB μV and the unit for the horizontal axis is the radius R of the electrode.

it is placed on a line, parallel to Ox, cutting Oy at abscissa $0.8R$. On this later figure, we can see an increase of sensitivity of the tripolar cuff in the vicinity of the rings, but this remains far lower than the sensitivity of any of the hexagonal patches.

4.4   External Sensitivity

For the evaluation of the rejection of parasitic signals, we must first recall that EMG are also AP, creating the same kind of electric field. But, in this case, we cannot make any assumption on the value of ψ in the eq. (7). So, to evaluate the external sensitivity of electrodes, we chosen to use only a $1/r^3$ model, unable to give voltages, but sufficient to compare the sensitivities of various electrodes.

The figure 9 show the external sensitivities of the two structures for an AP placed on the Ox or on the Oz axis of the electrode. As stated above, the quantity plotted is not a voltage, but is homogeneous to the reciprocal of the cube of a distance. Nevertheless, we can see on these two graphs that the hexagonal patches exhibit a better rejection of

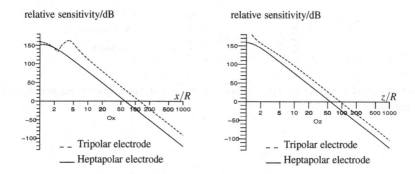

Fig. 9. External relative sensitivity along the Ox and Oz axes for a tripolar cuff electrode and a bent hexagonal patch. The vertical axis is in dB and the unit for the horizontal axis is the radius R of the cuff.

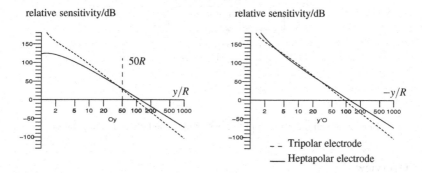

Fig. 10. External relative sensitivity along the two halves of the Oy axis for a tripolar cuff electrode and a bent hexagonal patch. The vertical axis is in dB and the unit for the horizontal axis is the radius R of the electrode.

parasitic signals than the tripolar cuff. This improvement is of 32 dB for Ox and 20 dB for Oz.

The same study conducted along the Oy axis (figure 10) shows that the wrapped hexagonal patch has a sensitivity decreasing slowly along this Oy axis. In fact, the bent hexagonal patch only begins to have larger sensitivity than the tripolar cuff for AP placed at more than fifty times the radius of the cuff, corresponding to approximately 7 cm. At this distance, the parasitic signal could be neglected in comparison to ENG signal.

5 ENG Amplifier ASIC

Because of the very low level of processed signals we propose to perform the maximum of signal processing as close as possible to the nerve. The more complex operations to be considered are those with the hexagonal electrode.

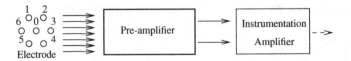

Fig. 11. Overview of the ASIC. Each channel is composed of a preamplifier and an instrumentation amplifier.

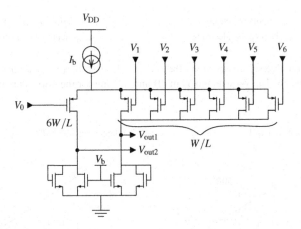

Fig. 12. Seven input preamplifier performing the laplacian computation

5.1 Overview

For this purpose, we have designed a seven channels ASIC. Each channel compute a weighted difference between the measurement point and the six closest surrounding points. This is done in the analog domain using the preamplifier shown on figure 12. This preamplifier is build around a differential pair whose negative input transistor was split into six transistors (six times smaller, of course). It has a voltage gain that is about 100 and it is followed by an instrumentation amplifier whose gain is configurable between 6 dB and 80 dB. Each channel is composed of a preamplifier followed by an instrumentation amplifier (fig. 11).

5.2 Simulation Results

This circuit was designed to give an input-referred noise below $1\,\mu V_{rms}$, a CMMR above 60 dB and a sufficient gain, i.e greater than 60 dB ; all these parameters in the bandwidth of interest ($1\,Hz \leq f \leq 3\,kHz$). The performances expected for this amplifier are given in table 1 (the noise is measured in the band 1 Hz-3 kHz).

A Microphotograph of the fabricated circuit is presented Fig. 14. This circuit was designed in CMOS AMS 0.35-μm technology and the prototype is currently under test.

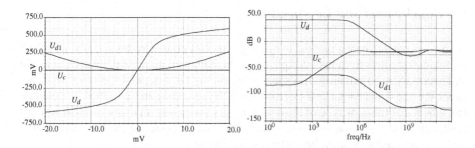

Fig. 13. Simulated static response and frequency response of the preamplifier. Uc stands for the common mode voltage (to be rejected), U_d is the component we want to amplify while U_d is a parasitic differential voltage.

Table 1. Amplifier characteristics (simulation)

Active area (7 channels)	$1.16\,\mathrm{mm}^2$
Supply voltage	3.3 V
DC Current (Preamp)	$20\,\mu$A
Voltage gain (Preamp)	100 (40 dB)
CMRR (Preamp)	80 dB (10 kHz)
Voltage gain (Inst amp)	$2 \leq G \leq 10000$
CMRR (Full amp)	80 dB (10 kHz)
Input-ref. noise (Preamp)	$0.672\,\mu$V RMS
Input-ref. noise (Full amp)	$0.677\,\mu$V RMS
Bandwidth (Full amp)	76 kHz

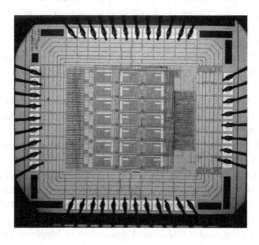

Fig. 14. Microphotograph of the seven-channel prototype

6   Conclusions and Perspectives

We have presented here a method to build multipolar cuff electrodes and how to extract useful informations from the multiple channels. Although numerical investigations are still necessary to an exhaustive comparison of multipolar structures, the comparison between the classical tripolar cuff electrode for ENG recording and a multipolar electrode has shown that this new type of the design is very promising. In every simulation, multipolar electrodes prove to be more sensitive to sources located inside the nerve, and in almost every case they show better far source rejection.

We hope the improvement of the recorded signal given by this new design will allow the use of signal processing techniques such as source separation. Then, multipolar configurations could make it possible to estimate more precise parameters like the speed and the direction of propagation of the AP [4,14].

References

1. Djilas, M., Azevedo Coste, C., Yoshida, K., Cathébras, G.: Interpretation of ENG signal for FES closed-loop control. In: IFESS 2006: 11th Annual Conference of the International Functional Electrical Stimulation Society, Miyagi-Zao, Japan, September 12-15, pp. 85–87 (2006)
2. Ramachandran, A., Sacristan, J., Lago, N., Dörge, T., Osès, M., Koch, K., Hoffmann, K.: Multipolar cuff electrodes with integrated pre-amplifier & filter to interface peripheral nerves for FES application. In: 10th Annual Conference of the International FES Society, Montreal, Canada (July 2005)
3. Winter, J., Rahal, M., Taylor, N., Donaldson, N., Struijk, J.: Improved spatial filtering of ENG signals using a multielectrode nerve cuff. In: 5th Annual Conference of the International Functional Electrical Stimulation Society, Aalborg, Denmark (June 2000)
4. Taylor, J., Donaldson, N., Winter, J.: Multiple-electrode nerve cuffs for low-velocity and velocity-selective neural recording. Medical and Biological Engineering and Computing 42(5), 634–643 (2004)
5. Haugland, M.K., Hoffer, J.A., Sinkjaer, T.: Skin contact force information in sensory nerve signals recorded by implanted cuff electrodes. IEEE Transactions on Rehabilitation Engineering 2(1), 18–28 (1994)
6. Jensen, W., Sinkjaer, T., Sepulveda, F.: Improving signal reliability for on-line joint angle estimation from nerve cuff recordings of muscle afferents. IEEE Transactions on Neural Systems and Rehabilitation Engineering 10(3), 133–139 (2002)
7. Andreasen, L.N.S., Struijk, J.J.: Signal strength versus cuff length in nerve cuff electrode recordings. IEEE J BME 49(9), 1045–1050 (2002)
8. Struijk, J.J., Thomsen, M.: Tripolar nerve cuff recording: stimulus artifact, EMG and the recorded nerve signal. In: IEEE 17th Annual Conference on Engineering in Medicine and Biology Society, Montreal, Que, September 20–23, vol. 2, pp. 1105–1106 (1995)
9. Pflaum, C., Riso, R.R., Wiesspeiner, G.: Performance of alternative amplifier configurations for tripolar nerve cuff recorded ENG. In: Engineering in Medicine and Biology Society. Bridging Disciplines for Biomedicine, Proceedings of the 18th Annual International Conference of the IEEE, Amsterdam, October 31–November 3, vol. 1, pp. 375–376 (1996)
10. Demosthenous, A., Triantis, I.F.: An adaptive ENG amplifier for tripolar cuff electrodes. IEEE J JSSC 40(2), 412–421 (2005)

11. Weisstein, E.W.: Tessellation. From MathWorld–A Wolfram Web Resource (2002), http://mathworld.wolfram.com/Tessellation.html
12. Gonzales, R.C., Woods, R.E.: Digital Image Processing. Addison-Wesley, Reading (1992)
13. Stein, R.B.: Nerve and Muscle. Plenum Press (1980)
14. Rieger, R., Schuettler, M., Pal, D., Clarke, C., Langlois, P., Taylor, J., Donaldson, N.: Very low-noise ENG amplifier system using CMOS technology. IEEE Transactions on Neural Systems and Rehabilitation Engineering 14(4), 427–437 (2006)

Part II

BIOSIGNALS

On a Chirplet Transform Based Method for Co-channel Voice Separation*

B. Dugnol, C. Fernández, G. Galiano, and J. Velasco

Dpto. de Matemáticas, Universidad de Oviedo, 33007 Oviedo, Spain
{dugnol,carlos,galiano,julian}@uniovi.es

Abstract. We use signal and image theory based algorithms to produce estimations of the number of wolves emitting howls or barks in a given field recording as an individuals counting alternative to the traditional trace collecting methodologies. We proceed in two steps. Firstly, we clean and enhance the signal by using PDE based image processing algorithms applied to the signal spectrogram. Secondly, assuming that the wolves chorus may be modelled as an addition of nonlinear chirps, we use the quadratic energy distribution corresponding to the Chirplet Transform of the signal to produce estimates of the corresponding instantaneous frequencies, chirp-rates and amplitudes at each instant of the recording. We finally establish suitable criteria to decide how such estimates are connected in time.

1 Introduction

Wolf is a protected specie in many countries around the world. Due to their predator character and to their proximity to human settlements, wolves often kill cattle interfering in this way in farmers' economy. To smooth this interference, authorities reimburse the cost of these lost to farmers. Counting the population of wolves inhabiting a region is, therefore, not only a question of biological interest but also of economic interest, since authorities are willing to estimate the budget devoted to costs produced by wolf protection, see for instance [1]. However, estimating the population of wild species is not an easy task. In particular, for mammals, few and not very precise techniques are used, mainly based on the recuperation of field traces, such as steps, excrements and so on. Our investigation is centered in what it seems to be a new technique, based on signal and image theory methods, to estimate the population of species which fulfill two conditions: living in groups, for instance, packs of wolves, and emitting some characteristic sounds, howls and barks, for wolves. The basic initial idea is to produce, from a given recording, some time-frequency distribution which allows to identify the different howls corresponding to different individuals by estimating the instantaneous frequency (IF) lines of their howls.

Unfortunately, the real situation is somehow more involved due mainly to the following two factors. On one hand, since natural sounds, in particular wolf howling,

---

* All authors are supported by Project PC07-12, Gobierno del Principado de Asturias, Spain. Third and fourth authors are supported by the Spanish DGI Project MTM2007-65088.

A. Fred, J. Filipe, and H. Gamboa (Eds.): BIOSTEC 2008, CCIS 25, pp. 163–175, 2008.

are composed by a fundamental pitch and several harmonics, direct instantaneous frequency estimation of the multi-signal recording leads to an over-counting of individuals since various IF lines correspond to the same individual. Therefore, more sophisticated methods are indicated for the analysis of these signals, methods capable of extracting additional information such as the slope of the IF, which allows to a better identification of the harmonics of a given fundamental tone. The use of a Chirplet type transform [2,3] is investigated in this article, although an equivalent formulation in terms of the Fourier fractional transform [4] could be employed as well. On the other hand, despite the quality of recording devices, field recordings are affected for a variety of undesirable signals which range from low amplitude broad spectrum long duration signals, like wind, to signals localized in time, like cattle bells, or localized in spectrum, like car engines. Clearly, the addition of all these signals generates an unstructured noise in the background of the wolves chorus which impedes the above mentioned methods to work properly, and which should be treated in advance. We accomplish this task by using PDE-based techniques which transforms the image of the signal spectrogram into a smoothed and enhanced approximation to the reassigned spectrogram introduced in [5,6] as a spectrogram readability improving method.

2 Signal Enhancement

In previous works [7,8], we investigated the noise reduction and edge (IF lines) enhancement on the spectrogram image by a PDE-based image processing algorithm. For a clean signal, the method allows to produce an approximation to the reassigned spectrogram through a process referred to as *differential reassignment*, and for a noisy signal this process is modified by the introduction of a nonlinear operator which induces isotropic diffusion (noise smoothing) in regions with low gradient values, and anisotropic diffusion (edge-IF enhancement) in regions with high gradient values.

Let $x \in L^2(\mathbb{R})$ denote an audio signal and consider the Short Time Fourier transform (STFT)

$$\mathcal{G}_\varphi(x; t, \omega) = \int_{\mathbb{R}} x(s)\varphi(s - t)e^{-i\omega s}ds, \qquad (1)$$

corresponding to the real, symmetric and normalized *window* $\varphi \in L^2(\mathbb{R})$. The energy density function or *spectrogram* of x corresponding to the window φ is given by

$$S_\varphi(x; t, \omega) = |\mathcal{G}_\varphi(x; t, \omega)|^2, \qquad (2)$$

which may be expressed also as [9]

$$S_\varphi(x; t, \omega) = \int_{\mathbb{R}^2} WV(\varphi; \tilde{t}, \tilde{\omega})WV(x; t - \tilde{t}, \omega - \tilde{\omega})d\tilde{t}d\tilde{\omega}, \qquad (3)$$

with $WV(y; \cdot, \cdot)$ denoting the Wigner-Ville distribution of $y \in L^2(\mathbb{R})$,

$$WV(y; t, \omega) = \int_{\mathbb{R}} y(t + \frac{s}{2})y(t - \frac{s}{2})e^{-i\omega s}ds.$$

The Wigner-Ville (WV) distribution has received much attention for IF estimation due to its excellent concentration and many other desirable mathematical properties, see [9]. However, it is well known that it presents high amplitude sign-varying cross-terms for multi-component signals which makes its interpretation difficult. Expression (3) represents the spectrogram as the convolution of the WV distribution of the signal, x, with the smoothing kernel defined by the WV distribution of the window, φ, explaining the mechanism of attenuation of the cross-terms interferences in the spectrogram. However, an important drawback of the spectrogram with respect to the WV distribution is the broadening of the IF lines as a direct consequence of the smoothing convolution. To override this inconvenient, it was suggested in [5] that instead of assigning the averaged energy to the geometric center of the smoothing kernel, (t, ω), as it is done for the spectrogram, one assigns it to the *center of gravity* of these energy contributions, $(\hat{t}, \hat{\omega})$, which is certainly more representative of the local energy distribution of the signal. As deduced in [6], the gravity center may be computed by the following formulas

$$\hat{t}(x; t, \omega) = t - \Re\left\{\frac{\mathcal{G}_{T\varphi}(x; t, \omega)}{\mathcal{G}_{\varphi}(x; t, \omega)}\right\},$$

$$\hat{\omega}(x; t, \omega) = \omega + \Im\left\{\frac{\mathcal{G}_{D\varphi}(x; t, \omega)}{\mathcal{G}_{\varphi}(x; t, \omega)}\right\},$$

where the STFT's windows in the numerators are $T\varphi(t) = t\varphi(t)$ and $D\varphi(t) = \varphi'(t)$. The reassigned spectrogram, $RS_{\varphi}(x; t, \omega)$, is then defined as the aggregation of the reassigned energies to their corresponding locations in the time-frequency domain. Observe that energy is conserved through the reassignment process. Other desirable properties, among which non-negativity and perfect localization of linear chirps, are proven in [10]. For our application, it is of special interest the fact that the reallocation vector, $\mathbf{r}(t, \omega) = (\hat{t}(t, \omega) - t, \hat{\omega}(t, \omega) - \omega)$, may be expressed through a potential related to the spectrogram [11],

$$\mathbf{r}(t, \omega) = \frac{1}{2}\nabla \log(S_{\varphi}(x; t, \omega)), \tag{4}$$

when φ is a Gaussian window of unit variance. Let $\tau \geq 0$ denote an artificial time and consider the dynamical expression of the reassignment given by $\Phi(t, \omega, \tau) = (t, \omega) + \tau\mathbf{r}(t, \omega)$ which, for $\tau = 0$ to $\tau = 1$, connects the initial point (t, ω) with its reassigned point $(\hat{t}, \hat{\omega})$. Rewriting this expression as

$$\frac{1}{\tau}(\Phi(t, \omega, \tau) - \Phi(0, \omega, \tau)) = \mathbf{r}(t, \omega),$$

and taking the limit $\tau \to 0$, we may identify the displacement vector $\mathbf{r}$ as the velocity field of the transformation Φ. In close relation with this approach is the process referred to as *differential reassignment* [11], defined as the transformation given by the dynamical system corresponding to such velocity field,

$$\begin{cases} \dfrac{d\chi}{d\tau}(t, \omega, \tau) = \mathbf{r}(\chi(t, \omega, \tau)), \\ \chi(t, \omega, 0) = (t, \omega), \end{cases} \tag{5}$$

for $\tau > 0$. Observe that, in a first order approximation, we still have that χ connects (t, ω) with some point in a neighborhood of $(\hat{t}, \hat{\omega})$, since

$$\chi(t, \omega, 1) \approx \chi(t, \omega, 0) + \mathbf{r}(\chi(t, \omega, 0))$$
$$= (t, \omega) + \mathbf{r}(t, \omega) = (\hat{t}, \hat{\omega}).$$

In addition, for $\tau \to \infty$, each particle (t, ω) converges to some local extremum of the potential $\log(S_\varphi(x; \cdot, \cdot))$, among them the maxima and ridges of the original spectrogram. The conservative energy reassignation for the differential reassignment is obtained by solving the following problem for $u(t, \omega, \tau)$ and $\tau > 0$,

$$\frac{\partial u}{\partial \tau} + \mathrm{div}(u\mathbf{r}) = 0, \tag{6}$$

$$u(\cdot, \cdot, 0) = u_0, \tag{7}$$

where we introduced the notation $u_0 = S_\varphi(x; \cdot, \cdot)$ and, consequently, $\mathbf{r} = \frac{1}{2}\nabla \log(u_0)$. Since in applications both signal and spectrogram are defined in bounded domains, we assume (6)-(7) to hold in a bounded time-frequency domain, Ω, in which we assume non energy flow conditions on the solution and the data

$$\nabla u \cdot \mathbf{n} = 0, \quad \mathbf{r} \cdot \mathbf{n} = 0 \quad \text{on } \partial\Omega \times \mathbb{R}_+, \tag{8}$$

being $\mathbf{n}$ the unitary outwards normal to $\partial\Omega$. Finally, observe that the positivity of the spectrogram [9] and the fact that it is obtained from a convolution with a C^∞ kernel implies the regularity $u_0, \mathbf{r} \in C^\infty$ and, therefore, problem (6)-(8) admits a unique smooth solution.

As noted in [11], differential reassignment can be viewed as a PDE based processing of the spectrogram image in which the energy tends to concentrate on the initial image ridges (IF lines). As mentioned above, our aim is not only to concentrate the diffused IF lines of the spectrogram but also to attenuate the noise present in our recordings. It is clear that noise may distort the reassigned spectrogram due to the change of the energy distribution and therefore of the gravity centers of each time-frequency window. Although even a worse situation may happen to the differential reassignment, due to its convergence to spectrogram local extrema (noise picks among them) an intuitive way to correct this effect comes from its image processing interpretation. As shown in [7,8], when a strong noise is added to a clean signal better results are obtained for approximating the clean spectrogram if we use a noise reduction edge enhancement PDE based algorithm than if we simply threshold the image spectrogram. This is due to the local application of gaussian filters in regions of small gradients (noise, among them) while anisotropic diffusion (in the orthogonal direction to the gradient) is applied in regions of large gradients (edges-IF lines). Therefore, a possible way to improve the image obtained by the differential reassigned spectrogram is modifying (6) by adding a diffusive term with the mentioned properties.

Let us make a final observation before writing the model we work with. In the derivation of both the reassigned and the differential reassigned spectrogram the property of energy conservation is imposed, implying that energy values on ridges increase. Indeed, let B be a neighborhood of a point of maximum for u_0, in which $\mathrm{div}\,\mathbf{r} = \Delta \log u_0 < 0$,

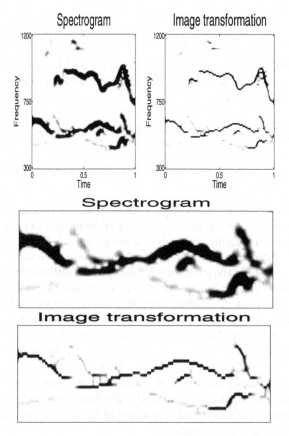

Fig. 1. First row: Spectrogram and its transformation with the PDE model. Subsequent plots: detail of the howl contained within the range $300 - 750$ Hz. We observe the IF concentration and smoothing effect of the PDE algorithm.

and let $(t_0, \omega_0) \in B$. Let $\chi_0(t, \omega, \tau)$ denote the characteristic defined by (5) starting at (t_0, ω_0). Evaluating Eq. (6) along χ_0 we obtain

$$\frac{d}{d\tau}u = \frac{\partial u}{\partial \tau} + \mathbf{r} \cdot \nabla u = -u \operatorname{div} \mathbf{r}, \qquad (9)$$

implying that u experiments exponential increase in B. For image processing, it is desirable the maximum principle to hold, i.e., that the bounds $\min u_0 \le u \le \max u_0$ hold for any $(t, \omega, \tau) \in \Omega \times \mathbb{R}_+$, ensuring that the processed image lies within the range of image definition ($[0, 255]$, usually). A simple way, which we shall address, to ensure this property is by dropping the right hand side term of Eq. (9), i.e., replacing Eq. (6) by the transport equation

$$\frac{\partial u}{\partial \tau} + \mathbf{r} \cdot \nabla u = 0. \qquad (10)$$

However, no energy conservation law will apply anymore (note that u is constant along the characteristics). The combination of the differential reassignment problem with the edge-detection image-smoothing algorithm [12] is written as

$$\frac{\partial u}{\partial \tau} + \frac{\varepsilon}{2} \nabla \log(u_0) \cdot \nabla u - g(|G_s * \nabla u|) A(u) = 0, \tag{11}$$

in $\Omega \times \mathbb{R}_+$, together with the boundary data (8) and the initial condition (7). Parameter $\varepsilon \geq 0$ allows us to play with different balances between transport and diffusion effects. The diffusion operator is given by

$$A(u) = (1 - h(|\nabla u|))\Delta u + h(|\nabla u|) \sum_{j=1,\ldots,n} f_j\left(\frac{\nabla u}{|\nabla u|}\right) \frac{\partial^2 u}{\partial x_j^2}.$$

Let us briefly remind the properties and meaning of the diffusive term components in equation (11):

- Function G_s is a Gaussian of variance s. The variance is a *scale parameter* which fixes the minimal size of the details to be kept in the processed image.
- Function g is non-increasing with $g(0) = 1$ and $g(\infty) = 0$. It is a *contrast* function, which allows to decide whether a detail is sharp enough to be kept.
- The composition of G_s and g on ∇u rules the speed of diffusion in the evolution of the image, controlling the *enhancement* of the edges and the noise smoothing.
- The diffusion operator A combines isotropic and anisotropic diffusion. The first smoothes the image by local averaging while the second enforces the diffusion only on the orthogonal direction to ∇u (along the edges). More precisely, for $\theta_j = (j-1) * \pi/n$, $j = 1, \ldots, n$ we define x_j as the orthogonal to the direction θ_j, i.e., $x_j = -t \sin \theta_j + w \cos \theta_j$. Then, smooth non-negative functions $f_j(\cos \theta, \sin \theta)$ are designed to be *active* only when θ is close to θ_j. Therefore, the anisotropic diffusion is taken in an approximated direction to the orthogonal of ∇u. The combination of isotropic and anisotropic diffusions is controlled by function $h(s)$, which is nondecreasing with

$$h(s) = \begin{cases} 0 \text{ for } s \leq h_0, \\ 1 \text{ for } s \geq 2h_0, \end{cases} \tag{12}$$

being h_0 the *enhancement* parameter.

In Fig. 1 we show an example of the outcome of our algorithm for a signal composed by two howls. See [8,13] for more details and other numerical experiments.

3   Howl Tracking and Separation

A wolf chorus is composed, mainly, by howls and barks which, from the analytical point of view, may be regarded as chirp functions. The former has a long time support and a small frequency range variation, while the latter is almost punctually localized in time but posses a large frequency spectrum. It is convenient, therefore, adopting a parametric model to represent the wolves chorus as an addition of chirps given by the function $f : [0, T] \to \mathbb{C}$,

$$f(t) = \sum_{n=1}^{N} a_n(t) \exp[i\phi_n(t)], \tag{13}$$

with T the length of the chorus emission, a_n and ϕ_n the chirps amplitude and phase, respectively, and with N, the number of chirps contained in the chorus. We notice that N is not necessarily the number of wolves since, for instance, harmonics of a given fundamental tone are counted separately.

To identify the unknowns N, a_n and ϕ_n we proceed in two steps. Firstly, for a time discretization of the time interval $[0, T]$, say t_j, for $j = 0, \ldots, J$, we produce estimates of the amplitude $a_n(t_j)$ and the phase $\phi_n(t_j)$ of the chirps contained at such discrete times. Secondly, we establish criteria which allow us deciding if the computed estimates at adjacent times do belong to the same global chirp or do not.

For the first step we use the Chirplet transform defined by

$$\Psi f (t_o, \xi, \mu; \lambda) = \int_{-\infty}^{\infty} f(t) \, \overline{\psi_{t_o, \xi, \mu, \lambda}(t)} dt, \tag{14}$$

with the complex window $\psi_{t_o, \xi, \mu, \lambda}$ given by

$$\psi_{t_o, \xi, \mu, \lambda}(t) = v_\lambda(t - t_o) \exp\left[i(\xi t + \frac{\mu}{2}(t - t_o)^2)\right]. \tag{15}$$

Here, $v \in L^2(\mathbb{R})$ denotes a real window, $v_\lambda(\cdot) = v(\cdot/\lambda)$, with $\lambda > 0$, and the parameters $t_o, \xi, \mu \in \mathbb{R}$, stand for time, instantaneous frequency and chirp rate, respectively. The quadratic energy distribution corresponding to the chirplet transform (14) is given by

$$P_\Psi f(t_o, \xi, \mu; \lambda) = |\Psi f(t_o, \xi, \mu; \lambda)|^2. \tag{16}$$

For a linear chirp of the form

$$f(t) = a(t) \exp[i(\frac{\alpha}{2}(t - t_o)^2 + \beta(t - t_o) + \gamma)],$$

it is straightforward to prove that the energy distribution (16) has a global maximum at (α, β), allowing us to determine the IF and chirp rate of a given linear chirp by localizing the maxima of the energy distribution. For more general forms of mono-component chirps we have the following localization result [14]

Theorem 1. *Let $f(t) = a(t) \exp[i\phi(t)]$, with $a \in L^2(\mathbb{R})$ non-negative and $\phi \in C^3(\mathbb{R})$. For all $\varepsilon > 0$ and $\xi, \mu \in \mathbb{R}$ there exists $L > 0$ such that if $\lambda < L$ then*

$$P_\Psi f(t_o, \xi, \mu; \lambda) \leqslant \varepsilon + P_\Psi f(t_o, \phi'(t_o), \phi''(t_o); \lambda). \tag{17}$$

In addition,

$$\lim_{\lambda \to 0} P_\Psi f(t_o, \phi'(t_o), \phi''(t_o); \lambda) = a(t_o)^2. \tag{18}$$

In other words, for a general mono-component chirp the energy distribution maximum provides an arbitrarily close approximation to the IF and chirp rate of the signal. More-over, its amplitude may also be estimated by shrinking the window time support at the maximum point.

Finally, for a multi-component chirp $f(t) = \sum_{n=1}^{N} a_n(t) \exp[i\phi_n(t)]$ the situation is somehow more involved since although the energy distribution still has maxima at $(\phi'_n(t_o), \phi''_n(t_o))$ for all n such that $a_n(t_o) \neq 0$, these are now of local nature, and in fact, spurious local maxima not corresponding to any chirp may appear due to the energy interaction among the actual chirps.

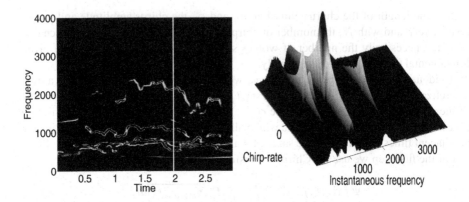

Fig. 2. Left: STFT of a field recorded signal. Right: quadratic energy distribution of the chirplet transform at $t_0 = 2$. Maxima correspond to IF and chirp rate chirps locations. We observe the different behavior in the ξ and μ directions at these maxima.

4   Numerical Experiments

According to the recording quality, we start our algorithm enhancing the signal with the PDE algorithm explained in Section 2 or directly with the separation algorithm introduced in Section 3. For details about the implementation of the former, we refer the reader to [13]. Following, we briefly comment about the separation algorithm implementation, see [14] for more details. We start by computing the energy distribution, $P_\Psi f (\tau_m, \xi, \mu; \lambda)$ at a set of discrete times $\tau_m = m * \tau$ of constant time step, τ, and for a fixed window width λ. Next we compute the maxima of the energy at each of these times. When the signal is mono-component or the various components of the signal are far from each other relative to the window width, the maxima of P_Ψ correspond to some $(\phi'_n (\tau_m), \phi''_n (\tau_m))$ which are then identified as the IF and chirp-rate of a chirp candidate. However, when multi-component signals are close to each other or are crossing, some spurious local maxima are produced which do not correspond to any actual chirp. Therefore, some criterium must be used to select the correct local maxima at each τ_m. Although we lack of an analytical proof, there are evidences suggesting that maxima produced by chirps, i.e., at points of the type $(\phi'_n (\tau_m), \phi''_n (\tau_m))$, decrease much faster in the ξ direction than in the μ direction, see Fig. 2, a phenomenon that does not occur at spurious maxima. We use this fact to choose the candidates first by selecting ξ_k, for $k = 1, \ldots, K$, which are maxima for

$$\sup_\mu P_\Psi f (\tau_m, \xi, \mu; \lambda),$$

and, among them, selecting the maxima with respect to μ of $P_\Psi f (\tau_m, \xi_k, \mu; \lambda)$. We finally establish a threshold parameter to filter out possible local maxima located at points that do not correspond to any $\phi'_n (\tau_m)$ but which are close to two of them. We set this threshold such that the existence of two consecutive maxima is avoided.

   In this way we obtain, for each τ_m, a set of points (μ_{i_m}, ξ_{i_m}), for $i_m = 1, \ldots, I_m$, which correspond to the IF's and chirp-rates of chirps with time support including τ_m.

The next step is the chirp separation. We note that if the time step $\tau = \tau_{m+1} - \tau_m$ is small enough, then

$$\xi_{j_{m+1}} - \tau \frac{\mu_{j_{m+1}}}{2} \approx \xi_{i_m} + \tau \frac{\mu_{i_m}}{2}.$$

Introducing a new parameter, ν, we test this property by imposing the condition

$$\frac{1}{\nu} < \frac{2\xi_{j_{m+1}} - \tau\mu_{j_{m+1}}}{2\xi_{i_m} + \tau\mu_{i_m}} < \nu, \tag{19}$$

for two points to be in the same chirp. In the experiments we take $\nu = 2^{1/13} \approx 1.0548$.

Finally, in the case in which test (19) is satisfied by more than one point, i.e., when there exist points $(\xi_{j_{m+1}}, \mu_{j_{m+1}})$ and $(\xi_{k_{m+1}}, \mu_{k_{m+1}})$ such that (19) holds for the same (ξ_{i_m}, μ_{i_m}), we impose a regularity criterium and choose the point with a closer chirp-rate to that of (ξ_{i_m}, μ_{i_m}). This is a situation typically arising at chirps crossings points.

Summarizing, the chirp separation algorithm is implemented as follows:

- Each point (ξ_{i_1}, μ_{i_1}), for $i_1 = 1, \ldots, I_1$, is assumed to belong to a distinct chirp.
- For $k = 2, 3, \ldots$, we use the described criteria to decide if (ξ_{i_k}, μ_{i_k}), for : $i_k = 1, \ldots, I_k$, belongs to an already detected chirp. On the contrary, it is established as the starting point of a new chirp.
- When the above iteration is finished and to avoid artifacts due to numerical errors, we disregard chirps composed by a unique point.

Finally, once the chirps are separated, we use the following approximation, motivated by Theorem 1, to estimate the amplitude

$$a(\tau_m)^2 \approx \frac{1}{\lambda[\hat{v}(0)]^2} P_\Psi f(t_o, \phi'(\tau_m), \phi''(\tau_m); \lambda).$$

Again, to avoid artifacts due to numerical discretization, we neglect portions of signals with an amplitude lower than certain relative threshold , $\epsilon \in (0, 1)$, of the maximum amplitude of the whole signal, considering that in this case no chirp is present.

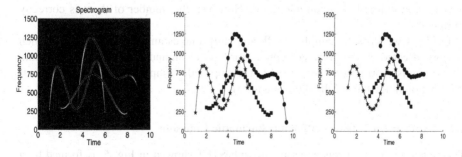

Fig. 3. Spectrogram of the clean signal and results of the chirp localization and separation algorithm for clean and noisy signals, respectively

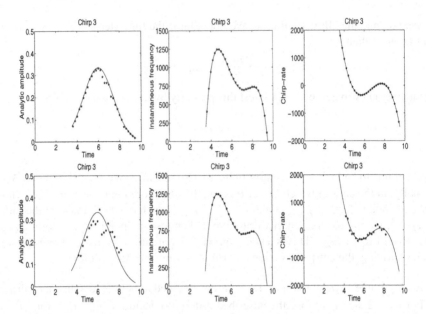

Fig. 4. Amplitude, IF and chirp rate for the clean signal (first row) and noisy signal (second row). Solid lines correspond to exact values and crosses to computed values.

4.1 Experiment 1. A Synthetic Signal

In this first experiment we test our algorithm with a synthetic signal, f, composed by the addition of three nonlinear chirps (spectrogram shown in Fig. 3) and with the same signal corrupted with an additive noise of similar amplitude than that of f, i.e., with $SNR = 0$.

We used the same time step, $\tau = 0.2$ sec, and window width, $\lambda = 0.1$ sec, to process both signals, while we set the relative threshold amplitude level to $\epsilon = 0.01$ for the clean signal and to $\epsilon = 0.1$ for the noisy signal. The results of our algorithm of denoising, detection and separation is shown in Fig. 3. We observe that for the clean signal all chirps are captured with a high degree of accuracy even at crossing points. We also observe that the main effect of noise corruption is the lose of information at chirps low amplitude range. However, the number of them is correctly computed.

In Fig. 4 we show the amplitude, IF and chirp-rate estimations of the chirp which is more affected by the noise corruption, for both clean and noisy signals. The main effect of noise corruption is observed in the amplitude computation and in the lose of information in the tails of the three quantities.

4.2 Experiment 2. One Wolf Multi-harmonic Emission

The signal analyzed in this experiment, with STFT showed in Fig. 5, is formed by a howl of a unique individual which is composed by a formant-chirp and its corresponding harmonic chirps among which only one is detected, due to the high intensity noise

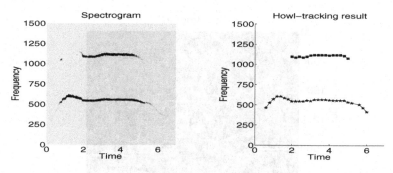

Fig. 5. Left: Experiment 2 signal's STFT. Right: Result of the separation algorithm.

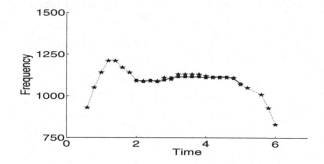

Fig. 6. The separation algorithm allows us to check the harmonicity of chirps. We plot the high frequency chirp and the diadic translation of the low frequency chirp of Fig. 5.

level. Before using the separation algorithm, we filtered the signal to the relevant frequency band $[250, 1500]$ Hz and then applied the noise reduction signal enhancement algorithm of [13].

Parameters were fixed as follows: time step to $\tau = 0.2$ sec, amplitude threshold to $\epsilon = 0.05$ and window width to $\lambda = 0.1$ sec. The result is shown in Fig. 5. We observe that a segment of the harmonic is not detected, as expected due to the high SNR. However, the algorithm detects and separates both chirps accurately enough to perform the computation plotted in Fig. 6, showing that both chirps are harmonics and, therefore, allowing us to conclude that only one individual is emitting.

4.3   Experiment 3. A Field Recorded Wolves Chorus

In this experiment we analyze a rather complex signal obtained from field recordings of wolves choruses in wilderness, [15]. Due to the noise present in the recording, we first use the PDE algorithm to enhance the signal and reduce the noise, see [13] for details. For the separation algorithm, we fixed the time step as $\tau = 0.03$ sec, the relative amplitude threshold as $\epsilon = 0.01$ and the window width as $\lambda = 0.0625$ sec.

The algorithm output is composed by 32 chirps which should correspond to the howls and barks (with all their harmonics) emitted by the wolves along the duration of the

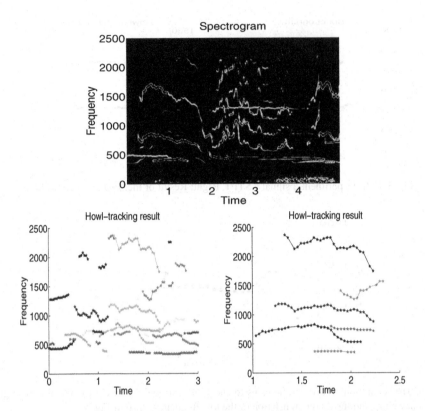

Fig. 7. First row: spectrogram of the field recorded signal utilized in Experiment 3. Second row. Left: result of the chirp localization and separation algorithm. Right: a zoom of the previous plot showing six separated chirps corresponding to five wolves howls.

recording (about five sec). The result is shown in Fig. 7. Since our aim is giving an estimate of how many individuals are emitting in a recording, we plot a zoom of the separating algorithm result for the time interval $(1, 2.5)$. Here, the number of chirps reduces to six. However, it seems that one couple of them are harmonics, the couples formed by the chirp around 1000 Hz and the highest IF chirp. Therefore, we may conclude that at least five wolves are emitting in this interval of time. A similar analysis is carried out with other time subintervals until all the recorded signal is analyzed.

5   Conclusions

A combined algorithm for signal enhancement and voice separation is utilized for wolf population counting. Although field recorded wolf chorus signals posses a complex structure due to noise corruption and nonlinear multi-component character, the outcome of our algorithm provides us with accurate estimates of the number of individuals emitting in a given recording. Thus, the algorithm seems to be a good complement or, even, an alternative to existent methodologies, mainly based in wolf traces collection or in the intrusive attaching of electronic devices to the animals. Clearly, our algorithm is not limited to wolves emissions but to any signal which may reasonably be modelled as an addition of chirps, opening its utilization to other applications. Drawbacks of the algorithm are related to the expert dependent election of some parameters, such as the amplitude threshold, or to the execution time when denoising and separation of long duration signals must be accomplished. We are currently working in the improvement of these aspects as well as in the recognition of components corresponding to the same emissor, such as harmonics of a fundamental chirp, pursuing the full automatization of the counting algorithm.

References

1. Skonhoft, A.: The costs and benefits of animal predation: An analysis of scandinavian wolf re-colonization. Ecol. Econ. 58(4), 830–841 (2006)
2. Mann, S., Haykin, S.: The chirplet transform: Physical considerations. IEEE Trans. Signal Process. 43(11), 2745–2761 (1995)
3. Angrisani, L., D'Arco, M.: A measurement method based on a modified version of the chirplet transform for instantaneous frequency estimation. IEEE Trans. Instrum. Meas. 51(4), 704–711 (2002)
4. Ozaktas, H.M., Zalevsky, Z., Kutay, M.A.: The Fractional Fourier Transform with Applications in Optics and Signal Processing. Wiley, Chichester (2001)
5. Kodera, K., Gendrin, R., de Villedary, C.: Analysis of time-varying signals with small bt values. IEEE Trans. Acoustics Speech Signal Process. 26(1), 64–76 (1978)
6. Auger, F., Flandrin, P.: Improving the readability of time-frequency and time-scale representations by the method of reassignment. IEEE Trans. Signal Process. 43(5), 1068–1089 (1995)
7. Dugnol, B., Fernández, C., Galiano, G.: Wolves counting by spectrogram image processing. Appl. Math. Comput. 186, 820–830 (2007)
8. Dugnol, B., Fernández, C., Galiano, G., Velasco, J.: On pde-based spectrogram image restoration. application to wolf chorus noise reduction and comparison with other algorithms. In: Damiani, E., Dipanda, A., Yetongnon, K., Legrand, L., Schelkens, P., Chbeir, R. (eds.) Signal processing for image enhancement and multimedia processing, vol. 31, pp. 3–12. Springer, Heidelberg (2008)
9. Mallat, S.: A wavelet tour of signal processing. Academic Press, London (1998)
10. Auger, F.: Représentation temps-fréquence des signaux non-stationnaires: Syntheèse et contributions. Thèse de doctorat, Ecole Centrale de Nantes (1991)
11. Chassandre-Mottin, E., Daubechies, I., Auger, F., Flandrin, P.: Differential reassignment. IEEE Signal Process. Lett. 4(10), 293–294 (1997)
12. Álvarez, L., Lions, P.L., Morel, J.M.: Image selective smoothing and edge detection by nonlinear diffusion. ii. SIAM J. Numer. Anal. 29(3), 845–866 (1992)
13. Dugnol, B., Fernández, C., Galiano, G., Velasco, J.: Implementation of a diffusive differential reassignment method for signal enhancement an application to wolf population counting. Appl. Math. Comput. 193, 374–384 (2007)
14. Dugnol, B., Fernández, C., Galiano, G., Velasco, J.: On a chirplet transform-based method applied to separating and counting wolf howls. Signal Process 887, 1817–1826 (2008)
15. LLaneza, L., Palacios, V.: Field recordings obtained in wilderness in Asturias (Spain) in the 2003 campaign. Asesores en Recursos Naturales, S.L. (2003)

Extraction of Capillary Non-perfusion from Fundus Fluorescein Angiogram

Jayanthi Sivaswamy[1], Amit Agarwal[1], Mayank Chawla[1], Alka Rani[2],
and Taraprasad Das[3]

[1] International Institute of Information Technology
`jsivaswamy@mail.iiit.ac.in`, `mayank_c@research.iiit.ac.in`
[2] Aravind Eye Institute, Hyderabad, India
`alka_bala@rediffmail.com`
[3] LV Prasad Eye Institute, Hyderabad, India
`tpd@lvpei.org`

Abstract. Capillary Non-Perfusion (CNP) is a condition in diabetic retinopathy where blood ceases to flow to certain parts of the retina, potentially leading to blindness. This paper presents a solution for automatically detecting and segmenting CNP regions from fundus fluorescein angiograms (FFAs). CNPs are modelled as valleys, and a novel technique based on extrema pyramid is presented for trough-based valley detection. The obtained valley points are used to segment the desired CNP regions by employing a variance-based region growing scheme. The proposed algorithm has been tested on 40 images and validated against expert-marked ground truth. In this paper, we present results of testing and validation of our algorithm against ground truth and compare the segmentation performance against two others methods. The performance of the proposed algorithm is presented as a receiver operating characteristic (ROC) curve. The area under this curve is 0.842 and the distance of ROC from the ideal point $(0, 1)$ is 0.31. The proposed method for CNP segmentation was found to outperform the watershed [1] and heat-flow [2] based methods.

1 Introduction

Diabetes is occurring in an ever increasing percentage in the world. Diabetes mellitus affects many organs of the body, and the eye is one of the organs that is affected relatively early (compared to the kidney). While diabetes affects all parts of the eye, the retina (retinopathy) is most commonly affected. Diabetic retinopathy progresses in phases. It starts with microaneurysms and superficial retinal hemorrhages (non-proliferative diabetic retinopathy; NPDR), progresses to accumulation of hard exudates in the posterior pole (diabetic maculopathy), and finally ends with new vessels in the surface of the retina and/or the optic disc (proliferative diabetic retinopathy; PDR). The underlying cause of the terminal event, the retinal new vessels, is retinal ischemia which manifests as areas of CNP that is most clearly seen in an FFA. These lesions appear as dark regions in the FFA images as shown in Fig. 1. If not treated in time, the CNP areas grow and spread across the entire retina. Large areas of non-perfusion lead to new vessel formation and bleeding into the vitreous cavity. These complications are

A. Fred, J. Filipe, and H. Gamboa (Eds.): BIOSTEC 2008, CCIS 25, pp. 176–188, 2008.
© Springer-Verlag Berlin Heidelberg 2008

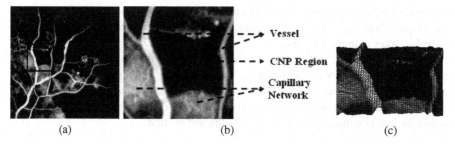

Fig. 1. (a) A sample FFA image with CNP. (b) Enlarged view and (c) surface plot of the CNP region in (a).

responsible for severe visual loss in most patients with PDR [3]. FFA guides the choice and extent of laser treatment in diabetic maculopathy and PDR.

An automatic identification of important events in FFA is objective and very useful both for referral and treatment. Automated analysis of FFA images for the purpose of extracting important structures as well as lesions have received some attention. Image conditioning solutions that have been proposed include illumination correction using a parametric bi-cubic model for the illumination function [4] and noise suppression for a sequence of angiogram images based on bilateral filtering [5]. In FFA segmentation, stochastic models have been proposed to segment the fovea, arteries and veins from the central (macular) view of FFAs [6] and among lesions, microaneurysms have received much attention. Several techniques ranging from morphological to model-based have been proposed for microaneurysm segmentation [7], [8] and [9]. An automated technique for measurement of blood flow in capillaries has been attempted from angiograms, for determining the effect of cardio-pulmonary bypass surgery [10]. The foveal region of the retinal image is processed to enhance the vascular structure and extract linear segments. The processed results from images taken before and after the bypass surgery are then compared (via a logical AND operation) to identify the differences. However, to our knowledge, there are no reports in the literature of any technique to detect the cause of PDR namely the presence of the CNP regions anywhere in the retina. Detecting and segmenting CNPs are the focus of this paper.

The clinical procedure to detect CNPs is a visual scan of an FFA image. In order to estimate the amount of area damaged, the scan is generally done on the composite image of the retina obtained after suitable mosaicing of several retinal segments. Such a procedure suffers from several drawbacks: the variable skills and subjectivity of the observer, which also depend on the quality of the images; a lack of precise understanding of the area of retina affected which helps in deciding the nature and extent of laser treatment. Automated image analysis techniques can be used to address these issues but there are several challenges in devising solutions for CNP segmentation. FFAs suffer from non-uniform illumination due to the eye geometry, imaging conditions and presence of other media opacity such as cataract. Inter-patient and intra-patient variability is also possible. The former is due to different pupil dilations and the latter is due to the time of image capture after injection of fluorescein dye. Another compounding factor is that the mean grey level of CNPs as well as their shape and size are variable, with the size ranging from very small to very large (from 100 to 55000 pixels). Often, the boundaries of CNPs are not well defined because of an inhomogeneous textured background. Thus, the only visually distinguishing characteristic of a CNP is that it is relatively darker than its surround.

In this paper, we propose a novel method to extract and quantify regions of CNP based on modeling CNPs as valleys in the image surface. The algorithm for CNP segmentation is developed and its details are presented in the next section. Section 3 provides implementation details and illustrative test results of the algorithm. Finally, some discussions and conclusions are presented in the last section.

2   Valley Based CNP Segmentation

2.1   Modelling CNP Regions

As discussed earlier, CNP occurs when the capillary network in a region of the human retina stops functioning and does not supply blood to the corresponding areas. In FFAs, regions receiving normal blood supply appear as bright white regions since they carry a fluorescent dye and regions lacking in blood (due to abnormal supply of blood) appear as dark regions. Hence, regions of CNP appear as dull/dark lesions bounded by healthy vasculature.

A sample FFA image and an enlarged view of a CNP region and its surroundings is shown in Fig. 1. Also, included in this figure is the surface plot of the corresponding CNP region from which we can observe that the prominent vessels, the healthy capillary network and the CNP have very different topographic characteristics: While the major vessel appears as a ridge, the CNP appears as a valley with the healthy capillary network appearing as a plateau in the image. Hence, one can conclude that CNPs can be modelled as valleys. Watershed-based solution to valley detection ([11], [12]) is possible, however, these result in oversegmentation which have been addressed by the use of markers [13]. The markers are either provided by a user [14] or extracted automatically from a given image [1]. In the case of CNP detection, since the size of a CNP and the nature of its surround is highly variable, obtaining such markers is quite challenging. A better alternative is to identify the trough (lowest point on a curve) and use it to segment a CNP. Hence, we have taken a different approach to the problem and propose a technique that detects trough points at multiple scales of the image and collates them across scales. We next present the details of our proposed algorithm for CNP segmentation comprising several steps.

2.2   CNP Detection Algorithm

The proposed CNP detection algorithm consists of these stages: Firstly, illumination correction (IC) is done to minimize the background intensity variation followed by denoising to eliminate noise that is frequently found in FFAs. Next, valley detection is performed to locate the seed points in the CNP regions which are used to extract the candidate CNP regions using a region growing algorithm. Finally, thresholding is done to reject false positives among the detected candidates. The processing in each of these stages are described next.

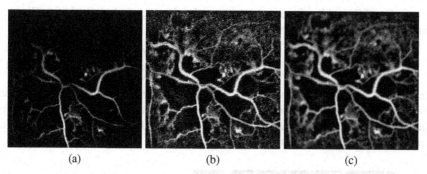

| (a) | (b) | (c) |

Fig. 2. (a)A sample FFA image,(b)Illumination corrected image and (c)denoised image

Illumination Correction. Nonuniform illumination is a problem in retinal colour images as well as angiograms. A camera-model based solution for illumination correction in angiograms, obtained with non-confocal imaging, is given in [4] which assumes a macula-centric view of the retina. Our images are not necessarily macula-centric and are obtained from a laser-based confocal imaging system. We modified a quotient based approach proposed for face images [15] and model the non-uniform illumination as a multiplicative degradation function which is estimated by blurring the corrupted image. Let $I(x, y)$, $I_s(x, y)$ and $I_0(x, y)$ denote the given, smoothed and corrected images, respectively and l_0 be the desired level of illumination. The corrected intensity value at location (x, y) is found as

$$I_0(x, y) = \begin{cases} I(x, y) \times \frac{l_0}{I_s(x,y)} & \text{if } I_s(x, y) < l_0 \\ I(x, y) & \text{if } I_s(x, y) \geq l_0. \end{cases} \tag{1}$$

As can be observed from eq. 1, a pixel where the estimated illumination is greater than the ideal illumination value is not corrected. This is to ensure that the regions which are inherently bright, like the optic-disk, haemorrhages, etc., are not wrongly classified as regions of excessive illumination and corrected accordingly. When the estimated illumination value is less than the ideal illumination value, multiplication by the fraction $\frac{l_0}{I_s(x,y)}$ ensures that regions with illumination less than the l_0 are elevated to the ideal illumination value. Moreover, contrast at such a pixel is improved by a factor of $\frac{l_0}{I_s(x,y)}$ thereby removing the need for subsequent brightness and contrast operations, as required in the case of quotient-image based technique. A sample FFA image and the resulting image after illumination correction are shown in Fig. 2 (a), (b) respectively.

Noise Removal. The laser-based imaging produces fine-grain speckle type of noise in the angiograms as can be seen in Fig. 2. A bilateral filter-based approach proposed for color and gray scale images in [16] has been successfully applied to denoise images in an angiogram sequence [17]. The strength of bilateral filter based denoising is its ability to denoise without compromising edge quality. This is due to the filter's nonlinear characteristic which permits one to take into account the spatial distance as well the photometric similarity of a pixel to its neighbors. The spatial context is provided by a domain filter while the photometric similarity is controlled by a range filter. We use a version of the bilater filter for our noise removal task which is described next. Given an input pixel $I(P)$, the output pixel $I_0(P)$ is found as

$$I_0(P) = \frac{\sum_w I(Q)W_d(P,Q)W_r(P,Q)}{\sum_w W_d(P,Q)W_r(P,Q)} \ . \tag{2}$$

where P and Q are position vectors, w is the current neighborhood and W_d, W_r are Gaussian kernels of the domain and range filters respectively. The edge preservation feature of the bilateral filter can be seen in the results of preprocessing (illumination correction + denoising) in Fig.6 (b).

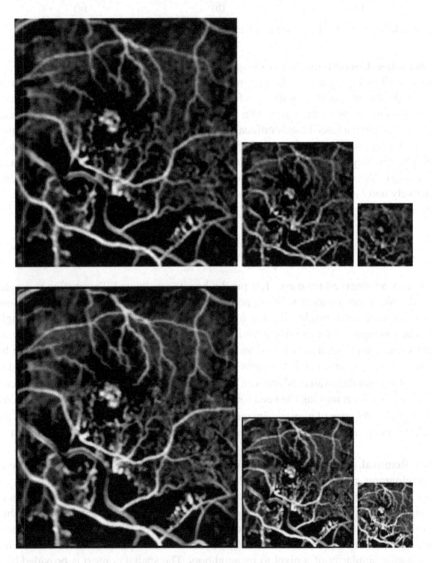

Fig. 3. An FFA image at levels 3, 4, 5 in the (first row) conventional and (second row) proposed pyramidal decomposition

CNP Segmentation. Now we turn to the main task of detecting and segmenting CNP regions. Since we have modelled CNPs as valleys, a valley detection algorithm is needed to detect seed points in the CNP regions. As the CNPs vary widely in size, the valleys can be extended. Hence, a multi-resolution approach is appropriate. The strategy we have adopted is to reduce the valleys to a single trough point via a pyramidal decomposition and then detect them using a trough detector at each level and collating them. Each of these steps is described next.

Extrema Pyramid Decomposition - A conventional pyramidal decomposition based on averaging and subsampling is inadequate for the problem at hand. This can be illustrated with an example shown in Fig. 3. It can be seen that the averaging process dulls the entire image and will therefore adversely affect CNP detection based on troughs. Another drawback with the averaging process is the difficulty in localising of the trough points in the full resolution image when performing the up-sampling process after trough detection. In the problem at hand, the CNP regions are generally bigger and darker relative to the brighter regions which are thin. Averaging and down-sampling will result in the bright regions to disappear faster than the CNP regions, whereas for locating troughs, it would help to more or less retain the bright regions across several levels while accepting some loss in the CNP area. Hence, to preserve the relation between a CNP and its surround, and maintain the depth of the valley across levels, we need a method for pyramidal decomposition that will minimize the CNP regions at a much faster rate compared to the brighter surrounding regions. Retaining intensity maxima rather than average will ensure that the thinner bright regions are largely preserved during downsampling. However, this is detrimental to the relative contrast between a CNP and its surround as it elevates the average intensity of the CNP regions. The end result is a lowering of the depth of the troughs, which is undesireable. A better alternative is to generate the pyramid through an adaptive selection of pixels. The solution we propose is a novel technique for decomposition which is based on intensity *extrema*. The decomposition obtained through this procedure is an extrema pyramid representation of the given image. We now describe the technique for generating an extrema pyramid. Given an image I_1 of size $M \times N$ a L-level decomposition is found as follows:

$$I_l(m,n) = \left\{ \begin{array}{ll} min\{g_{i,j}(m,n)\} & \text{if } g_{i,j}(m,n) \leq t \\ max\{g_{i,j}(m,n)\} & \text{otherwise .} \end{array} \right\}. \tag{3}$$

$$\forall\, i,j = 0,1$$

where $g_{i,j}(m,n) = I_{l-1}(2m+i, 2n+j)$, with $l = 2, ...L$ and t is a suitable threshold, taken to be the global mean in our experiments. An illustration of the equation is given in Fig. 5 for $t = 100$. Based on the above definition, it is worthwhile to note that the extrema pyramid is not strictly a multi-resolution representation for a given image. This is because, given an image at some level in the pyramid it is not possible to generate a finer level image through an interpolation process. However, every pixel at a given level in an extrema pyramid corresponds exactly to a pixel at the finer level. This is a useful property in the problem at hand as locating seeds in the CNP region is of main interest.

In an extrema pyramidal decomposition of an angiogram, the CNP regions diminish in size at a much faster rate than non-CNP regions across the levels. This is illustrated

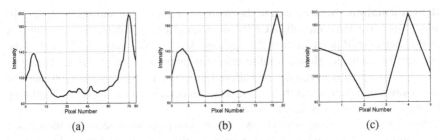

Fig. 4. Intensity profile of a CNP and its surround, at levels (a) 1, (b) 3 and (c) 5 of the image pyramid

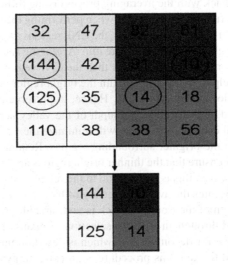

Fig. 5. Example demonstrating how pixel values are computed in the extrema pyramid using Equation 3

in Fig. 3 (b) where the thickness of vessels are more or less preserved but the CNPs are reduced to near-dots in the lowest level image. This effect is also seen from the intensity profiles shown in Fig. 4, along a horizontal line passing through a CNP region at different levels. The width of the valley reduces from 50 pixels at the first level, to about 1 pixel in the fifth level whereas the image has been downsampled by 16 between these levels. The relative brightness value (130) of the ridge and valley regions is preserved as a result of not performing a smoothing operation.

Trough Detection - A trough is defined as the lowest point on a curve. Alternatively, the brightness at a trough is a local minimum. Since the context in which CNPs, and hence troughs, occur is variable in an angiogram, two parameters can be used to characterise a trough: μ, the mean brightness of the surround and P, the peak factor which represents the depth of the trough. These two parameters are used to develop the following trough detection algorithm in which the image is denoted by $I(x, y)$.

For every pixel (x, y) do the following:

1. *Initialize a Boolean variable isTrough = False.*
2. *Check if I(x,y) is a local minimum in a M × M neighborhood.*
3. *If yes, then calculate the mean (μ) of a N × N neighborhood, with N > M. Else, do nothing.*
4. *Let T = μ*P and check if $I(x,y) < T$.*
5. *If yes, then isTrough = True.*
6. *If isTrough = True, then mark I(x,y) as a trough pixel. Else, do nothing.*

The threshold T represents the depth of the valley from the mean μ. Since the image pyramid retains extrema, this threshold value has to be carefully chosen to ensure that enough seed pixels are captured in a valley while minimising the possibility of false alarms. A region with low μ is likely to be a CNP region and hence the required depth for that region is less whereas the same may not be true if μ is high and hence, a stricter condition is required in the latter case. Thus, choosing T proportional to μ is appropriate. Furthermore, since trough detection is carried out at multiple levels a peak factor has to be chosen for each level. A guiding factor in this choice is that due to retention of extremas, the likelihood of the local minima being a CNP region will be higher at upper levels. Hence, the peak factor should be progressively increased with the levels in the pyramid.

After performing trough detection at all levels, the results are combined with a simple logical *OR* operation. For locating the seed pixels in the original image, the fact that the extrema of four pixels is selected at every level is used iteratively.

CNP Region Extraction - The detected trough points can serve as seed points for region based approach to segmenting the CNP regions. Although geometric methods can potentially yield better results, as an initial experiment we chose to use a simple region growing technique for extracting the CNP regions as it was computationally simpler. Given the variability of the appearance of the CNP regions within and across images, the traditionally used intensity-based homogeneity criterion for region growing is not suitable. Instead, by noting that CNP regions are smooth, the better alternative is to perform the pixel aggregation in the variance space.

In our experiments, the range for the variance was taken to be ± 4. In order to reject false candidates, a final thresholding operation was performed. A threshold based on the global mean intensity was applied since the global mean is always lowered with the presence of CNPs.

3   Implementation and Results

The proposed algorithm was implemented as follows. In the illumination correction stage, the ideal illumination l_0 in (1) was set to be roughly half the maximum grey value in the image or 120. In order to approximate the illumination function with a Gaussian smoothed image, the level of smoothing required is quite large (roughly 49 × 49). Since this is computationally intensive, for faster processing, we subsampled the original image (by 8 × 8) and smoothed the subsampled image by by 5 × 5) mask and upsampled the smoothed result to the full resolution. For denoising, a filter kernel size of 9 × 9 was used and σ for the domain and range filters were fixed at 3 and 10 respectively. For valley detection, a 5-level pyramid was generated; M, N were fixed at

5 and 7 respectively and the peak factor was incremented by 0.02 at each level in the pyramid. In region growing, the variance was calculated over a 5×5 neighborhood.

The proposed CNP segmentation algorithm was tested on 40 images of size 512×512, each of which contained many CNPs. These were acquired from the digital confocal scanning laser ophthalmoscope of Heidelberg Retina Angiograph. The images were of retinal segments for which the ground truth, in the form of boundaries of CNPs, were prepared manually by a retina expert (a co-author). Some sample test images along with corresponding ground truth and results of our CNP segmentation algorithm, with a peak factor of 0.41, are shown in Fig. 6.

CNP regions are shown in black in both ground truth and segmented results. The five sample test images indicate the variability in images in terms of quality, size of CNPs and presence of other structures such as optic disk, macula and microaneurysms. A quantitative assessment of the algorithm was done using a ROC curve and not a FROC curve since the area of CNP is of clinical interest. A comparison between computed and marked CNP segments was done on a pixel by pixel basis. By using the peak factor as a control parameter, the obtained ROC curve, shown in Fig. 7, was found to have an area under the curve (AUC) of 0.842 and a distance (D_i) to the ideal point (1,0) of 0.35. The ideal values for AUC and D_i are 1 and 0 respectively.

In order to asses the proposed method against other segmentation methods, we conducted some experiments by choosing two candidate methods. The first method was watershed-based segmentation with automatic marker extraction, as described in [1]. The markers are extracted in this method by locating local minima in the smoothed gradient image. This method requires no parameter tuning. In our experiments the input images were corrected and denoised prior to applying the watershed segmentation algorithm. The illumination correction and denoising steps were identical to the first two stages in our algorithm. The second method selected was heat-flow based segmentation described in [2]. This method essentially is a shape extraction scheme using heat conduction and geometric flow and results of testing on medical images (small in size) can be found in [2]. The heat-flow method requires a manual seed selection, which in our experiments were taken to be the detected trough points. Further, the number of iterations and number of modes in the histogram of image need to be specified. We chose 20 iterations and unimodal distribution to describe the CNP images. Heat flow-based segmentation was applied on test images after illumination correction and denoising.

CNP segmentation using the above two methods was done on all 40 test images. Five test images and the corresponding results are shown in Fig. 8. Of these 5 test images, the top 2 are from the set shown in Fig. 6.

4   Discussion and Conclusions

An unsupervised algorithm for automatically segmenting CNPs from FFA images has been presented. Its overall performance is quite good as indicated by the ROC curve and the AUC, D_i metrics. Since there is no reported work on this problem it is not possible to do any benchmarking. A visual inspection of segmented results in Fig. 6 indicates that the algorithm successfully detects CNPs of all sizes, however, it tends to undersegment large CNPs because the illumination correction stage intensifies the variability within

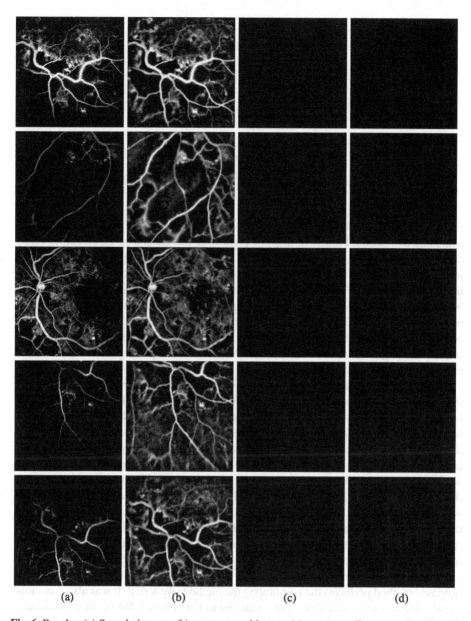

Fig. 6. Results: (a) Sample images, (b) preprocessed image, (c) corresponding ground truths and (d) segmented results with CNP regions shown in black

CNPs. A failure analysis indicates that the macula region gets mislabled as a CNP (as seen in the bottom row of Fig. 6) since the two have similar characteristics, and CNPs in the image periphery tend to be missed since the valley model is weak in this region.

The results of using a watershed algorithm for CNP segmentation were overall very unsatisfactory. The tendency of watershed segmentation to oversegment is clearly vis-

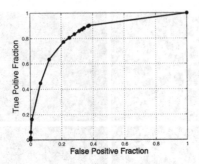

Fig. 7. ROC plot

ible from the results shown in Fig. 8. The reason for the poor performance can be explained as follows. Firstly, unlike our algorithm, the watershed algorithm is based on the gradient image, which makes it very sensitive to noise, some of which remains in the denoised image that forms the input to this algorithm. Secondly, marker extraction is difficult on CNP images because the boundary between a CNP region and its background is often quite fuzzy. As a consequence, there were several spurious minima leading to false CNP regions (see for instance, the top row in Fig. 8). Thirdly, to address the oversegmentation problem, markers internal and external to the region of interest are required whereas the method we used for comparison, used only internal markers (as it picks local minima).

The heat-flow based CNP segmentation yielded a better performance compared to the watershed-based method as can also be observed from the results in Fig. 8. A reason for this is that the seed selection was via our proposed method. However, in comparison with our proposed method, which uses a variance-image based region growing technique, the performance was still found to be poor in a majority of the cases. An exception was the image in the second row of Fig. 6 and Fig. 8. Here, it can be seen that our algorithm fails to detect the CNP regions in the corners of the image unlike the heat-flow method. The results also indicate that the heat-flow method tends to oversegment, whereas our algorithm tends to undersegment CNP regions. In the heat-flow algorithm, the number of iterations is a variable that ideally depends on the image condition. It is difficult to determine the optimal number adaptively. It was found that increasing the number of iterations leads to unstable solutions as also mentioned in [2].

Overall, we conclude that, for the CNP detection and segmentation problem, our proposed method performs the best among the methods we tested. It was also computationally less intensive and hence faster than the tested methods. While the performance of our current implementation is quite good, there is scope for improvement of the algorithm's performance: geometric techniques such as fast marching method in [18] can be used to more accurately extract the CNP region boundaries while a pixel-based classifier will help improve the rejection of the false alarms. Likewise, incorporation of a macula detection stage will help avoid a misclassification of macula as a CNP region.

Finally, it was observed that the ground truth generation process for CNPs is a laborious one. Retina experts found it challenging to draw precise boundaries because they

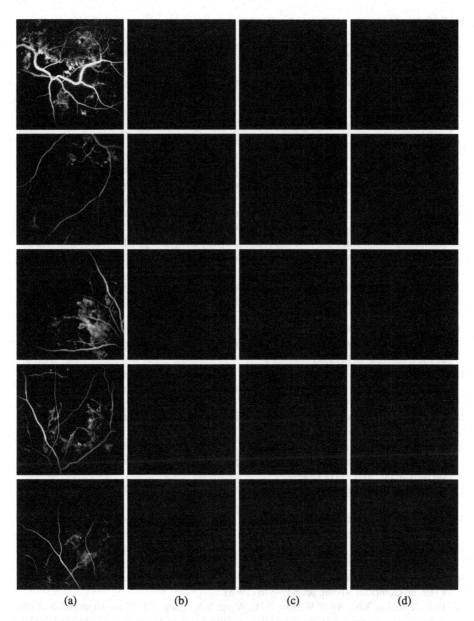

Fig. 8. Results: (a) Sample images, (b) corresponding ground truths, segmented results with (c) Watershed [1] and with (d) Heat Flow methods [2] (CNP regions shown in black)

often appear to be intricate or ill-defined. This points to the need for the use of multiple expert-markings for a fuller evaluation of the algorithm. Such an exercise might also shed light on the degree of observer bias in CNP segmentation.

References

1. Gao, L., Yang, S., Xia, J., Liang, J., Qin, Y.: A new marker-based watershed algorithm. In: Proc. ISCAS, vol. 2, pp. 11–84–4 (2004)
2. Direkoäÿlu, C., Nixon, M.S.: Shape extraction via heat flow analogy. In: Blanc-Talon, J., Philips, W., Popescu, D., Scheunders, P. (eds.) ACIVS 2007. LNCS, vol. 4678, pp. 553–564. Springer, Heidelberg (2007)
3. Kohner, E.M.: Diabetic retinopathy. BMJ 307, 1195–1199 (1993)
4. Cree, M.J., Olson, J.A., McHardy, C.K., Sharp, P.F., Forresters, J.V.: The preprocessing of retinal images for the detection of fluorescein leakage. Phys. Med. Biol. 44, 293–308 (1999)
5. Guo, X.X., Lu, Y.N., Xu, Z.W., Pang, Y.J.: An eawa filter for denoising of filtering of fluorescein angiogram sequences. In: Proc. International Conf. on Computer and Information Technology, pp. 614–618 (2005)
6. Simó, A., de Ves, E.: Segmentation of macular fluorescein angiographies. a statistical approach. Pattern Recognition 34, 795–809 (2001)
7. Fleming, A.D., Philip, S., Goatman, K.A., Olson, J.A., Sharp, P.F.: Automated microaneurysm detection using local contrast normalization and local vessel detection. IEEE Trans. on Medical Imaging 25, 1223–1232 (2006)
8. Hafez, M.: Using adaptive edge technique for detecting microaneurysms in fluorescein angiograms of the ocular fundus. In: Proc. Mediterranean Electrotechnical Conf., pp. 479–483 (2002)
9. Mendonça, A.M., Campilho, A.J.: Automatic segmentation of microaneurysms in retinal angiograms of diabetic patients. In: Proc. International Conf. on Image Analysis and Processing, pp. 728–733 (1999)
10. Jagoe, R., Arnold, J., Blauth, C., Smith, P., Taylor, K.M., Wootton, R.: Measurement of capillary dropout in retinal angiograms by computerised image analysis. Pattern Recognition Letters 13, 143–151 (1992)
11. Gauch, J.M.: Image segmentation and analysis via multiscale gradient watershed hierarchies. IEEE Trans. on Image Processing 8, 69–79 (1999)
12. Karvelis, P.S., Fotiadis, D.I., Georgiou, I., Syrrou, M.: A watershed based segmentation method for multispectral chromosome images classification. In: Proc. International Conf. on EMBS, pp. 3009–3012 (2006)
13. Sonka, M., Fitzpatrick, J.: Hand Book of Medical Imaging. Medical Image Processing and Analysis (2004)
14. Beuchen, S.: The watershed transformation applied to image segmentation. Scanning microscopy international 34, 795–809 (2001)
15. Wang, H., Li, S.Z., Wang, Y.: Generalized quotient image. In: Proc. Conf. on Computer Vision and Pattern Recognition, vol. 2, pp. 498–505 (2004)
16. Tomasi, C., Manduchi, R.: Bilateral filtering for gray and color images. In: Proc. International Conf. on Computer Vision, pp. 839–846 (1998)
17. Guo, X.X., Lu, Y.N., Xu, Z.W., Liu, Z.H., Wang, Y.X., Pang, Y.J.: Noise suppression of fluorescein angiogram sequences using bilateral filter. In: Proc. International Conf. on Machine Learning and Cybernetics, vol. 9, pp. 5366–5371 (2005)
18. Malladi, R., Sethian, J.A.: Fast methods for shape extraction in medical and biomedical imaging. In: Malladi, R. (ed.) Geometric Methods in Bio-medical Image Processing, pp. 49–61. Springer, Berlin (2006)

Content-Based Retrieval of Medical Images with Elongated Structures

Alexei Manso Correa Machado* and Christiano Augusto Caldas Teixeira

Pontifical Catholic University of Minas Gerais, Belo Horizonte MG, Brazil

Abstract. In this paper we propose a set of methods to describe, register and retrieve images of elongated structures from a database based on their shape content. Registration is performed based on an elastic matching algorithm that jointly takes into account the gross shape of the structure and the shape of its boundary, resulting in anatomically consistent deformations. The method determines a medial axis that represents the full extent of the structure with no branches. Discriminative anatomic features are computed from the results of registration and used as variables in a content-based image retrieval system. A case study on the morphology of the corpus callosum in the chromosome 22q11.2 deletion syndrome illustrates the effectiveness of the method and corroborates the hypothesis that retrieval systems may also act as knowledge discovery tools.

1 Introduction

Many important problems in medical imaging applications are related to elongated structures such as vessels, bones and brain ventricles [20,18] which can be efficiently represented by centerlines or medial axes. Those structures have in common the fact that their gross shape is, if not more, as important as the shape of the boundary. Contour may present important anatomical features, but the overall shape is of greater interest.

In this work, a set of methods is proposed to describe, register and ultimately retrieve images of elongated structures from a database based on their shape content. Image registration techniques have been widely used in morphometry, as they provide detailed description of the anatomy, taking a reference image as a basis for comparison. Registration algorithms are nevertheless computationally intensive procedures and, when applied to the whole image or to the boundary of elongated structures, may yield unsatisfactory results. Figure 1 shows a schematic of a content-based image retrieval (CBIR) system that follows this approach. A set of images of elongated structures is segmented and the structures represented by their boundaries and medial axes. Another image, taken as a common reference, is deformed through elastic registration so as to align its anatomy with the anatomy of the images in the dataset. The result of registration is a mapping function from each point in the reference to a point in the target image that enable detailed shape description. After the structures have been described, e.g. based on the curvature of their boundaries and medial axes, they are stored in the database

---

* This work was partially supported by PUC Minas and CNPq grant 20043054198. The authors are grateful to the University of Pennsylvania and Children's Hospital of Philadelphia for sharing the corpus callosum data.

A. Fred, J. Filipe, and H. Gamboa (Eds.): BIOSTEC 2008, CCIS 25, pp. 189–201, 2008.

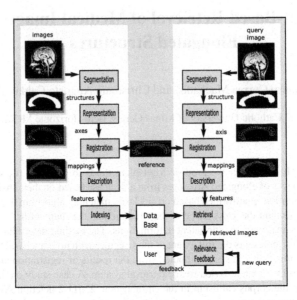

Fig. 1. Schematic of a CBIR system based on registration. The left part of the scheme shows the steps performed off-line for each image in the database. The on-line part of the retrieval process is shown in the right. The link between the on-line and off-line phases is the reference image that is registered to the query and to the database, establishing a basis for shape comparison.

for future searching. The querying phase follows the same steps used to convert the images into descriptive features. The query image converted to the corresponding feature vector is searched in the database, the most similar images are retrieved and presented to the user. The user may rank the results according to their relevance, choose one of the retrieved images as a new query or redefine a region of interest that should be given greater priority in the next retrieving iteration. The query vector is therefore updated taking into account the user's feedback.

A contribution of this work is a registration algorithm that takes into account both the gross shape of the structure and the shape of its boundary, with emphasis to the former aspect. The characterization of the gross shape is also critical to the registration and retrieval of elongated structures. A semi-automatic solution to the extraction of a medial axis is presented, that yields a representation of the full extent of the structure, with no branches. Finally, discriminative anatomic features are computed from the results of registration and used as variables in a CBIR system. A case study on the morphology of the corpus callosum in the chromosome 22q11.2 deletion syndrome illustrates the effectiveness of the method and corroborates the hypothesis that CBIR systems may also act as knowledge discovery tools.

2   Background

The representation of elongated structures through single sequences of connected points that describe their intrinsic geometry has been extensively studied. Pioneered by Blum and Nagel [2], the use of medial axes to describe 2D shapes is based on the removal of points in the boundary until the gross shape is minimally represented. Many skeleton and thinning algorithms can be found in the literature, revealing the difficulty on determining a standard definition for medial axis [6]. Other more complex models include the medial representations [14,22], in which the medial axis and a radial scalar field are parametrically described such that the boundary can be further reconstructed, and the medial profiles [9], that provide a shape representation and deformation operators that can be used to derive shape distributions.

Registration is considered one of the most important approaches to provide detailed description of shape. Automatic registration algorithms [12,19] may be applied to the contour [3,4] or medial axis [15,7] of specific structures. Registration is also used together with the medial axis transform [21] to align the anatomy of structures based on their skeletons.

Retrieval of images based on their content is still in its infancy. Smeulders [17] and Lew [10] present comprehensive discussions on the main aspects and challenges of image retrieval. Muller [13] shows how CBIR systems can be used to retrieve images in general medical databases. In the next section, we discuss the specific issues related to the registration and retrieval of images depicting elongated structures and propose a registration algorithm that jointly considers the axis and boundaries of such structures.

3   Methods

An image retrieval solution can be divided into four steps: midline extraction, registration, description and retrieval.

3.1   Midline Extraction

The definition of a midline is not a consensus in the image processing literature. A midline can be informally defined as a curve that splits the structure into dorsal and ventral regions, such that, at any point, the perpendicular line segments connecting the midline to dorsal and ventral parts of the boundary have roughly the same length (properties of perpendicularity and congruency). The difficulty in providing a formal definition for the midline is that, for many structures, the properties of perpendicularity and congruency cannot be jointly guaranteed in all points of the representation. Furthermore, most of the alternative definitions will either fail to guarantee a representation without branches or the observation of all properties.

The process of midline extraction starts by determining a skeleton based on a variation of the thinning algorithm described by Gonzalez and Woods [8], for 8-connected objects. Object points are labeled as 1 and the background is set to 0. In order for the curve to fully extend from one extremity to the other, two object points are manually chosen and forced to be respectively the starting and ending points of the skeleton. Additionally, the thinning algorithm is modified so as to prune any other branches of the structure's skeleton. The final curve is, therefore, a single sequence of pixels, each one connected to two neighbors, with the exception of the starting and ending points.

The following algorithm summarizes skeleton extraction, where p_1 and p_2 are the endpoints; the neighbors of p are denoted as n_i, numbered counterclockwise from 0 (east) to 7 (southeast); function N returns the number of neighbors of p that belong to the object, i.e., $N(p) = \sum_i n_i$; and function S returns the number of connected sequences of object points in the neighborhood of p, i.e., read as an 8-bit string, the neighbors of p must match the regular expression $0^+1^+0^*\bigcup 1^+0^+1^*$. It can be shown that only 42 neighborhood configurations satisfy the condition to mark a point, so that the algorithm can be efficiently implemented using look-up tables:

Repeat
    For each point p of the object, $p \notin \{p_1, p_2\}$, do
        If $N(p) < 7$ and $S(p) = 1$ and $n_0 n_6 (n_2 + n_4) = 0$
            Mark p to be removed;
    Remove marked points;
    For each point p of the object, $p \notin \{p_1, p_2\}$, do
        If $N(p) < 7$ and $S(p) = 1$ and $n_2 n_4 (n_0 + n_6) = 0$
            Mark p to be removed;
    Remove marked points;
until no more points can be removed.

The linear length of the skeleton is computed considering the distances between each pair of consecutive pixels: pixels connected by a face with distance equals to 1 and the ones connected by a vertex with distance equals to $\sqrt{2}$. The coordinates of the pixels are smoothed and interpolated so as to yield an isotropic rotation-invariant representation of the midline. The derivative of this curve, taken at equidistant points, guides the computation of perpendicular segments that link the dorsal and ventral boundaries of the structure. Problems may occur in regions where the midline presents increased curvature. In this case, it may be impossible to jointly satisfy the requirements of perpendicularity and congruency for the segments. Figure 2 shows an example where two consecutive segments intersect each other as the result of increased midline curvature. A solution for this problem is to violate the property of perpendicularity so that points with increasing coordinates at the midline will be connected to points of non-decreasing coordinates at both boundaries. It is however expected that elongated structures will not frequently incur in this problem.

The curvature (second derivative) of the midline can be determined based on the k-curvature metric, that is defined in each point $p_i = (x_i, y_i)$ as the difference between the average of the derivatives at the k next points and the average of the derivatives at the k previous points (including p_i):

$$kcurv(p_i) = \frac{1}{k}\left(\sum_{j=i+1}^{i+k} d(p_j) - \sum_{j=i-k+1}^{i} d(p_j)\right), \qquad (1)$$

$$d(p_j) = \tan^{-1}(x_j - x_{j-1}, y_j - y_{j-1}).$$

Parameter k should be empirically chosen so as to provide enough smoothness. The midline curve should be extrapolated at the extremities (e.g. based on autoregression), so that the curvatures will be computed over all the midline extension. Analogously, the curvature at the dorsal and ventral boundaries should be computed at the intersection of the segments. The curvatures at the midline and boundaries will play a fundamental role as a measure of similarity during registration.

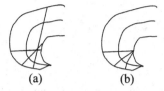

(a)   (b)

Fig. 2. Example where consecutive segments intersect each other as the result of increased midline curvature (a) and the solution to the problem (b)

3.2 Image Registration

Image registration is a key step to morphometric analysis. The images in the database should be registered to a reference in order to establish a common basis for comparison. Image registration can be stated as the process of determining a correspondence between each point p in the midline of the reference image to a point $u(p)$ in the midline of the subject image. Let $C_M(p) = kcurv(p) - kcurv(u(p))$ be the difference between the k-curvature taken at point p in the reference midline and the k-curvature taken at point $u(p)$ in the subject midline. Analogously, let C_D and C_V be the same difference function computed respectively at the intersection points of the perpendicular segments emanating from the midline with the dorsal and ventral boundaries. Image registration aims at determining a correspondence that minimizes a cost function given as

$$cost = D - S, \tag{2}$$

where D is the deformation penalty and S is the similarity between the curvatures of registered points of the midline, dorsal and ventral boundaries, given as

$$D = \alpha \int_0^1 (\frac{du(p)}{dp})^2 dp + \beta \int_0^1 (\frac{d^2u(p)}{dp^2})^2 dp,$$

$$S = \sum_{i\in\{M,D,V\}} \gamma_i \int_0^1 C_i(p)^2 dp. \tag{3}$$

Parameters α and β weight the amount and smoothness of deformation, respectively. Parameters γ_M, γ_D and γ_V are negative and weight the importance of the similarity terms computed respectively for the midline, dorsal and ventral boundaries.

Registration is performed through dynamic programming, in which equidistant points in the reference midline are mapped to points in the midlines of the database by minimizing the cost function in (2). After registering the midlines and corresponding boundaries, thin plate splines [1] are used to interpolate the warping applied to these curves to the whole structure, so that each pixel in the reference image is assigned a displacement vector.

An advantage of the proposed registration algorithm is that it will always map a segment perpendicular to the reference midline to a segment perpendicular to the midline of the subject image. This is a very important constraint to be observed when dealing with elongated structures. Figure 3 shows two examples where an image registration

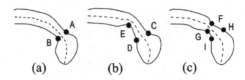

Fig. 3. Examples of unsatisfactory registration of the segment $\overline{AB}$ in the reference structure (a) to segment $\overline{CE}$ in (b) and to segment $\overline{HI}$ in (c). An algorithm based on both the boundary and midline would correctly map $\overline{AB}$ to $\overline{CD}$ and $\overline{FG}$.

algorithm based only on the boundary or only on the midline would fail to provide satisfactory deformations. The structure in (a) is the reference, whose boundary points A and B must be found correspondence in the other structures. A registration algorithm that takes into account only the boundaries would map point A to C (correctly), but B to E instead of D, since the boundary curvature in B is more similar to the curvature in E than it is in D. If, on the other hand, the algorithm is based only on the curvature of the midline, the registration of the reference to the structure in (c) would probably map the segment $\overline{AB}$ to $\overline{HI}$ instead of $\overline{FG}$, ignoring the similarity between the curvatures at the boundaries. The similarity function proposed in (3) avoid both mistakes, since the curvatures at the midline, dorsal and ventral boundaries are jointly taken into account.

Evaluating the effectiveness of registration methods is always a difficult task, as ground truth data is usually inexistent, particularly when the structure being registered does not present well-defined landmarks. Alternatively, landmarks may be chosen by experts, but in this case human subjectivity and lack of repeatability should be considered in the analysis. In this work, we designed an interactive interface in which an expert chooses a set of landmarks in the reference structure and the corresponding loci in the subjects. The procedure is repeated after 2 weeks, in order to evaluate repeatability. The results achieved by automatic registration are compared to the mapping provided by the expert: if the result falls within the interval of values provided by the expert, it is considered satisfactory, otherwise the distance in millimeters to nearest value is stored and averaged.

3.3   Description

The output of registration is a displacement field that maps each pixel of the reference image to a point in the subject. From this set of vectors, it is possible to obtain diverse measurements that describe the imaged objects, such as point-wise area and length variation, curvature of axes and contours, relationships between axes of orientation, moments and other shape descriptors. Feature selection is a fundamental step in image retrieval systems, as it is critical to the effectiveness and efficiency of many algorithms. The set of features that will represent the objects should be concise and discriminative, as distinguishing features facilitates the retrieval of relevant images, while non-relevant characteristics are confounders. Feature selection and information retrieval are synergetic steps: while the choice of distinguishing features increases the relevance of retrieval results, retrieval itself act as a "mining" tool, selecting the features that discriminate between classes of images. This is the fundamental relationship that characterizes image retrieval as a potential knowledge discovery methodology. In this work, structures are described as vectors of k-curvatures (1) taken at each matched point of the subjects, after being registered to the reference.

3.4   Image Retrieval

In a CBIR system, the user presents an image as a query, which is registered to the reference image. The features obtained from the resulting mapping function are compared to the features of the images in the database, which have been previously processed and registered to the same reference. Following a measure of similarity, the most similar images are retrieved and presented to the user.

A common assumption in image retrieval is that the images can be represented in an Euclidean space [5]. If $\mathbf{q}$ is the feature vector representing the query and $\mathbf{v}_k$ is the feature vector representation of image k in the database, the similarity between them can be computed as

$$sim(\mathbf{v}_k, \mathbf{q}) = ((\mathbf{v}_k - \mathbf{q})^T (\mathbf{v}_k - \mathbf{q}))^{1/2}.$$

The performance of an image retrieval system can be evaluated by computing two metrics [5]: The recall of the system is the ability to retrieve relevant images. It is defined as the ratio between the number of retrieved images considered relevant and the total number of relevant images in the database. The precision reflects the ability of the system to retrieve only relevant images. It is defined as the ratio between the number of retrieved images considered relevant and the total number of retrieved images. The plot of recall × precision gives an estimate of the overall effectiveness of a CBIR system, as a compromise between both performance metrics is expected.

4   A Case Study on the Morphology of the Corpus Callosum

We illustrate the proposed registration-based retrieval system with a case study on the morphology of the corpus callosum in the chromosome 22q11.2 deletion syndrome (DS22q11.2). The DS22q11.2 is an example of genetic abnormality for which many hypotheses on anatomical differences have been recently stated [11]. This syndrome is the result of a 1.5 - 3Mb microdeletion on the long arm of chromosome 22 and is characterized by a range of medical manifestations that include cardiac, palatal and immune disorders, as well as particular problems in cognitive domains associated with the orienting and executive attention systems and with numerically related processing. Recent studies have drawn particular attention to changes in the corpus callosum — the largest bundle of axons connecting the two hemispheres of the brain, as differences in the shape of this structure may indicate changes in brain connectivity that may be related to the observed cognitive impairments [16]. We hypothesized that an image retrieval system would be able to retrieve images of subjects sharing the same diagnosis, based on a shape representation of the corpus callosum, if the features used to index the images could be considered discriminative for the syndrome. In this sense, the system would reveal the most distinguishing features associated with the disease.

Participants in this study were 18 children with chromosome 22q11.2 deletion syndrome, ranging in age from 7.3 to 14.0 years (mean,S.D.=9.9,1.4 years) and 18 typically developing control children, ranging in age from 7.5 to 14.2 years (mean,S.D.=10.4,2.0

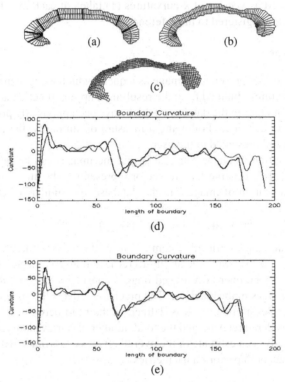

Fig. 4. An example of registration. The midline and boundary of the reference (a) is registered to the subject (b) and the result interpolated to the whole structure (c). The original plot of the boundary curvatures (d) and result of registration (e) are also shown, where the curvatures of the template and the subject are represented by thick and thin lines, respectively. The 7 landmarks used for registration evaluation, numbered from left to right, are depicted in (a) with thick lines.

years) [16]. Magnetic resonance imaging was performed on a 1.5 Tesla Siemens MAG-NETOM Vision scanner (Siemens Medical Solutions, Erlangen, Germany). For each subject, a high-resolution three-dimensional structural MRI was obtained using a T1-weighted magnetization prepared rapid gradient echo (MP-RAGE) sequence with the following parameters: repetition time (TR) = 9.7 ms, echo time (TE) = 4 ms, flip angle = 12(, number of excitations = 1, matrix size = 256x256, slice thickness = 1.0 mm, 160 sagittal slices, in-plane resolution = 1x1 mm. The midsagittal slice of each brain image volume was manually extracted as the best plane spanning the interhemispheric fissure, and on which the anterior and posterior commissures and the cerebral aqueduct were visible.

The callosa in the midsagittal images were segmented by manual thresholding and delineation. The boundaries of the callosa were automatically determined using the Rosenfeld algorithm for 8-connected contours [8]. The midlines of the callosa were also extracted based on the algorithm proposed in Section 3.1 and interpolated so as to yield an isotropic rotation-invariant representation, in which any two consecutive sampled points were 1 mm apart. The pointwise curvature of the callosum midline was computed for each subject, using the k-curvature metric (1), where k was empirically chosen to be 10% of the length of the midline, so as to provide enough smoothness.

Shape measurement was performed, by aligning a reference image of the callosum to subject callosa. One of the control subjects was arbitrarily chosen as the reference. The midline of the reference, sampled at 87 equidistant points, was registered to the subjects' midlines based on the cost function described in (2) with parameters $\alpha=0.001$, $\beta=1000.0$ and $\gamma_i=-1.0$ mm^2/degree^2 for $i \in \{M,D,V\}$, which were empirically determined. The midline curves of the subject callosa were interpolated to provide sub-pixel precision (0.5 mm). The result of registration was a mapping from each of the 87 points in the reference to corresponding points in the subjects. Registration took 7.78 seconds to compute. All methods were implemented in IDL language (Research Systems) and run in a 1.1 GHz Intel Celeron processor computer with 256 MB of RAM, under Windows XP operating system.

An example of registration is shown in Fig. 4, where the reference image described through its midline and perpendicular segments (a) is deformed to match the subject (b). The resulting deformation is shown as a warped grid (c). A plot of the original k-curvatures (in degrees/mm) at the boundaries of both images (in mm), taken counterclockwise from the leftmost endpoint of the midline, is given in (d) and the resulting registration is depicted in (e).

The effectiveness of registration was evaluated based on 2 sets of landmarks provided by an expert, taken in an interval of 2 weeks. Seven landmarks were defined at the reference, from anterior to posterior callosum (Fig. 4a), and the expert was asked to determine their corresponding loci at each of the 36 subjects. The set of 504 landmarks were compared to the results of registration. Table 1 summarizes the results, where it is possible to compare the average error of the method with the variability of measures provided by the expert, for each landmark. The average error of the method for the whole set of landmarks was 1.7 mm, a satisfactory result considering that the average variability of the expert's measures was 1.2 mm. Larger errors were observed at landmarks 3 and 4 (callosal body) where the subjects present larger variability with respect to curvature. The best results were achieved at landmarks 5 and 6 (posterior callosum) where the errors obtained with automatic registration were smaller than the average variability observed in manual registration.

The results of image retrieval were evaluated with the aid of a simple retrieval environment. Initially, the user browses the database and chooses an image that will represent the query. The system ranks the remaining images, showing the n most relevant ones to the user appraisal. In this study, we considered as relevant the images that shared the same diagnosis of the query (with or without the deletion). Following the recent findings on anatomic differences in the callosum of these populations [11](see Fig. 5), an effective CBIR would be able to retrieve images sharing the same diagnosis, unless outliers would be present in the database.

Table 1. Average error (mm) for each landmark, considering manual and automatic registration

Landmark	0	1	2	3	4	5	6
Manual	0.6	0.4	0.7	1.7	1.3	2.6	1.0
Automatic	1.0	0.9	0.8	3.7	2.4	1.6	0.9

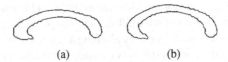

(a)   (b)

Fig. 5. Mean callosal shape for the typically developing children (a) and children with the deletion (b). Controls have shorter, more curved anterior callosum (rostrum and genu) and less curved midbody. Children with the deletion present more arched callosum (larger height/length ratio).

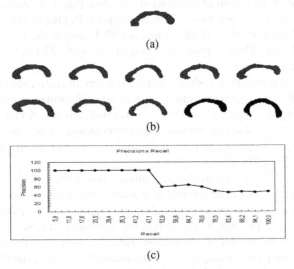

(a)

(b)

(c)

Fig. 6. Example of a query image (a) and the result of retrieval (b). The plot of recall × precision is shown in (c).

An example of the results of image retrieval is shown in Fig. 6. The query image presented by the user (a) is registered to the same reference used in the registration of the images stored in the database. The 10 images that yield greater similarity with respect to the curvature of the midline and boundary are retrieved and displayed (b). Images of controls are shown in gray and images of children with the deletion are shown in black. A plot of the recall × precision computed after the retrieval of each of the 17 relevant images in the database is presented in (d). In this case, the query is a typical control, yielding high precision.

An example in which an outlier is retrieved is given in Fig. 7. The third retrieved image is a control with arched callosum, whereas the query is a child with the deletion. In this case, the precision is affected. Worse result occurs when the query itself is an outlier, as exampled in Fig. 8. In this case, the query is a control with longer, less curved rostrum (left-most end of the midline) that is more common in children with the deletion. As a consequence, the precision is drastically affected, staying bellow 50% from the second retrieved image, a level that would be expected by pure chance.

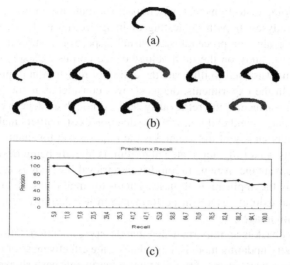

Fig. 7. Example of a query image (a) and the result of retrieval (b). In this case, the third best-ranked image is an outlier. The plot of recall × precision is shown in (c).

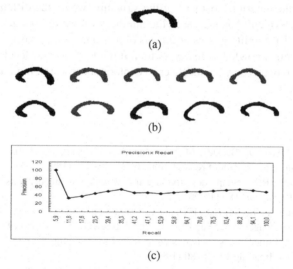

Fig. 8. Example of a query image (a) and the result of retrieval (b). In this case, the query is an outlier, yielding poor performance (c).

5 Conclusions

We have addressed the problem of registering and retrieving images of elongated structures. Traditional registration methods may yield anatomically inconsistent results while applying warping models only to the structure's contour or medial axis. The method proposed in this paper jointly registers the medial axis, dorsal and ventral boundaries, avoiding distortions that may impact substantially in the results of further morphometric analyses, hypothesis testing or image retrieval. The method deserves more systematic evaluation procedures, as visual inspection is subjective and difficult to quantify. A case study on the morphology of the corpus callosum in the 22q11.2 deletion syndrome was used to illustrate the ability of registration to provide effective image retrieval. In the experiments, diagnosis was considered as the ground truth to evaluate the performance of the retrieval system. Although evidences of shape differences between controls and children with the deletion exist, outliers make evaluation a difficult task. A deficiency of the method is the requirement for manual choice of the midline endpoints, so a fully automated algorithm is already being designed. Another well-known disadvantage of registration-driven retrieval methods is its inadequacy to indexing, limiting the application of these systems to small datasets. Furthermore, the vector model that exhibits excellent performance in text retrieval is not a consensus when dealing with images.

Relevance feedback is an important step that deserves attention. Different similarity functions and query updating models may enhance the effectiveness of image retrieval, as the user's preferences are more rapidly met. Experiments have shown that when the set of features is restricted to specific regions of interest, the precision is enhanced. In the case of the study on the corpus callosum morphometry, restricting the computation of similarity to the anterior-most part of the structure, where the differences between groups are more evident, has increased the number of retrieved images that share the same diagnosis. This ability to cluster images of the same group may qualify image retrieval as a potential knowledge discovery tool. It implements new levels of supporting environments and opens new perspectives to exploratory research in image databases.

References

1. Barrodale, I., Skea, D., Berkley, M., Kuwahara, R., Poeckert, R.: Warping digital images using thin plate splines. Pattern Recognition 26(2), 375–376 (1993)
2. Blum, H., Nagel, R.N.: Shape description using weighted symmetric axis features. Pattern Recognition 10(3), 167–180 (1978)
3. Cootes, T., Hill, A., Taylor, C.J., Haslam, J.: Use of active shape models for locating structures in medical images. Image and Vision Computing 12(6), 355–365 (1994)
4. Davatzikos, C., Prince, J.: An active contour model for mapping the cortex. IEEE Transactions on Medical Imaging 14, 65–80 (1995)
5. Del Bimbo, A.: Visual Information Retrieval. Morgan Kaufmann, San Francisco (1999)
6. Dvies, E.R., Plummer, A.P.: Thinning algorithms: a critique and a new methodology. Pattern Recognition 14, 53–63 (1981)
7. Golland, P., Grimson, W.E.L., Kikinis, R.: Statistical shape analysis using fixed topology skeletons: Corpus callosum study. In: Kuba, A., Sámal, M., Todd-Pokropek, A. (eds.) IPMI 1999. LNCS, vol. 1613, pp. 382–387. Springer, Heidelberg (1999)
8. Gonzalez, R.C., Woods, R.E.: Digital Image Processing. Prentice-Hall, Upper Saddle River (2002)
9. Hamarneh, G., Abu-Gharbieh, R., McInerney, T.: Medial profiles for modeling deformation and statistical analysis of shape and their use in medical image segmentation. International Journal of Shape Modeling 10(2), 187–209 (2004)

10. Lew, M., Sebe, N., Djeraba, C., Jain, R.: Content-based multimedia information retrieval: State of the art and challenges. ACM Transactions on Multimedia Computing, Communications, and Applications 2(1), 1–19 (2006)
11. Machado, A., Simon, T., Nguyen, V., McDonald-McGinn, D., Zackai, E., Gee, J.: Corpus callosum morphology and ventricular size in chromosome 22q11.2 deletion syndrome. Brain Research 1131, 197–210 (2007)
12. McInerney, T., Terzopoulos, D.: Deformable models in medical image analysis: A survey. Medical Image Analysis 1(2), 91–108 (1996)
13. Muller, H., Michoux, N., Bandon, D., Geissbuhler, A.: A review of content-based image retrieval systems in medical applications-clinical benefits and future directions. International Journal of Medical Informatics 73(1), 1–23 (2004)
14. Pizer, S.M., Fletcher, P.T., Joshi, S., Thall, A., Chen, J.Z., Fritsch, D.S., Gash, A.G., Glotzer, J.M., Jiroutek, M.R., Lu, C., Muller, K.E., Tracton, G., Yushkevich, Chaney, E.: Deformable m-reps for 3D medical image segmentation. International Journal of Computer Vision 55(2), 85–106 (2003)
15. Pizer, S.M., Fritsch, D.S., Yushkevich, P., Johnson, V., Chaney, E.: Segmentation, registration and measurement of shape variation via image object shape. IEEE Transactions on Medical Imaging 18(10), 851–865 (1996)
16. Simon, T.J., Ding, L., Bish, J.P., McDonald-McGinn, D.M., Zackai, E.H., Gee, J.C.: Volumetric, connective, and morphologic changes in the brains of children with chromosome 22q11.2 deletion syndrome: an integrative study. Neuroimage 25, 169–180 (2005)
17. Smeulders, A., Worring, M., Santini, S., Gupta, A., Jain, R.: Content-based image retrieval at the end of the early years. IEEE Transactions on Pattern Analysis and Machine Intelligence 22(12), 1349–1380 (2000)
18. Staal, J.: Segmentation of elongated structures in medical imaging. PrintPartners Ipskamp, Enschede (2004)
19. Toga, A.W.: Brain Warping. Academic Press, New York (1999)
20. Toledo, R., Orriols, X., Binefa, X., Radeva, P., Vitria, J., Villanueva, J.: Tracking of elongated structures using statistical snakes. In: Proceedings of the CVPR, Hilton Head Island, pp. 157–162 (2000)
21. Xie, J., Heng, P.A.: Shape modeling using automatic landmarking. In: Duncan, J.S., Gerig, G. (eds.) MICCAI 2005. LNCS, vol. 3750, pp. 709–716. Springer, Heidelberg (2005)
22. Yushkevich, P., Fletcher, P.T., Joshi, S., Thall, A., Pizer, S.: Continuous medial representations for geometric object modeling in 2d and 3d. Image and Vision Computing 21(1), 17–28 (2003)

Cardiovascular Response Identification Based on Nonlinear Support Vector Regression

Lu Wang[1], Steven W. Su[1,2], Gregory S.H. Chan[1], Branko G. Celler[1,*],
Teddy M. Cheng[1], and Andrey V. Savkin[1]

[1] Biomedical System Lab, School of Electrical Engineering & Telecommunications, Faculty of Engineering, University of New South Wales, UNSW Sydney, N.S.W. 2052, Australia
b.celler@unsw.edu.au,
{LuWang,Steven.Su,Gregory.Chan,B.Celler,T.Cheng,
A.Savkin}@unsw.edu.au
[2] Key University Research Centre for Health Technologies, Faculty of Engineering, University of Technology, Sydney, Broadway, NSW 2007, Australia

Abstract. This study experimentally investigates the relationships between central cardiovascular variables and oxygen uptake based on nonlinear analysis and modeling. Ten healthy subjects were studied using cycle-ergometry exercise tests with constant workloads ranging from 25 Watt to 125 Watt. Breath by breath gas exchange, heart rate, cardiac output, stroke volume and blood pressure were measured at each stage. The modeling results proved that the nonlinear modeling method (Support Vector Regression) outperforms traditional regression method (reducing Estimation Error between 59% and 80%, reducing Testing Error between 53% and 72%) and is the ideal approach in the modeling of physiological data, especially with small training data set.

Keywords: Cardiovascular system, Nonlinear modeling, Cardiovascular responses to Exercise, Machine learning.

1 Introduction

The relationships between central cardiovascular variables and oxygen uptake during steady state of graded exercise have been widely examined by numerous investigators [1] [2] [3] [4] [5] [6] [7] [8] [9]. Most of them investigated the relationship between cardiac output (CO) and oxygen uptake ($\dot{V}O_2$) using linear regression methods and found the slope between the two variables to be approximately $5-6$ in normal and athletic subjects [10]. Beck et al [11] in contrast, investigated this relationship in healthy humans using polynomial regression. Turley [9] described both the relationship of stroke volume (SV) and the total peripheral resistance (TPR) to oxygen uptake during steady state of sub-maximal exercise using linear regression. However, from the point view of modeling, the regression methods used by the previous researchers have several limitations. First the empirical risk minimization (ERM) principle used by traditional regression models does not guarantee good generalization performance and may

---

* Corresponding author.

A. Fred, J. Filipe, and H. Gamboa (Eds.): BIOSTEC 2008, CCIS 25, pp. 202–213, 2008.

produce models that over-fit the data [12]. Secondly, most of the regression models developed from early research based on a small sample set with limited subjects during three or four exercise intensities. Traditional regression approaches are particularly not recommended for modeling small training sets. Determination of the size of the training set is a main issue to be solved in the modeling performance because the sufficiency and efficiency of the training set is one of the most important factors to be considered.

This study presents a novel machine learning approach, Support Vector Regression (SVR) [13] to model the central cardiovascular response to exercise. SVR, developed by Vapnik and his co-workers in 1995, has been widely applied in forecasting and regression [14] [15] [16] [17]. The following characteristics of SVR make it an ideal approach in modeling of cardiovascular system. Firstly, SVR avoids the over-fitting problem which exists in the traditional modeling approaches. Second, SVR condenses information in the training data and provide a sparse representation by using a small number of data points [18]. Thirdly, SVR is insensitive to modeling assumption due to its being a non-parametric model structure. Finally, the SVR model is unique and globally optimal, unlike traditional training which can risk converging to local minima.

The rest of this paper is organized as follows: Section 2 describes the experimental design for the data collection. Section 3 applies SVR for modeling the relationships between central cardiovascular variables and oxygen uptake. Finally, some conclusions are drawn in Section 4.

2 Experimental Design

2.1 Subjects

We studied 12 normal male subjects. They are all active, but do not participate in formal training or organized sports. However, since two of them could not complete 6 minutes of higher level exercise, only the data recorded from 10 subjects (aged 25 ± 4yr, height 177 ± 5cm, body weight 73 ± 11kg) are used for this study. All the subjects knew the protocol and the potential risks, and had given their informed consent.

2.2 Experimental Procedure

All tests were conducted in the afternoon in an air-conditioned laboratory with temperature maintained between 23-24 °C. The subjects were studied during rest and a series of exercise in an upright position on an electronically braked cycle ergometer. Exercise was maintained at a constant workload for 6 minutes, followed by a period of rest. The initial exercise level was 25W and each successive stint of exercise was increased in 25W steps until a workload of 125W was reached. The rest periods were increased progressively from 10 to 30 minutes after each stint of exercise. Six minutes of exercise was long enough to approach a steady state since the values of oxygen uptake and the A-V oxygen difference had become stable by the 5th and 6th minutes even for near maximum exertion [19].

2.3 Measurement and Data Processing

Heart rate was monitored beat by beat using a single lead ECG instrument, while ventilation and pulmonary exchange were measured on a breath by breath basis.

Minute ventilation was measured during inspiration using a Turbine Flow Transducer model K520-C521 (Applied Electrochemistry, USA). Pulmonary gas exchange was measured using S-3A and CD-3A gas analyzers (Applied Electrochemistry, USA). Before each individual exercise test, the turbine flow meter was calibrated using a 3.0 liters calibration syringe. Before and after each test, the gas analyzers were calibrated using reference gases with known O_2 and CO_2 concentrations. The outputs of the ECG, the flow transducer and the gas analyzers were interfaced to a laptop through an A/D converter (NI DAQ 6062E) with a sampling rate of 500 Hz. Programs were developed in Labview 7.0 for breath by breath determination of pulmonary gas exchange variables but with particular reference to $\dot{V}O_2$ ($\dot{V}O_2$ STPD). Beat by beat stroke volume and cardiac outputs were measured noninvasively using the ultrasound based device (USCOM, Sydney, Australia) at the ascending aorta. This device has previously been reported to be both accurate and reproducible [20]. In order to keep consistent measurements, all CO/SV measurements were conducted by the same person. An oscillometric blood pressure measurement device (CBM-700, Colin, France) was used to measure blood pressure.

The measurement of $\dot{V}O_2$ and HR were conducted during the whole exercise and recovery stage. The static values ($\dot{V}O_2$ and HR) were calculated for each workload from data collected in the last minute of the six minute exercise protocol. The measurements of SV, CO and BP (blood pressure) were similarly conducted during the last minute of the six minute exercise for each workload with the additional requirement that subjects keep their upper body as still as possible to minimize artifacts caused by the movement of the chest during exercise. We then, calculated their static values (CO, SV and BP) based on the measurement in the last minute for each workload.

2.4   Results

We found that the percentage changes of cardiovascular variables relative to their rest values more uniform than when absolute values are used. This may be because using relative values diminish the variability between subjects.

Based on the above finding, we model CO, SV and TPR to $\dot{V}O_2$ by modeling the percentage changes in CO, SV and TPR with respect to their corresponding rest values to percentage change in $\dot{V}O_2$ with respect to its rest value. We use CO%, SV%, TPR% and $\dot{V}O_2$ % to represent their relative values (expressed as percentage), respectively.

3   Application of SVR for Modeling

We selected radial basic function (RBF) kernels for this study, that is $K(x, x_i) = \exp(-\dfrac{\left\| x - x_i \right\|^2}{2\sigma^2})$ where σ is the kernel parameter, x_i is the ith input support value and x is the input value.

Detailed discussion about SVR, such as the selection of regularization constant C, radius ε of the tube and kernel function, can be found in [12] [21]. In order to show the effectiveness of SVR, we applied both SVR and traditional linear regression (Least-Square linear regression (LS)) to investigate the relationships between percentage change of cardiovascular variables (CO%, SV% and TPR%) and $\dot{V}O_2\%$.

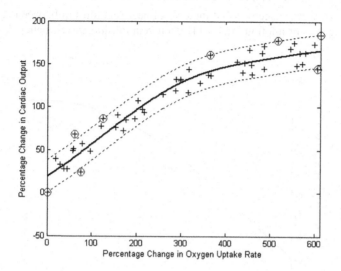

(a) Estimation of percentage change in CO from percentage change in $\dot{V}O_2$ using SVR

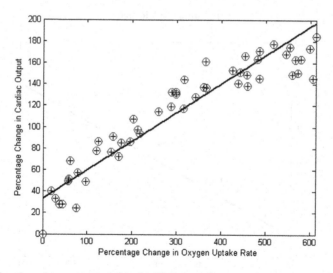

(b) Estimation of percentage change in CO from percentage change in $\dot{V}O_2$ using LS

Fig. 1. Comparison of estimation results of CO% between using SVR and using LS

3.1 The Relationship between CO% and $\dot{V}O_2$%

3.1.1 Model Identification

A SVR model was developed to estimate CO% from $\dot{V}O_2$% (Table 1 and Fig. 1). Although it is widely accepted that there is a linear relationship between cardiac output and oxygen consumption [1] [2] [4], their relationship can be better described by

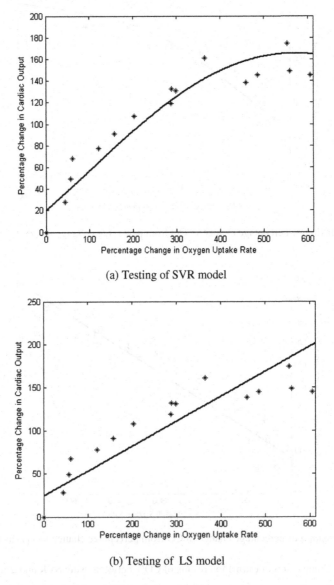

(a) Testing of SVR model

(b) Testing of LS model

Fig. 2. Comparison of models of CO% against % change in oxygen uptake using SVR and using LS methods

the nonlinear SVR model in terms of reducing the errors (MSE) from 418 to 171 (Table 2), an improvement of 59% comparing with that of LS method. The results in Table 1 also show the efficiency of SVR. Unlike traditional regression method where the solution of the model depends on the whole training data points, in SVR, the solution to the problem is only dependent on a subset of training data points which are referred to as support vectors. Using only support vectors, the same solution can be obtained as using all the training data points. SVR uses just 13% of the total points available to model their nonlinear behavior efficiently.

Table 1. Fitting data for the model of cardiovascular variables and oxygen uptake rate using SVR

Relation	CO% vs $\dot{V}O_2$%	SV% vs $\dot{V}O_2$%	TPR% vs $\dot{V}O_2$%
Kernel	RBF	RBF	RBF
Parameter	$\sigma = 200$	$\sigma = 500$	$\sigma = 500$
Regularization Constant C	5000	5000	5000
ε-insensitivity	19	3	8
Support vector number	8 (13.3%)	8 (13.3%)	8 (13.3%)
Estimation error	171	5	30

Table 2. Comparison of the estimation errors (MSE) between using SVR and using linear regression method

Relation	CO% vs $\dot{V}O_2$%	SV% vs $\dot{V}O_2$%	TPR% vs $\dot{V}O_2$%
SVR	171	5	30
LS	418	15	151

3.1.2 Model Validation

To further evaluate the feasibility of this proposed SVR model, the whole data set is divided into two parts: the first part (70% of the data) is used to design the model and the second part (30% of the data) is used to test its performance. Because we do not have large sample of data, we separated the data set into two parts randomly five

Table 3. Comparison of the model fitting errors (MSE) using SVR and linear regression methods (N=5)

Relation	CO% vs $\dot{V}O_2$%	SV% vs $\dot{V}O_2$%	TPR% vs $\dot{V}O_2$%
SVR testing error	245 ± 15	8 ± 2	36 ± 5
LS Testing error	521 ± 19	22 ± 7	130 ± 12

times. Each time we use 70% of the data for training and the rest for testing. We established the SVR model with the three design parameters (kernel function, capacity (C) and the radius of insensitivity (ε) based on the training set, and test its goodness on the testing set. In Fig. 2, we present the results for one of the 5 tests. As shown in Table 3, the averaged results (MSE) for the 5 times testing for SVR is 245±15. However, the averaged error for traditional linear regression is as high as 521±19. It indicates that SVR can build more robust models to predict CO% from

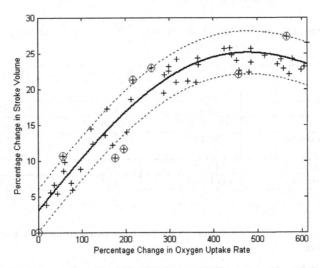

(a) Estimation of percentage change in SV from percentage change in $\dot{V}O_2$ using SVR

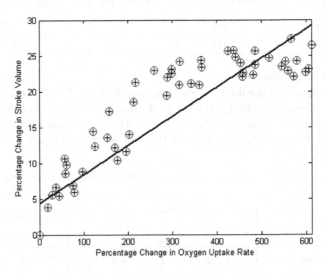

(b) Estimation of percentage change in SV with percentage change in $\dot{V}O_2$ using linear regression

Fig. 3. Comparison of estimation results for SV% between using SVR and using LS

$\dot{V}O_2\%$ using only a small training set. It also demonstrates that SVR can overcome the over-fitting problem, even though SVR has more model parameters than the traditional linear regression method.

3.2 The Relationship between SV% and $\dot{V}_{O_2}\%$

Fig. 3 shows the models for estimating SV%. The SVR model gives more precisely estimation than the LS does and decreases estimation errors (MSE) by 67% (Table 2).

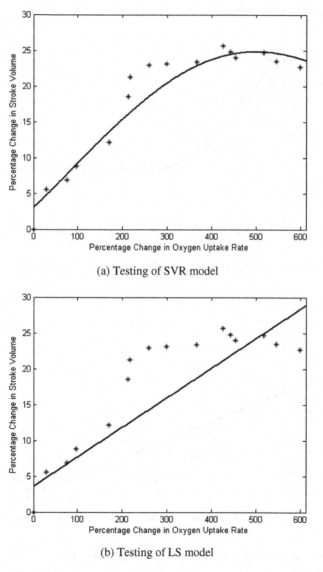

(a) Testing of SVR model

(b) Testing of LS model

Fig. 4. Comparison of the testing results for Stroke Volume using SVR and using traditional linear regression

The testing models are given in Fig. 4 and the testing errors are in Table 3. As indicated, the SVR model decreases the testing error by 64%.

3.3 The Relationship between TPR% and $\dot{V}O_2$%

As shown in Fig. 5, the SVR model describes a rapid fall in TPR% at low workloads which remains relatively constant even with increasing $\dot{V}O_2$%. SVR uses just 13%

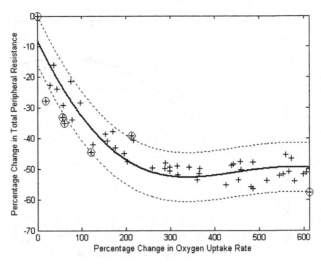

(a) Estimation of percentage change in TPR from percentage change in $\dot{V}O_2$ using SVR

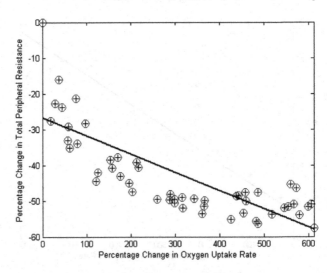

(b) Estimation of percentage change in TPR with percentage change in $\dot{V}O_2$ using linear regression

Fig. 5. Comparison of the estimation results of TPR% between using SVR and LS

(Table 1) of the total points to get an efficient nonlinear model. Compared with linear regression, the SVR model decreases MSE from 151 to 30, an improvement of 80%.

The testing results for this SVR model and the equivalent LS model are given in Fig. 6 and Table 3, respectively. Both of these (Fig. 6 and Table 3) demonstrate that SVR outperforms the traditional linear regression method by reducing testing errors significantly, from 130 to 36.

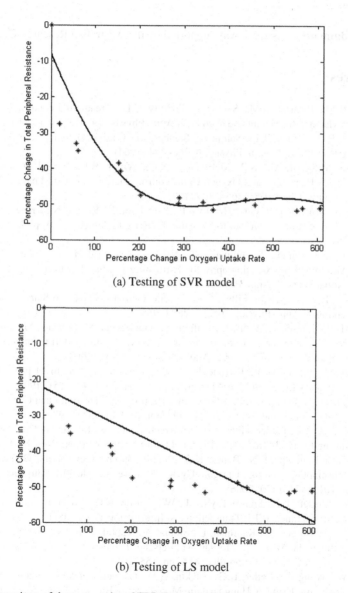

(a) Testing of SVR model

(b) Testing of LS model

Fig. 6. Comparison of the test results of TPR% against % change on Oxygen uptake using SVR and using LS

4 Conclusions

This is the first time that SVR has been applied to experimentally investigate the steady state relationships between key central cardiovascular variables and oxygen consumption during incremental exercise. The impressive results obtained prove that SVR is an effective approach that can be recommended for the modeling of physiological data.

Acknowledgments. This work was supported by the Australian Research Council.

References

1. Allor, K.M., Pivarnik, J.M., Sam, L.J., Perkins, C.D.: Treadmill Economy in Girls and Women Matched for Height and Weight. J. Appl. Physiol. 89, 512–516 (2000)
2. Astrand, P.O., Cuddy, T.E., Saltin, B., Stenberg, J.: Cardiac Output During Submaximal and Maximal Work. J. Appl. Physiol. 9, 268–274 (1964)
3. Fairbarn, M.S., Blackie, S.P., McElvaney, N.G., Wiggs, B.R., Pare, P.D., Pardy, R.L.: Prediction of Heart Rate and Oxygen Uptake during Incremental and Maximal Exercise in Healthy Adults. Chest 105, 1365–1369 (1994)
4. Freedman, M.E., Snider, G.L., Brostoff, P., Kimelblot, S., Katz, L.N.: Effects of Training on Response of Cardiac Output to Muscular Exercise in Athletes. J. Appl. Physiol. 8, 37–47 (1955)
5. Kobayashi, Y., Andoh, Y., Fujinami, T., Nakayama, K., Takada, K., Takeuchi, T., Okamoto, M.: Impedance Cardiography for Estimating Cardiac Output during Submaximal and Maximal Work. J. Appl. Physiol. 45, 459–462 (1978)
6. Reeves, J.T., Grover, R.F., Filley, G.F.: Cardiac Output Response to Standing and Treadmill Walking. J. Appl. Physiol. 16, 283–288 (1961)
7. Richard, R., Lonsdorfer-Wolf, E., Dufour, S., Doutreleau, S., Oswald- Mammosser, M., Billat, V.L., Lonsdorfer, J.: Cardiac Output and Oxygen Release during Intensity Exercise Performed until Exhaustion. Eur. J. Appl. Physiol. 93, 9–18 (2004)
8. Rowland, T., Popowski, B., Ferrone, L.: Cardiac Response to Maximal Upright Cycle Exercise in Healthy Boys and Men. Med. Sci. Sport Exer. 29, 1146–1151 (1997)
9. Turley, K.R., Wilmore, J.H.: Cardiovascular Responses to Treadmill and Cycle Ergometer Exercise in Children and Adults. J. Appl. Physiol. 83, 948–957 (1997)
10. Rowell, L.B.: Circulatory adjustments to dynamic exercise. In: Human Circulation Regulation during Physical Stress, pp. 213–256. Oxford University Press, New York (1986)
11. Beck, K.C., Randolph, L.N., Bailey, K.R.: Relationship between Cardiac Output and Oxygen Consumption during Upright Cycle Exercise in Healthy Humans. J. Appl. Physiol. 101, 1474–1480 (2006)
12. Guo, Y., Bartlett, P.L., Shawe-Taylor, J., Williamson, R.C.: Covering Numbers for Support Vector Machines. IEEE Transactions on Information Theory 48(1), 239–250 (2002)
13. Drucker, H., Burges, C., Kaufman, L., Smola, A., Vapnik, V.N.: Support Vector Regression Machines. In: Mozer, M., Jordan, M., Petsche, T. (eds.) Advances in Neural Information Procession Systems, Cambridge, MA, pp. 155–161 (1997)
14. Su, S.W., Wang, L., Celler, B.G., Savkin, A.V.: Oxygen Uptake Estimation in Humans during Exercise Using a Hammerstein Model. Ann. Biomed. Eng. 35(11), 1898–1906 (2007)

15. Su, S.W., Wang, L., Celler, B.G., Savkin, A.V., Guo, Y.: Identification and Control for Heart Rate Regulation during Treadmill Exercise. IEEE Transactions on Biomedical Engineering 54(7), 1238–1246 (2007)
16. Su, S.W., Wang, L., Celler, B., Savkin, A.V.: Estimation of Walking Energy Expenditure by Using Support Vector Regression. In: Proceedings of the 27th Annual International Conference of the IEEE Engineering in Medicine and Biology Society (EMBS), Shanghai, China, pp. 3526–3529 (2005)
17. Valerity, V.G., Supriya, B.G.: Volatility Forecasting from Multiscale and High-dimensional Market Data. Neurocomputing 55, 285–305 (2003)
18. Girosi, F.: An Equivalence between Sparse Approximation and Support Vector Machines. Neural Computation 20, 1455–1480 (1998)
19. Reeves, J.T., Grover, R.F., Filley, G.F., Blount Jr., S.G.: Circulatory Changes in Man during Mild Supine Exercise. J. Appl. Phsiol. 16, 279–282 (1961)
20. Knobloch, K., Lichtenberg, A., Winterhalter, M., Rossner, D., Pichlmaier, M., Philips, R.: Non-invasive Cardiac Output Determination by Two-dimensional Independent Doppler during and after Cardiac Surgery. Ann. Thorac. Surg. 80, 1479–1483 (2005)
21. Vapnik, V.: Statistical Learning Theory. Wiley, New York (1998)

Vessel Cross-Sectional Diameter Measurement on Color Retinal Image

Alauddin Bhuiyan, Baikunth Nath, Joey Chua, and Ramamohanarao Kotagiri

Department of Computer Science and Software Engineering
NICTA Victoria Research Laboratory, The University of Melbourne, Australia 3010
{bhuiyanm,bnath,jjchua,rao}@csse.unimelb.edu.au

Abstract. Vessel cross-sectional diameter is an important feature for analyzing retinal vascular changes. In automated retinal image analysis, the measurement of vascular width is a complex process as most of the vessels are few pixels wide or suffering from lack of contrast. In this paper, we propose a new method to measure the retinal blood vessel diameter which can be used to detect arteriolar narrowing, arteriovenous (AV) nicking, branching coefficients, etc. to diagnose various diseases. The proposed method utilizes the vessel centerline and edge information to measure the width for a vessel cross-section. Using the Adaptive Region Growing (ARG) segmentation technique we obtain the edges of the blood vessels, and then applying the unsupervised texture classification method we segment the blood vessels from where the vessel centerline is obtained. The potential pixels pairs for each centerline pixel are obtained from the edge image that pass through this centerline pixel. We apply a rotational invariant mask to search the pixel pairs from the edge image, and calculate the shortest distance pair which provides the vessel width (or diameter) for that cross-section. The method is evaluated with manually measured width for different vessels' cross-sectional area. For the automated measurement of vascular width we achieve an average accuracy of 95.8%.

Keywords: Vessel Cross-section, Gradient Operator, Adaptive Region Growing Technique, Texture Classification, Gabor Energy Filter Bank, Fuzzy C-Means Clustering.

1 Introduction

Accurate measurement of retinal vessel diameter is an important part in the diagnosis of many diseases. A variety of morphological changes occur to retinal vessels in different disease conditions. The change in width of retinal vessels within the fundus image is believed to be indicative of the risk level of diabetic retinopathy; venous beading (unusual variations in diameter along a vein) is one of the most powerful predictor of proliferate diabetic retinopathy. Generalized and focal retinal arteriolar narrowing and arteriovenous nicking have been shown to be strongly associated with current and past hypertension reflecting the transient and persistent structural effects of elevated blood pressure on the retinal vascular network. In addition, retinal arteriolar bifurcation diameter exponents have been shown to change significantly in patients with peripheral

A. Fred, J. Filipe, and H. Gamboa (Eds.): BIOSTEC 2008, CCIS 25, pp. 214–227, 2008.

vascular disease and arteriosclerosis, and a variety of retinal microvascular abnormalities have been shown which are related to the risk of stroke [12]. Therefore, an accurate measurement of vessel diameter and its geometry is necessary for effective diagnosis of such diseases.

The measurement of the vascular diameter is critical and a challenging task whose accuracy depends on the accuracy of the segmentation method. The study of vessel diameter measurement is still an open area for improvement. Zhou et al. [10] have applied a model-based approach for tracking and to estimating widths of retinal vessels. Their model assumes that image intensity as a function of distance across the vessel displays a single Gaussian form. However, high resolution fundus photographs often display a central light reflex [11]. Intensity distribution curves is not always of single Gaussian form, so that using a single Gaussian model for simulating intensity profile of vessel could produce poor fit and subsequently provide inaccurate diameter estimation [9]. Gao et al. [9] model the intensity profiles over vessel cross section using twin Gaussian functions to acquire vessel width. This technique may produce poor results in case of minor vessels where the contrast is less. Lowell et al. [12] have proposed an algorithm based on fitting a local 2D vessel model, which can measure vascular width to an accuracy of about one third of a pixel. However, the technique is biased on smooth data (image) and suffers from measuring the width of minor vessels where the contrast is poor.

The rest of the paper is organized as follows: Section 2 introduces the proposed method for measuring the blood vessel width. Edge based blood vessel segmentation technique is described in section 3. Section 4 details the vessel centerline detection procedure. The vessel width measurement method is described in section 5. The experimental results on vessel width measurement accuracy are provided in section 6 and finally the conclusion and future research directions are drawn in section 7.

2 Proposed Method

We propose the blood vessels' width measurement algorithm based on the vessel edge and centerline. The major advantage of our technique is that it is less sensitive to noise and works equally for the low contrast vessels (particularly for minor vessels). Another advantage of our technique is that it can calculate the vessel width even when it is one pixel wide. We adopt segmented images that are produced from the original RGB image. At first, we apply the ARG segmentation technique on the gradient image to obtain the vessel edges, then we apply the unsupervised texture classification method to segment the blood vessels from where we obtain the vessel centerline. We map the vessel centerline image and pick any of the vessel centerline pixel. For that particular pixel we apply a rotational invariant mask whose center is that pixels position and searches the potential pixels from the edge image using a continuous increment of lower to higher distance and orientation. For each case, if the gray scale value of that pixel position is 255 or white it finds the mirror of this pixel by searching through a fixed angle (exactly incrementing 180 degree) but in variable distance. This is to give the flexibility and consistency to our method as the centerline pixels may not be in the exact position of vessel center. In this way, we can obtain all the potential pairs (line end

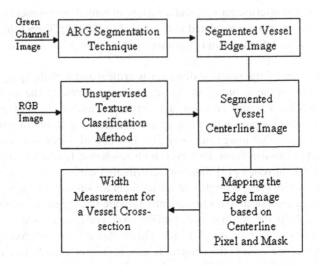

Fig. 1. The overall system for measuring blood vessel width

points) which pass through that centerline pixel. From those pairs we calculate the minimum distance/length pair which is the width of that cross-section of the blood vessel. Figure 1 depicts the overall technique of our proposed method.

3   Vessel Edge Detection

We implemented the vessel segmentation technique based on vessel edges. In the following subsections, we provide a brief illustration of this method.

3.1   Preprocessing and Conversion of Retinal Image

Adaptive Histogram Equalization (AHE) method is implemented, using MATLAB, to enhance the contrast of the image intensity by transforming the values using contrast-limited adaptive histogram equalization (Fig. 2). The enhanced retinal image is converted into gradient image (Fig. 2) using first order partial differential operator. The gradient of an image $f(x, y)$ at location (x, y) is defined as the two dimensional vector [8]

$$G[f(x,y)] = [G_x, G_y] = \left[\frac{\partial f}{\partial x}, \frac{\partial f}{\partial y}\right] \tag{1}$$

For edge detection, we are interested in the magnitude $M(x, y)$ and direction $\alpha(x, y)$ of the vector $G[f(x, y)]$, generally referred to simply as the gradient and and commonly takes the value of

$$M(x,y) = mag(G[f(x,y)]) \approx |G_x| + |G_y| \\ \alpha(x,y) = \tan^{-1}(G_y/G_x) \tag{2}$$

where the angle is measured with respect to the x axis.

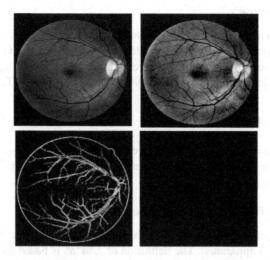

Fig. 2. Original retinal image, its Adaptive Histogram Equalized image (top; left to right), the Gradient Image and final ARG output image (bottom; left to right)

3.2   Adaptive Region Growing Technique

The edges of vessels are segmented using region growing procedure [7] that groups pixels or sub regions into larger regions based on gradient magnitude. As the gradient magnitude is not constant for the whole vessel we need to consider an adaptive gradient value that gradually increases or decreases to append the pixel to a region. We call it an adaptive procedure, as the difference of neighboring pixels intensity value is always adapted for the region growing process. The region growing process starts with appending the pixels that pass certain threshold value. For region growing we find the intensity difference between a pixel belonging to a region and its neighboring potential region growing pixels. The pixel is considered for appending in that region if the difference is less than a threshold value. The threshold value is calculated by considering the maximum differential gradient magnitude for any neighboring pixels with equal (approximately) gradient direction. Region growing stops when no more pixels satisfy the criteria for inclusion in that region.

Table 1. Measuring the detection accuracy blood vessel detection

Image number	Total number of vessels	Number of detected vessels	Accuracy (%)	Overall accuracy (%)
Image 1	93	92	98.92	
Image 2	74	70	94.59	
Image 3	85	79	92.94	94.98
Image 4	81	76	93.82	
Image 5	75	71	94.66	

3.3 Detection Accuracy

Using DRIVE database [6] we applied our technique on five images for initial assessment. To evaluate performance of our procedure we employed an expert to determine the number of vessels in the original image and detected output image (Fig. 2). We achieve an overall detection accuracy of 94.98% which is shown on Table 1.

4 Vessel Centerline Detection

We implement the unsupervised texture classification based vessel segmentation method from which we detect the vessel centerline. The steps in the method are described in the following subsections.

4.1 Color Space Transformation and Preprocessing

Generally image data is given in RGB space (because of the availability of data produced by the camera apparatus). The definition of $L^*a^*b^*$ is based on an intermediate system, known as the CIE XYZ space (ITU-Rec 709). This space is derived from RGB as follows [3],[2].

$$
\begin{aligned}
X &= 0.412453R + 0.357580G + 0.180423B \\
Y &= 0.212671R + 0.715160G + 0.072169B \\
Z &= 0.019334R + 0.119193G + 0.950227B
\end{aligned}
\tag{3}
$$

$L^*a^*b^*$ color space is defined as follows:

$$
\begin{aligned}
L^* &= 116f(Y/Y_n) - 16 \\
a^* &= 500[f(X/X_n) - f(Y/Y_n)] \\
b^* &= 200[f(Y/Y_n)] - f(Z/Z_n)
\end{aligned}
\tag{4}
$$

where $f(q) = q^{1/3}$ if $q < 0.008856$ and is constant $7.87+16/116$ otherwise. X_n, Y_n and Z_n represent a reference white as defined by a CIE standard illuminant, D_{65} in this case. This is obtained by setting $R = G = B = 100$ in (1), $q \in \{X/X_n, Y/Y_n, Z/Z_n\}$.

Gaussian color model can also be well approximated by the RGB values. The first three components $\hat{E}$, $\hat{E}_\lambda$ and $\hat{E}_{\lambda\lambda}$ of the Gaussian color model (Taylor expansion of the Gaussian weighted spectral energy distribution at Gaussian central wavelength and scale) can be approximated from the CIE 1964 XYZ basis (Fig. 3) when taking $\lambda_0 = 520nm$ (Gaussian central wavelength) and $\sigma_\lambda = 55nm$ (scale) [3]. The Adaptive Histogram Equalization method was implemented, using MATLAB, to enhance the contrast of the image intensity.

4.2 Texture Feature Extraction

Texture generally describes second order property of surfaces and scenes, measured over image intensities. A Gabor filter has weak responses along all orientations on the smooth (background) surface. However, when positioned on a linear pattern object (like a vessel) it produces relatively large differences in its responses when the orientation parameter changes [13]. Hence, the use of Gabor filters to analyze the texture of the retinal images is very promising.

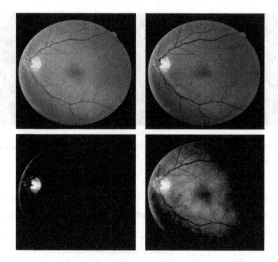

Fig. 3. Original RGB and its Green channel image (top; left to right), Gaussian transformed first and second component image (bottom; left to right)

Gabor Filter. An input image $I(x, y)$, $(x, y) \in \Omega$ where Ω is the set of image points, is convolved with a 2D Gabor function $g(x, y)$, $(x, y) \in \omega$, to obtain a Gabor feature image $r(x, y)$ (Gabor filter response) as follows [4]

$$r(x, y) = \iint\limits_{\Omega} I(\xi, \eta) g(x - \xi, y - \eta) d\xi d\eta \tag{5}$$

We use the following family of 2D Gabor functions to model the spatial summation properties of an image [4]

$$g_{\xi, \eta, \lambda, \Theta, \phi}(x, y) = \exp(-\frac{x'^2 + \gamma^2 y'^2}{2\sigma^2}) \cos(2\pi \frac{x'}{\lambda} + \phi)$$

$$x' = (x - \xi) \cos \Theta - (y - \eta) \sin \Theta$$
$$y' = (x - \xi) \cos \Theta - (y - \eta) \sin \Theta \tag{6}$$

where the arguments x and y specify the position of a light impulse in the visual field and $\xi, \eta, \sigma, \gamma, \lambda, \Theta, \phi$ are parameters. The pair (ξ, η) specifies the center of a receptive field in image coordinates. The standard deviation σ of the Gaussian factor determines the size of the receptive filed. Its eccentricity is determined by the parameter γ called the spatial aspect ratio. The parameter λ is the wavelength of the cosine factor which determines the preferred spatial frequency $\frac{1}{\lambda}$ of the receptive field function $g_{\xi, \eta, \lambda, \Theta, \phi}(x, y)$. The parameter Θ specifies the orientation of the normal to the parallel excitatory and inhibitory stripe zones - this normal is the axis x' in (5). Finally, the parameter $\phi \in (-\pi, \pi)$, which is a phase offset argument of the harmonic factor $\cos(2\pi \frac{x'}{\lambda} + \phi)$, determines the symmetry of the function $g_{\xi, \eta, \lambda, \Theta, \phi}(x, y)$.

Gabor Energy Features. A set of textures was obtained based on the use of Gabor filters (6) according to a multichannel filtering scheme. For this purpose, each image

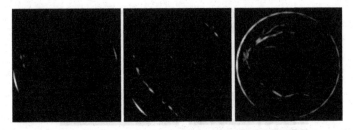

Fig. 4. Texture analyzed image with the orientations of 15, 45 degrees and maximum response of all twenty-four orientations (left to right)

was filtered with a set of Gabor filters with different preferred orientation, spatial frequencies and phases. The filter results of the phase pairs were combined, yielding the Gabor energy quantity [4]:

$$E_{\xi,\eta,\Theta,\lambda} = \sqrt{r^2_{\xi,\eta,\Theta,\lambda,0} + r^2_{\xi,\eta,\Theta,\lambda,\pi/2}} \qquad (7)$$

where $r^2_{\xi,\eta,\Theta,\lambda,0}$ and $r^2_{\xi,\eta,\Theta,\lambda,\pi/2}$ are the outputs of the symmetric and antisymmetric filters. We used Gabor energy filters with twenty-four equidistant preferred orientations ($\Theta = 0, 15, 30, .., 345$) and three preferred spatial frequencies ($\lambda = 6, 7, 8$). In this way an appropriate coverage was performed of the spatial frequency domain.

We considered the maximum response value per pixel on each color channel to reduce the feature vector length and complexity of training on data for the classifier. In addition, we constructed an image (Fig. 4) on each color channel which was used for histogram analysis to determine the cluster number. From these images we constructed a twelve element length feature vector for each pixel in each retinal image for classifying them into vessel and non-vessel using the Fuzzy C-Means (FCM) clustering algorithm.

4.3   Texture Classification and Image Segmentation

he FCM is a data clustering technique wherein each data point belongs to a cluster to some degree that is specified by a membership grade. Let $X = x_1, x_2, , x_N$ where $x \in R^N$ present a given set of feature data. The objective of the FCM clustering algorithm is to minimize the Fuzzy C-Means cost function formulated as [5]

$$J(U,V) = \sum_{j=1}^{C} \sum_{i=1}^{N} (\mu_{ij})^m \|x_i - v_j\|^2 \qquad (8)$$

$V = \{v_1, v_2, , v_C\}$ are the cluster centers. $U = (\mu_{ij})_{N \times C}$ is fuzzy partition matrix, in which each member is between the data vector x_i and the cluster j. The values of matrix U satisfy the following conditions:

$$\mu_{ij} \in [0,1], i = 1, .., N, j = 1, .., C \qquad (9)$$

$$\mu_{ij} = 1, i = 1, .., N \qquad (10)$$

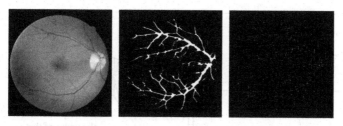

Fig. 5. Original RGB image, vessel segmented image, and its centerline image (from left to right)

The exponent $m \in [1, \infty]$ is the weighting exponent, which determines the fuzziness of the clusters. The most commonly used distance norm is the Euclidean distance $d_{ij} = ||x_i - v_j||$.

We used the Matlab Fuzzy Logic Toolbox for clustering 253440 vectors (the size of the retinal image is 512x495) in length twelve for each retinal image. In each retinal image clustering procedure, the number of clusters was assigned after analyzing the histogram of the texture image. We obtained the membership values on each cluster for every vector, from which we choose the cluster number that belonged to the highest membership value for each vector and converted it into a 2D matrix. From this matrix we produced the binary image considering the cluster central intensity value which identifies the blood vessels only.

4.4 Segmentation Accuracy

Using the DRIVE database [6] we applied our method on five images for vessel segmentation. For performance evaluation, we detected the vessel centerline in our output segmented images and hand-labeled ground truth segmented (GT) images applying the morphological thinning operation (Fig. 5). We achieved an overall 84.37% sensitivity $(TP/(TP+FN))$ and 99.61% specificity $(TN/(TN+FP))$ where TP, TN, FP and FN are true positive, true negative, false positive and false negative respectively. We note, Hoover et al. [1] method applied on the same five segmented images provided 68.23% sensitivity and 98.06% specificity. Clearly, our method produces superior results.

5 Vessel Width Measurement

After obtaining the vessels edge image and centerline image, we mapped these images to find the vessel width for a particular vessel centerline pixel position. To do this we first pick a pixel from the vessel centerline image, then we apply a mask considering this centerline pixel as its center. The purpose of this mask is to find the potential edge pixels (which may fall in width or cross-section of the vessels) in any side of that centerline pixel position. Therefore, we will apply the mask to the edge image only. For searching all the pixel positions inside the mask, we calculate the pixel position by shifting by one up to the size of the mask and rotating each position from 0 to 180 degrees at the same time. For increasing the rotation angle we use the step size (depending on the size of the mask) less than $180°/(\text{mask length})$. Hence, we can access every cell in the mask using this angle.

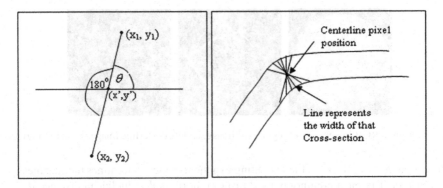

Fig. 6. Finding the mirror of an edge pixel(left) and width or minimum distance from potential pairs of pixels (right)

For each obtained position we search the edge image gray scale value to check whether it is an edge pixel or not. Once we find an edge pixel we then find it's mirror by shifting the angle of 180 degree and increasing the distance from one to the maximum size of the mask (Fig. 6). In this way we produce a rotational invariant mask and pick all the potential pixel pairs to find the width or diameter of that cross sectional area.

$$x1 = x' + r * \cos \theta$$
$$y1 = y' + r * \sin \theta \qquad (11)$$

where (x', y') is the vessel centerline pixel position, r=1,2,..(mask size)/2 and $\theta = 0, .., 180°$. For any pixel position, if the gray scale value in the edge image is 255 (white or edge pixel) then we find the pixel $(x2, y2)$ in the opposite edge (mirror of this pixel) considering $\theta = (\theta + 180)$ and varying r.

After applying this operation we obtain the pairs of pixels which are on the opposite edges (at line end points) giving imaginary lines passing through the centerline pixels (Fig. 6). From these pixels pairs we find the minimum Euclidian distance $\sqrt{(x_1 - x_2)^2 + (y_1 - y_2)^2}$, the width of that cross-section. This enables us to measure the width for all vessels including the vessels' one pixel wide (for which we have the edge and the centreline itself).

6   Experimental Results and Discussion

We used the centreline images and edge images for measuring the width of the blood vessels. We measure the accuracy qualitatively by comparing with the width measured by plotting the centreline pixel and its surround edge pixels. We considered ten different vessel cross-sections of these images and observed that our method is working very accurately. Figure 7 depicts the detected width for some cross-sectional points indicating by white lines (enlarged).

For quantitative evaluation we considered ten images (each 3072×2048 captured with the Canon D-60 digital fundus camera) with manually measured width on different

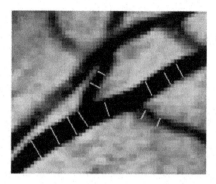

Fig. 7. Measured vessel width showing by the white lines in an image portion

Table 2. Measuring the accuracy of the automatic width measurement

Cross-section	Centreline pixel X_c	Y_c	Detected width end points X_1	Y_1	X_2	Y_2	Auto. width (A)	Error (%)	Accuracy (%)
1	2055	629	2068	632	2046	628	22.361	0.86	99.14
2	1859	519	1871	519	1850	520	21.024	2.50	97.50
3	2259	815	2259	811	2259	824	13.000	0.54	99.46
4	2350	1077	2350	1070	2350	1084	14.000	12.39	87.61
5	2233	1317	2239	1314	2239	1322	11.314	6.51	93.49
6	2180	1435	2189	1431	2172	1440	19.235	4.61	95.39
7	2045	1451	2055	1452	2042	1452	13.000	14.45	85.55
8	1683	1500	1691	1509	1680	1496	17.029	12.48	87.52
9	1579	617	608	1593	630	1573	23.409	1.52	98.48
10	1434	855	853	1436	859	1432	7.211	14.52	85.48
11	1443	1000	999	1446	1004	1440	7.810	8.77	91.23
12	1618	1331	1335	1623	1330	1617	7.810	10.46	89.54
13	1475	1164	1169	1479	1162	1474	8.602	16.80	83.20

cross-sections povided by the Eye and Ear Hospital, Victoria, Australia. For each cross-section, we received the graded width by five different experts who are trained retinal vessel graders of that institution. For manual grading a computer program was used where the graders could zoom in and out at will, moving around the image and selecting various parts.

We applied our technique on these images to produce the edge image and vessel centreline image. We considered these images and randomly picked ninety-six cross-sections of vessels of varying width from one to twenty-seven pixels. We measured the width for each cross-section by our automatic width measurement technique (we call it automatic width, A) and considered the five manually measured width (we call it manual width) by experts. We calculated the average of the manual width (μ), the standard deviation on manual widths (σ_m) and considered the following formula to find the error,

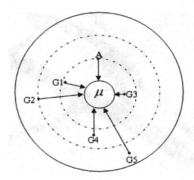

Fig. 8. Distances of individual Graders and Automatic width from the mean value

$$E = \left| \frac{\frac{(\mu - \sigma_m) - A}{(\mu - \sigma_m)} + \frac{(\mu + \sigma_m) - A}{(\mu + \sigma_m)}}{2} \right|$$

$$= \left| 1 - \frac{\mu \times A}{\mu^2 - \sigma_m^2} \right|$$

(12)

In equation (12), we considered $(\mu \pm \sigma_m)$ to normalize it. This formula is a good measure as the error rate will be less if it is within one standard deviation interval. Using this formula, we calculated the error and accuracy in all ninety-six cross-section and achieve an average of 95.8% accuracy (maximum accuracy is 99.58% and minimum accuracy is 83.20%) in the detection of vessel width. We found the maximum error is 16.80% which is 2.04 pixel and the minimum error is 0.698% which is 0.139 pixel. Tables 2 and 3 depict the manual and automatic width measurement accuracy on different cross-sections in an image.

We compared the accuracy of automatically measured width with each grader using the same equation (12) with a simple modification, by replacing the automatic width by

Table 3. Manually measured widths for an image cross-sections

Cross-section	Manually measured width (in micron)					Avg. width (in micron)	Mean width in pixel, μ	Standard Deviation, σ_m
	One	Two	Three	Four	Five			
1	112.42	117.53	107.31	117.53	112.42	113.44	22.2	0.8366
2	107.31	112.42	107.31	117.53	107.31	110.38	21.6	0.8944
3	66.43	76.65	61.32	71.54	61.32	67.45	13.2	1.3088
4	61.32	71.54	61.32	71.54	56.21	64.39	12.6	1.3416
5	56.21	66.43	56.21	66.43	66.43	62.32	12.2	1.0954
6	107.31	107.31	102.2	102.2	97.09	103.22	20.2	0.8366
7	56.21	66.43	45.99	61.32	66.43	59.28	11.6	1.6733
8	86.87	107.31	102.2	107.31	97.09	100.16	19.6	1.6733
9	132.86	127.75	112.42	132.86	107.31	122.64	24	2.3452
10	45.99	51.1	35.77	56.21	35.77	44.97	8.8	1.7889
11	40.88	56.21	35.77	45.99	45.99	44.97	8.8	1.4832
12	35.77	51.1	45.99	56.21	40.88	37.99	9	1.5811
13	35.77	45.99	35.77	45.99	30.66	38.84	7.6	1.3416

Table 4. Comparison of automatic width measurement accuracy with an individual Grader (Grader 4)

Manual width	Manual accuracy	Manual error	Automatic width	Automatic accuracy	Automatic error
22	95.3103	4.6896	22.3610	**98.2189**	1.7811
21	91.7047	8.2952	21.0240	**98.9913**	1.0087
13	91.0179	8.9820	13.0000	**98.8024**	1.1976
12	99.6200	0.3800	14.0000	88.2100	11.7900
11	90.6542	9.3457	11.3140	**95.1645**	4.8355
21	98.9867	1.0132	19.2350	95.2005	4.7995
11	92.6766	7.3233	13.0000	83.7330	16.2670
17	90.0436	9.9563	17.0290	89.1641	10.8359
26	88.2146	11.7853	23.4090	**99.3545**	0.64550
9	62.1082	37.8917	7.2110	**90.3943**	9.6057
8	93.0693	6.9306	7.8102	92.7950	7.2050
7	67.5324	32.4675	7.8102	**94.0550**	5.9450
7	72.0065	27.9934	8.6023	**77.6621**	22.3379

the width measured by individual grader. We considered this mean width (μ) and standard deviation (σ_m) for all the graders except the one which we want to compare with. We consider this mean value as the Oracle (the correct width for that cross-section). For automatically measured width comparison we use the same values of the mean width and the standard deviation in the widths measured by the four graders. We calculate the number of winner that has the higher accuracy in each cross-section, width measured by the individual grader (Table 4) or automatic method (bold in Table 4). The ratios' of higher accuracy width measurement between the automatic and the individual graders are shown in Table 5. Clearly, our method produced better results for most of the image cross-sections, and is also performing better than the majority of the graders. The results from Table 5 are produced in Fig. 8, which shows the distance between the mean

Table 5. Ratio of the higher accuracy on automatic vs. individual Grader

Image number	Number of Cross-sections	Automatic (A) vs. Manual (M)					Overall performance (A:M)
		Grader 1	Grader 2	Grader 3	Grader 4	Grader 5	
1	13	7:6	12:1	8:5	8:5	10:3	**45:20**
2	10	6:4	2:8	3:7	6:4	9:1	**26:24**
3	11	7:4	3:8	7:4	5:6	6:5	**28:27**
4	9	4:5	6:3	3:6	3:6	4:5	**20:25**
5	7	2:5	4:3	1:6	3:4	4:3	**14:21**
6	12	5:7	9:3	5:7	6:6	8:4	**33:27**
7	10	4:6	7:3	4:6	6:4	8:2	**29:21**
8	7	2:5	3:4	2:5	3:4	3:4	**13:22**
9	8	4:4	6:2	5:3	5:3	6:2	**26:14**
10	9	6:3	5:4	6:3	7:2	4:5	**28:17**
Total	**96**	**47:49**	**57:39**	**44:52**	**52:44**	**62:34**	**262:218**

width and the width measured by the individual graders and automatic method. Figure 8 demonstrates the consistency of our method.

We also compared our technique with Lowell et al. [12] that achieved the maximum accuracy of 99% (did not mention the average accuracy for all cross-sections) with minimum pixel error of 0.34. Using the same formula, $|(\mu - A)/\mu|$, we achieved 100% accuracy for a particular cross-section which indicates the superiority of our method.

7   Conclusions

In this paper we proposed a new and efficient technique for blood vessels width measurement. Our approach provides a robust estimator of vessel width in the presence of low contrast and noise. The results obtained are encouraging, and the detected width can be used to measure different parameters (nicking, narrowing, branching coefficients, etc.) for diagnosing various diseases. Currently, we are working on the blood vessels' bifurcation and cross-over detection where the measured width is contributing as an important information for perceptual grouping process.

Acknowledgements. We would like to thank David Griffiths (Research Assistant, The University of Melbourne and Eye and Ear Hospital, Melbourne, Australia) for providing us with the manually measured width images and data.

References

1. Hoover, A., Kouznetsova, V., Goldbaum, M.: Locating Blood Vessels in Retinal Images by Piece-wise Threshold Probing of a Matched Filter Response. IEEE Transactions on Medical Imaging 19(3), 203–210 (2000)
2. Wyszecki, G.W., Stiles, S.W.: Color Science: Concepts and Methods, Quantitative Data and Formulas. Wiley, New York (1982)
3. Geusebroek, J., Boomgaard, R.V.D., Smeulders, A.W.M., Geerts, H.: Color Invariance. IEEE Transactions on Pattern Analysis and Machine Intelligence 23(2), 1338–1350 (2001)
4. Kruizinga, P., Petkov, N.: Nonlinear Operator for Oriented Texture. IEEE Transactions on Image Processing 8(10), 1395–1407 (1999)
5. Bezdek, J.: Pattern Recognition with Fuzzy Objective Function Algorithms. Plenum Press, USA (1981)
6. DRIVE-database: Image Sciences Institute, University Medical Center Utrecht, The Netherlands (2004), http://www.isi.uu.nl/Research/Databases/DRIVE/
7. Gonzalez, R.C., Woods, R.E., Eddins, S.L.: Digital Image Processing Using MATLAB. Prentice-Hall, Englewood Cliffs (2004)
8. Gonzalez, R.C.: Woods: Digital Image Processing, 3rd edn. Prentice Hall, New Jersey (2008)
9. Gao, X., Bharath, A., Stanton, A., Hughes, A., Chapman, N., Thom, S.: Measurement of Vessel Diameters on Retinal Images for Cardiovascular Studies. In: Proceedings of Medical Image Understanding and Analysis, pp. 1–4 (2001)
10. Zhou, L., Rzeszotarsk, M.S., Singerman, L.J., Chokreff, J.M.: The Detection and Quantification of Retinopathy Using Digital Angiograms. IEEE Transactions on Medical Imaging 13(4), 619–626 (1994)

11. Brinchman-Hansen, O., Heier, H.: Theoritical Relations Between Light Streak Characterstics and Optical Properties of Retinal Vessels. Acta Ophthalmologica 179(33), 33–37 (1986)
12. Lowell, J., Hunter, A., Steel, D., Basu, D., Ryder, R., Kennedy, R.L.: Measurement of Retinal Vessel Widths From Fundus Images Based on 2-D Modeling. IEEE Transactions on Medical Imaging 23(10), 1196–1204 (2004)
13. Wu, D., Zhang, M., Liu, J.: On the Adaptive Detetcion of Blood Vessels in retinal Images. IEEE Transactions on Biomedical Engineering 53(2), 341–343 (2006)

Automatic Detection of Laryngeal Pathology on Sustained Vowels Using Short-Term Cepstral Parameters: Analysis of Performance and Theoretical Justification

Rubén Fraile[1], Juan Ignacio Godino-Llorente[1], Nicolás Sáenz-Lechón[1],
Víctor Osma-Ruiz[1], and Pedro Gómez-Vilda[2]

[1] Department of Circuits & Systems Engineering, Universidad Politécnica de Madrid
Carretera de Valencia Km 7, 28031 Madrid, Spain
{rfraile,igodino,vosma}@ics.upm.es, nicolas.saenz@upm.es
[2] Department of Computer Systems' Architecture and Technology
Universidad Politécnica de Madrid
Campus de Montegancedo s/n, Boadilla del Monte, 28660 Madrid, Spain
pedro@pino.datsi.fi.upm.es

Abstract. The majority of speech signal analysis procedures for automatic detection of laryngeal pathologies mainly rely on parameters extracted from time-domain processing. Moreover, calculation of these parameters often requires prior pitch period estimation; therefore, their validity heavily depends on the robustness of pitch detection. Within this paper, an alternative approach based on cepstral - domain processing is presented which has the advantage of not requiring pitch estimation, thus providing a gain in both simplicity and robustness. While the proposed scheme is similar to solutions based on Mel-frequency cepstral parameters, already present in literature, it has an easier physical interpretation while achieving similar performance standards.

1 Introduction

Analysis of recorded speech is an attractive method for pathology detection since it is a low-cost non-invasive diagnostic procedure [1]. Although there is a wide range of causes for pathological voice (functional, neural, laryngeal, etc.) and a correspondingly wide range of acoustic parameters has been proposed for its detection (see [2] for summarising tables and typical values), these intend to detect speech signal features that may be roughly classified in only three classes [3]:

- *Short-term Frequency Perturbations*: both in fundamental frequency and in formants.
- *Short-term Amplitude Perturbations*.
- *Noise* or, more specifically, speech-to-noise ratio.

Calculation of above-mentioned acoustic parameters requires previous and reliable detection of speech fundamental frequency (pitch) [4] [1]. Nevertheless, pitch detection

A. Fred, J. Filipe, and H. Gamboa (Eds.): BIOSTEC 2008, CCIS 25, pp. 228–241, 2008.

is not an easy task due to its sensitiveness to noise, signal distorsion, speech formants, etc. [5].

An alternative approach to speech signal analysis is doing it in cepstral domain, more specifically in Mel-frequency cepstral domain. Such approach, consisting in classifying patterns of so-called Mel-frequency cepstral coefficients (MFCC), does not require prior pitch estimation and has proven to be fairly robust against different kinds of speech distortion [6], including that of the telephone channel [7], and reasonably independent of the particular way in which computations may be implemented [8]. For these reasons, their application to automatic voice pathology detection has been proposed during the last years [9]. Yet, to authors' knowledge, up to now no physical explanation exists on the meaning of MFCC and their relevance on pathology detection.

Within this paper, a new scheme for automatic voice pathology detection is proposed. This lies half-way between usual cepstral domain and Mel-frequency cepstral domain. Namely, it takes profit from the conceptual interpretation of cepstral processing of speech signals [10], the pattern separation capability of cepstral distances [11] and the smoother spectrum estimation provided by the filter banks in MFCC calculation [11]. The mathematical formulation of both cepstrum and MFCC parameters is revised in Sect. 2, while the newly proposed set of parameters is introduced in Sect. 3. The results from the application of these features to the detection of pathologies on voices belonging to a commercial database are reported in Sect. 4. Last, the conclusions are presented in Sect. 5.

2 Mathematical Formulation

2.1 Short-Time Fourier Transform

As stated in previous section, the variability of speech signal is a key feature for pathology detection. The need for detecting such variability leads to the convenience of employing short-time techniques for speech processing. For this reason, in the following lines the mathematical framework for short-time processing of speech provided in [10] is revised.

Let $x[n]$ be a speech signal composed by N samples ($n = 0 \cdots N - 1$) obtained at a sampling frequency equal to f_s; then it can be segmented in frames defined by:

$$f[n;m] = x[n] \cdot w[m-n] \ , \tag{1}$$

where $w[n]$ is the framing window:

$$w[n] = 0 \ \text{if} \ n < 0 \ \text{or} \ n \geq L \tag{2}$$

and L is the frame length. Consequently, $f[n;m]$ has non-zero values only for $n \in [m - L + 1, m]$. If consecutive speech frames are overlapped a number of l_0 samples, then m may have the following values:

$$m = L + p \cdot (L - l_0) - 1 \ , \tag{3}$$

where p is the frame index and it is an integer such that:

$$0 \leq p \leq \frac{N - L}{L - l_0} \ . \tag{4}$$

Considering the relation between the frame shift m and the frame index p, frames without time shift reference may be renamed as:

$$g_p[n] = f[n+m-L+1;m] = f[n+p\cdot(L-l_0);m] =$$
$$= x[n+p\cdot(L-l_0)]\cdot w[(L-1)-n] \ , \tag{5}$$

where $n = 0\cdots L-1$. From these speech frames, the short-term Discrete Fourier Transform (stDFT) is computed as:

$$S_p(k) = \sum_{n=0}^{N_{DFT}-1} \tilde{g}_p[n]\cdot e^{-j\cdot\frac{2\pi}{N_{DFT}}\cdot kn} \ , \tag{6}$$

where N_{DFT} is the number of points of the stDFT, $k = 0\cdots N_{DFT}-1$ and:

$$\tilde{g}_p[n] = \begin{cases} g_p[n] & \text{if } 0 \le n < L \\ 0 & \text{otherwise} \end{cases} \ . \tag{7}$$

Thus, if $N_{DFT} \ge L$ then (6) is equal to:

$$S_p(k) = \sum_{n=0}^{L-1} g_p[n]\cdot e^{-j\cdot\frac{2\pi}{N_{DFT}}\cdot kn} \tag{8}$$

and the frequency values that correspond to each stDFT coefficient are:

$$f_k = \begin{cases} f_s\cdot\frac{k}{N_{DFT}} & \text{if } k \le \frac{N_{DFT}}{2} \\ f_s\cdot\frac{k-N_{DFT}}{N_{DFT}} & \text{if } k > \frac{N_{DFT}}{2} \end{cases} \ . \tag{9}$$

2.2   Short-Time Cepstrum

In [10], an algorithm for computing the short-time cepstrum from the stDFT is given, under the assumption that $N_{DFT} \gg L$:

$$c_p[q] = \frac{1}{N_{DFT}}\cdot\sum_{k=0}^{N_{DFT}-1} \log|S_p(k)|\cdot e^{j\cdot\frac{2\pi k}{N_{DFT}}\cdot q} \ . \tag{10}$$

A physical interpretation of cepstrum can be derived from the discrete-time model for speech production that can also be found in [10]. This model may be written in frequency domain as:

$$S\left(e^{j\Omega}\right) = E\left(e^{j\Omega}\right)\cdot G\left(e^{j\Omega}\right)\cdot H\left(e^{j\Omega}\right) \ , \tag{11}$$

where $S\left(e^{j\Omega}\right)$ is the speech, $E\left(e^{j\Omega}\right)$ is the impulse train corresponding to the pitch and its harmonics, $G\left(e^{j\Omega}\right)$ is the glottal pulse waveform that modulates the impulse train and $H\left(e^{j\Omega}\right)$ is, herein, the combined effect of vocal tract and lip radiation. These components can be appreciated in Fig. 1, which corresponds to the average modulus of

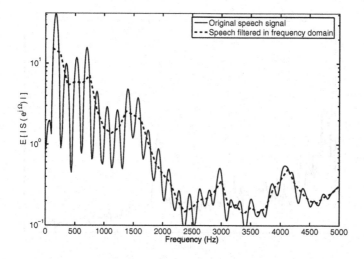

Fig. 1. Average modulus of the short-term DFT for one voice record

the short-term DFT calculated from one of the voice records belonging to the database referred in Sect. 4.1.

The quick impulse-like variations in Fig. 1 correspond to the fundamental frequency and its harmonics $E\left(e^{j\Omega}\right)$ and the evolution of the impulse amplitude envelope is related both to the glottal waveform $G\left(e^{j\Omega}\right)$ and to the formants induced by the vocal tract $H\left(e^{j\Omega}\right)$. These formants correspond to the three envelope peaks with a decreasing level of energy that are centered at 750 Hz, 1375 Hz and 3000 Hz. In fact, these center frequencies are coherent with the range of typical values given in [2].

The logarithm operation in (10) converts the products in (11) into sums. Consequently, it allows the cepstrum to separate fast from slow signal variations in frequency domain. This widely known fact is illustrated in Fig. 2, where the peak around 5.7 ms clearly identifies the fundamental frequency (175 Hz) and the values below 2 ms correspond to the spectrum envelope (dashed line in Fig. 1).

2.3 Short-Time MFCC

Once the stDFT of a speech signal is available, another option for further processing, as mentioned in Sect. 1, is the calculation of short-time MFCC (stMFCC) parameters. For stMFCC computation, only the positive part of the frequency axis is considered [11], that is, $f_k \geq 0$ and, therefore, $k \leq N_{\mathrm{DFT}}/2$ (recall (9)). In order to calculate stMFCC coefficients, a transformation is applied to the frequencies so as to convert them to Mel-frequencies f_k^{mel} [9]:

$$f_k^{\mathrm{mel}} = 2595 \cdot \log_{10}\left(1 + \frac{f_k}{700}\right) \tag{12}$$

and the stDFT is further processed through band-pass integration along M equally long Mel-frequency intervals, being $M = \lfloor 3 \cdot \log f_s \rfloor$ ($\lfloor \cdot \rfloor$ means rounding to the previous integer). Namely, the ith interval ($i = 1 \cdots M$) in Mel-domain is defined by:

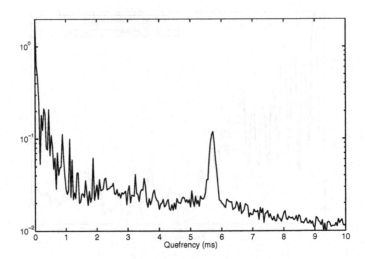

Fig. 2. Short term cepstrum averaged for all frames of the same voice record as used for figure 1

$$I_i^{\text{mel}} = \left[F^{\text{mel}} \cdot \frac{i-1}{M+1}, F^{\text{mel}} \cdot \frac{i+1}{M+1} \right] \, , \tag{13}$$

where F^{mel} is the maximum Mel-frecuency:

$$F^{\text{mel}} = \max_k f_k^{\text{mel}} = 2595 \cdot \log_{10} \left(1 + \frac{f_s/2}{700} \right) \tag{14}$$

and the interval length in Mel-domain is given by:

$$L\left(I_i^{\text{mel}} \right) = \frac{2}{M+1} \cdot F^{\text{mel}} \, . \tag{15}$$

According to the previous equations, the N_{DFT} coefficients of the stDFT are transformed to M frequency components as follows:

$$\tilde{S}_p(i) = \sum_{f_k \in I_i} \left(1 - \frac{\left| f_k^{\text{mel}} - F^{\text{mel}} \cdot \frac{i}{M+1} \right|}{L\left(I_i^{\text{mel}} \right)/2} \right) \cdot |S_p(k)| \, . \tag{16}$$

Last, the qth ($q = 1 \cdots Q$) stMFCC of the pth speech frame, where Q is the desired length of the Mel-cepstrum, is given by cosine transform of the logarithm of the smoothed "Mel-spectrum" [11]:

$$\tilde{c}_p[q] = \sum_{i=1}^{M} \log \left| \tilde{S}_p(i) \right| \cdot \cos \left[q \cdot \left(i - \frac{1}{2} \right) \cdot \frac{\pi}{M} \right] \, . \tag{17}$$

3 Cepstral Coefficients Based on Smoothed Spectrum

3.1 Justification

As stated in Sect. 1, while MFCC parameters exhibit both good performance and robustness in feature extraction from speech, they lack a clear physical interpretation. On the opposite, cepstrum has a physical meaning (recall Sect. 2.2), yet raw cepstrum coefficients are not as useful for speech parametrisation. In the next paragraphs, the reasons for these facts are exposed.

Cepstrum calculation, as formulated in (10), is based on the spectrum estimate provided by the absolute value of the stDFT. Due to the logarithm, this gives a result that is proportional to the case of periodogram-based spectrum estimation. However, such estimation is very dependent on the specific values of the original speech frame. A more robust spectrum estimate can be obtained if the periodogram is smoothed (Blackman and Tukey method, [12]). In fact, this is what (16) expresses in the calculation of MFCC. Therefore, filtering of the stDFT may be assumed to be one of the sources of MFCC robustness.

In contrast, an explanation for the lack of clear interpretation of MFCC also lies in the meaning of (16). According to that equation, stDFT smoothing for MFCC computation is carried out with a variable-length filter, that is, a Bartlett window whose length decreases for lower frequency bands. Moreover, the smoothed stDFT is downsampled to obtain only M samples in the interval $[0, f_s/2]$ that are not uniformly spaced [11]. While the downsampling is positive in the sense that it reduces the dimensionality of the problem, its non-uniformness, together with the previous variable-length filtering, obscures the interpretation of the output of the cosine transform in (17).

From the previous reasoning, if stDFT is smoothed with a fixed-length filter and its output is uniformly decimated prior to the logarithm computation, the cepstral coefficients in (10) can be transformed to a more robust parameter set. Moreover, this is achieved while keeping the physical meaning of cepstrum, since the output of the first operation gives an improved spectrum estimate and the second one only limits the length of the cepstrum in quefrency domain.

3.2 Formulation

Starting from (8), if the stDFT modulus is smoothed with a Bartlett window of constant length equal to Δf then the following output is obtained:

$$S'_p(i) = \sum_{f_k \in I_i} \left(1 - \frac{|f_k - i \cdot \Delta f/2|}{\Delta f/2}\right) \cdot |S_p(k)| \ , \tag{18}$$

where $I_i = [\Delta f \cdot (i-1)/2, \Delta f \cdot (i+1)/2]$ and the Bartlett window has been chosen for similarity with (16). Herein, only the positive part of the frequency axis has been considered, as in Sect. 2.3.

If the filtered stDFT is decimated so as to keep only the outputs of consecutive windows with a 50% overlap, this is equivalent to decimation by a factor:

$$D = \lfloor \Delta f \cdot N_{\text{DFT}}/(2 \cdot f_s) \rfloor \ . \tag{19}$$

The short-term modified cepstrum (stMC) then becomes:

$$c'_p[q] = \frac{D}{N_{DFT}} \cdot \sum_{k=0}^{\frac{N_{DFT}}{2 \cdot D}} \log\left|S'_p(k \cdot D)\right| \cdot \cos\left((k-1) \cdot \frac{2\pi D}{N_{DFT}} \cdot q\right) , \qquad (20)$$

where only the positive frequencies have been considered, hence computing the inverse DFT as a cosine transform as in (17). $c'_p[q]$ has the twofold advantage over $c_p[q]$ of being based on a smoother spectrum estimate $S'_p(i)$ and having a period length that has been reduced by a factor D, thus providing some dimensionality reduction.

3.3 Cepstral Distances

Differences in cepstrum can be used for speech signal classification. An example of such usage is the definition of the cepstral distance in [11] as the norm of the vector resulting form substraction of the two cepstra to be compared. This, if directly applied to pathology detection, would result in comparing the cepstrum of consecutive speech frames so as to assess the variability of the signal. Mathematically:

$$d_p^2 = \sum_{q=0}^{\frac{N_{DFT}}{D}-1} \left|c'_{p+1}[q] - c'_p[q]\right|^2 . \qquad (21)$$

However, bearing in mind the physical interpretation of cepstrum, this definition has the drawback of mixing pitch variations with formant and glottal pulse variations. To overcome this problem an individual frame-to-frame cepstral parameter variation analysis is proposed:

$$d_p[q] = \left|c'_{p+1}[q] - c'_p[q]\right| . \qquad (22)$$

This way, analysis of the distribution of $d_p[q]$ related to speech formant and glottal pulse variability (low values of q) can be isolated from pitch changes associated to values of q around the pitch period.

4   Application and Results

For the purpose of performance analysis, the stMC parametrisation presented in previous section has been applied to the problem of automatic pathology detection on recorded voice. The results have been compared to those produced using stMFCC. Within this section, first the voice database is presented, second the classifier is described and, last, the results from two different experiments are shown and commented.

4.1   Database

The voice records used in this investigation are the same as in [13]. They belong to a database distributed by the company Kay Elemetrics [14]. The recorded sounds correspond to sustained phonations (1-3 s long) of the vowel /ah/ from patients with either normal or disordered voice. Such voice disorders belong to a wide variety of organic, neurological, traumatic and psychogenic classes. The sampling rate of speech records has been made uniform for all of them and equal to 25 KHz, while the coding has a resolution of 16 bits. The subset taken from the database contains 53 normal and 173 pathological speakers which are uniformly distributed in age and gender [13].

4.2 Classifier Description

For the classification stage of the pathology detector, a Multilayer Perceptron (MLP) with two hidden layers, each consisting of 4 neurons, and a two-neuron output layer has been used. All neurons have logistic activation functions. An MLP with a single hidden layer having 50 neurons was utilised in [9]. The structure herein proposed, in contrast, has less free parameters, thus allowing a faster learning, and the reduced number of neurons is compensanted by the introduction of an additional hidden layer that permits learning of more complex relations [15].

The MLP classifier has been trained with 60% of available speech records in such a way that the outcome of one neuron of the output layer is expected to be "1" for pathological voices and "0" for normal voices while the second output neuron is expected to have an opposite behaviour. 10% of the records have been used for cross-validation during the training phase as a criterion to stop training. The remaining 30% of records have been used for testing. The experiment has been repeated 200 times, each of them with different, randomly chosen, training and cross-validation sets.

4.3 Results for Short-Time Parameters

Within the previous classification scheme, each feature vector, corresponding to one speech frame, is assigned a pair of likelihoods, one coming from each output neuron:

- likelihood of belonging to the phonation of a healthy person (l_{nor}^p) and
- likelihood of that person having pathology (l_{pat}^p);

being p the frame index. From the pair (l_{nor}^p, l_{pat}^p), the classification decision at frame level is taken based on the value of the log-likelihood ratio (LLR^p) and comparing it to a threshold θ:

$$LLR^p = \log \frac{l_{nor}^p}{l_{pat}^p} \gtrless \theta \ . \tag{23}$$

For a record consisting of P frames, decision at record level is taken based on the mean log-likelihood:

$$LLR = \frac{1}{P} \cdot \sum_{p=1}^{P} \log \frac{l_{nor}^p}{l_{pat}^p} \gtrless \theta \ . \tag{24}$$

In this first experiment, for each speech record stMC coefficients, as defined in (20), have been calculated. Specifically, a filter length $\Delta f = 200$ Hz has been chosen for sfDFT smoothing. Consequently, this results in a cepstrum length equal to $(f_s - \Delta f/2)/(\Delta f/2) = 124$ samples. The choice of Δf is consistent to the approximate length of the low-band filters used for MFCC calculation (recall (16)). At first sight, however, it has the drawback of loosing pitch information of the signal spectrum. This is illustrated in Fig. 1 where the filtered DFT has been plotted with a dashed line and also in Fig. 3 where

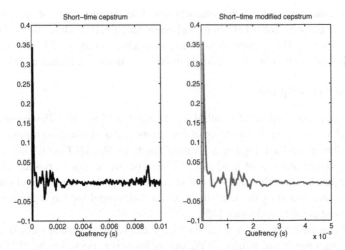

Fig. 3. Short-time cepstrum of a speech frame taken from the database, without spectrum filtering (left) and after spectrum filtering (right). It can be noticed that the limitation on the length of cepstrum has produced a loss of information about pitch (peak on the right part of the left graph).

the cepstrum obtained with and without spectral smoothing is represented. Nevertheless, such filtered spectrum contains information on both harmonic-to-noise ratio (HNR) and glottal pulse waveform [16] and HNR is a useful parameter for pathology detection that is closely related to both frequency and amplitude perturbations of pitch [2].

For the sake of comparison, another classifier based on a parameter vector consisting of 20 stMFCC calculated using 31 Mel-band filters ($M = \lfloor 3 \cdot \log f_s \rfloor = 31$) has also been

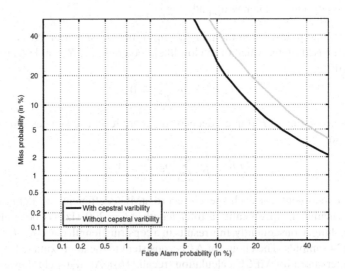

Fig. 4. DET plot for short-time cepstral parameters including information on cepstral variability (black) and lacking that information (gray)

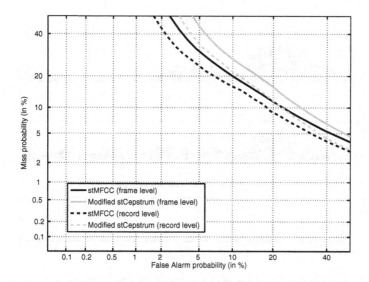

Fig. 5. DET plot for the stMFCC (black) and the short-time modified cepstrum (gray) parametrisation schemes.

tested. Figure 5 shows the detection-error-tradeoff (DET) plots [17] for both parametrisation schemes at frame level and at record level. It can be noticed that while the stMFCC-based system provides better performance (14.96 % equal error rate - EER), possibly due to the higher dimensionality reduction, the herein presented stMC-based scheme has a performance within the same range (17.79 % EER) with a clearer physical interpretation. Another observation that can be drawn from the plot is that results at record level (respectively, 13.36% and 15.40% EER) are better that at frame level. Such fact indicates the presence of a certain degree of variability among speech frames belonging to the same record. This is intended to be confirmed in the second experiment, as reported next.

4.4   Results for Averaged Parameters

In order to assess the relevance of cepstral variability, a second experiment has been carried out. In this case, the input vectors for the pathology detector are averaged for each record, thus working directly at record level, instead of getting record-level likelihoods as combinations of frame-level likelihoods. The first 124 elements correspond to the average values of the stMC, as calculated before. The rest of the input vector contains information about the variability of stMC around those average values. More specifically, the mean and variance of $d_p[q]$ for every value of q are used as descriptors of the cepstrum variability. Therefore, on the whole, a parameter vector of $124 \times 3 = 372$ elements is produced.

Figure 4 shows the results of using this scheme for speech record parametrisation compared to those obtained only with the first 124 components of the feature vectors, that is, without including information on cepstral variability. It should be noted that the structure of the MLP classifier for this experiment has been simplified by removing one

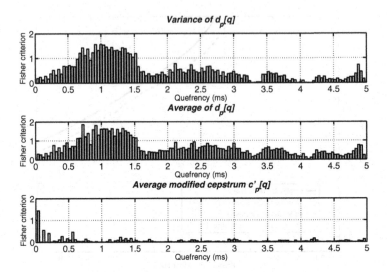

Fig. 6. Value of Fisher criterion for each cepstral parameter

of the hidden layers. The reason for this is that when passing from the frame level to the record level a great reduction on the number of feature vectors is obtained. Still, a similar performance at record-level classification is achieved for the case of the feature vectors including information on cepstral variability (14.70 % EER). However, if this information is removed from the classifier input, the performace is significantly degraded (19.17 % EER). This confirms the relevance of cepstral variability for pathology detection.

In order to acquire a deeper understanding of the reasons for these results, an analysis of the relevance of these stMC-based parameters for speech classification as either pathological or not has been realised. Such analysis is based on the evaluation of the Fisher criterion [18] for each individual parameter of the above-described 372-element feature vectors. The results, differentiated for the three subsets of parameters (average stMC, variance of differences and average of absolute differences) are plotted in Fig. 6.

According to this plot, the most relevant cepstral parameters for pathology detection maybe roughly classified into two groups:

– The modified cepstrum values with lowest indices (plot at the bottom of Fig. 6): these are related to the slowest components of the spectrum envelope in Fig. 1, which, on their side, are associated to spectral noise levels and HNR [16].
– The frame-to-frame variations in stMC-based coefficients whose quefrencies are within the interval [0.5, 1.5] miliseconds approximately. The coefficients within that interval correspond to the short frequency range components of the spectrum envelope. These components, as justified in Sect. 2.2, are related to glottal waveform and speech formants. However, this information itself does not help to discriminate the presence of pathology, as indicated by the low values of the Fisher criterion in the bottom plot of Fig. 6. Instead, frame-to-frame variations of these factors are much more relevant, as depicted in the other two plots of the same figure.

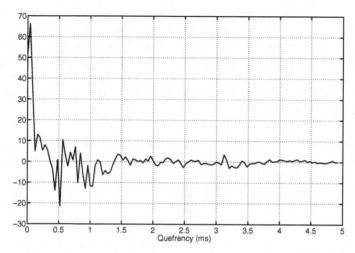

Fig. 7. 124 modified cepstral parameters from one of the database's voice records

To be more specific, since the voice records of the database used for this experiment correspond to sustained vowel phonations, it can be assumed that the vocal tract has very little variations, hence formants do not change and the second group of parameters should be more closely related to changes in the glottal waveform. As for the limits of the quefrency interval in which parameters from the second group are relevant, the lower limit of 0.5 ms corresponds to the quefrency band that separates slow components of the spectrum envelope (first group of parameters) from faster components (associated to the second set); on the other hand, the upper limit of 1.5 ms corresponds to the highest quefrency range at which the modified cepstrum $c'_p[q]$ has significant values. This is shown in Fig. 7, where a plot of the frame-averaged stMC of one voice record is depicted.

5   Conclusions

Speech parametrisation in cepstral domain is a useful technique for automatic pathology detection. Specifically, MFCC have been successfully used for this purpose. While the computation of these parameters has an intrinsic robustness due to its independency from pitch extraction and the spectrum filtering, their physical interpretation is obscure because of the non-linear Mel-frequency transformation.

Within this paper an alternative set of cepstrum-based parameters has been proposed. Such parameters share the robustness of MFCC since they do not require pitch estimation and filtering of the estimated speech spectrum is also performed. In contrast to MFCC, the calculation of these newly proposed parameters does not involve any non-linear frequency transformation and, consequently, their physical interpretation remains clear. Namely, their values have been shown to be related to the amount of noise energy present in speech and the glottal waveform variability. Both factors are directly associated to laringeal pathologies.

Finally, the performance of the proposed cepstral parameters for pathology detection has been tested using a MLP classifier and results have been compared to those of MFCC. The obtained misclassification rates indicate that the performances of both sets of parameters are similar. Moreover, a deeper analysis on the individual impact of each parameter on the classification task has revealed that the most relevant parameters are those more closely linked to the above-mentioned two factors: noise energy and glottal wave variations.

Acknowledgements. This research was carried out within projects funded by the Ministry of Science and Technology of Spain (TEC2006-12887-C02) and the Universidad Politécnica de Madrid (AL06-EX-PID-033). The work has also been realised within the framework of European COST action 2103.

References

1. Boyanov, B., Hadjitodorov, S.: Acoustic analysis of pathological voices. A voice analysis system for the screening of laryngeal diseases. IEEE Engineering in Medicine and Biology 16, 74–82 (1997)
2. Jackson-Menaldi, M.C.A.: La voz patológica. Editorial Médica Panamericana, Buenos Aires (Argentina) (2002)
3. Godino-Llorente, J.I., Sáenz-Lechón, N., Osma-Ruiz, V., Aguilera-Navarro, S., Gómez-Vilda, P.: An integrated tool for the diagnosis of voice disorders. Medical Engineering & Physics 28, 276–289 (2006)
4. Deliyski, D.D.: Acoustic model and evaluation of pathological voice production. In: Proceedings of the 3rd Conference on Speech Communication and Technology (EUROSPEECH 1993), Berlin (Germany), pp. 1969–1972 (1993)
5. Boyanov, B., Ivanov, T., Hadjitodorov, S., Chollet, G.: Robust hybrid pitch detector. IEE Electronics Letters 29, 1924–1926 (1993)
6. Bou-Ghazale, S.E., Hansen, J.H.L.: A comparative study of traditional and newly proposed features for recognition of speech under stress. IEEE Transactions on Speech and Audio Processing 8, 429–442 (2000)
7. Fraile, R., Godino-Llorente, J.I., Sáenz-Lechón, N., Osma-Ruiz, V., Gómez-Vilda, P.: Analysis of the impact of analogue telephone channel on mfcc parameters for voice pathology detection. In: 8th INTERSPEECH Conference (INTERSPEECH 2007), Antwerp (Belgium), pp. 1218–1221 (2007)
8. Ganchev, T., Fakotakis, N., Kokkinakis, G.: Comparative evaluation of various MFCC implementations on the speaker verification task. In: Proceedings of the 10th International Conference on Speech and Computer (SPECOM 2005), Patras (Greece), pp. 191–194 (2005)
9. Godino-Llorente, J.I., Gómez-Vilda, P.: Automatic detection of voice impairments by means of short-term cepstral parameters and neural network based detectors. IEEE Transactions on Biomedical Engineering 51, 380–384 (2004)
10. Deller, J.R., Proakis, J.G., Hansen, J.H.L.: Discrete-time processing of speech signals. Macmillan Publishing Company, New York (1993)
11. Rabiner, L., Juang, B.H.: Fundamentals of speech recognition. Prentice-Hall, Englewood Cliffs (1993)
12. Proakis, J.G., Manolakis, D.G.: Digital Signal Processing. Principles, Algorithms and Applications, 3rd edn. Prentice-Hall International, New Jersey (1996)

13. Godino-Llorente, J.I., Gómez-Vilda, P., Blanco-Velasco, M.: Dimensionality reduction of a pathological voice quality assessment system based on gaussian mixture models and short-term cepstral parameters. IEEE Transactions on Biomedical Engineering 53, 1493–1953 (2006)
14. Kay Elemetrics Corp.: Disordered voice database. version 1.03 (1994)
15. Haykin, S.: Neural Networks: a comprehensive foundation, 1st edn. Macmillan College Publishing Company, New York (1994)
16. Murphy, P.J., Akande, O.O.: Quantification of glottal and voiced speech harmonics-to-noise ratios using cepstral-based estimation. In: Faundez-Zanuy, M., Janer, L., Esposito, A., Satue-Villar, A., Roure, J., Espinosa-Duro, V. (eds.) NOLISP 2005. LNCS, vol. 3817, pp. 224–232. Springer, Heidelberg (2006)
17. Martin, A., Doddington, G., Kamm, T., Ordowski, M., Przybocki, M.: The DET curve in assessment of detection task performance. In: Proceedings of the 5th Conference on Speech Communication and Technology (EUROSPEECH 1997), Rhodes (Greece), pp. 1895–1898 (1997)
18. Duda, R.O., Hart, P.E., Stork, D.G.: Pattern classification, 2nd edn. John Wiley & Sons, New York (2001)

MM-Correction: Meta-analysis-Based Multiple Hypotheses Correction in Omic Studies

Christine Nardini[1], Lei Wang[1], Hesen Peng[1], Luca Benini[2], and Michael D. Kuo[3]

[1] CAS-MPG PICB, Yue Yuan Road 320, Shanghai, PRC
christine@picb.ac.cn
www.healomic.org
[2] DEIS, Università di Bologna, Viale Risorgimento 2, Bologna, Italy
[3] UCSD Medical Center HillCrest, 200 West Arbor Drive, San Diego, CA, U.S.A.

Abstract. The post-Genomic Era is characterized by the proliferation of high-throughput platforms that allow the parallel study of a complete body of molecules in one single run of experiments (*omic* approach). Analysis and integration of *omic* data represent one of the most challenging frontiers for all the disciplines related to *Systems Biology*. From the computational perspective this requires, among others, the massive use of automated approaches in several steps of the complex analysis pipeline, often consisting of cascades of statistical tests. In this frame, the identification of statistical significance has been one of the early challenges in the handling of *omic* data and remains a critical step due to the multiple hypotheses testing issue, given the large number of hypotheses examined at one time. Two main approaches are currently used: *p*-values based on random permutation approaches and the False Discovery Rate. Both give meaningful and important results, however they suffer respectively from being computationally heavy -due to the large number of data that has to be generated-, or extremely flexible with respect to the definition of the significance threshold, leading to difficulties in standardization. We present here a complementary/alternative approach to these current ones and discuss performances, properties and limitations.

Keywords: Statistical testing, statistical significance, multiple hypothesis testing, false discovery rate, statistical resampling methods, statistical meta-analysis, omic data.

1 Introduction

In recent times high-throughput devices for genome-wide analyses have greatly increased in size, scope and type. In the post-Genomic Era, several solutions have been devised to extend the successful approach adopted for gene expression analyses with microarray technology to other bodies of data such as proteomes, DNA copy number, single nucleotide polymorphisms, promoter sites and many more [1]. These data supports, and notably their integration, represent the future of molecular biology; for this reason the elucidation and definition of tools and methods suited to handle the data produced by these high-throughput devices is of great importance.

Early methods for such analyses were mainly dealing with gene expression data, their goal being to extract items that appear to have coherent trends among themselves (in this

A. Fred, J. Filipe, and H. Gamboa (Eds.): BIOSTEC 2008, CCIS 25, pp. 242–255, 2008.

context commonly called *unsupervised* methods) or with respect to external features, such as clinical markers (*supervised* methods). Both types of approaches have been used for example for the classification of subtypes of poorly understood diseases with unpredictable outcomes [2,3]. Currently, other approaches, that take advantage of larger and diverse sources of information are being devised to address questions of varying complexity in different areas of research rooted in molecular biology. These methods cover a broad variety of applications, from the study of complex hereditary diseases [4] to the identification of radiological traits' *surrogate markers* (the molecular origin of a clinical trait) for enabling non-invasive personalized medicine [5,6]. Overall, besides the variety and complexity of the analyses and methods adopted, some invariants can be identified. The most common atomic step is the identification on the large scale of similarities or associations among molecular behaviors. Such association measures consist for example of scores that evaluate similarities across several samples of genes' expression profiles, or genetic coherence in genes copy number or deletion, and more. Coherence among expression profiles and other association measures can be assessed by means of statistical techniques, namely, by computing a measure of trend similarity (test score, θ) and evaluating the likelihood of this measure to occur by chance (α-level or p-value). The test score is then assumed to be either a measure of actual similarity or only a random effect, based on the value of the associated p-value. The p-value represents the probability of being wrong when assuming that the score represents an actual similarity. This error (type I error) can happen for non-extreme values of the test θ that are difficult to classify as *good* or *bad* and results in erroneously refuting the null hypothesis ($H_0 : \theta = 0$) which assumes that there is no relationship, when actual facts show that the items are tightly related.

The scientific community typically assumes to be meaningful (i.e. *statistically significant*) test scores that are coupled to p-values lower or equal to one of the following nominal p-values: $0.05, 0.01, 0.001$. These values represent the probability of committing type I errors. Given these definitions, the highly dimensional nature of genome-wide data has posed problems and challenges to conventional biostatistical approaches. Indeed, when performing in parallel such a large number of tests, type I errors inherently rise in number, since over a large number of items, the possibility of faults increases.

For this reason, p-values need to be readjusted in a more conservative way, accounting for the so called *multiple hypothesis testing* issue. The most classical technique to account for this problem is the Bonferroni correction [7] that simply multiplies the actual p-value of every single test by the total number of tests observed. However, this approach is not considered viable in *omic* studies, as in fact it often leads to the rejection of too many tests, since none of the corrected p-value is smaller than any of the nominal p-values. An alternative and less conservative approach to this problem is the generation of a random distribution, based on random resampling or on the generation of scores obtained from the randomization of the data. Such approaches allow to build a distribution that represents the population's behavior, and can thus be used to test the hypothesis of interest. When operating with *omic* data, another statistic, the False Discovery Rate (FDR) has been introduced [8,9,10]. Like the p-value, the FDR measures the false positives, however while the p-value controls the number of false positive over the number of truly null tests, the FDR controls the number of false positive over the fraction of significant tests. The utility of this statistic is undeniable, however, its interpretation is far less standardized than the better known p-value, and thus, very often, the value of acceptance of a test based on FDR is much more flexible and dependent on the investigator experience. Globally, these characteristics make the results assessed by FDR highly dependent on the rejection level the investigator chooses. This makes it difficult to automate with high parallelism the identification of statistically significant hypotheses. This problem can becomes relevant due to the increasingly common necessity to merge different sources of information to assess the validity of a given biological hypothesis. Examples of such circumstances arise whenever, for example, the analysis aims at refining, by means of cascades of statistical tests, a set of genes candidate to explain a biological assumption. The hypothesis in fact is refined collecting information across various databases or other forms of *a priori* knowledge, that progressively filter out the spurious data -only as an example see various tools presented in [11,4]. To be efficient, the analysis requires the result of each filtering step to be automatically sent to the following one. Thus the possibility to assess significance by mean of universally accepted values of significance becomes relevant. This latter observation was one of the stimuli motivating the search for an alternative/integrative approach to the multiple hypotheses problem encountered when dealing with genomic datasets.

We also wanted this method to be reasonably efficient to be computed. We thus approached the problem based on techniques that allow the intrinsic correction of p-values in case of multiple tests (*meta analyses* approaches) used for the combination of various statistical tests. Among them, we turned our attention to the category of the *omnibus tests* [12]. These approaches are non-parametric, meaning that they do not depend on the distribution of the underlying data, as long as the test statistic is continuous. In fact, p-values derived from such tests have a uniform distribution under the null hypothesis, regardless of the test statistic or the distribution they have been derived from. However, omnibus tests suffer from a strong limitation: they can be used to assess whether there is a superior outcome in *any* of the studies performed. This means that the combined significance is not a measure of the average significance of the studies performed. An omnibus test therefore cannot be used *as is*, to assess the global statistical validity of the number of tests considered simultaneously. Thus, we manipulated this approach to make it applicable to the definition of a significance threshold.

The main advantage of our solution is twofold. On one side the p-values can be computed in very reasonable times and can thus help managing the computational issues related to permutations techniques; on the other side they represent p-values for which nominal threshold of significance (e.g. $0.05, 0.01, 0.001$) can be applied, and can overcome the threshold selection issue faced when using FDR approaches. Additionally, this method appears to perform better in avoiding the selection of false positives. However, this is coupled to a partially diminished ability in identifying correctly true positives, limited -in our experience- only to the discovery of complex patterns of association, whose biological validity remains to be assessed. These consideration support the findings of several authors that strongly suggest to validate the results obtained from *omic* studies through the use of different techniques and threshold of significance, given the highly noisy nature of the data [13].

2 Related Work

Two main methodologies are currently being used to approach the multiple hypothesis testing issue. The first is based on the principles that define the resampling statistical approaches [7]. In particular we adopted the permutation method that requires the construction of a null distribution to which to compare the actual data. This distribution must be built from the generation of a large number of random data. When the distribution is built using the randomized data generated by *all* the tests, the corresponding p-value is corrected for these same multiple hypotheses. This represents a structurally simple, robust, but computationally intensive approach, given the large numbers involved in the analysis of *omic* data. The computational efficiency issue can become extremely relevant, since most of the interpreted languages commonly used for their large libraries of bioinformatics related functions (notably R and the Bioconductor Project [14], and Matlab), cannot reasonably handle such approaches. Even with the recent improvements for (implicit) parallelization of the computation, time lags for the evaluation of the results remain large. Moreover, for large datasets, compiled languages such as C also require intensive and long lasting computational efforts, unless specific architectures are adopted to enhance efficiency. The second approach consists of novel methods purposely introduced to handle *omic* data that defines the concept of False Discovery Rate. This statistic comes in a number of flavors, and relies on complex statistical assumption. A full description is beyond the scope of this paper, here we briefly describe three of the most used approaches: (i) the pioneering work of Benjamini [8]; (ii) the definition of the q-value [9]; (iii) the FDR adopted in the tool Significance Analysis of Microarray -SAM, [10]- a widespread software used for the analysis of microarray data.

Benjamini FDR: This approach controls the FDR by modifying the p-values obtained on a single test, rescaling it in the following way: $FDR_{BEN} = \frac{K p_i}{i \sum_{i=1}^{K} i^{-1}}$, where p_i represents the i-th of the K single p-values.

q-value: The q-value is the minimum false discovery rate. This measure can be approximated by the ratio of the number of false positives over the number of significant tests, the implementation of the q-value provides several options to evaluate this estimate and to compare it to the corresponding p-values.

SAM FDR: SAM is a tool that allows the extraction of significant genes that help differentiate 2 or more sample classes by means of various scores suited to answer different questions (i.e. depending on the number of sample classes observed and on the meaning of the scores defining the classes, such as survival times, experimental points in time course experiments etc.). Statistical validation of the score value produced by SAM is performed by the generation of a distribution of random score values. These scores are evaluated by means of random permutations of the class labels. These new values, along with the ones from the original classification are used to evaluate the FDR as the average of falsely significant items: $FDR_{SAM} = \frac{\frac{\#signif.\ permuted\ scores}{\#permutations}}{\#signif.\ actual\ scores}$ i.e. the number of items with permuted test scores called significant divided by the number of permutations over the number of items called significant in actual data.

The q-value approach is one of the most widespread, both because of its quality and because of the various and user-friendly implementations the authors have made available. For this reason we choose this method for comparison to ours. In general, FDR scores represent an extremely valuable information while dealing with *omic* data, however, the main issue to the fully automated use of these techniques lies in the flexible acceptance of the threshold values for significance. In other words the investigator can set his threshold for the acceptance of the False Discovery Rate, but no universally accepted thresholds have been recognized. This issue has been pointed out for example in [15]. In this work the authors designed three other statistical scores to help in the choice of the threshold for significance. Among these scores, two are designed to assess general significance threshold criteria for large-scale multiple tests and one is based on existing biological knowledge. Our method does not represent a novel way to evaluate FDR, but it defines a *p*-value, for this reason universally accepted thresholds for significance can be adopted.

More recently and independently from our approach another method based on omnibus tests has been designed to improve the identification of the FDR [16]. Again, one of our goals is to provide an efficient way to evaluate a *p*-value that takes into account the multiple hypotheses tested, in order to be able to adopt the thresholds of significance accepted by the scientific community (0.05,0.01,0.001), easier to automate in long pipelines of tests, and not to improve FDR definition. In this paper we show that the *p*-value obtained with manipulation of the inverse χ^2 method (one of the omnibus tests) can also be used directly as a measure of significance for the identification of statistically significantly tests.

3 Method

We chose as the base for our approach the inverse χ^2 method [12], an *omnibus* statistical test used to ascertain if at least one among several tests is significant, by evaluation of the following statistics: $S(k) = -2\sum_{i=1}^{k} ln(p_i)$ and $s(k) = \chi^2(S, 2k)$ where $k = 1...K$ are the tests performed and p_i the *p*-value of the *i*-th test. S has a $\chi^2_{(s,2k)}$ distribution, where s is the *p*-value of the χ^2 distribution with $2k$ degrees of freedom, and represents the significance of the combined tests, meaning that it can assess if *any* of the tests can be considered significant, accounting for the total number of K tests performed. Thus, in the following, s will indicate the *p*-value we can use for assessing the statistical significance of the tests taking into account the multiple hypothesis issue, while p will indicate the significance of the single test. The score θ is the value resulting from the statistical test. Making use of the χ^2 inverse method means testing the null hypothesis $H_0 : H_{0,1} = ... = H_{0,K} = 0$. Values of $s > 0.05$ indicate that H_0 cannot be rejected and thus that it holds for all the subhypotheses $H_{0,i} = 0, i \in [1, K]$. Conversely, more than one combination of rejection and non-rejection of single hypotheses $H_{0,i}$ is possible to justify the rejection of the global null hypothesis H_0. For example, all but one of the subhypotheses could be null, or only one could be null etc. Evaluating s on all the tests performed would be of no interest in terms of defining a global threshold for significance. In fact, while a non significant value of s would indicate that none of the items has a score value that allows the rejection of the null hypothesis, a low value of s (< 0.05) would only mean that at least one item's score is relevant to the rejection of the null hypothesis, with no indication on which one(s) are the relevant items. To overcome this limitation we ranked the tests scores θ in ascending order (assuming that significant values of the test are represented by high values of the score), and ordered the p-values consistently. We then evaluated s for sets of p-values of increasing size, starting from a set made of only the p-value corresponding to the worse test score, then adding at each iteration of this algorithm another p-value coupled to the immediately higher or equal (better) score (θ), and closing the last iteration with all the p-values. By induction (Equation 1) we can show that whenever the value of s drops below any of the standard values of significance $(0.05, 0.01, 0.001)$ the score corresponding to the last p-value added is the threshold for significance, since it represents the specific test that accounts for the impossibility to reject the global null hypothesis H_0. By construction, at each iteration, the p-value added is always smaller, and correspondingly, due to the logarithm properties, S shows a fast growth $(S(k) = -2\sum_{i=1}^{k} ln(p_i))$. At the same time the parameter of the χ^2 function k, grows linearly $(2 \cdot k)$. Because of the shape of the χ^2 function and because of the logarithm properties, if there are *enough* small p-values, S becomes quickly and abruptly very large, and moves to behaviors typical of the ones on the right hand side of Figure 1(c), $\chi_k^2(S) \rightarrow_{k \rightarrow inf, S \rightarrow inf} 0$. This gives s its typical shape (shown in Figure 1(b)), with a very abrupt drop from values very close to 1 to values very close to 0.

$$
\begin{aligned}
For \quad i = 1 \quad & s(i) > 0.05 \Rightarrow H_0 \ not \ rej. \ H_{0,1} \ not \ rej. \\
Let \quad i = n \quad & s(i) > 0.05 \Rightarrow H_0 \ not \ rej. \ H_{0,i} \ not \ rej., \forall i \in [1,n] \\
Then \ i = n+1 \ & s(i) > 0.05 \Rightarrow H_0 \ not \ rej. \ H_{0,i} \ not \ rej., \forall i \in [1,n+1] \\
& s(i) \leq 0.05 \Rightarrow H_0 \ rej. \quad H_{0,i} \ not \ rej., \forall i \in [1,n], H_{0,n+1} \ rej.
\end{aligned}
\tag{1}
$$

Figure 1 shows an example of the trends of the variables involved in the evaluation of global significance: the statistics S and s that define the global significance, the test score θ and the corresponding single p-value that are the basic units of the analysis. The statistic S represents the argument of the χ^2 function and is associated to a given

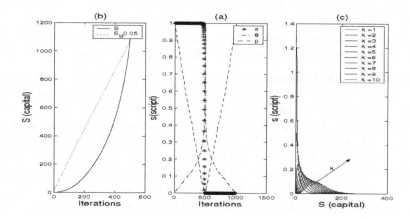

Fig. 1. Graphical representation of the different scores involved in the analysis. Figure(a) deals with the statistic S and S_{id}. Figure(b) plots the corrected p-value s, the absolute value of the correlation score θ and the single p-value p. Figure(c) shows the χ^2 probability density function.

degree of freedom (k). For any given degree of freedom it is possible to identify the minimum value (here called $S_{id\alpha}$) for which the inverse χ^2 function returns the suited probability α. Since S_{id} is the minimum value, the p-value that represents the threshold for significance is associated to $k_{sign\alpha}$ and can be conveniently visualized as the point in which $k_{sign\alpha} = k | S_{id\alpha}(k) = S(k)$. Equivalently for s the threshold for significance at a given nominal level α can be defined as $k_{sign} = min_{k\in[1,K]} |s(k) \leq \alpha$. In our experiments θ is the Spearman correlation score [7]. Before processing the test values we separated positive from negative scores, and then performed the previously described operations on the absolute values. This sign segregation of the data has a two-fold objective. On one side this fulfills the requirement for the applicability of the test since one tailed p-values are required. On the other side it satisfies the biological necessity to discern between significantly over and under expressed genes, based on positive and negative values of the test scores. As far as the permutation approach is involved we generated 1000 random permutations (as well as a run of 10,000 for Dataset 2) of each trait values as it was done in other applications with this same goal [17]. We then re-evaluate the θ scores for all 1000 (10,000) randomized instance of each trait, these constitute the null distribution. For the FDR approach, we used the q-value R package with default settings. For the identification of significant items, we adopted as threshold the same values we used for the p-value. The method was implemented in R.

3.1   Data

To test our method, we simulated the typical set up of a common genomic experiment. Namely, we generated a random expression matrix $1000x100$ (i.e. 1000 genes and 100 samples) and we defined a number of *external traits* for which we search the *surrogate markers*. In other words, these external traits mimic any clinical trait or molecular marker. The goal of the experiment is to identify the genes associated to the external traits, to define the traits' surrogate markers. This approach is then used to investigate the molecular etiology of commonly used clinical markers. Several examples of such approaches can be found in literature, only as a sample see [3,17,6]. We choose to perform two levels of validation: the first (Dataset 1) to assess more in general the limitations of the proposed approach with respect to the existing methods, the second (Dataset 2) to investigate in more details the characteristics of the results.

Dataset 1 - Methods Comparisons. The surrogate markers were obtained either by simple copy of expression profiles (in varying number of copies, namely 0, 1, 5), or by sum of varying numbers of profiles (namely 5, 30). The first group of external traits (#1, #2, #3) provides both the negative control (0 copies, obtained by elimination of a randomly chosen expression profile, and exported as external trait) and helps measuring the comparative ability of the 3 different approaches (FDR, permutations and MM-correction) in extracting small cluster of correlated profiles (1, 5 copies). The second set of traits (#4, #5) tests the approach with more challenging data (sums of 5, 30 copies). To each expression value we added varying levels of gaussian noise (0%, 50%, 100%) proportional to the expression value, to asses the limits of applicability of the approaches. The method was also tested to assess its reliability with variable numbers of genes.

Dataset 2 - Testing Results Properties. In order to explore in more details the characteristics of the trait-related genes that the MM-correction produces, we generated a second set of data, to expand the characteristics of traits #1-#3, that are better captured by the 3 approaches under study. Noise level was reduced to 0%, 1% and 10% only [18]. Again, the surrogate markers were obtained by simple copy of expression profiles (in varying number of copies, namely 1, 5, 30), and also including a negative control (0 copies, obtained by elimination of a randomly chosen expression profile, and exported as external trait). In total we defined 4 associations (3 different sized linear associations, one negative control). For this data set both 1000 and 10,000 permutations were run, resulting in the same classification ability performances.

In both data sets, to avoid specific case results, we replicated our the data generation 3 times per each noise level and averaged the results of specificity, sensitivity, positive and negative predictive value. We observed the approach for the 3 levels of significance 0.05, 0.01, 0.001.

3.2 Multiclass Statistical Scores

To compare the results of different methods we evaluated the specificity, sensitivity, negative and positive predictive value. These statistics are used in combination to quantify different aspects of the accuracy of a binary test, evaluating different proportions of correctly and incorrectly classified items, when compared to a known classification, considered the gold standard. In this context the *test* is the ensemble of all the operations performed to classify each items; *positive* and *negatives* label the items according to the two classes $c = N, P = 0, 1$ they belong to; *true* (T) and *false* (F) represent the ability of the test to classify coherently or not a given item in the test classification with respect to the gold standard classification. Thus, for example, in classical definitions TN (true negative) labels items belonging to class 0 (N) correctly classified by the test, and FP (false positive) labels items incorrectly classified as 1 (P) by the test. Given these definitions, positive and negative predictive value (PPV, NPV), sensitivity (Se) and specificity (Sp) are usually formalized with the relationships in the first part of Equations 2.

When the test classifies $n > 2$ categories, these definitions become more complex to apply. However, it still remains important to be able to characterize the performances of the test in terms of its ability to distinguish between items that belong and do not belong to any category (in our case between genes that constitute and do not constitute any molecular surrogate). To reach this goal and preserve the meaning of the 4 scores (PPV, NPV, Se, Sp) some caution must be used. In fact the meaning of *positive* and *negative* is not relevant anymore, since there are now *positives*. Then, while the definition of *true* remains straightforward, as it indicates coherence between the classification of the test and the gold standard, the definition of *false* can be cumbersome, since there are $n - 1$ ways to misclassify an item. Additionally, the possibly intuitive definition of *false positives* (or *negatives* as items that are non-zero in the test (or in the gold standard) classification leads to ambiguity, since items happen to be contemporary false positives *and* false negatives. To avoid confusion and ambiguities the actual values of all false can be identified by rewriting the problem in terms of a system of equation based on the relationships indicated in Table 1. Here P_t, N_t represent the total number of positive and negative items that can be found in the test (t) categorization, and P_{gs}, Ngs in the

Table 1. Classical definition and generalization to 3 classes for *true, false, negatives, positives*

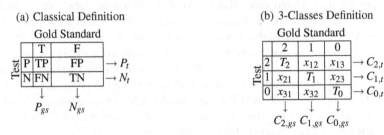

(a) Classical Definition

Gold Standard

	T	F	
P	TP	FP	→ P_t
N	FN	TN	→ N_t
	↓	↓	
	P_{gs}	N_{gs}	

(b) 3-Classes Definition

Gold Standard

	2	1	0	
2	T_2	x_{12}	x_{13}	→ $C_{2,t}$
1	x_{21}	T_1	x_{23}	→ $C_{1,t}$
0	x_{31}	x_{32}	T_0	→ $C_{0,t}$
	↓	↓	↓	
	$C_{2,gs}$	$C_{1,gs}$	$C_{0,gs}$	

gold standard (*gs*) classification. The definitions can be generalized to $n > 2$ classes changing the term negative and positive with the indices of the corresponding classes $c = 0, 1, ..., n$, and having C_c that designs the total number of positives for each given class. The system of equations obtained from the relationships in the rows and columns of Table 1 contains $2 \cdot n$ equations (i.e. $TP + FP = P_t$) and $2 \cdot n$ unknown (x_{ij}), thus it is completely specified. It is worth noticing, that with these general definitions, in case of 2-classes test, Se and Sp appear to be dual scores. Thus, when generalizing to n-classes it is possible to define the predictive ability of the test for each given class $c \in 0, 1, .., n$ as $PV_c = T_c/C_t$ and the Sensitivity/Specificity (now called Sep) for the same class c as $Sep_c = T_c/C_{gs}$. To clarify the situation it is extremely useful to rewrite the definitions as they are written on the left hand side of Equation 2, namely:

$$
\begin{aligned}
PPV &= TP/TP + FP) = TP/P_t \\
PPN &= TN/(TN + FN) = TN/N_t \\
Se &= TP/(TP + FN) = TP/P_{gs} \\
Sp &= TN/(TN + FP) = TN/N_{gs}
\end{aligned}
\tag{2}
$$

For n classes this gives:

$$
\begin{aligned}
PPV &= \Sigma_c T_c / \Sigma_c C_{c,t}, c = 1, .., n \\
PPN &= T_0/N_t = T_0/C_{0,t} \\
Se &= \Sigma_c T_c / \Sigma_c C_{c,gs}, c = 1, .., n \\
Sp &= T_0/N_{gs} = T_0/C_{0,gs}
\end{aligned}
\tag{3}
$$

4 Results and Discussion

Provided that the single test p-values can be generated making use of known distribution, the MM-correction is much more efficient than the permutation method, since MM-correction's computational complexity is $O(g \cdot t)$ while the bootstrapping one is $O(g \cdot t \cdot p)$, with g indicating the number of genes, t the number of external traits, and p the number of permutations. The comparison with FDR in these terms is not relevant, since this method is computationally efficient.

Dataset 1. As far as the first comparison is involved, all methods performed with varying good degrees of specificity ($Sp > 0.95$), but none had satisfactory sensitivity ($Se < 0.5$ to $Se \ll 0.5$) except the permutation method for only the threshold 0.05,

Table 2. Table 2(a) shows the statistics of the performances of the 3 methods compared on Dataset 1: MM-correction, permuted p-values and FDR. The comparison is done on expression matrices 1000x100 and 5 traits as they are described in Section 3.1. Results are averaged over 3 instances of the random data generated with the same specifics. Standard deviations of these averages are below 10^{-2}. The first column indicates the noise level (n), the second the threshold of significance chosen (α) and then all the scores for the 3 methods. Because of space constraints only values for noise 0.5 are shown. Table 2(b) shows class by class comparison of the algorithms performances. MM-correction performs better in terms of avoiding false positive ans worse with false negatives. Data are shown as averages across the random replicates and across the 3 different levels of significance, for 3 different levels of noise (n). Figures in italic were inferred from NANs.

(a) Global Performances

n	α	MM-correction		Permutations		FDR - q-value	
		Se	Sp	Se	Sp	Se	Sp
0.5	.05	.1905	.9998	.6746	.9512	.1667	.9948
	.01	.1667	.9999	.4603	.9898	.1667	.9948
	.001	.1667	1.000	.3175	.9981	.1667	.9948

(b) Trait By Trait Performances

n	Method	PV (classes)						Sep (classes)					
		0	1	2	3	4	5	0	1	2	3	4	5
0	MM-corr.	.9998	1.000	*1.000*	1.000	.3111	.0556	.9936	1.000	*1.000*	1.000	*0.000*	*0.000*
	Perm.	.9797	1.000	*1.000*	1.000	.9556	.2852	.9956	.2510	0.000	.3846	.4325	.3494
0.5	MM-corr.	.9999	1.000	*1.000*	1.000	.0444	.0037	.9931	1.000	*1.000*	1.000	*0.000*	*0.000*
	Perm.	.9797	1.000	*1.000*	1.000	.9556	.2852	.9956	.2510	0.000	.3846	.4325	.3494
1	MM-corr.	.9999	.3333	*1.000*	.7333	.0000	.0037	.9925	*0.000*	*1.000*	.9506	*0.000*	*0.000*
	Perm.	.9797	1.000	*1.000*	1.000	.9556	.2852	.9956	.2510	0.000	.3846	.4325	.3494

$Se_{perm,\alpha=0.05} = .67$ (see Table 2(a)). In particular, MM-correction has intermediate sensitivity (better than FDR) and specificity (better than permutations). Since the FDR method at the chosen thresholds for significance appears to behave in extreme ways, i.e. with better specificity and worse sensitivity with respect to both methods, we focused our attention to a more refined comparison between the bootstrapping and the MM-correction method, and did not pursue the goal, out of our scope here, to evaluate FDR results with other thresholds for significance. Namely, we performed the second experiment, on a trait by trait basis, with two goals: to investigate the reasons of the improved performances of our method in terms of specificity; to assess the reasons for the poor global performances in terms of sensitivity. For this we evaluated PV and Sep for each one of the 6 classes ($c = 0, 1, .., n$). In general MM-correction seems to have more problems with false negatives, while the bootstrapping method collects a much larger number of false positives (Table 2(b)). These characteristics depend on the intrinsic properties of s as they have been described in Section 3. The abrupt drop in value of s is responsible for an almost binary behavior of this score. This leaves very little *gray* areas for spurious classification, thus ambiguous θ values are quickly coupled to high s values and discarded from the significant tests set. Overall, trait #5 defines a too complex pattern (sum of 30 profiles), and none of the method can treat it correctly, conversely, trait #4 (sum of 5 profiles) can be superiorly handled by the permutation method and trait #1, #2 and #3 ($1, 0, 5$ correlated profiles) are better recognized with our method. It is difficult to speculate on whether surrogate markers of type #3 are more or less common than the ones of type #4 in actual biology, we can state however that our method is able to identify the surrogate markers of trait #3 with profiles that have as little correlation as 0.33 (100% noise addedd).

To summarize the results on Dataset 1 we evaluated ROC curves to assess if any of the methods was strikingly outperforming the other (ROC curves in this case are not used to evaluate the relationship between sensitivity and specificity, but to compare two

Table 3. Class by class comparison on Dataset 2 of the algorithms performances averaged over 3 random instances, 3 levels of noise and 3 levels of significance for space constraints and given the overall small variance of the data. MM-correction seems to have overall better performances. The variance in the permutation approach is due to poor performances of the approach at the 0.005 level of significance. In this case, in fact, the method collects a large number of false positives.

Avg (Stdev)	PV Control	Sep Control	PV Lin. 1G	Sep Lin 1G	PV Lin 5G	Sep Lin 5G	PV Lin 30G	Sep Lin 30G
Permut	0.998(.001)	0.980(.022)	0.253(.282)	1.000(.000)	0.433(.339)	1.000(.000)	0.740(.245)	1.000(.000)
MM-Correct	0.996(.000)	0.998(.002)	1.000(.000)	1.000(.000)	0.931(.093)	1.000(.000)	1.000(.000)	1.000(.000)

populations of data, that happen to be PV and Sep scores). Namely, we compared: (i) PV and Sep for each method, (ii) Sep only, (iii) PV only. Namely, sensitivity and specificity combined, as well as specificity alone lead to $AUC \approx 0.5$, while the sensitivity test leads to $AUC \approx 0.6$, slightly better, but not statistically significant ($AUC = 0.5$ indicates tests with comparable performances). Finally, we tested our method for the same hypotheses for varying numbers of genes, from 100 to 2000 (steps of 100 genes). Across 20 samples we obtained median values that reproduce the findings of the two previous experiments (global and trait by trait performances) with very small variances across the 20 samples ($\approx 10^{-2}$ for sensitivity and $\approx 10^{-3}$ for specificity). Thus, the method appears to be stable with respect to the number of items tested.

Dataset 2. Because the trait by trait performances of Dataset 1 proved traits #4 and #5 to be too difficult to recognize with all the approaches tested, Dataset 2 was designed to investigate in more details linear associations with genes sets of varying size. Performances are shown in Table 3.

As a final investigation, we also tried to shed light on the characteristics of the trait-related genes sets found. We speculated that a desirable characteristic of a supervised method should be to guarantee that not only the genes are indeed statistically significantly associated with the external trait of interest, but also that they form a cluster, i.e. they share association among them. The clustering coefficient [19] used in network theory [20] is specifically designed to measure this characteristics.

Briefly, for each trait we built a graph where the nodes (a set of n vertices $V = v_1, ..., v_n$) represent genes and the edges (a set E, where $e_{i,j}$ connects v_i and v_j) are the links between these nodes, and represent the correlation among the corresponding genes' expression (we call it *self*-correlation to distinguish it from the *trait*-correlation evaluated above). Depending on the value chosen to represent a significant self-correlation, the graph will present more or less interconnections among the genes. This property can be measured with the clustering coefficient: once we define the neighborhood of a node i as $N_i = \{v_j\}|e_{i,j} \in E$, C_i measures the ratio between the number of edges that actually exist between the neighboring nodes, and the maximum possible number of $(k_i \cdot (k_i - 1))/2$ edges, where $k_i = |N_i|$: $C_i = \frac{2 \cdot |\{e_{j,k}\}|}{k_i(k_i-1)} |v_j, v_k \in N_i, e_{j,k} \in E$ The clustering coefficient of the whole cluster is given by the average value of all C_i's, $i = 1, .., n_C$, where n_C is the size of the cluster.

Since our goal was to assess how trait-correlation could also be a measure of self-correlation (i.e. if and how *surrogate markers* represent *gene clusters*) we performed the following test. For each trait we selected the genes identified by changing values of the trait-correlation. We then computed the clustering coefficient of these genes. The

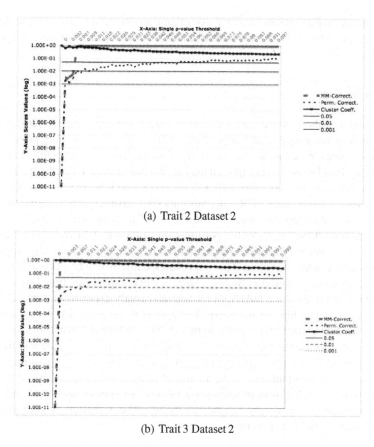

(a) Trait 2 Dataset 2

(b) Trait 3 Dataset 2

Fig. 2. Plot of *p*-values corrected with permutations and MM-correction compared with the Clustering Coefficient. Only the traits from Dataset 2 associated with 5 (Trait #2) and 30 (Trait #3) genes are presented, since they mimic better biological associations. Statistical significance is met for higher values of the Clustering Coefficient when MM-correction is used, compared to permutations correction.

clustering coefficient was computed as the average of the clustering coefficient of the nodes in the trait-associated set, however each clustering coefficient could involve the connection to other genes, connected to the trait associated ones due to high self-correlation. We performed the same process for both permutations and MM-correction. Since each trait-correlation threshold is associated to a corrected p-value, we could plot corrected *p*-values for both MM-correction and permutations method versus the Clustering Coefficient (see Figure 2), and observe how the Clustering Coefficient is associated to the significance of the trait association. Interestingly, due to the 'step function'-like behavior of the MM-corrected p-values, the surrogate markers found with this method appear to guarantee a higher clustering property of the genes. This characteristic is particularly useful and desirable when the trait-associated genes do not appear to be enriched for (i.e. statistically significantly associated with) any known molecular function (i.e. Gene Ontology analyses [21]). In this case, in fact, high values of the Clustering

Coefficient can give a measure of actual association among genes, when no previous information is available, this can represent a valuable hypothesis for further experimental investigation.

5 Conclusions

We presented a method for the identification of p-values in *omic* studies. This approach is based on a meta-analysis and has two main advantages. On one side, provided that the single p-values are computed using known distributions, it is computationally efficient, and can thus be used in interpreted languages such as R and Matlab that offer rich libraries of functions for *omic* analyses. On the other side it is based on the identification of a p-value rather than FDR, and can thus take advantage of nominal threshold for significance, allowing for an easier automation of filtering steps in analyses based on statistical tests. Conversely to the permutation technique, that remains a computationally intensive but very robust reference method, our approach, globally, appears to be more specific but less sensitive. This improved specificity can be extremely advantageous in the practice of Systems Biology, since novel compact functional subunits can emerge or remain uncovered and require longer and costly experimental investigations to be extracted, depending on the noise they appear to be identified with. Along the same line, the data extracted using this approach share a higher Clustering Coefficient, this is again an information that can help in the speculative process of the identification of genomic functional units. Application of this method to real data can provide insightful and useful information, during the initial analysis of complex *omic* data, since it will allow the identification of interesting targets, for further analytical and experimental investigations. For these reasons we believe the definition of alternative and complementary method is appropriate.

Acknowledgements. The authors would like to thank Diego di Bernardo and Mukesh Bansal for constructive discussion.

References

1. Nardini, C., Benini, L., Micheli, G.D.: Circuits and systems for high-throughput biology. Circuits and Systems Magazine, IEEE 6(3), 10–20 (2006)
2. Ramaswamy, S., Ross, K.N., Lander, E.S., Golub, T.R.: A molecular signature of metastasis in primary solid tumors. Nat. Genet. 33(1), 49–54 (2003)
3. Lapointe, J., Li, C., Higgins, J.P., van de Rijn, M., Bair, E., Montgomery, K., Ferrari, M., Egevad, L., Rayford, W., Bergerheim, U., Ekman, P., DeMarzo, A.M., Tibshirani, R., Botstein, D., Brown, P.O., Brooks, J.D., Pollack, J.R.: Gene expression profiling identifies clinically relevant subtypes of prostate cancer. Proc. Natl. Acad. Sci. 101, 811–816 (2004)
4. Rossi, S., Masotti, D., Nardini, C., Bonora, E., Romeo, G., Macii, E., Benini, L., Volinia, S.: TOM: a web-based integrated approach for efficient identification of candidate disease genes. Nucleic Acids Res. 34, 285–292 (2006)
5. Segal, E., Sirlin, C.B., Ooi, C., Adler, A.S., Gollub, J., Chen, X., Chan, B.K., Matcuk, G., Barry, C., Chang, H.Y., Kuo, M.D.: Decoding global gene expression programs in liver cancer by noninvasive imaging. Nature Biotechnology 25, 675–680 (2007)

6. Diehn, M., Nardini, C., Wang, D.S., McGovern, S., Jayaraman, M., Liang, Y., Aldape, K., Cha, S., Kuo, M.D.: Identification of non-invasive imaging surrogates for brain tumor gene expression modules. PNAS 105(13), 5213–5218 (2008)
7. Sokal, R.R., Rohlf, F.J.: Biometry. Freeman, New York (2003)
8. Benjamini, Y., Hochberg, Y.: Controlling the false discovery rate: a practical and powerful approach to multiple testing. J. R. Stat. Soc. B 57, 289–300 (1995)
9. Storey, J.D., Tibshirani, R.: Statistical significance for genomewide studies. PNAS 10(16), 9440–9445 (2003)
10. Tusher, V.G., Tibshirani, R., Chu, G.: Significance analysis of microarrays applied to the ionizing radiation response. Proc. Natl. Acad. Sci. 98, 5116–5121 (2001)
11. Tiffin, N., Adie, E., Turner, F., Brunner, H., van Drielnd, M., Oti, M.A., Lopez-Bigas, N., Ouzunis, C., Perez-Iratxeta, C., Andrade-Navarro, M.A., Adeyemo, A., Patti, M.E., Semple, C.A.M., Hide, W.: Computational disease gene identification: a concert of methods prioritizes type 2 diabetes and obesity candidate genes. Nucleic Acids Res. 34 (2006)
12. Hedges, L.B., Olkin, I.: Statistical Methods in Meta-Analysis. Academic Press, New York (1985)
13. Pan, K.H., Lih, C.J., Cohen, S.N.: Effects of threshold choice on biological conclusions reached during analysis of gene expression by DNA microarrays. Proc. Natl. Acad. Sci. 102(25), 8961–8965 (2005)
14. Gentleman, R., Carey, V., Huber, W., Irizarry, R., Dudoit, S.: Bioinformatics and Computational Biology Solutions Using R and Bioconductor. Springer, Heidelberg (2005)
15. Cheng, C., Pounds, S., Boyett, J., Pei, D., Kuo, M., Roussel, M.F.: Statistical significance threshold criteria for analysis of microarray gene expression data. Stat. Appl. Genet. Mol. Biol. 3, Article36 (2004)
16. Yang, J.J., Yang, M.C.: An improved procedure for gene selection from microarray experiments using false discovery rate criterion. BMC Bioinformatics 7, 15 (2006)
17. Liang, Y., Diehn, M., Watson, N., Bollen, A.W., Aldape, K.D., Nicholas, M.K., Lamborn, K.R., Berger, M.S., Botstein, D., Brown, P.O., Israel, M.A.: Gene expression profiling reveals molecularly and clinically distinct subtypes of glioblastoma multiforme. Proc. Natl. Acad. Sci. 102(16), 5814–5819 (2005)
18. Bansal, M., Belcastro, V., Ambesi-Impiombato, A., di Bernardo, D.: How to infer gene networks from expression profiles. Mol. Syst. Biol. 3 (2007)
19. Watts, D.J., Strogatz, S.: Collective dynamics of 'small-world' networks. Nature 393, 440–442 (1998)
20. Lauritzen, S.L.: Graphical Models. Oxford University Press, New York (1996)
21. The Gene Ontology Consortium. Creating the gene ontology resource: Design and implementation. Genome Res. 11(8), 1425–1433 (2001)

A Supervised Wavelet Transform Algorithm for R Spike Detection in Noisy ECGs*

G. de Lannoy, A. de Decker, and M. Verleysen

Université Catholique de Louvain, Machine Learning Group
Place du levant 3, 1348 Louvain-La-Neuve, Belgium
{gael.delannoy,arnaud.dedecker,michel.verleysen}@uclouvain.be

Abstract. The wavelet transform is a widely used pre-filtering step for subsequent R spike detection by thresholding of the coefficients. The time-frequency decomposition is indeed a powerful tool to analyze non-stationary signals. Still, current methods use consecutive wavelet scales in an a priori restricted range and may therefore lack adaptativity. This paper introduces a supervised learning algorithm which learns the optimal scales for each dataset using the annotations provided by physicians on a small training set. For each record, this method allows a specific set of non consecutive scales to be selected, based on the record's characteristics. The selected scales are then used for the decomposition of the original long-term ECG signal recording and a hard thresholding rule is applied on the derivative of the wavelet coefficients to label the R spikes. This algorithm has been tested on the MIT-BIH arrhythmia database and obtains an average sensitivity rate of 99.7% and average positive predictivity rate of 99.7%.

1 Introduction

In the framework of biomedical engineering, the analysis of the electrocardiogram (ECG) is one of the most widely studied topics. The easy recording and visual interpretation of the non-invasive electrocardiogram signal is a powerful way for medical professionals to extract important information about the clinical condition of their patients.

The ECG is a measure of the electrical activity associated with the heart. It is characterized by a time-variant cyclic occurrence of patterns with different frequency content (QRS complexes, P and T waves). The P wave corresponds to the contraction of the atria, the QRS complex to the contraction of the ventricles and the T wave to their repolarization. Because the ventricles contain more muscle mass than the atria, the QRS complex is more intensive than the P wave. The QRS wave is therefore the most representative feature of the ECG. Furthermore, once the QRS complex has been identified, other features of interest can be more easily detected.

Analyzing ECGs for a long time can lead to errors and misinterpretations. This is the reason why automatic feature extraction of the ECG signal can help physicians in their diagnosis for early detection of cardiac troubles. The feature extraction mainly consists in the automatic annotation of the different waves in the recording, the most important of them being the QRS. One of the main application of the QRS detection is the

---

* This paper is an extended version of a conference paper.

A. Fred, J. Filipe, and H. Gamboa (Eds.): BIOSTEC 2008, CCIS 25, pp. 256–264, 2008.

heart rate variability (HRV) analysis [29]. HRV measures have been proven successful in diagnosing cardiac abnormalities and neuropathies or evaluating the actions of the autonomic nervous system on the heart [1]. However, HRV measures heavily rely on the accuracy of the QRS feature detection on the digitalized ECG signal.

Automatic feature extraction and especially R spike detection is thus a milestone for ECG analysis. However, it is a difficult task in real situations: (1) The physiological variations due to the patient and its disease make the ECG a non-stationary signal. (2) Other ECG components such as the P or T wave looking like QRS complexes often lead to wrong detections. (3) There are many sources of noise that pollute the ECG signal such as power line interferences, muscular artifacts, poor electrode contacts and baseline wanderings due to respiration. These three problems highly compromise the detection of R spikes.

The detection of QRS complexes in the ECG has been conducted by many researchers in the past years. However, to our knowledge, none of the current algorithms are able to automatically learn their parameters using pre-labeled beats provided by physicians. The aim of this paper is to introduce a new algorithm for R peak detection that does not blindly detect beats but learns and propagates the annotations provided by physicians on a small portion of the signal, which is often wanted in real situations. Our contribution consists in the design and experiment of a supervised learning algorithm for an optimal and automatic signal decomposition and subsequent optimal R spike detection. The associated detection method by hard thresholding rule is also presented. The algorithm does not require any pre-processing of the signal and can also be adapted for the detection of other features such as the P or T wave.

The following of this paper is structured as follows. After this introduction, Sect. 2 gives a brief literature review about the state of the art on ECG feature detection and especially the QRS detection. Section 3 provides a summary of the theory about the continuous wavelet transform used in this paper. Section 4 introduces the methodology followed by the algorithm and Sect. 5 shows the experiments and results obtained on a real public database.

2   State of the Art

Due to the non-stationarity of the ECG signal, the physiological conditions and the presence of many artifacts, finding a robust and general algorithm for ECG feature detection is a tough task. A lot of work has been published in the literature about the detection of various interesting ECG features such as P waves, QRS waves, T waves, QT intervals or abnormal beats by numerous techniques [2,25,26]. This paper focuses on R spike detection only.

For this purpose, several approaches using different signal processing methods have been reported previously: template matching [12], mathematical models [24], signal envelop [23], matched filters [18], ECG slope criterion [4], dynamic time warping [30], syntactic methods [19], hidden Markov models [10], neural networks [31,27], adaptive thresholding [16,9] and geometrical approach [28].

In all cases, using a pre-filtering step prior to beat detection is required for the elimination of artifacts in the signal. The time-frequency decompositions by wavelet transform

(WT) seem the most intuitive tool for ECG filtering [2]. The WT is naturally appropriate for analyzing non-stationary signals because it allows precise time-frequency representation of the signal with a low computational complexity. A lot of work has been published in past years on the use of the WT for QRS detection. In 1995, [20] used an algorithm based on finding the maxima larger than a threshold obtained from the pre-processed initial beats. Later, [17] produced a method allocating a R peak at a point being the local maxima of several consecutive dyadic wavelet scales. In both these methods, a post-processing allowed to eliminate false R detections. Based on these two publications, a lot of other researches were published on the beat detection using a WT filtering step [27,13,22,2,8,7].

The main problem of the WT is that one has to choose the mother wavelet and the scales used to analyze the signal on an empirical basis. While the mother wavelet can easily be chosen based on its characteristics and resemblance with a QRS wave, the ideal scale(s) at which the QRS are matched is harder to guess a priori. Current algorithms blindly search for QRS complexes in a limited number of consecutive scales selected in a range of a priori fixed scales. However, the shape of the QRS pattern can be varying between patients but also with time. One or several consecutive fixed wavelet scales may not be enough to match all complexes at once in a dataset. In this paper, we propose a new supervised learning algorithm based on the continuous wavelet transform that overcomes these issues. It only relies on the annotations provided by physicians on a small portion of the signal in order to select the optimal subset of non-consecutive scales for each dataset.

3   Theoretical Background

The continuous wavelet transform (CWT) is a tool which produces a time-frequency decomposition of a signal $x(t)$ by the convolution of this signal with a so-called *wavelet function*.

A wavelet function $\psi(t)$ is a function with several properties. It must be a function of finite energy, that is

$$E = \int_{-\infty}^{+\infty} |\psi(t)|^2 dt < \infty , \tag{1}$$

and it must have a zero mean.

From a wavelet function, one can obtain a family of time-scale waveforms by translation and scaling

$$\psi_{a,b}(t) = \frac{1}{\sqrt{a}} \psi\left(\frac{t-b}{a}\right) , \tag{2}$$

where $a > 0$ represents the scale factor, b the translation and $a, b \in \mathbf{R}$. When $a = 1$ and $b = 0$, the wavelet is called the *mother wavelet*.

The *wavelet transform* of a function $x(t) \in L^2(\mathbf{R})$ is a projection of this function on the wavelet basis $\{\psi_{a,b}\}$:

$$T(a,b) = \int_{-\infty}^{+\infty} x(t)\psi_{a,b}(t)dt . \tag{3}$$

For each a, the wavelet coefficients $T(a,b)$ are signals (that depend on b) which represent the matching degree between wavelet $\psi_{a,b}(t)$ and the analyzed function $x(t)$.

The signal energy at a specific scale and position can be calculated as

$$E(a,b) = |T(a,b)|^2 \ . \tag{4}$$

The two-dimensional wavelet energy density function is called the *scalogram*.

The CWT is a suitable tool for ECG analysis because of this time-frequency representation of the signal. With the multiscale feature of WTs, the QRS complex can be distinguished from high P or T waves, noise, baseline drift, and artifacts. The important time aspect of the non-stationary ECG signal is kept. Moreover, very efficient implementations of the algorithm exist and a low computational complexity is required, allowing real-time analysis. With the aim of a QRS detection, an appropriate mother wavelet must be chosen. It must match nicely with a QRS complex, in order to emphasizes these complexes and to filter the useless noise. For more details on the wavelet transform and on the standard wavelet functions available, the interested reader can consult [21,2,11].

4   Methodology

4.1   General Description

The detection of R spikes is a tough task due to the complexity of the ECG signal. The aim of the algorithm introduced here is to automatically find the best subset of wavelet scales for subsequent optimal R detection. For each dataset, this subset is selected on a short training sample by a supervised learning procedure. The CWT at the selected scales is then computed on the complete dataset. Finally, R spikes are detected by a hard thresholding rule on the selected wavelet coefficients.

4.2   Supervised Pre-filtering

The algorithm uses a supervised learning approach: it will use the labeled information that is provided and learn the best way to adapt to the problem. Here, the labeled information that is provided is the location of the R peaks in a training dataset.

Each dataset consists in a long-term ECG signal recording (for example 24 hours). With such long recording, the problem is that a manual extraction of the R peaks cannot be performed, as detailed in the Introduction. However, asking a specialist to annotate a small part of the signal by indicating the R peaks is perfectly feasible; this annotated part will consist in labeled segments of one minute each, taken at random locations over the entire dataset. Choosing random locations along the signal is a way to obtain a representative training set maximizing the probability to include all types of beats contained in the recordings. The CWT is then computed on the training set in a wide (therefore non restrictive) range of 50 fixed scales defined as $\{s_i\}$, $1 \leq i \leq 50$. The mother wavelet $\psi(t)$ that was used in our experiments is the mexican hat wavelet, for its similarity with the regular morphology of the QRS complex. The wavelet is represented Fig. 1. It is defined as the second derivative of the gaussian probability density function:

$$\psi(t) = \frac{1}{\sqrt{2\pi\sigma^3}} \left(1 - \frac{t^2}{\sigma^2}\right) e^{\frac{-t^2}{2\sigma^2}} \ . \tag{5}$$

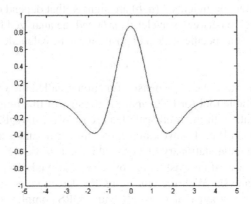

Fig. 1. The Mexican hat mother wavelet

In order to select the appropriate scales among the wide range of wavelet scales, one needs a criterion. A natural criterion is the percentage of correct R peaks detection on the annotated parts of the signal using the coefficients of the wavelet transform at the trial scales in the set $\{\psi(t)_i\}$. A *stepwise forward* method automatically selects the best subset $\{a_k\} \subset \{s_i\}$ of scales on the basis of the detection rate. It involves starting with an empty subset, trying out at each step the trial scales one by one and including them to the model if the detection rate is improved. The procedure stops when no scale left in $\{s_i\}$ can improve the detection rate. In addition, at each step, the scales previously selected in $\{a_k\}$ are individually challenged: if their removal does not decrease the detection rate, the scale is now useless and therefore removed from the model.

The set $\{a_k\}$ of scales coming from the selection is thus made of the scales giving the best R detection when combined together. The selected set of scales is then used for R spike detection on the complete original long-term recording. Figure 2 shows an original ECG segment and the coefficients of the first selected wavelet scale.

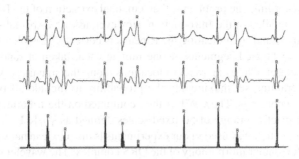

Fig. 2. Example of an original ECG segment (upper plot), the first selected wavelet scale (middle plot) and its squared derivative (lower plot)

4.3 Detection of R Spikes

The supervised pre-filtering procedure extracted $\{a_k\} \subset \{s_i\}$, the best subset of scales on the training set. Note that the scales in the subset are not necessarily consecutive, which means that different QRS shapes can be matched at different scale levels. The CWT is computed on the whole signal at the scales $\{a_k\}$.

Because of the non-stationarity of the signal, a moving window of 5 seconds length with an overlap of one second is used to cut $T(a_k, b)$ into J parts, $1 \leq j \leq J$. For each a_k and b_j, the R spikes are detected on $T(a_k, b_j)$ by a hard thresholding rule, where index b_j scans the jth window. Let us define

$$D(a_k, b_j) = \left(\frac{d|T(a_k, b_j)|^2}{db_j} \right)^2 . \tag{6}$$

In order to emphasize the importance of the pre-filtering step by the CWT over the detection procedure itself, the thresholding method is a very simple non-parametric one. The threshold $th(a_k, j)$ is estimated as the mean of $D(a_k, b_j)$. As $D(a_k, b_j)$ has sharp peaks in the slopes of the QRS complexes, the intervals $I(a_k, j)$ satisfying

$$D(a_k, b_j) > th(a_k, j) \tag{7}$$

belong to QRS complexes. The R spikes are then defined as the maxima of $|T(a_k, b_j)|^2$ in each of the $I(a_k, j)$ intervals. All the R spikes obtained at each scale k are then merged together.

4.4 Post-processing

A last step of post-processing makes sure that T waves or Q and S spikes have not been wrongly labeled as a R. If two or more R spikes were detected in a window smaller than 250ms (two heartbeats cannot physiologically happen in less than 250ms [9]), the algorithm keeps only the peak which has the highest value on the original ECG.

5   Results and Validation

The learning of the model on the training set and the assessment of performances must be done using an annotated database. The public standard MIT-BIH arrhythmia database [14] was used in this work. It contains 48 half-hour recordings of annotated ECG with a sampling rate of 360Hz and 11-bit resolution over a 10-mV range. The recorded signals contain different wave types and only a robust algorithm can perform well on all datasets together. Some datasets include very few artifacts and clear R peaks, but others make the detection of the R spike more difficult because of (1) abnormal QRS shapes or P and T waves, (2) low signal-to-noise ratio, (3) heavy baseline drifts, (4) lots of non normal beats such as premature ventricular contraction, left bundle branch block beat, atrial premature contraction etc... Among the 48 available datasets, the four ones including paced beats were a priori rejected because they consist in a special case. After visual inspection of the data, datasets 207 and 208 were also rejected. The reason is that a representative training set of five times one minute would be hard to extract randomly as several minutes of these two datasets contain only non-labeled parts looking like a sinus wave.

The performances were assessed by evaluating two parameters as suggested in [19]. The sensitivity is measured as

$$\frac{TP}{TP + FN} \tag{8}$$

and the positive predictivity as

$$\frac{TP}{TP + FP} , \tag{9}$$

where TP is the number of true positive detections, FN the number of false negatives and FP the number of false positives. The error rate is also reported. It is computed by

$$\frac{FN + FP}{n_{QRS}} , \tag{10}$$

where n_{QRS} is the total number of QRS labeled in a dataset. On the database, the algorithm obtains an average sensitivity rate of 99.7% and average positive predictivity rate of 99.7%. The average error rate is below one percent. To our knowledge, only the R spike detectors based on Li's algorithm obtained comparable results with a sensitivity and a positive predictivity between 99.7% and 99.9% [22,20,6]. However, this algorithm makes use of several heuristic rules and requires the setting of many empirical parameters. Our algorithm achieves comparable performances with a very simple non-parametric thresholding method and without the need for a more advanced post-processing stage such as those used in these articles. These results demonstrate the importance of an optimal pre-filtering step over the detection method.

6   Conclusions

In this paper, a supervised learning algorithm for the automatic detection of R peaks in ECG is introduced. It uses the multiscale feature of the continuous wavelet transform (CWT) to emphasize the QRS complex over high P or T waves, noise, baseline drift and artifacts. The CWT keeps the important time aspect of the non-stationary ECG signal. Moreover, very efficient implementations of the CWT exist and a low computational complexity is required, allowing real-time analysis. This algorithm learns and propagates the annotations provided by a physician on a small annotated segment. For this purpose, the method selects the best subset of wavelet scales on a representative training set by a stepwise forward procedure. The forward procedure allows to select scales that are not necessarily consecutive and it does not a priori restrict the range of computed scales on an empirical basis. It allows a complete different set of scales to be selected for each ECG signal, based on its characteristics. The selected scales are then used on the original long-term ECG signal recording and a hard thresholding rule is applied on the derivative of the wavelet coefficients to label the R spikes. The method is robust and does not require any pre-processing stage. The selection procedure can be generalized in order to detect other ECG features such as the P and T wave.

Experiments on the public annotated MIT-BIH database lead to a sensitivity of 99.7% and a positive predictivity of 99.7% without the need of an advanced post-processing stage on the detected peaks. To our knowledge, only three R spike detectors based on WT reported in the literature obtained comparable results, while requiring a more complex thresholding and post-processing stage.

Further works will include: (1) The development of a more advanced thresholding rule that takes the peaks detected so far into account; (2) the use of a more advanced post-processing stage to eliminate wrong detections; (3) the design of an automatic selection of the best mother wavelet by the same learning methodology; (4) the generalization of the method for the detection of other ECG features such as P or T wave.

Acknowledgements. This work was partly supported by the Belgian "Région Wallonne" ADVENS convention 4994 project and by the Belgian "Région de Bruxelles-Capitale" BEATS project. G. de Lannoy and A. de Decker are funded by a Belgian F.R.I.A. grant.

References

1. Rajendra Acharya, U., Paul Joseph, K., Kannathal, N., Lim, C.M., Suri, J.S.: Heart rate variability: a review. Medical and Biological Engineering and Computing, November 17 (2006)
2. Addison, P.D.: Wavelet transform and the ecg: a review. Physiological Measurement 26, 155–199 (2005)
3. Afonso, V.X., Tompkins, W.J., Nguyen, T.Q., Luo, S.: Ecg beat detection using filter banks. IEEE Transactions on Biomedical Engineering 2, 192–201 (1999)
4. Algra, A., Zeelenberg, H.: An algorithm for computer measurement of qt intervals in the 24h ecg. In: Proceedings of the IEEE Computer Society Press, p. 1179 (1987)
5. Brychta, R.J., Tuntrakool, S., Appalsamy, M., Keller, N.R., Robertson, D., Shiavi, R.G., Diedrich, A.: Wavelet methods for spike detection in mouse renal symathetic nerve activity. IEEE Transactions on Biomedical Engineering 54(1), 82–93 (2007)
6. Chen, S.M., Chen, H.C., Chan, H.L.: Dsp Implementation of wavelet Transform for Real Time ecg Waveforms detection and heart rate analysis. Computer Methods and program in Biomedicine 55(1), 35–44 (1997)
7. Chen, S.W., Chen, H.C., Chan, H.L.: A real-time qrs detection method based on moving-averaging incorporating with wavelet denoising. Comput. Methods Programs Biomed. 82(3), 187–195 (2006)
8. Chen, Y., Yan, Z., He, W.: Detection of qrs-wave in electrocardiogram: Based on kalman and wavelets. Conf. Proc. IEEE Eng. Med. Biol. Soc. 3, 2758–2760 (2005)
9. Christov, I.I.: Real time electrocardiogram qrs detection using combined adaptive threshold. BioMedical Engineering OnLine 3(1), 28 (2004)
10. Clavier, L., Boucher, J.-M., Lepage, R., Blanc, J.-J., Cornily, J.-C.: Automatic p-wave analysis of patients prone to atrial fibrillation. Med. Biol. Eng. Comp. 40, 63–71 (2002)
11. Daubechies, I.: Ten Lectures on Wavelets (C B M S - N S F Regional Conference Series in Applied Mathematics). Soc. for Industrial & Applied Math. (December 1992)
12. Dobbs, S.E., Schmitt, N.M., Ozemek, H.S.: Qrs detection by template matching using real-time correlation on a microcomputer. Journal of clinical engineering 9(3), 197–212 (1984)
13. Jafari Moghadam Fard, P., Moradi, M.H., Tajvidi, M.R.: A novel approach in r peak detection using hybrid complex wavelet (HCW). International Journal of Cardiology (March 27, 2007) (in press, 2007)

14. Goldberger, A.L., Amaral, L.A.N., Glass, L., Hausdorff, J.M., Ivanov, P.C., Mark, R.G., Mietus, J.E., Moody, G.B., Peng, C.-K., Stanley, H.E.: PhysioBank, PhysioToolkit, and PhysioNet: Components of a new research resource for complex physiologic signals. Circulation 101(23), e215–e220 (2000)
15. Huber, P.J.: Robust Statistics. Wiley, New York (1981)
16. Madeiro, J., Cortez, P., Oliveira, F., Siqueira, R.: A new approach to qrs segmentation based on wavelet bases and adaptive threshold technique. Medical Engineering and Physics 29, 2637 (2007)
17. Kadambe, S., Murray, R., Boudreaux-Bartels, G.F.: Wavelet transform-based qrs complex detector. IEEE Transactions on Biomedical Engineering 46, 838–848 (1999)
18. Koeleman, A.S.M., Ros, H.H., van den Akker, T.J.: Beat-to-beat interval measurement in the electrocardiogram. Med. Biol. Eng. Comp. 23, 2139 (1985)
19. Kohler, B.U., Hennig, C., Orglmeister, R.: The principles of software qrs detection. IEEE Eng. Med. Biol. Mag. 2(1), 42–57 (2002)
20. Li, C., Zheng, C., Tai, C.: Detection of ecg Characteristic Points using Wavelet Transform. IEEE Trans. Biomed. 42(1), 21–28 (1995)
21. Mallat, S.: A Wavelet Tour of Signal Processing. In: Wavelet Analysis and Its Applications, 2nd edn. IEEE Computer Society Press, San Diego (1999)
22. Martinez, J.P., Almeida, R., Olmos, S., Rocha, A.P., Laguna, P.: A wavelet-based ECG delineator: evaluation on standard databases. IEEE Transactions on Biomedical Engineering 51, 570–581 (2004)
23. Nygards, M., Sornmo, L.: Delineation of the qrs Complex using the Envelope of the ecg. Med. Biol. Eng. Comp. 21, 53847 (1983)
24. Pahlm, O., Sornmo, L.: Software QRS Detection in Ambulatory Monitoring a review. Med. Biol. Eng. Comp. 22, 28997 (1984)
25. Sahambi, J.S., Tandon, S.N., Bhatt, R.K.P.: An automated approach to beat-by-beat qt-interval Analysis. IEEE Eng. Med. Biol. Mag. 19(3), 97–101 (2000)
26. Senhadji, L., Carrault, G., Bellanger, J.J., Passariello, G.: Comparing wavelet transforms for recognizing cardiac patterns. IEEE Eng. Med. Biol. Mag. 149(2), 167–173 (1995)
27. Shyuand, L.-Y., Wu, Y.-H., Hu, W.: Using wavelet Transform and Fuzzy Neural Network for vpc detection from the Holter ecg. IEEE Transactions on Biomedical Engineering 51, 1269–1273 (2004)
28. Suárez, K.V., Silva, J.C., Berthoumieu, Y., Gomis, P., Najim, M.: Ecg beat Detection using a Geometrical Matching Approach. IEEE Transactions on Biomedical Engineering 54(4), 641–650 (2007)
29. Task Force of the European Society of Cardiology and The North American Society of Pacing and Electrophysiology. Heart-rate variability: Standards of measurement, physiological interpretation, and clinical use. Circulation 93(5), 1043–1065 (1996)
30. Vullings, H., Verhaegen, M., Verbruggen, H.: Automated ecg segmentation with dynamic time warping. In: Proceedings of the 20th Annual International Conference on IEEE Engineering in Medicine and Biology Society, p. 1636 (1998)
31. Xue, X., Hu, Y.H., Tompkins, W.J.: Neural-network-based Adaptive matched Filtering for QRS Detection. IEEE Transactions on Biomedical Engineering 32(4), 317–329 (1992)

Four-Channel Biosignal Analysis and Feature Extraction for Automatic Emotion Recognition

Jonghwa Kim and Elisabeth André

Institute of Computer Science, University of Augsburg
Eichleitnerstr. 30, D-86159 Augsburg, Germany
{kim,andre}@informatik.uni-augsburg.de
http://mmwerkstatt.informatik.uniaugsburg.de

Abstract. This paper investigates the potential of physiological signals as a reliable channel for automatic recognition of user's emotial state. For the emotion recognition, little attention has been paid so far to physiological signals compared to audio-visual emotion channels such as facial expression or speech. All essential stages of automatic recognition system using biosignals are discussed, from recording physiological dataset up to feature-based multiclass classification. Four-channel biosensors are used to measure electromyogram, electrocardiogram, skin conductivity and respiration changes. A wide range of physiological features from various analysis domains, including time/frequency, entropy, geometric analysis, subband spectra, multiscale entropy, etc., is proposed in order to search the best emotion-relevant features and to correlate them with emotional states. The best features extracted are specified in detail and their effectiveness is proven by emotion recognition results.

Keywords: Emotion recognition, physiological measures, skin conductance, electrocardiogram, electromyogram, respiration, affective computing, human-com-puter interaction, musical emotion, autonomic nervous system, arousal, valence, feature extraction, pattern recognition.

1 Introduction

In human communication, expression and understanding of emotions facilitate to complete the mutual sympathy. To approach it in human-machine interaction, we need to equip machines with the means to interpret and understand human emotions without input of user's translated intention. Hence, one of the most important prerequisites to realize such an advanced user interface is a reliable emotion recognition system which guarantees acceptable recognition accuracy, robustness against any artifacts, and adaptability to practical applications. It is about to model, analyze, process, train, and classify emotional features measured from the implicit emotion channels of human communication, such as speech, facial expression, gesture, pose, physiological responses, etc. In this paper we concentrate on finding emotional cues from various physiological measures.

Recently many works on engineering approaches to automatic emotion recognition have been reported. For an overview we refer to [1]. Particularly, most efforts have been

A. Fred, J. Filipe, and H. Gamboa (Eds.): BIOSTEC 2008, CCIS 25, pp. 265–277, 2008.

taken to recognize human emotions using audiovisual channels of emotion expression, facial expression, speech, and gesture. Relatively little attention, however, has been paid so far to using physiological measures. Reasons are some significant limitations resulting from the use of physiological signals for emotion recognition. The main difficulty lies in the fact that it is a very hard task to uniquely map subtle physiological patterns onto specific emotional states. As an emotion is a function of time, context, space, culture, and person, physiological patterns may also widely differ from user to user and from situation to situation.

In this paper, we treat all essential stages of automatic emotion recognition system using physiological measures, from data collection up to classification of four typical emotions (joy, anger, sadness, pleasure) using four-channel biosignals. The work in this paper is novel in trying to recognize naturally induced musical emotions using physiological changes, in acquiring the physiological dataset through everyday life recording over many weeks from multiple subjects, in finding emotion-relevant ANS (autonomic nervous system) specificity through various feature contents, and in designing an emotion-specific classification method. After the calculation of a great number of features (a total of 110 features) from various feature domains, we try to identify emotion-relevant features using the backward feature selection method combined with a linear classifier. These features can be directly used to design affective human-machine interfaces for practical applications.

2   Related Research

A significant amount of work has been conducted by Picard and colleagues at MIT Lab who showed that certain affective states may be recognized by using physiological data including heart rate, skin conductivity, temperature, muscle activity and respiration velocity [2]. They used personalized imagery to elicit target emotions from a single subject who had two years' experience in acting, and achieved overall 81% recognition accuracy in eight emotions by using hybrid linear discriminant classification [3]. Nasoz et al. [4] used movie clips based on the study by Gross and Levenson [5] for eliciting target emotions from 29 subjects and achieved best emotion classification accuracy of 83% through the Marquardt Backpropagation algorithm (MBP). More recently, interesting user-independent emotion recognition system is reported by Kim et al. [6]. They developed a set of recording protocols using multimodal stimuli (audio, visual, and cognitive) to evoke targeted emotions (sadness, stress, anger, and surprise) from the 175 children aged from five to eight years. Classification ratio of 78.43% for three emotions (sadness, stress, and anger) and 61.76% for four emotions (sadness, stress, anger, and surprise) has been achieved by adopting support vector machine as pattern classifier.

Note that the recognition rates in the privious works should be strongly dependent on the datasets they used and context of subjects. Moreover, the physiological datasets used in most of the previous works are gathered by using visual elicitation materials in a lab setting. The subjects then "tried and felt" or "acted out" the target emotions while looking at selected photos or watching movie clips that are carefully prearranged to the emotions. It means, extremely speaking, that the recognition results were achieved for specific users in specific contexts with the "forced" emotional states.

3   Musical Emotion Induction

A well established mechanism of emotion induction would be either to imagine or to recall from individual memory. Emotional reaction can be triggered by a specific cue and be evoked by an experimental instruction to imagine certain events. On the other hand, it can spontaneously be resurged in memory. Music is a pervasive element accompanying many highly significant events in human social life and particular pieces of music are often connected to significant personal memories. This claims that certain music can be a powerful cue in bringing emotional experiences from memory back into awareness. Since music listening is often done by an individual in isolation, the possible artifacts by social masking and social interaction may be minimized in the experiment. Furthermore, like odors, music may be treated at lower levels of the brain that are particularly resistant to modifications by later input, contrary to cortically based episodic memory [7].

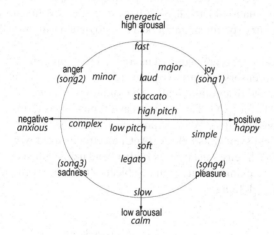

Fig. 1. Reference emotional cues in music based on the 2D emotion model. Metaphoric cues for song selection: song1 (enjoyable, harmonic, dynamic, moving), song2 (noisy, loud, irritating, discord), song3 (melancholic, sad memory), song4 (blissful, pleasurable, slumberous, tender).

To collect a database of physiological signals in which the targeted emotions (joy, anger, sadness, pleasure)[1] can be *naturally* reflected without any deliberate expression, we decided to use musical induction method, recording physiological signals while the subjects are listening to different music songs. The subjects were three males (two students and an academic employee) between 25-38 years old and enjoyed listening to music everyday. The subjects individually handpicked four music songs by themselves that should spontaneously evoke their emotional memories and certain moods corresponding to the four target emotions. Figure 1 shows the musical emotion model referred

---

[1] We note that these four expression words are used to cover each quadrant in the 2D emotion model, i.e. joy should represent *all* emotions with high arousal and positive valence, anger with high arousal and negative, sadness with low arousal and negative, and pleasure with low arousal and positive valence.

to for the selection of their songs. Generally, emotional responses to music would vary greatly from individual to individual depending on their unique past experiences. Moreover, cross-cultural comparisons in literature suggest that the emotional responses can be quite differentially emphasized by different musical cultures and training. This is why we advised the subjects to choose themselves the songs that recall their individual special memories with respect to the target emotions.

For the experiment, we prepared a quiet listening room in our institute in order to ensure the subjects to unaffectedly feel the emotions from the music. For the recording, the subject needs to position himself the sensors by instruction posters in the room, to apply the headphones, and to select a song from his song list saved in the computer. When he does mouse-click just at the start of recording, the recording and music systems are automatically setting up by preset values for each song. Recording schedule was determined by the subjects themselves too, at any time when they will listen to music and which song they choose. It means, different from methods used in other studies, that the subjects were not forced to participate in a lab setting scenario and to use prespecified stimulation materials. We believe that this voluntary participation of the subjects during our experiment might be a help to obtain a high-quality dataset with natural emotions.

The physiological signals are acquired using the Procomp Infiniti™ (www.mindmedia.nl) with four biosensors, electromyogram (EMG), skin conductivity (SC), electrocardiogram (ECG), and respiration (RSP). The sampling rates are 32 Hz for EMG, SC, and RSP, and 256 Hz for ECG. The positions and typical waveforms of the biosensors we used are illustrated in Fig. 2. We used pre-gelled single Ag/AgCl electrodes for ECG and EMG sensors and standard single Ag/AgCl electrodes fixed with two finger bands for SC sensor. A stretch sensor using latex rubber band fixed with velcro respiration belt is used to capture breathing activity of the subjects. It can be worn either thoracically or abdominally, over clothing.

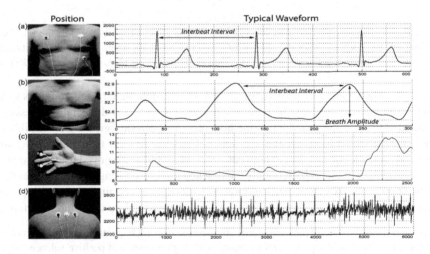

Fig. 2. Position and typical waveforms of the biosensors: (a) ECG, (b) RSP, (c) SC, (d) EMG

During the three months, a total of 360 samples (90 samples for each emotion) from three subjects is collected. Signal length of each sample is between 3-5 minutes depending on the duration of the songs.

4   Methodology

Overall structure of our recognition system is illustrated in Figure 3.

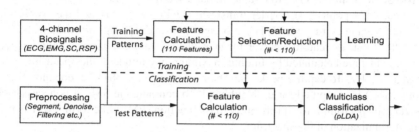

Fig. 3. Block diagram of supervised statistical classification system for emotion recognition

4.1   Preprocessing

Different types of artifacts were observed in all the four channel signals, such as transient noise due to movement of the subjects during the recording, mostly at the begin and end of the each recording. Thus, uniformly for all subjects and channels, we segmented the signals into final samples with fixed length of 160 seconds by cutting out from the middle part of each signal. Particularly to the EMG signal, we needed to pay closer attention because the signal contains artifacts generated by respiration and heart beat (Fig. 4). It was due to the position of EMG sensor at the nape of the neck. For other signals we used pertinent lowpass filters to remove the artifacts without loss of information.

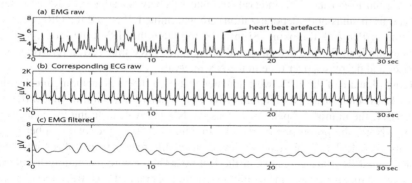

Fig. 4. Example of EMG signal with heart beat artifacts and denoised signal

4.2   Measured Features

From the four channel signals we calculated a total of 110 features from various analysis domains including conventional statistics in time series, frequency domain, geometric analysis, multiscale sample entropy, subband spectra, etc. For the signals with non-periodic characteristics, such as EMG and SC, we focused on capturing the amplitude variance and localizing the occurrences (number of transient changes) in the signals.

Electrocardiogram (ECG). To obtain subband spectrum of the ECG signal we used the typical 1024 points fast Fourier transform (FFT) and partitioned the coefficients within the frequency range 0-10 Hz into eight non-overlapping subbands with equal bandwidth. First, as features, power mean values of each subband and fundamental frequency (F0) are calculated by finding maximum magnitude in the spectrum within the range 0-3 Hz. To capture peaks and their locations in subbands, subband spectral entropy (SSE) is computed for each subband. To compute the SSE, it is necessary to convert each spectrum into a probability mass function (PMF) like form. Eq. 1 is used for the normalization of the spectrum.

$$x_i = \frac{X_i}{\sum_{i=1}^N X_i}, \qquad \text{for } i = 1 \ldots N \tag{1}$$

where X_i is the energy of ith frequency component of the spectrum and $\tilde{x} = \{x_1 \ldots x_N\}$ is to be considered as the PMF of the spectrum. In each subband the SSE is computed from $\tilde{x}$ by

$$H_{\text{sub}} = -\sum_{i=1}^N x_i \cdot \log_2 x_i \tag{2}$$

By packing the eight subbands into two bands, i.e., subbands 1-3 as low frequency (LF) band and subbands 4-8 as high frequency (HF) band, the ratios of LF/HF bands are calculated from the power mean values and the SSEs.

To obtain the HRV (heart rate variability) from the continuous ECG signal, each QRS complex is detected and the RR intervals (all intervals between adjacent R waves) or the normal-to-normal (NN) intervals (all intervals between adjacent QRS complexes resulting from sinus node depolarization) are determined. We used the QRS detection algorithm of Pan and Tompkins [8] in order to obtain the HRV time series. Figure 5 shows example of R wave detection and interpolated HRV time series, referring to the increases and decreases over time in the NN intervals.

In time-domain of the HRV, we calculated statistical features including mean value, standard deviation of all NN intervals (SDNN), standard deviation of first difference of the HRV, the number of pairs of successive NN intervals differing by greater than 50 ms (NN50), the proportion derived by dividing NN50 by the total number of NN intervals. By calculating the standard deviations in different distances of RR interbeats, we also added Poincaré geometry in the feature set to capture the nature of interbeat (RR) interval fluctuations. Poincaré plot geometry is a graph of each RR interval plotted against the next interval and provides quantitative information of the heart activity by calculating the standard deviations of the distances of the $R - R(i)$ to the lines $y = x$ and $y = -x + 2 * R - R_m$, where $R - R_m$ is the mean of all $R - R(i)$, [9]. Figure 5.(e) shows

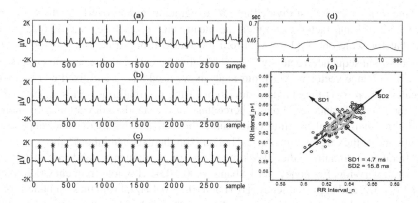

Fig. 5. Example of ECG Analysis: (a) raw ECG signal with respiration artifacts, (b) detrended signal, (c) detected RR interbeats, (d) interpolated HRV time series, (e) Poincaré plot of the HRV time series

an example plot of the Poincaré geometry. The standard deviations SD_1 and SD_2 refer to the fast beat-to-beat variability and longer-term variability of $R - R(i)$ respectively.

Entropy-based features from the HRV time series are also considered. Based on the so-called *approximate entropy* and *sample entropy* proposed in [10], a multiscale sample entropy (MSE) has been introduced [11] and successfully applied to physiological data, especially for analysis of short and noisy biosignal. Given a time series $\{X_i\} = \{x_1, x_2, ..., x_N\}$ of length N, the number $(n_i^{(m)})$ of similar m-dimensional vectors $y^{(m)}(j)$ for each sequence vectors $y^{(m)}(i) = \{x_i, x_{i+1}, ..., x_{i+m-1}\}$ is determined by measuring their respective distances. The relative frequency to find the vector $y^{(m)}(j)$ within a tolerance level δ is defined by

$$C_i^{(m)}(\delta) = \frac{n_i^{(m)}}{N - m + 1} \tag{3}$$

The approximate entropy, $h_A(\delta, m)$, and the sample entropy, $h_S(\delta, m)$ are defined as

$$h_A(\delta, m) = \lim_{N \to \infty} [H_N^{(m)}(\delta) - H_N^{(m+1)}(\delta)], \tag{4}$$

$$h_S(\delta, m) = \lim_{N \to \infty} -\ln \frac{C^{(m+1)}(\delta)}{C^{(m)}(\delta)}, \tag{5}$$

where

$$H_N^{(m)}(\delta) = \frac{1}{N - m + 1} \sum_{i=1}^{N-m+1} \ln C_i^{(m)}(\delta), \tag{6}$$

Because of advantage of being less dependent on time series length N, we applied the sample entropy h_S to coarse-grained versions $(y_j^{(\tau)})$ of the original HRV time series $\{X_i\}$,

$$y_j(\tau) = \frac{1}{\tau} \sum_{i=(j-1)\tau+1}^{j\tau} x_i, \quad 1 \leq j \leq N/\tau, \quad \tau = 1, 2, 3, ... \tag{7}$$

The time series $\{X_i\}$ is first divided into N/τ segments by non-overlapped windowing with length of scale factor τ and then the mean value of each segment is calculated. Note that for scale one $y_j(1) = x_j$. From the scaled time series $y_j(\tau)$ we obtain the m-dimensional sequence vectors $y^{(m)}(i,\tau)$. Finally, we calculate the sample entropy h_S for each sequence vector $y_j(\tau)$. In our analysis we used $m = 2$ and fixed $\delta = 0.2\sigma$ for all scales, where σ is the standard deviation of the original time series x_i. Note that using the fixed tolerance level δ as a percentage of the standard deviation corresponds to initial normalizing of the time series and it thus enables that h_S does not depend on the variance of the original time series, but only on their sequential ordering.

In frequency-domain of the HRV time series, three frequency bands are of interest in general; very-low frequency (VLF) band (0.003-0.04 Hz), low frequency (LF) band (0.04-0.15 Hz), and high frequency (HF) band (0.15-0.4 Hz). From these subband spectra, we computed dominant frequency and power of each band by integrating the power spectral densities (PSD) obtained by using Welch's algorithm, and ratio of power within the low-frequency and high-frequency band (LF/HF). Since the parasympathetic activity dominates at high frequency, the ratio of LF/HF is generally thought to distinguish sympathetic effects from parasympathetic effects [12].

Respiration (RSP). Including the typical statistics of the raw RSP signal we calculated similar types of features like the ECG features, power mean values of three subbands (obtained by dividing the Fourier coefficients within the range 0-0.8 Hz into non-overlapped three subbands with equal bandwidth), and the set of subband spectral entropies (SSE).

In order to investigate inherent correlation between respiration rate and heart rate, we considered a novel feature content for the RSP signal. Since RSP signal exhibits quasi periodic waveform with sinusoidal property, it is not unreasonable to process HRV like analysis for the RSP signal, i.e. to estimate breathing rate variability (BRV). After detrending with mean value of the entire signal and lowpass filtering, we calculated the BRV by detecting the peaks in the signal using the maxima ranks within each zero-crossing. From the BRV time series, similar to the ECG signal, we calculated mean value, SD, SD of first difference, MSE, Poincaré analysis, etc. In the spectrum of the BRV, peak frequency, power of two subbands, low-frequency band (0-0.03Hz) and high-frequency band (0.03-0.15 Hz), and the ratio of power within the two bands (LF/HF) are calculated.

Skin Conductivity (SC). The mean value, standard deviation, and mean of first and second derivations are extracted as features from the normalized SC signal and the low-passed SC signal using 0.2 Hz of cutoff frequency. To obtain a detrended SCR (skin conductance response) waveform without DC-level components, we removed continuous, piecewise linear trend in the two low-passed signals, i.e., very low-passed (VLP) with 0.08 Hz and low-passed (LP) signal with 0.2 Hz of cutoff frequency, respectively (see Fig. 6 (a)-(e)).

The baseline of the SC signal was calculated and subtracted to consider only relative amplitudes.

By finding two consecutive zero-crossings and the maximum value between them, we calculated the number of SCR occurrences within 100 seconds from each LP and

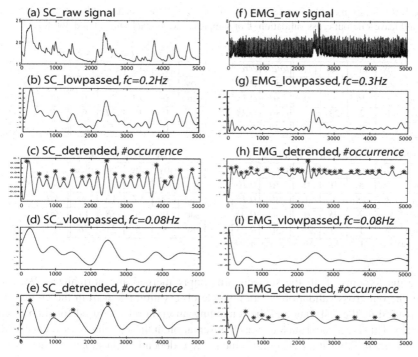

Fig. 6. Analysis Examples of SC and EMG signals

VLP signal, mean of the amplitudes of all occurrences, and ratio of the SCR occurrences within the low-passed signals (VLP/LP).

Electromyography (EMG). For the EMG signal we calculated similar types of features as in the case of the SC signal. From normalized and low-passed signals, the mean value of entire signal, the mean of first and second derivations, and the standard deviation are extracted as features. The occurrence number of myo-responses and ratio of that within VLP and LP signals are also added in feature set by similar manner used for detecting the SCR occurrence but with 0.08 Hz (VLP) and 0.3 Hz (LP) of cutoff frequency (see Fig. 6.(f)-(j)). *Finally we obtained a total of 110 features from the 4-channel biosignals; 53 (ECG) + 37 (RSP) + 10 (SC) + 10 (EMG).*

5 Classification Result

For classification we used the pseudoinverse linear discriminant analysis (pLDA) [13], combined with the sequential backward selection (SBS) [14] to select significant feature subset. The pLDA is a natural extension of classical LDA by applying eigenvalue decomposition to the scatter matrices, in order to deal with the sigularity problem of LDA.

Table 1 with confusion matrix presents the correct classification ratio (CCR) of subject-dependent (Subject A, B, and C) and subject-independent (All) classification

Table 1. Recognition results in rates (*error 0.00 = CCR 100%*) achieved by using pLDA with SBS and leave-one-out cross validation. # of samples: 120 for each subject and 360 for All.

Subject A (*CCR % = 81%*)

	joy	ang	sad	plea	*total**	*error*
joy	22	4	1	3	30	0.27
ang	3	26	1	0	30	0.13
sad	1	2	23	4	30	0.23
plea	3	0	1	26	30	0.13

Subject C (*CCR % = 89%*)

	joy	ang	sad	plea	*total**	*error*
joy	28	0	2	0	30	0.07
ang	0	30	0	0	30	0.00
sad	0	0	24	6	30	0.20
plea	0	0	5	25	30	0.17

Subject B (*CCR % = 91%*)

	joy	ang	sad	plea	*total**	*error*
joy	27	3	0	0	30	0.10
ang	3	25	1	1	30	0.17
sad	0	2	28	0	30	0.07
plea	0	1	0	29	30	0.03

All (*CCR % = 65%*)

	joy	ang	sad	plea	*total**	*error*
joy	62	9	8	11	90	0.31
ang	15	57	13	5	90	0.37
sad	9	6	58	17	90	0.36
plea	8	5	21	56	90	0.38

*: Actual total # of samples,   All: Subject-independent.

where the features of all the subjects are simply merged and normalized. We used leave-one-out cross-validation where a single observation taken from the entire samples is used as the test data and the remaining observations are used for training the classifier. This is repeated such that each observation in the samples is used once as the test data. In the Table, it turned out that the *CCR* is depending on subject to subject. For

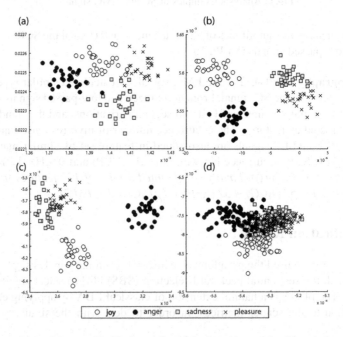

Fig. 7. Comparison of feature distributions of subject-dependent and subject-independent case. (a) Subject A, (b) Subject B, (c) Subject C, (d) Subject-independent.

example, best accuracy was 91% from subject B and lowest was 81% from subject A. Not only the overall accuracy but the CCR of the single emotions differs from subject to subject. On the other side, it is very meaningful that relatively robust recognition accuracy is achieved for the classification between emotions that are reciprocal with each other regarding the diagonal quadrants in the 2D emotion model, i.e., joy vs. sadness and anger vs. pleasure. Moreover, the accuracy is much better than that of arousal classification.

The CCR of subject-independent classification was not comparable to that obtained for subject-dependent classification. As shown in Figure 7, merging the features of all subjects does not refine the discriminating information related to the emotions, but rather leads to scattered class boundaries.

We also tried to differentiate the emotions based on the two axes, arousal and valence, in the 2D emotion model. The samples of four emotions are divided into groups of negative valence (anger/sadness) and positive valence (joy/pleasure) and into groups of high arousal (joy/anger) and low arousal (sadness/pleasure). By using the same methods, we then performed two-class classification of the divided samples for arousal and valence separately. Finally, it turned out that emotion-relevant ANS specificity can be observed more conspicuously in the arousal axis regardless of subject-dependent or independent case. Classification of arousal achieved an acceptable CCR of 97-99% for the subject-dependent recognition and 89% for the subject-independent recognition, while the results for valence were 88-94% and 77%, respectively.

6   Best Emotion-Relevant ANS Features

In Table 2, the best emotion-relevant features, that we determined by ranking the features selected for all subjects (including Subject All) in each classification problem, are listed in detail by specifying their values and domains. One interesting result is that each classification problem respectively links together with certain feature domain. The features obtained from time/frequency analysis of HRV time series are decisive for arousal and four emotions classification, while the features from MSE domain of ECG signals are a predominant factor for correct valence differentiation. Particularly, mutually sympathizing correlate between HRV and BRV which is firstly proposed in this paper has been clearly observed in all the classification problems by the features from their time/frequency analysis and Poincaré domain, *_PoincareHRV* and *_PoincareBRV*. This reveals a manifest cross-correlation between respiration and cardiac activity with respect to emotional state. Furthermore, the correlation between heart rate and respiration is obviously captured by the features from HRV power spectrum (*_HRVspec*), the fast/long-term HRV/BRV analysis using Poincaré method, and the multiscale variance analysis of HRV/BRV (*_MSE*), and that the peaks of high frequency range in HR subband spectrum (*_SubSpectra*) provide information about how the sinoatrial node responds to vagal activity at certain respiration frequency.

In addtion, we analyzed the number of selected features for the three classification problems, arousal, valence, and four emotion states. For the arousal classification, relatively few features are used but achieved higher recognition accuracy compared to the other class problems. After the ratio of number of selected features to the total feature

Table 2. Best emotion-relevant features extracted from four channel physiological signals. Arousal classes: joy+anger/sadness+pleasure, Valence classes: joy+pleasure/anger+sadness, Four classes: joy/anger/sadness/pleasure.

Classes	Best Emotion-relevant Features *(Ch_value_domain, C: ECG, R: RSP, S: SC, M: EMG)*
Arousal	*C_std(diff)_HRVtime*, *C_sd2_PoincareHRV*, *C_powerLow_HRVspec*, ***R_mean_MSE*** , ***R_meanEnergy_SubSpectra***, *R_sd2_PoincareBRV*, *S_mean_RawLowpassed*, *S_std_RawLowpassed*, *M_#occurrenceRatio_RawLowpassed*, *M_mean_RawNormed*
Valence	*C_sd2_PoincareHRV*, *C_meanEnergy_SubSpectra*, *C_ratioLH_HRVspec*, *C_mean_MSE*, *C_mean(diff)_MSE*, ***R_meanEnergy_SubSpectra***, *R_mean(diff)_SubSpectra*, *R_sd1_PoincareBRV*, *R_sd2_PoincareBRV*, ***R_mean_MSE***, *S_mean(diff)_RawNormed*, *M_mean(diff)_RawNormed*
Four Emotions	*C_mean_HRVtime*, *C_std_HRVtime*, *C_std(diff)_HRVtime*, *C_mean(diff)_MSE*, *C_mean_MSE*, *C_mean_SSE*, ***C_sd2_PoincareHRV***, *C_mean_SubSpectra*, ***R_meanEnergy_SubSpectra***, *R_mean_SSE*, *R_mean_BRVtime*, *R_sd1_PoincareBRV*, ***R_sd2_PoincareBRV***, ***R_mean_MSE***, *R_power_BRVspec*, *S_std_RawLowpassed*, *S_mean(diff)_RawNormed*, *S_mean(diff(diff))_RawLowpassed*, *S_mean_RawNormed*, *S_#occurrence_RawLowpassed*, *M_mean(diff)_RawNormed*

: overall selected features are printed in bold.

number of each channel, it was obvious that the SC and EMG activities reflected in both *_RawLowpassed* and *_RawNormed* domains (see Table 2) are more significant for arousal classification than the other channels. This supports also the experimental elucidation in previous works that the SCR is linearly correlated with the intensity of arousal. On the other side, we could observe a remarkable increase of number of the ECG and RSP features for the case of valence classification.

7   Conclusions

In this paper, we treated all essential stages of automatic emotion recognition system using multichannel physiological measures, from data collection up to classification process, and analyzed the results from each stage of the system. For four emotional states of three subjects, we achieved average recognition accuracy of 91% which connotes more than a prima facie evidence that there exist some ANS differences among emotions.

A wide range of physiological features from various analysis domains including time, frequency, entropy, geometric analysis, subband spectra, multiscale entropy, and HRV/BRV has been proposed to search the best emotion-relevant features and to correlate them with emotional states. The selected best features are specified in detail and their effectiveness is proven by classification results. We found that SC and EMG are linearly correlated with arousal change in emotional ANS activities, and that the features in ECG and RSP are dominant for valence differentiation. Particularly, the HRV/BRV analysis revealed the cross-correlation between heart rate and respiration.

As we humans use several modalities jointly to interpret emotional states since emotion affects almost all modes, one most challenging issue in near future work is to explore multimodal analysis for emotion recognition. Toward the human-like analysis and finer resolution of recognizable emotion classes, an essential step would be therefore to find innate priority among the modalities to be preferred for each emotional state. In this sense, physiological channel can be considered as a "baseline channel" in designing a multimodal fashion of emotion recognition system, since it provides several advantages over other external channels and acceptable recognition accuracy, as we presented in this work.

Acknowledgements. This research was partially supported by the European Commission (HUMAINE NoE: FP6 IST-507422).

References

1. Cowie, R., Douglas-Cowie, E., Tsapatsoulis, N., Votsis, G., Kollias, S., Fellenz, W., Taylor, J.G.: Emotion recognition in human-computer interaction. IEEE Signal Processing Mag. 18, 32–80 (2001)
2. Healey, J., Picard, R.W.: Digital Processing of Affective Signals. In: Proc. IEEE Int. Conf. Acoust., Speech, and Signal Proc., Seattle, WA, pp. 3749–3752 (1998)
3. Picard, R., Vyzas, E., Healy, J.: Toward Machine Emotional Intelligence: Analysis of affective physiological state. IEEE Trans. Pattern Anal. and Machine Intell. 23(10), 1175–1191 (2001)
4. Nasoz, F., Alvarez, K., Lisetti, C., Finkelstein, N.: Emotion Recognition from Physiological Signals for Presence Technologies. International Journal of Cognition, Technology, and Work - Special Issue on Presence 6(1) (2003)
5. Gross, J.J., Levenson, R.W.: Emotion Elicitation using Films. Cognition and Emotion 9, 87–108 (1995)
6. Kim, K.H., Bang, S.W., Kim, S.R.: Emotion Recognition System using Short-term Monitoring of Physiological Signals. Medical & Biological Engineering & Computing 42, 419–427 (2004)
7. LeDoux, J.E.: Emotion and the Amygdala. In: The Amygdala: Neurobiological Aspects of Emotion, Memory, and Mental Dysfunction, pp. 339–351. Wiley-Liss, New York (1992)
8. Pan, J., Tompkins, W.: A real-time qrs detection algorithm. IEEE Trans. Biomed. Eng. 32(3), 230–323 (1985)
9. Kamen, P.W., Krum, H., Tonkin, A.M.: Poincare Plot of Heart Rate Variability allows Quantitative Display of Parasympathetic Nervous Activity. Clin. Sci. 91, 201–208 (1996)
10. Richmann, J., Moorman, J.: Physiological Time Series Analysis using Approximate Entropy and Sample Entropy. Am. J. Physiol. Heart Circ. Physiol. 278, H2039 (2000)
11. Costa, M., Goldberger, A.L., Peng, C.K.: Multiscale Entropy Analysis of Biological Signals. Phys. Rev. E 71(021906) (2005)
12. Malliani, A.: The Pattern of Sympathovagal Balance explored in the Frequency Domain. News Physiol. Sci. 14, 111–117 (1999)
13. Ye, J., Li, Q.: A Two-stage Linear discriminant Analysis via QR-decomposition. IEEE Trans. Pattern Anal. and Machine Intell. 27(6) (June 2005)
14. Kittler, J.: Feature Selection and Extraction, p. 5983. Academic Press, London (1986)

Fast, Accurate and Precise Mid-Sagittal Plane Location in 3D MR Images of the Brain

Felipe P.G. Bergo[1], Alexandre X. Falcão[1], Clarissa L. Yasuda[2], and Guilherme C.S. Ruppert[1]

[1] LIV, Institute of Computing, University of Campinas (UNICAMP)
C.P. 6176, 13083-970, Campinas, SP, Brazil
[2] Dept. of Neurology, Faculty of Medical Sciences, University of Campinas (UNICAMP)
C.P. 6111, 13083-970, Campinas, SP, Brazil

Abstract. Extraction of the mid-sagittal plane (MSP) is a key step for brain image registration and asymmetry analysis. We present a fast MSP extraction method for 3D MR images, based on automatic segmentation of the brain and on heuristic maximization of the cerebro-spinal fluid within the MSP. The method is robust to severe anatomical asymmetries between the hemispheres, caused by surgical procedures and lesions. The method is also accurate with respect to MSP delineations done by a specialist. The method was evaluated on 64 MR images (36 pathological, 20 healthy, 8 synthetic), and it found a precise and accurate approximation of the MSP in all of them with a mean time of 60.0 seconds per image, mean angular variation within a same image (precision) of $1.26°$ and mean angular difference from specialist delineations (accuracy) of $1.64°$.

1 Introduction

The human brain is not perfectly symmetric [1,2,3]. However, for the purpose of analysis, it is paramount to define and distinguish a *standard of asymmetry*, considered as normal for any given measurement, from abnormal asymmetry, which may be related to neurological diseases, cerebral malformations, surgical procedures or trauma. Several works sustain this claim. For example, accentuated asymmetries between left and right hippocampi have been found in patients with Schizophrenia [4,5,6,7,8,9], Epilepsy [10,11] and Alzheimer Disease [12,13].

The brain has two hemispheres, and the structures of one side should have their counterpart in the other side with similar shapes and approximate locations [1]. The hemispheres have their boundaries limited by the longitudinal (median) fissure, being the corpus callosum their only interconnection.

The ideal separation surface between the hemisferes is not perfectly planar, but the mid-sagittal plane (MSP) can be used as a reference for asymmetry analysis, without significant loss in the relative comparison between normal and abnormal subjects. The MSP location is also important for image registration. Some works have used this operation as a first step for intra-subject registration, as it reduces the number of degrees of freedom [14,15], and to bring different images into a same coordinate system [16], such as in the Talairach [17] model.

A. Fred, J. Filipe, and H. Gamboa (Eds.): BIOSTEC 2008, CCIS 25, pp. 278–290, 2008.

The longitudinal fissure forms a gap between the hemispheres filled with cerebrospinal fluid (CSF). Given that there is no exact definition of the MSP, we define it as a large intersection between a plane and an *envelope* of the brain (a binary volume whose surface approximates the convex hull of the brain) that maximizes the amount of CSF. This definition leads to an automatic, precise, accurate and efficient algorithm for MSP extraction. We have evaluated the precision of the method with respect to random tilts applied to normal and abnormal images. Its accuracy evaluation has also been included, with respect to manual delineations done by a specialist, extending a previous version of this work [18].

The paper is organized as follows. In Section 2, we review existing works on automatic location of the mid-sagittal plane. In section 3, we present the proposed method. In section 4, we show experimental results and validation with simulated and real MR-T1 images. Section 5 states our conclusions.

2   Related Works

MSP extraction methods can be divided in two groups: (i) methods that define the MSP as a plane that maximizes a symmetry measure, extracted from both sides of the image [19,20,21,14,22,16,23,24,25], and (ii) methods that detect the longitudinal fissure to estimate the location of the MSP [26,27,28,29]. Table 1 summarizes these works, and extensive reviews can be found in [28], [29], [23] and [16].

Methods in the first group address the problem by exploiting the rough symmetry of the brain. Basically, they consist in defining a symmetry measure and searching for the plane that maximizes this score. Methods in the second group find the MSP by detecting the longitudinal fissure. Even though the longitudinal fissure is not visible in some modalities, such as PET and SPECT, it clearly appears in MR images. Particularly,

Table 1. Summary of existing MSP methods

Method	Features	Measure
(Brummer, 1991) [26]	Fissure, 2D; MR	Edge Hough Transform
(Guillemaud, 1996) [27]	Fissure, 2D; MR	Active contours
(Hu, 2003) [28]	Fissure, 2D; MR, CT	Local symmetry of fissure
(Volkau, 2006) [29]	Fissure, 3D; MR, CT	Kullback-Leibler's measure
(Junck, 1990) [19]	Symmetry, 2D; PET, SPECT	Intensity cross correlation
(Miroshima,1992) [20]	Symmetry, 3D; PET	Stochastic sign change
(Ardekani, 1997) [14]	Symmetry, 3D; MR, PET	Intensity cross correlation
(Sun, 1997) [21]	Symmetry, 3D; MR, CT	Extended Gaussian image
(Smith, 1999) [22]	Symmetry, 3D; MR, CT, PET, SPECT	Ratio of intensity profiles
(Liu, 2001) [16]	Symmetry, 2D; MR, CT	Edge cross correlation
(Prima, 2002) [23]	Symmetry, 3D; MR, CT, PET, SPECT	Intensity cross correlation
(Tuzikov, 2003) [24]	Symmetry, 3D; MR, CT, SPECT	Intensity cross correlation
(Teverovskiy, 2006) [25]	Symmetry, 3D; MR	Edge cross correlation

we prefer these methods because patients may have very asymmetric brains and we believe this would affect the symmetry measure and, consequently, the MSP detection.

The aforementioned approaches based on longitudinal fissure detection present some limitations that we are circumventing in the proposed method. In [27], the MSP is found by using snakes and orthogonal regression for a set of points manually placed on each slice along the longitudinal fissure, thus requiring human intervention. Other method [26] uses the Hough Transform to automatically detect straight lines on each slice [26], but it does not perform well on pathological images. The method in [28] assumes local symmetry near the plane, which is not verified in many cases (see Figs. 2, 5 and 8). Volkau et al. [29] propose a method based on the Kullback and Leibler's measure for intensity histograms in consecutive candidate planes (image slices). The method presents excellent results under a few limitations related to rotation, search region of the plane, and pathological images.

3   Methods

Our method is based on detection of the longitudinal fissure, which is clearly visible in MR images. Unlike some previous works, our approach is fully 3D, automatic, and applicable to images of patients with severe asymmetries.

We assume that the mid-sagittal plane is a plane that contains a maximal area of cerebro-spinal fluid (CSF), excluding ventricles and lesions. In MR T1 images, CSF appears as low intensity pixels, so the task is reduced to the search of a sagittal plane that minimizes the mean voxel intensity within a mask that disregards voxels from large CSF structures and voxels outside the brain.

The method is divided in two stages. First, we automatically segment the brain and morphologically remove thick CSF structures from it, obtaining a brain mask. The second stage is the location of the plane itself, searching for a plane that minimizes the mean voxel intensity within its intersection with the brain mask. Our method uses some morphological operations whose structuring elements are defined based on the image resolution. To keep the method description independent of image resolution, we use the notation S_r to denote a spherical structuring element of radius r mm.

3.1   Segmentation Stage

We use the tree pruning approach to segment the brain. Tree pruning [30] is a segmentation method based on the Image Foresting Transform [31], which is a general tool for the design of fast image processing operators based on connectivity. In tree pruning, we interpret the image as a graph, and compute an optimum path forest from a set of seed voxels inside the object. A gradient-like image with high pixel intensities along object borders must be computed to provide the edge weights of the implicit graph. A combinatorial property of the forest is exploited to prune tree paths at the object's border, limiting the forest to the object being segmented.

To segment the brain (white matter (WM), gray matter (GM) and ventricles), we compute a suitable gradient image, a set of seed voxels inside the brain and apply the tree pruning algorithm. A more detailed description of this procedure is given in [30]. Note that any other brain segmentation method could be used for this purpose.

Gradient Computation. MR-T1 images of the brain contain two large clusters: the first with air, bone and CSF (lower intensities), and the second, with higher intensities, consists of GM, WM, skin, fat and muscles. Otsu's optimal threshold [32] can separate these clusters (Figs. 1a and 1b), such that the GM/CSF border becomes part of the border between them. To enhance the GM/CSF border, we multiply each voxel intensity $I(p)$ by a weight $w(p)$ as follows:

$$
w(p) = \begin{cases} 0 & I(p) \leq m_1 \\ 2\left(\frac{I(p)-m_1}{m_2-m_1}\right)^2 & m_1 < I(p) \leq \tau \\ 2 - 2\left(\frac{I(p)-m_2}{m_2-m_1}\right)^2 & \tau < I(p) \leq m_2 \\ 2 & I(p) > m_2 \end{cases} \tag{1}
$$

where τ is the Otsu's threshold, and m_1 and m_2 are the mean intensities of each cluster. We compute a 3D gradient at each voxel as the sum of its projections along 26 directions around the voxel, and then use its magnitude for tree pruning (Fig. 1c).

Seed Selection. The brighter cluster contains many voxels outside the brain (Fig. 1b). To obtain a set of seeds inside the brain, we apply a morphological erosion by S_5 on the binary image of the brighter cluster. This operation disconnects the brain from adjacent structures. We then select the largest connected component as the seed set (Fig. 1d).

Morphological Closing. The brain object obtained by tree pruning (Fig. 1e) might not include the entire longitudinal fissure, especially when the fissure is too thick. To ensure its inclusion, we apply a morphological closing by S_{20} to the binary brain image (Fig. 1f).

Thick CSF Structure Removal. The last step of this phase is the removal of thick CSF structures (such as the ventricles, lesions and post-surgery cavities) from the brain object, to avoid the MSP from snapping to a dark structure other than the longitudinal fissure. We achieve this with a sequence of morphological operations: we start from a binary image obtained by thresholding at Otsu's optimal threshold (Fig. 1b). We apply a morphological opening by S_5 to connect the thick ($> 5~mm$) CSF structures (Fig. 1g), and then dilate the result by S_2 to include a thin ($2~mm$) wall of the CSF structures (Fig. 1h). This dilation ensures the reinclusion of the longitudinal fissure, in case it is removed by the opening. The binary intersection of this image with the brain object is then used as brain mask (Fig. 1i) by the next stage of our method. Only voxels within this mask are considered by stage 2. Figs. 2a and 2b show how the computed brain mask excludes the large cavity in a post-surgery image, and figures 2c and 2d show how the mask excludes most of the ventricles in patients with large ventricles.

3.2 Plane Location Stage

To obtain the CSF score of a plane, we compute the mean voxel intensity in the intersection between the plane and the brain mask (Figs. 3a and 3b). The lower the score, the more likely the plane is to contain more CSF than white matter and gray matter. The plane with a sufficiently large brain mask intersection and minimal score is the most likely to be the mid-sagittal plane.

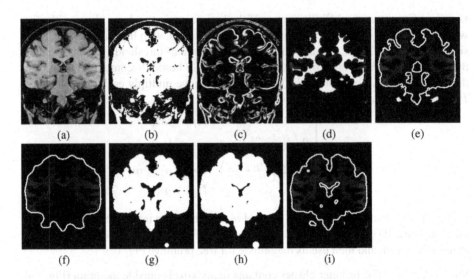

Fig. 1. Sample slice of the intermediary steps in stage 1: (a) original coronal MR slice; (b) binary cluster mask obtained by thresholding; (c) gradient-like image used for tree pruning; (d) seed set used for tree pruning (white); (e) border of the brain object obtained by tree pruning (white); (f) border of the brain object after morphological closing; (g) CSF mask after opening; (h) CSF mask after dilation; (h) brain mask (intersection of (f) and (h))

To find a starting candidate plane, we compute the score of all sagittal planes in 1 mm intervals (which leads to 140–180 planes in usual MR datasets), and select the plane with minimum score. Planes with intersection area lower than 10 000 mm^2 are not considered to avoid selecting planes tangent to the surface of the brain. Planes with small intersection areas may lead to low scores due to alignment with sulci and also due to partial volume effect between gray matter and CSF (Figs. 3c and 3d).

Once the best candidate plane is found, we compute the CSF score for small transformations of the plane by a set of rotations and translations. If none of the transformations

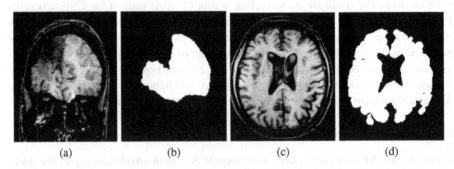

Fig. 2. Examples of thick CSF structure removal: (a) coronal MR slice of a patient with post-surgical cavity; (b) brain mask of (a); (c) axial MR slice of a patient with large ventricles; (d) brain mask of (c)

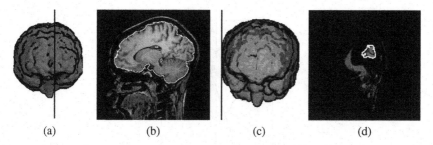

(a)	(b)	(c)	(d)

Fig. 3. Plane intersection: (a–b) sample plane, brain mask and their intersection (white outline). (c–d) example of a plane tangent to the brain's surface and its small intersection area with the brain mask (delineated in white), overlaid on the original MR image.

lead to a plane with lower CSF score, the current plane is the mid-sagittal plane and the algorithm stops. Otherwise, the transformed plane with lower CSF score is considered the current candidate, and the algorithm is repeated. The algorithm is finite, since each iteration reduces the CSF score, and the CSF score is limited by the voxel intensity domain.

We use a set of 42 candidate transforms at each iteration: translations on both directions of the X, Y and Z axes by $10\,mm$, $5\,mm$ and $1\,mm$ (18 translations) and rotations on both directions around the X, Y and Z axes by $10°$, $5°$, $1°$ and $0.5°$ (24 rotations). All rotations are about the central point of the initial candidate plane. Rotations by less than $0.5°$ are useless, as this is close to the limit where planes fall over the same voxels for typical MR datasets, as discussed in Section 4.1.

4   Evaluation and Discussion

4.1   Error Measurement

The discretization of $\mathbb{R}^3$ makes planes that differ by small angles to fall over the same voxels. Consider two planes A and B that differ by an angle Θ (Fig. 4). The minimum angle that makes A and B differ by at least 1 voxel at a distance r from the rotation center is given by Equation 2.

$$\Theta = \arctan\left(\frac{1}{r}\right) \tag{2}$$

An MR dataset with $1\,mm^3$ voxels has a typical maximum dimension of $256\,mm$. For rotations about the center of the volume, the minimum angle that makes planes A and B differ by at least one voxel within the volume (point p_i in Fig. 4) is approximately $\arctan\left(\frac{1}{128}\right) = 0.45°$. For most MSP applications, we are only concerned about plane differences within the brain. The largest length within the brain is usually longitudinal, reaching up to 200 mm in adult brains. The minimum angle that makes planes A and B differ by at least one voxel within the brain (point p_b in Fig. 4) is approximately $\arctan\left(\frac{1}{100}\right) = 0.57°$.

Therefore, we can consider errors around $1°$ excellent and equivalent results.

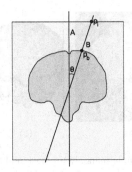

Fig. 4. Error measurement in discrete space: points and angles

4.2 Experiments

We evaluated the method on 64 MR datasets divided in 3 groups: A control group with 20 datasets from subjects with no anomalies, a surgery group with 36 datasets from patients with significant structural variations due to brain surgery, and a phantom group with 8 synthetic datasets with varying levels of noise and inomogeneity, taken from the BrainWeb project [33].

All datasets in the control group and most datasets in the surgery group were acquired with a voxel size of $0.98 \times 0.98 \times 1.00 \ mm^3$. Some images in the surgery group were acquired with a voxel size of $0.98 \times 0.98 \times 1.50 \ mm^3$. The images in the phantom group were generated with an isotropic voxel size of $1.00 \ mm^3$. All volumes in the control and surgery groups were interpolated to an isotropic voxel size of $0.98 \ mm^3$ before applying the method.

We performed two sets of experiments: random tilt evaluation (precision) and comparison with expert delineations of the MSPs (accuracy).

4.3 Random Tilt Evaluation

For each of the 64 datasets, we generated 10 variations (tilted datasets) by applying 10 random transforms composed of translations and rotations of up to 12 mm and 12° in all axes. The MSP location method was applied to the 704 datasets (64 untilted, 640 tilted), and visual inspection showed that the method correctly found acceptable approximations of the MSP in all of them. Fig. 5 shows sample slices of some datasets and the computed MSPs.

Table 2. Angles between computed MSPs in the random tilt evaluation

Group	Datasets	Angles Mean	σ
Control	20	1.33°	0.85°
Surgery	36	1.32°	1.03°
Phantom	8	0.85°	0.69°
Overall	64	1.26°	0.95°

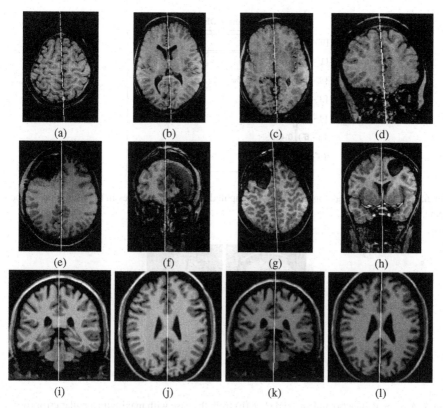

Fig. 5. Examples of planes computed by the method: (a–d): sample slices from a control dataset; (e–f) sample slices from a surgery dataset; (g–h) sample slices from another surgery dataset; (i–j): sample slices from a phantom dataset; (k–l): sample slices from a tilted dataset obtained from the one in (i–j)

For each tilted dataset, we applied the inverse transform to the computed mid-sagittal plane to project it on its respective untilted dataset space. Thus, for each untilted dataset we obtained 11 planes which should be similar. We measured the angle between all $\binom{11}{2} = 55$ distinct plane pairs. Table 2 shows the mean and standard deviation (σ) of these angles within each group. The low mean angles (column 3) and low standard deviations (column 4) show that the method is precise and robust with regard to rigid transformations of the input. The similar values obtained for the 3 groups indicate that

Table 3. Angles between computed MSPs and expert delineations

Group	Datasets	Angles Mean	σ
Control	20	1.95°	1.23°
Surgery	36	1.62°	1.17°
Phantom	8	0.99°	0.27°
Overall	64	1.64°	1.16°

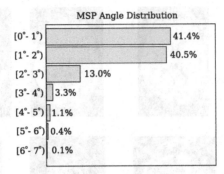

Fig. 6. Distribution of the angles between computed mid-sagittal planes in the random tilt experiment

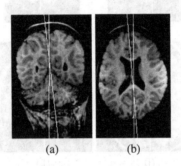

Fig. 7. A coronal slice (a) and an axial slice (b) from the case with maximum angular error (6.9°), with planes in white: The fissure was thick at the top of the head, and curved in the longitudinal direction, allowing the MSP to snap either to the frontal or posterior segments of the fissure, with some degree of freedom

the method performs equally well on healthy, pathological and synthetic data. The majority (94.9%) of the angles were less than 3°, as shown in the histogram of Fig. 6. Of $64 \times 55 = 3520$ computed angles, only 5 (0.1%) were above 6°. The maximum measured angle was 6.9°. Even in this case (Fig. 7), both planes are acceptable in visual

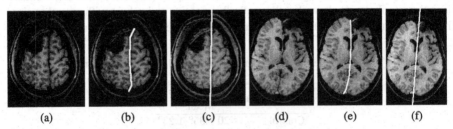

Fig. 8. Non-planar fissures: (a) irregular fissure, (b) non-planar fissure delineation of (a) by an expert and (c) MSP computed by our method. (d) Curved fissure, (e) non-planar fissure delineation of (d) by an expert and (f) MSP computed by our method.

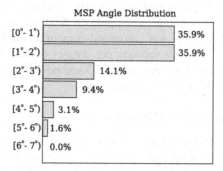

Fig. 9. Distribution of the angles between computed mid-sagittal planes and expert delineations

inspection, and the large angle between different two computations of the MSP can be related to the non-planarity of the fissure, which allows different planes to match with similar optimal scores. The lower mean angle in the phantom group (column 3, line 3 of Table 2) can be related to the absence of curved fissures in the synthetic datasets. Fig. 8 shows some examples of non-planar fissures.

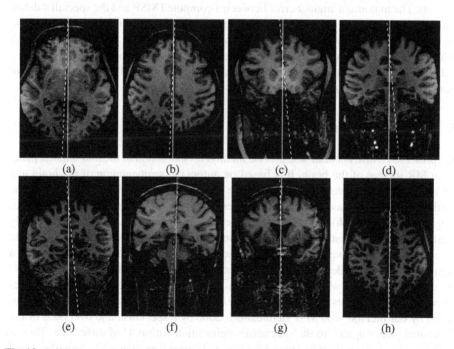

Fig. 10. Solid lines represent computed MSPs and dashed lines represent the specialist's delineations. (a)–(e) Sample slices from the single worst case where the computed MSP differed from the specialist's delineation by 5.7° (subject belongs to the control group). (f)–(h): Example slices from 3 distinct images in the surgery group, with angular errors 3.5° (f), 1.6° (g) and 0.3° (h).

These experiments were performed on a 2.0 GHz Athlon64 PC running Linux. The method took from 41 to 78 seconds to compute the MSP on each MR dataset (mean: 60.0 seconds). Most of the time was consumed computing the brain mask (stage 1). Stage 1 required from 39 to 69 seconds per dataset (mean: 54.8 seconds), while stage 2 required from 1.4 to 20 seconds (mean: 5.3 seconds). The number of iterations in stage 2 ranged from 0 to 30 (mean: 7.16 iterations).

4.4   Comparison with Expert Delineations

The MSPs of all 64 images were manually delineated by a specialist (Dr. Yasuda, neurosurgeon). In each image, the specialist selected 10 points over the longitudinal fissure, and the MSP was computed as the minimum squares fit of the points. The fitted plane was displayed interactively as the points were selected, and the specialist was able to inspect the proper location of the plane on any slice (in 3 different orientations), and each of these control points could be dragged with the mouse until the specialist was satisfied with the fitted plane. The specialist took about 2 minutes to delineate the plane on each image. We computed the angles between the manually delineated planes and the MSP located by our method. Table 3 shows the results, and Fig. 9 shows the angle distribution histogram.

About 85% of the computed planes matched the specialist's delineation within 3° (Fig. 9). The maximum angular error between a computed MSP and the specialist delineation was 5.7°, which occurred in one of the subjects of the control group. Figs. 10a–e show sample slices from this worst case, and Figs. 10f–h show typical slices from 3 other subjects. The majority of the computed planes closely matched the specialist's delineations. The small angular errors in these cases lead to small offsets in voxel location, considering the discrete image domain.

5   Conclusions and Future Work

We presented a fast, accurate and precise method for extraction of the mid-sagittal plane from MR images of the brain. It is based on automatic segmentation of the brain and on a heuristic search based on maximization of CSF within the MSP. We evaluated the method on 64 MR datasets, including images from patients with large surgical cavities (Fig. 2a and Figs. 5e–h). The method succeeded on all datasets and performed equally well on healthy and pathological cases. Rotations and translations of the datasets led to mean MSP variations around 1° (high accuracy and repeatability), which is not a significant error considering the discrete space of MR datasets. MSP variations over 3° occurred only in cases where the longitudinal fissure was not planar, and multiple planes fitted different segments of the fissure with similar scores. We compared the MSPs computed by our method with MSP delineations done by a specialist, and over 85% of the computed planes agreed to the specialist's delineation within 3° of difference. The remaining cases were misled by curved fissures and surgical cavities adjacent to the actual fissure. The mean angle variation within a same image (precision) was 1.26°, and the mean angle difference between the computed MSPs and the specialist's delineations (accuracy) was 1.64°. The method required a mean time of 60 seconds to extract the MSP from each MR dataset on a common PC.

Previous fissure-based works were either evaluated on images of healthy patients, on images with small lesions [29], or relied on local symmetry measurements [28]. As future work, we intend to implement some of the previous works and compare their accuracy, precision and performance with our method on the same datasets. Brain mask computation is responsible for most of the computing time. We also plan to evaluate how the computation of the brain mask on lower resolutions affect the accuracy, precision and efficiency of the method.

Acknowledgements. The authors thank CAPES (Proc. 01P-05866/2007), CNPq (Proc. 302617/2007-8), and FAPESP (Procs. 03/13424-1, 05/56578-4, 05/59258-0).

References

1. Davidson, R.J., Hugdahl, K.: Brain Asymmetry. MIT Press/Bradford Books (1996)
2. Crow, T.J.: Schizophrenia as an anomaly of cerebral asymmetry. In: Maurer, K. (ed.) Imaging of the Brain in Psychiatry and Related Fields, pp. 3–17. Springer, Heidelberg (1993)
3. Geschwind, N., Levitsky, W.: Human brain: Left-right asymmetries in temporale speech region. Science 161, 186–187 (1968)
4. Wang, L., Joshi, S.C., Miller, M.I., Csernansky, J.G.: Statistical analysis of hippocampal asymmetry in schizophrenia. NeuroImage 14, 531–545 (2001)
5. Csernansky, J.G., Joshi, S., Wang, L., Haller, J.W., Gado, M., Miller, J.P., Grenander, U., Miller, M.I.: Hippocampal morphometry in schizophrenia by high dimensional brain mapping. Proceedings of the National Academy of Sciences of the United States of America 95, 11406–11411 (1998)
6. Styner, M., Gerig, G.: Hybrid boundary-medial shape description for biologically variable shapes. In: Proc. of IEEE Workshop on Mathematical Methods in Biomedical Imaging Analysis (MMBIA), pp. 235–242. IEEE, Los Alamitos (2000)
7. Mackay, C.E., Barrick, T.R., Roberts, N., DeLisi, L.E., Maes, F., Vandermeulen, D., Crow, T.J.: Application of a new image analysis technique to study brain asymmetry in schizophrenia. Psychiatry Research 124, 25–35 (2003)
8. Highley, J.R., DeLisi, L.E., Roberts, N., Webb, J.A., Relja, M., Razi, K., Crow, T.J.: Sex-dependent effects of schizophrenia: an MRI study of gyral folding, and cortical and white matter volume. Psychiatry Research: Neuroimaging 124, 11–23 (2003)
9. Barrick, T.R., Mackay, C.E., Prima, S., Maes, F., Vandermeulen, D., Crow, T.J., Roberts, N.: Automatic analysis of cerebral asymmetry: na exploratory study of the relationship between brain torque and planum temporale asymmetry. NeuroImage 24, 678–691 (2005)
10. Hogan, R.E., Mark, K.E., Choudhuri, I., Wang, L., Joshi, S., Miller, M.I., Bucholz, R.D.: Magnetic resonance imaging deformation-based segmentation of the hippocampus in patients with mesial temporal sclerosis and temporal lobe epilepsy. J. Digital Imaging 13, 217–218 (2000)
11. Wu, W.C., Huang, C.C., Chung, H.W., Liou, M., Hsueh, C.J., Lee, C.S., Wu, M.L., Chen, C.Y.: Hippocampal alterations in children with temporal lobe epilepsy with or without a history of febrile convulsions: Evaluations with MR volumetry and proton MR spectroscopy. AJNR Am. J. Neuroradiol. 26, 1270–1275 (2005)
12. Csernansky, J.G., Wang, L., Joshi, S., Miller, J.P., Gado, M., Kido, D., McKeel, D., Morris, J.C., Miller, M.I.: Early DAT is distinguished from aging by High-dimensional Mapping of the hippocampus. Neurology 55, 1636–1643 (2000)
13. Liu, Y., Teverovskiy, L.A., Lopez, O.L., Aizenstein, H., Meltzer, C.C., Becker, J.T.: Discovery of biomarkers for alzheimer's disease prediction from structural MR images. In: 2007 IEEE Intl. Symp. on Biomedical Imaging, pp. 1344–1347. IEEE, Los Alamitos (2007)

14. Ardekani, B., Kershaw, J., Braun, M., Kanno, I.: Automatic detection of the mid-sagittal plane in 3-D brain images. IEEE Trans. on Medical Imaging 16, 947–952 (1997)
15. Kapouleas, I., Alavi, A., Alves, W.M., Gur, R.E., Weiss, D.W.: Registration of three dimensional MR and PET images of the human brain without markers. Radiology 181, 731–739 (1991)
16. Liu, Y., Collins, R.T., Rothfus, W.E.: Robust midsagittal plane extraction from normal and pathological 3D neuroradiology images. IEEE Trans. on Medical Imaging 20, 175–192 (2001)
17. Talairach, J., Tournoux, P.: Co-Planar Steriotaxic Atlas of the Human Brain. Thieme Medical Publishers (1988)
18. Bergo, F.P.G., Ruppert, G.C.S., Pinto, L.F., Falcão, A.X.: Fast and robust mid-sagittal plane location in 3D MR images of the brain. In: Proc. BIOSIGNALS 2008 – Intl. Conf. on Bio-Inspired Syst. and Sig. Proc., pp. 92–99 (2008)
19. Junck, L., Moen, J.G., Hutchins, G.D., Brown, M.B., Kuhl, D.E.: Correlation methods for the centering, rotation, and alignment of functional brain images. The Journal of Nuclear Medicine 31, 1220–1226 (1990)
20. Minoshima, S., Berger, K.L., Lee, K.S., Mintun, M.A.: An automated method for rotational correction and centering of three-dimensional functional brain images. The Journal of Nuclear Medicine 33, 1579–1585 (1992)
21. Sun, C., Sherrah, J.: 3D symmetry detection using the extended Gaussian image. IEEE Trans. on Pattern Analysis and Machine Intelligence 19, 164–168 (1997)
22. Smith, S.M., Jenkinson, M.: Accurate robust symmetry estimation. In: Taylor, C., Colchester, A. (eds.) MICCAI 1999. LNCS, vol. 1679, pp. 308–317. Springer, Heidelberg (1999)
23. Prima, S., Ourselin, S., Ayache, N.: Computation of the mid-sagittal plane in 3D brain images. IEEE Trans. on Medical Imaging 21, 122–138 (2002)
24. Tuzikov, A.V., Colliot, O., Bloch, I.: Evaluation of the symmetry plane in 3D MR brain images. Pattern Recognition Letters 24, 2219–2233 (2003)
25. Teverovskiy, L., Liu, Y.: Truly 3D midsagittal plane extraction for robust neuroimage registration. In: Proc. of 3rd IEEE Intl. Symp. on Biomedical Imaging, pp. 860–863. IEEE, Los Alamitos (2006)
26. Brummer, M.E.: Hough transform detection of the longitudinal fissure in tomographic head images. IEEE Trans. on Medical Imaging 10, 66–73 (1991)
27. Guillemaud, R., Marais, P., Zisserman, A., McDonald, B., Crow, T.J., Brady, M.: A three dimensional mid sagittal plane for brain asymmetry measurement. Schizophrenia Research 18, 183–184 (1996)
28. Hu, Q., Nowinski, W.L.: A rapid algorithm for robust and automatic extraction of the midsagittal plane of the human cerebrum from neuroimages based on local symmetry and outlier removal. neuroimage 20, 2153–2165 (2003)
29. Volkau, I., Prakash, K.N.B., Ananthasubramaniam, A., Aziz, A., Nowinski, W.L.: Extraction of the midsagittal plane from morphological neuroimages using the Kullback-Leibler's measure. Medical Image Analysis 10, 863–874 (2006)
30. Bergo, F.P.G., Falcão, A.X., Miranda, P.A.V., Rocha, L.M.: Automatic image segmentation by tree pruning. J. Math. Imaging and Vision 29, 141–162 (2007)
31. Falcão, A.X., Stolfi, J., Lotufo, R.A.: The image foresting transform: Theory, algorithms and applications. IEEE Trans. on Pattern Analysis and Machine Intelligence 26, 19–29 (2004)
32. Otsu, N.: A threshold selection method from gray level histograms. IEEE Trans. Systems, Man and Cybernetics 9, 62–66 (1979)
33. Collins, D.L., Zijdenbos, A.P., Kollokian, V., Sled, J.G., Kabani, N.J., Holmes, C.J., Evans, A.C.: Design and construction of a realistic digital brain phantom. IEEE Trans. on Medical Imaging 17, 463–468 (1998)

Facing Polychotomies through Classification by Decomposition: Applications in the Bio-medical Domain

Paolo Soda

Facoltà di Ingegneria, Università Campus Bio-Medico di Roma, Italy
`p.soda@unicampus.it`

Abstract. Polychotomies are recognition tasks with a number of categories grea-ter than two, consisting in assigning patterns to a finite set of classes. Although many of the learning algorithms developed so far are capable of handling polychotomies, most of them were designed by nature for dichotomies, that is, for binary learning. Therefore, various methods that decompose the multiclass recognition task in a set of binary learning problems have been proposed in the literature. After addressing the different dichotomies, the final decision is recon-structed according to a given criterion. Among the decomposition approaches, one of them is based on a pool of binary modules, where each one distinguishes the elements of one class from those of the others. For this reason, it is also known as *one-per-class* method. Under this decomposition scheme, we propose a novel reconstruction criterion to set the final decision on the basis of the single binary classifications. It looks at the quality of the current input and, more specifically, it is a function of the reliability of each classification act provided by the binary modules. The approach has been tested on six biological and medical datasets (two private, four public) and the achieved performance has been compared with the one previously reported in the literature, showing that the method improves the accuracies achieved so far.

1 Introduction

The problem of assigning input patterns to a finite set of categories or classes is a typical pattern recognition task. They are usually named as dichotomies, or binary learning, when they aim at distinguishing instances of two categories, whereas they are referred to as polychotomies, or multiclass learning, if there are more classes.

There is a huge number of applications that require multiclass categorisation, such as the support to medical diagnosis, text classification, object recognition and face identi-fication, to name a few.

In the literature numerous learning algorithms have been devised for multiclass prob-lems, such as neural networks or decision trees. However it exists a different approach that is based on the reduction of the multiclass task to multiple binary problems, and is referred to as *decomposition method*. The problem complexity is therefore reduced trough the decomposition of the polychotomy in less complex subtasks. The basic observation that supports such an approach is that in the literature most of the avail-able algorithms, which handle classification problems, are best suited to learning binary function [1,2]. Different dichotomizers, i.e. the discriminating functions that subdivide

A. Fred, J. Filipe, and H. Gamboa (Eds.): BIOSTEC 2008, CCIS 25, pp. 291–304, 2008.

the input patterns into two separate classes, perform the corresponding recognition task. To provide the final classification, their outputs are combined according to a given rule, usually referred to as *decision* or *reconstruction rule*.

In the framework of decomposition methods for classification, the various methods proposed to-date can be traced back to the following three categories [1,2,3,4,5,6,7,8,9].

The first one, called *one-per-class*, is based on a pool of binary learning functions, where each one separates a single class from all the others. The assignment of a new input to a certain class is typically performed looking at the function that returns the highest activation [1,4].

The second approach, commonly referred to as *distributed output code*, assigns a unique codeword, i.e. a binary string, to each class. If we assume that the string has n bit, the recognition system is composed by n binary classification functions. Given an unknown pattern, the classifiers provide an n-bit string that is compared with the codeword to set the final decision. For example, the input sample is assigned to the class with the closest codeword, according to a distance measure, such as the Hamming one. In this framework, in [1] the authors proposed an approach, known as *error-correcting techniques* (ECOC), where the use of error correcting codes as distributed output representation yielded a recognition system less sensitive to noise. This result could be achieved via the implementation of an error-recovering capability that belongs to the coding theory.

Recently, other researchers investigated ECOC approach proposing diversity measures between the codewords and the output of dichotomizers that differ from the Hamming distance, that is, the measures presented by Dietterich and Bakiri in [1]. For example, Kuncheva in [8] presented a measure that accounted for the overall diversity in the ensemble of binary classifiers, whereas Windeatt described two techniques for correlation reduction between the different codes [26].

The last approach is called n^2 classifier or *pairwise coupling*. In this case the recognition system is composed of $(n^2 - n)/2$ base dichotomizers, where each one is specialized in discriminating between pair of classes. Then, their predictions are aggregated to a final decision using a voting criterion. For example, in [3,9] the authors proposed a voting scheme adjusted by the credibilities of the base classifiers, which were calculated during the learning phase of the classification.

This short description of the methods proposed so far shows that the recognition systems based on decomposition methods are constituted by an ensemble of binary discriminating functions. Based on this observation, for brevity such systems are referred to as Multy Dichotomies System (MDS) in the following.

In the framework of the one-per-class approach, we present here a novel reconstruction rule that relies upon the quality of the input pattern and looks at the reliability of each classification act provided by the dichotomizers. Furthermore, the classification scheme that we propose allows employing either a single expert or an ensemble of classifiers internal to each module that solves a dichotomy. Finally, the effectiveness of the recognition system has been evaluated on six different datasets that belongs to biological and medical applications.

The rest of the paper is organized as follows: in the next section we introduce some notations and we present general considerations related to system configuration.

Section 3 details the reconstruction method and section 4 describes and discusses the experiments performed on six different medical datasets. Finally section 5 offers a conclusion.

2   Problem Definition

2.1   Background

Let us consider a classification task on c data classes, represented by the set of labels $\Omega = \{\omega_1, \ldots, \omega_i, \ldots, \omega_c\}$, with $c > 2$. The decomposition of the c classes polychotomy generates a pool of L dichotomizers, $M = \{M_1, \ldots, M_j, \ldots, M_L\}$, with the value of L depending upon the decomposition approach adopted. The dichotomizer M_j is a discriminating function that classifies each input sample x in two separates superclasses, represented by the label set $\Omega_j = \{0, 1\}$, each label grouping a subset of polychotomy classes. Therefore, the overall decomposition scheme can be specified by a *decomposition matrix* $D \in \Re^c \times \Re^L$, whose elements are defined as:

$$d_{ij} = \begin{cases} 1 & \text{if class } i \text{ is in the subgroup associated to label 1 of } M_j \\ 0 & \text{if class } i \text{ is in the subgroup associated to label 0 of } M_j \\ -1 & \text{if class } i \text{ is in neither groups associated to label 0 or 1 of } M_j \end{cases} \quad (1)$$

Hence, the dichotomizer M_j is trained to associate patterns belonging to classe ω_i with values d_{ij} of decomposition D. For instance, figure 1 represents the decomposition matrices of one-per-class, ECOC[1] and pairwise coupling when $c = 4$. Each row corresponds to one class, and each column to one dichotomy. Each class is therefore assigned a unique binary string, usually referred to as codeword, given by the corresponding row of the decomposition matrix.

$$
\begin{pmatrix} 1 & 0 & 0 & 0 \\ 0 & 1 & 0 & 0 \\ 0 & 0 & 1 & 0 \\ 0 & 0 & 0 & 1 \end{pmatrix}
\qquad
\begin{pmatrix} 1 & 1 & 1 & 1 & 1 & 1 & 1 \\ 0 & 0 & 0 & 0 & 1 & 1 & 1 \\ 0 & 0 & 1 & 1 & 0 & 0 & 1 \\ 0 & 1 & 0 & 1 & 0 & 1 & 0 \end{pmatrix}
\qquad
\begin{pmatrix} 1 & 0 & -1 & -1 \\ 1 & -1 & 0 & -1 \\ 1 & -1 & -1 & 0 \\ -1 & 1 & 0 & -1 \\ -1 & 1 & -1 & 0 \\ -1 & -1 & 1 & 0 \end{pmatrix}
$$

$$\qquad\qquad (a) \qquad\qquad\qquad\qquad (b) \qquad\qquad\qquad\qquad (c)$$

Fig. 1. Typical decomposition matrices D of four classes recognition problem for three different decomposition schemes. (a) one-per-class method, (b) ECOC and (c) pairwise coupling. Each row corresponds to one class, and each column to one dichotomy.

In this paper we propose a recognition approach that belongs to the one-per-class framework. Therefore, the multiclass task is reduced into c binary problems, each one

---

[1] The code matrix has been computed applying an exhaustive code generation approach [1].

addressed by one module of the pool M. Hence, L is equal to c. In the following, we say that the dichotomizer M_j is specialized in the jth class when it aims at recognizing if the sample x belongs either to the jth class ω_j or, alternatively, to any other class ω_i, with $i \neq j$. Therefore, each module assigns to the input pattern $x \in \Re^n$ a binary label:

$$M_j(x) = \begin{cases} 1 \text{ if } x \in \omega_j \\ 0 \text{ if } x \in \omega_i, i \neq j \end{cases} \tag{2}$$

where $M_j(x)$ indicates the output of the jth module on the pattern x. On this basis, the codeword associated to the class ω_j has a bit equal to 1 at the jth position, and 0 elsewhere.

Notice that we have just used the term module and not classifier to emphasize that each dichotomy can be solved not only by a single expert, but also by an ensemble of classifiers. To our knowledge, the dichotomizers of decomposition systems typically adopt the former approach, i.e. they are composed by one classifier per specialized module. For example, for their experimental assessments in [2] and [4] the authors used a a decision tree and a multi layer perceptrons with one hidden layer, respectively. The same functions were employed by Dietterich and Bakiri for the evaluation of their proposal in [1], whereas Allwein et al. used a Support Vector Machine [5]. Rather than using a single expert, a viable alternative consists of combining the outputs of classifiers that solve the same recognition task. The idea is that the classification performance attainable by their combination should be improved by taking advantage of the strength of the single classifiers. Classifier selection and fusion are the two main combination strategies reported in the literature. The former presumes that each classifier has expertise in some local area of the feature space [10,11,12]. For example, when an unknown pattern is submitted for classification, the more accurate classifier in the vicinity of the input is selected to label it [10]. The latter algorithms assume that the classifiers are applied in parallel and their outputs are combined to attain somehow a group of "consensus" [13,14,15]. Typical fusion techniques include weighted mean, voting, correlation, probability, etc..

It is worth noticing that the dichotomizers, e.g. a single classifier or an ensemble of them, besides labelling each pattern, may supply other information typically related to the degree that the sample belongs to that class. In this respect, the various classification algorithms are divided into three categories, on the basis of the output information that they are able to provide [12]. The classifiers of type 1 supply only the label of the presumed class and, therefore, they are also known as experts that work at the abstract level. Type 2 classifiers work at the rank level, i.e. they rank all classes in a queue where the class at the top is the first choice. Learning functions of type 3 operate at the measurement level, i.e. they attribute each class a value that measures the degree that the input sample belongs to that class. If a crisp label of the input pattern is needed, we can use the maximum membership rule that assigns x to the class for which the degree of support is maximum (ties are resolved arbitrarily). Although abstract classifiers provide a n-bit string that can be compared with the codewords, decision schemes that exploit information derived from the classifiers working at the measurement level permit us to define reconstruction rules that are potentially more effective. Indeed, in the literature most of the one-per-class scheme use measurements classifiers and assign the input sample corresponding to the classifier that returns the highest activation [4,1]. Furthermore, if the module is constituted by a multi-experts system, the information supplied by the single classifiers can be used to compute a measure similar to that provided by measurement classifiers.

It worth observing that the activation of a measurement classifier is a value that measures the degree that a sample belongs to the output class. However, this quantity represents only a raw information: in this respect, in different fields of pattern recognition, it has been proven that classifier outputs can be combined more effectively using the reliability of the current decision on sample x rather than its activation [13,15,16,17]. The contribution of this paper resides in combing the reliability of decisions provided by each binary module M_j with the analysis of the output codeword. Using these two quantities permits us to define a novel reconstruction rule, as presented in the following sections.

2.2   The Reconstruction Method

The reconstruction method addresses the issues of determining the final label of the input pattern x on the basis of the modules' decisions and, eventually, of information directly derived from their outputs.

To present our method, let us introduce two auxiliaries quantities. The first, named *binary profile*, represents the state of the module outputs. It is a c-bit vector defined by:

$$\mathbf{M}(x) = [M_1(x), \cdots, M_j(x), \cdots, M_c(x)] \tag{3}$$

whose entries are the crisp labels provided by each module in the classification of sample x (see equation 2). Note that this quantity represents the codeword associated to the input sample.

Since each block has a binary output, the 2^c possible bit combinations of the binary profile can be grouped into the following three categories:

(i) only one module classifies the sample in the class in which it is specialized;
(ii) more modules classify the sample in its own class;
(iii) none module classifies the sample in its own class.

In the first case, only one entry of $\mathbf{M}(x)$ is one; in the second more elements are one (at least two and no more than c), whereas in the last situation all the elements are zero. Such an observation naturally leads to distinguish these three cases using the summation over the binary profile. Indeed,

$$m = \sum_{j=1}^{c} M_j(x) = \begin{cases} 1, & \text{in case (i)} \\ [2,c], & \text{in case (ii)} \\ 0, & \text{in case (iii)} \end{cases} \tag{4}$$

where m therefore represents the number of modules whose outputs are 1.

The second quantity that we introduce is referred to as *reliability profile* and it is described by:

$$\psi(x) = [\psi_1(x), \cdots, \psi_j(x), \cdots, \psi_c(x)] \tag{5}$$

where each element $\psi_j(x)$ measure the reliability of the classification act on pattern x provided by the jth module. Note that the reliability varies in the interval $[0, 1]$, and a value near 1 indicates a very reliable classification.

Assuming that we determined both the binary and the reliability profiles, i.e. $\mathbf{M}(x)$ and $\psi(x)$ respectively, in the next section we will present the reconstruction rule.

3   Reliability Based Reconstruction

In this section we introduce the novel reconstruction strategy we propose in the paper. It chooses an output in any of the 2^c combinations of the binary profile. We deem that an accurate final decision can be taken if the reconstruction rule looks at the quality of the classification provided by the modules, i.e. at the reliability of their specific decisions. To our knowledge the application of such a parameter can not be found in the literature related to decomposition methods. Indeed, the papers of this field that used the information directly derived from the outputs of the base classifiers typically considered only the highest activation among the experts, e.g. the maximum output from a pool of neural networks. However, this measure cannot be regarded as a reliability parameter, since it has been demonstrated that it should be computed considering not only the winner output neurons but also the losers [16].

Therefore, differently from the past, we propose a criterion that makes use of the reliability measure, i.e. of the reliability profile, named as *Reliability-based-Reconstruction (RbR)*. Denoting by s the index of the module that sets the final output $O(x) \in \Omega$, referred to as selected module for brevity in the following, the final decision is given by:

$$O(x) = \omega_s \tag{6}$$

with

$$s = \begin{cases} \arg\max_j (M_j(x) \cdot \psi_j(x)), \text{ if } m \in [1, c] \\ \arg\min_j (\overline{M_j(x)} \cdot \psi_j(x)), \text{ if } m = 0 \end{cases} \tag{7}$$

where $\overline{M_j(x)}$ indicates the negate output of the jth block.

The first row of this equation considers both cases (i) and (ii). Indeed, since in the first case all the modules agree in their decision, as a final output is chosen the class of the module whose output is 1. Conversely, in cases (ii) and (iii) the final decision is performed looking at the reliability of each modules' classifications. In case (ii), m modules vote for their own class, whereas the others $(c - m)$ indicate that x does not belong to their own class. To solve the dichotomy between the m conflicting modules we look at the reliability of their classification and choose the class associated to the more reliable one. In case (iii) $m = 0$, suggesting that all modules classify x as belonging to another class than the one they are specialized. In this case, the bigger is the reliability parameter $\psi_j(x)$, the less is the probability that x belongs to ω_j, and the bigger is the probability that it belongs to the other classes. These observations suggest finding out which module has the minimum reliability and then choosing the class associated to it as a final output.

Panel A of figure 2 shows the architecture of the proposed recognition system. The decision $M_j(x)$ and the reliability $\psi_j(x)$ supplied by each of the c modules are

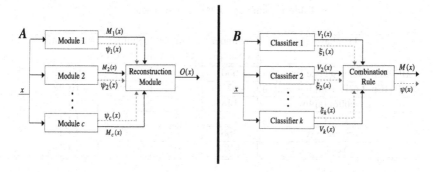

Fig. 2. The system architecture, which is based on the aggregation of binary modules (panel *A*), according to the one-per-class approach. Note that each module can be constituted by a multi-experts system, as depicted in panel *B*.

aggregated in the *reconstruction module* to provide the final decision $O(x)$. As observed in section 2.1, the use of an ensemble of classifiers in each module is a way to improve its discrimination capability. In this respect, panel *B* of the same figure depicts a typical configuration of a multi-experts system. Notice that both the output of the kth classifier and its reliability, denoted as $V_k(x)$ and $\xi_k(x)$, respectively, can be given to the combination rule in order to label the input sample.

4 Experimental Evaluation

In this section we first describe the datasets used to assess the performance of the reconstruction method and, second, we briefly discuss the configuration of the MDS modules. Third, we present a strategy to evaluate the classification reliability when the modules are constituted both by a single classifier and by an ensemble of experts, respectively. Finally, we report the experimental results.

4.1 Datasets

For our tests we use six datasets, described in the following and summarized in table 1.

Indirect Immunofluorescence Well Fluorescence Intensity. Connective tissue diseases are autoimmune disorders characterized by a chronic inflammatory process involving connective tissues. When they are suspected in a patient, the Indirect Immunofluorescence (IIF) test based on HEp-2 substrate is usually performed, since it is the recommended method. The interested reader may find a wide explanation of the IIF and its issues in [18,19]. The dataset consists of 14 features extracted from 600 patients sera collected at Università Campus Bio-Medico di Roma. The samples are distributed over three classes, namely positive (36.0%), negative (32.5%) and intermediate

Table 1. Summary of the datasets used

Database	Number of Samples	Number of Classes	Number of features	Avalaibility
IIF Well Fluorescence Intensity	600	3	14	Private
IIF HEp-2 cells staining pattern	573	5	159	Private
Lymphography	148	4	18	UCI
Ecoli	336	8	9	UCI
Dermatology	366	6	33	UCI
Yeast	1484	10	8	UCI

(31.5%). Previous results are reported in [20] where the authors employed a multiclass approach, achieving an accuracy of 76% approximately.

Indirect Immunofluorescence HEp-2 Cells Staining Pattern. This is a dataset with 573 instances represented by 159 statistical and spectral features. The samples are distributed in five classes that are representative of the main staining patterns exhibited by HEp-2 cells, namely homogeneous (23.9%), peripheral nuclear (21.8%), speckled (37.0%), nucleolar (8.2%) and artefact (9.1%). These patterns are related to one of the different autoantibodies that give rise to a connective tissue disease. For the details on these classes see [19]. On this dataset, we performed some tests adopting a multiclass approach, which exhibits a hit rate of 63.6% approximately, evaluated using the eight-fold cross validation.

Lymphography. A database of lymph diseases was obtained from the University Medical Centre, Institute of Oncology, Ljubljana. It is composed by 148 instances described by 18 numeric attributes. There are four classes, namely normal (1.4%), metastases (54.7%), malign lymph (41.2%) and fibrosis (2.7%). The data are available within the UCI Machine Learning Repository[2] [21]. Different approaches were used in the literature to address the recognition task. For instance, for Naive Bayes classifier and C4.5 decision tree the achieved performance was 79% and 77% respectively [22], whereas induction techniques correctly classified the 83% of samples [23].

Ecoli. The database is composed by 336 samples, described by a nine-dimensional vector and distributed in eight classes. Each class represents a localization site, which can be cytoplasm (42.5%), inner membrane without signal sequence (22.9%), periplasm (15.5%), inner membrane, uncleavable signal sequence (10.4%), outer membrane (6.0%), outer membrane lipoprotein (1.5%), inner membrane lipoprotein (0.6%) and inner membrane, cleavable signal sequence (0.6%). Again, the data are available within the UCI Machine Learning Repository [21]. In [3], the authors reported an accuracy that ranges from 79.7% up to 83.0%, achieved employing both a decision tree and a Multi Layer Perceptrons, respectively. In [5], using many popular classification

---

[2] For each dataset of this repository the users have access to a description of the application domain, to the features and to the ground truth.

algorithms, such as the Support Vector Machines, AdaBoost, regression and decision-tree algorithms, the hit rate varies from 78.5% up to 86.1%.

Dermatology. This is a dataset with 366 instances represented by 33 features, available within the UCI Machine Learning Repository [21]. Each class represents an erythemato-squamous disease, such as psoriasis (30.6%), seboreic dermatitis (16.6%), lichen planus (19.7%), pityriasis rosea (13.4%), cronic dermatitis (14.2%), and pityriasis rubra pilaris (5.5%).

Their differential diagnosis is a real problem in dermatology. Usually a biopsy is necessary for the diagnosis but unfortunately these diseases share many histopathological features as well. Another difficulty for the differential diagnosis is that a disease may show the features of another disease at the beginning stage and may have the characteristic features at the following stages. Patients were first evaluated clinically with 12 features. Afterwards, skin samples were taken for the evaluation of 22 histopathological features. The values of the histopathological features are determined by an analysis of the samples under a microscope. To solve this recognition problem, in [5] the experimentally evaluated accuracy ranged between 95.0% and 96.9%. In [24] a multiobjective genetic algorithm was proposed, which exhibited an accuracy that varied between 94.5% and 95.8%.

Yeast. A dataset of localization site of protein was achieved from the UCI Machine Learning Repository [21]. It is composed by 1484 instances distributed over 10 classes and described by 8 features. Each class represents a localization sites, namely cytoskeletal (31.2%), nuclear (28.9%), mitochondrial (16.4%), membrane protein with no N-terminal signal (11.0%), membrane protein with uncleaved signal (3.5%), membrane protein with cleaved signal (3.0%), extracellular (2.4%), vacuolar (2.0%), peroxisomal (1.3%) and endoplasmic reticulum lumen (0.3%). Previous results reported in [25] show an accuracy of 55.0% using a probabilistic classification system, whereas in [9] the authors achieved an hit rate that varied between 52.8% and 55.7% applying a pairwise coupling decomposition scheme in conjunction neural network and decision trees. In [5] the best measured accuracy was 58.2%.

4.2 MDS Configuration

The modules of the MDS are essentially composed by a single classifier or by an ensemble of classifiers. In both cases, as single expert we use k-Nearest Neighbour (kNN) or Multi-Layer Perceptron (MLP). For each dichotomy, we first select a subset of features that simplifies both the pattern representation and the classifier complexity as well as the risk of the incurring in the peaking phenomenon[3]. Then we carry out some preliminary tests to determine the best configuration of experts parameters, e.g. the number of neighbours for kNN classifier or the number of hidden layers, neurons per layer, etc., for the MLP network. Furthermore, when the module is constituted by an ensemble of experts we adopt a fusion technique to combine their outputs, namely the Weighted

---

[3] The peaking phenomenon is a paradoxical behaviour in which the added features may actually degrade the performance of a classifier if the number of training samples that are used to design the classifier is small relative to the number of features.

Voting (WV). In such a method the opinion of each expert about the class of the input pattern is weighted by the reliability of its classification. Since each expert deals with a binary learning task, to further present this scheme we can simplify the notation as follows. Denoting as $V_k(x)$ and as $\xi_k(x)$ the output and the classification reliability of kth classifier on sample x, the weighted sum of the votes for each of the two classes is given by:

$$W_h(x) = \sum_{k:V_k(x)=h} \xi_k(x), \text{ with } h = \{0,1\} \qquad (8)$$

The output of the fusion of the jth module, $M_j(x)$, is defined by[4]:

$$M_j(x) = \begin{cases} 1 & \text{if } W_1(x) > W_0(x) \\ 0 & \text{otherwise} \end{cases} \qquad (9)$$

Turning our attention to the configuration of the system in the experimental tests, notice that the modules that label the samples of the IIF well fluorescence intensity, lymphography, dermatology and yeast datasets are composed by one classifiers. The modules that classify the samples of HEp-2 cells and ecoli databases are constituted by kNN and MLP classifiers combined by the WV criterion.

4.3   Reliability Parameter

The approach described for deriving the final decision according to the RbR rule requires the introduction of a reliability parameter that evaluates the quality of the classification performed by each module, which can be composed by a single classifier or by an aggregation of classifiers (figure 2). In the former case its reliability ψ_j coincides with the one of the single classifier, i.e. ξ. In the latter case, each entry of the reliability profile generally depends on the combination rule adopted in the module, on the number k of composing experts and on their individual reliabilities ξ. Formally,

$$\psi_j(x) = \begin{cases} \xi(x), & \text{if } k = 1 \\ f(\xi_1(x), \cdots, \xi_i(x), \cdots, \xi_k(x)), & \text{if } k > 1 \end{cases} \qquad (10)$$

where all the reliabilities are reported as function of the input pattern to emphasize that they are computed for each classification act.

In the rest of this section we first present two techniques to measure the reliability of kNN and MLP decisions, and then we introduce a novel method that estimates such parameter in the case of the application of the WV criterion.

A typical approach that measures the reliability of the decision taken by the single expert, i.e. ξ, makes use of the confusion matrix[5] estimated on the learning set. The drawback of this method is that all the patterns with the same label have equal reliability, regardless of the quality of the sample. Indeed, the average performance on the learning set, although significant, does not necessarily reflect the actual reliability of

---

[4] In case of tie, i.e. if $W_1(x)$ is equal to $W_0(x)$, the output $M_j(x)$ is set arbitrarily to zero. Note that it never occurred in all tests we performed.

[5] The confusion matrix reports for each entry (p, q) the percentage of samples of the class C_p assigned to the class C_q.

each classification act. To overcome such limitations we adopt an approach that relies upon the quality of the current input. To this end, we refer to the work presented in [16], where the quality of the sample is related to its position in the feature space. In this respect, the low reliability of a recognition act can be traced back to one of the following situations: (a) in the feature space x is located in a region that is far from those associated with the various classes, i.e. the sample is significantly different from those present in the training set, (b) the point representing x lies in a region of the feature space where two or more classes overlap. These observations lead to introduce the parameters ξ_a and ξ_b that distinguish between the two situations of unreliable classification. Then, a comprehensive parameter ξ can be derived adopting the following conservative choice:

$$\xi = min(\xi_a, \xi_b) \tag{11}$$

Indeed, it implies that a low value for only one of the parameters is sufficient to consider unreliable the classification.

In the case of kNN classifiers, following [16], the two parameters are defined are given by:

$$\xi_a = max\left(1 - D_{min}/D_{max}, 0\right) \tag{12}$$
$$\xi_b = 1 - D_{min}/D_{min2} \tag{13}$$

where D_{min} is the smallest distance of x from a reference sample belonging to the same class of x, D_{max} is the highest among the values of D_{min} obtained for samples taken from the training-test set, i.e. a set that is disjoint from both the reference and the test set, D_{min2} is the distance between x and the reference sample with the second smallest distance from x among all the reference set samples belonging to a class that is different from that determining D_{min}.

In the case of MLP classifier, the two quantities are defined as follows:

$$\xi_a = N_{win} \tag{14}$$
$$\xi_b = N_{win} - N_{2win} \tag{15}$$

where N_{win} is the output of the winner neuron, N_{2win} is the output of the neuron with the highest value after the winner. From this definition, it is straightforward that $\xi = \xi_b$. For further details see [13].

When the jth module is composed by more than one classifier combined according to the WV rule, the reliability estimator considers again the situations which can give rise to an unreliable classification. In this respect, we need to introduce the following two auxiliary quantities:

$$\pi_1(x) = max\left(\{\xi_k(x)|k : V_k(x) = M_j(x)\}\right) \tag{16}$$
$$\pi_2(x) = max\left(\{\xi_k(x)|k : V_k(x) \neq M_j(x)\} \cup \{0\}\right) \tag{17}$$

where $\pi_1(x)$ and $\pi_2(x)$ represent the maximum reliabilities of experts voting for the winning class and for other classes (0 if all the experts agree on the winner class), respectively. Given these definitions, the reliability of the WV rule can be evaluated according to the following conservative choice:

$$\psi(x) = min\left(\pi_1(x), max\left(0, 1 - \pi_2(x)/\pi_1(x)\right)\right) \tag{18}$$

Table 2. Testing accuracy achieved on the used datasets

Database	Past Usage	MDS using RbR
IIF Well Fluorescence Intensity	75.9%	94.3%
IIF HEp-2 cells staining pattern	63.6%	75.9%
Lymphography	77% − 83.0%	89.9%
Ecoli	78.5% − 86.1%	87.9%
Dermatology	94.5% − 96.9%	98.7%
Yeast	52.8% − 58.2%	59.7%

4.4   Results and Discussion

This section presents the experimental results that we achieved using the system described so far. To evaluate and then compare the results of this approach with those reported in the literature we perform eightfold and tenfold cross validation on the two IIF datasets, i.e. well fluorescence intensity and HEp-2 cells staining pattern, and on the other four databases, i.e. lymphography, ecoli, dermatology and yeast, respectively.

The third column of table 2 shows the testing accuracies achieved on the six databases. To simply compare them with the past results, the second column of the same table summarizes the performance reported in literature. Turning our attention to the tests carried out on the first and on the second datasets, a relevant accuracy improvement can be observed. Indeed, the hit rate increases of 18.4% and of 12.3% in the case of the well fluorescence intensity and HEp-2 cells staining pattern databases, respectively. In our opinion, such an improvement is twofold motivated. On the one hand, the set of extracted features is more stable and more effective when we adopt a decomposition approach rather than a multiclass one. On the other hand, the reconstruction rule exhibits a very good capability of solving the disagreements between the specialized modules. Indeed, when the binary profile of the input sample $M(x)$ differs from one of the possible codewords (i.e. $m = 0$ or $2 \leq m \leq c$), the decision is taken looking at the reliability profile $\psi(x)$, as presented in the formula 7.

These considerations are strengthened by the observation of the performance attained in the classification of samples belonging to the four UCI datasets. Indeed, since they are benchmark datasets, any reported improvement is due to the recognition approach rather than to the use of a different features set. The tests on both the lymphography and ecoli datasets exhibit an accuracy better than the one reported to date. Indeed, for the former dataset the improvement ranges both from 6.9% up to 12.9% , whereas for the latter one it varies from 1.8% up to 9.4%. Similar advances are observed on dermatology and yeast data. Indeed, the accuracy improves of 1.9% − 3.3% in case of dermatology samples, and of 2.5% − 11.5% for yeast data. Therefore, also in these cases the MDS in combination with the RbR rule improves the recognition performance.

Furthermore, it is worth noting that the approach seems independent of the modules' arrangement. The rationale lies in observing that in two of the six tests the MDS modules are constituted by a multi-experts system, whereas in the others they are composed by a single classifier (section 4). Consequently, the reliability ψ_j is measured according to a method that varies with the module configuration, as previously presented (see equations 11-18). Nevertheless, these variations do not affect the effectiveness of the recognition system. Therefore, we deem that the reconstruction rule is robust with respect to different reliability estimators.

5 Conclusions

In the framework of decomposition methods, we have presented a classification approach that reconstructs the final decision looking at the reliability of each classification act provided by all dichotomizers. Furthermore, the reconstruction rule does not depend on the configuration of each module, i.e. on its architecture. Such an observation is strengthened by the good performance achieved when both a single classifier and a fusion of experts constitute each module, respectively.

For all the six tested databases, the experimental results show that the proposed system outperforms the performance reported in the literature.

Future works are directed towards two issues. First, the test of the system on other public datasets and, second, the definition of reliability parameter of each decision taken by the MDS.

Acknowledgements. The author would like to thank Prof. Giulio Iannello for his support and DAS s.r.l {www.dasitaly.com}, which has funded this work.

References

1. Dietterich, T.G., Bakiri, G.: Solving multiclass learning problems via error-correcting output codes. Journal of Artificial Intelligence Research 2, 263–286 (1995)
2. Mayoraz, E., Moreira, M.: On the decomposition of polychotomies into dichotomies. In: ICML 1997: Proceedings of the Fourteenth International Conference on Machine Learning, pp. 219–226 (1997)
3. Jelonek, J., Stefanowski, J.: Experiments on solving multiclass learning problems by n^2 classifier. In: Nédellec, C., Rouveirol, C. (eds.) ECML 1998. LNCS, vol. 1398, pp. 172–177. Springer, Heidelberg (1998)
4. Masulli, F., Valentini, G.: Comparing decomposition methods for classication. In: KES 2000, Fourth International Conference on Knowledge-Based Intelligent Engineering Systems & Allied Technologies, pp. 788–791 (2000)
5. Allwein, E.L., Schapire, R.E., Singer, Y.: Reducing multiclass to binary: a unifying approach for margin classifiers. Journal of Machine Learning Research 1, 113–141 (2001)
6. Crammer, K., Singer, Y.: On the algorithmic implementation of multiclass kernel-based vector machines. Journal of Machine Learning Research 2, 265–292 (2002)
7. Hastie, T., Tibshirani, R.: Classification by pairwise coupling. In: NIPS 1997: Proceedings of the 1997 conference on Advances in neural information & processing systems, vol. 10, pp. 507–513. MIT Press, Cambridge (1998)
8. Kuncheva, L.I.: Using diversity measures for generating error-correcting output codes in classifier ensembles. Pattern Recognition Letters 26, 83–90 (2005)
9. Stefanowski, J.: Multiple and hybrid classifiers, 174–188 (2001)
10. Woods, K., Kegelmeyer, W.P., Bowyer, K.: Combination of multiple classifiers using local accuracy estimates. IEEE Transactions on Pattern Analysis and Machine Intelligence 19, 405–410 (1997)

11. Kuncheva, L.I.: Switching between selection and fusion in combining classifiers: anexperiment. IEEE Transactions on Systems, Man and Cybernetics, Part B 32, 146–156 (2002)

12. Suen, C.Y., Lam, L.: Multiple classifier combination methodologies for different output levels. In: Kittler, J., Roli, F. (eds.) MCS 2000. LNCS, vol. 1857, pp. 52–66. Springer, Heidelberg (2000)

13. De Stefano, C., Sansone, C., Vento, M.: To reject or not to reject: that is the question: an answer in case of neural classifiers. IEEE Transactions on Systems, Man, and Cybernetics–Part C 30, 84–93 (2000)

14. Kuncheva, L.I., Bezdek, J.C., Duin, R.P.W.: Decision templates for multiple classifier fusion: an experimental comparison. Pattern Recognition 34, 299–314 (2001)

15. Kittler, J., Hatef, M., Duin, R.P.W., Matas, J.: On combining classifiers. IEEE Transactions On Pattern Analysis and Machine Intelligence 20, 226–239 (1998)

16. Cordella, L.P., Foggia, P., Sansone, C., Tortorella, F., Vento, M.: Reliability parameters to improve combination strategies in multi-expert systems. Pattern Analysis & Applications 2, 205–214 (1999)

17. Cordella, L.P., Sansone, C., Tortorella, F., Vento, M., De Stefano, C.: Neural networks classification reliability. In: Academic Press theme volumes on Neural Network Systems, Techniques and Applications, vol. 5, pp. 161–199. Academic Press, London (1998)

18. Kavanaugh, A., Tomar, R., Reveille, J., Solomon, D.H., Homburger, H.A.: Guidelines for clinical use of the antinuclear antibody test and tests for specific autoantibodies to nuclear antigens. American College of Pathologists, Archives of Pathology and Laboratory Medicine 124, 71–81 (2000)

19. Rigon, A., Soda, P., Zennaro, D., Iannello, G., Afeltra, A.: Indirect immunofluorescence in autoimmune diseases: Assessment of digital images for diagnostic purpose. Cytometry B (Clinical Cytometry) 72, 472–477 (2007)

20. Soda, P., Iannello, G.: A multi-expert system to classify fluorescent intensity in antinuclear autoantibodies testing. In: Computer Based Medical Systems, pp. 219–224. IEEE Computer Society, Los Alamitos (2006)

21. Asuncion, A., Newman, D.J.: UCI machine learning repository (2007)

22. Clark, P., Niblett, T.: Induction in noisy domains. In: Proc. of Progress in Machine Learning, pp. 11–30 (1987)

23. Cheung, N.: Machine learning techniques for medical analysis. Master's thesis, University of Queensland (2001)

24. Pappa, G.L., Freitas, A.A., Kaestner, C.A.A.: Attribute selection with a multi-objective genetic algorithm, 280–290 (2002)

25. Horton, P., Nakai, K.: A probabilistic classification system for predicting the cellular localization sites of proteins 4, 109–115 (1996)

26. Windeatt, T., Ghaderi, R.: Binary labelling and decision-level fusion. Information Fusion 2(2), 103–112 (2001)

Automatic Speech Recognition Based on Electromyographic Biosignals

Szu-Chen Stan Jou[1] and Tanja Schultz[1,2]

[1] International Center for Advanced Communication Technologies
Carnegie Mellon University, Pittsburgh, PA, U.S.A.
[2] Cognitive Systems Laboratory, Karlsruhe University, Karlsruhe, Germany
tanja@cs.cmu.edu

Abstract. This paper presents our studies of automatic speech recognition based on electromyographic biosignals captured from the articulatory muscles in the face using surface electrodes. We develop a phone-based speech recognizer and describe how the performance of this recognizer improves by carefully designing and tailoring the extraction of relevant speech feature toward electromyographic signals. Our experimental design includes the collection of audibly spoken speech simultaneously recorded as acoustic data using a close-speaking microphone and as electromyographic signals using electrodes. Our experiments indicate that electromyographic signals precede the acoustic signal by about 0.05-0.06 seconds. Furthermore, we introduce articulatory feature classifiers, which had recently shown to improved classical speech recognition significantly. We describe that the classification accuracy of articulatory features clearly benefits from the tailored feature extraction. Finally, these classifiers are integrated into the overall decoding framework applying a stream architecture. Our final system achieves a word error rate of 29.9% on a 100-word recognition task.

1 Introduction

Computers have become an integral part of our daily lives and consequently require user-friendly interfaces for efficient human-computer interaction. Automatic speech recognition (ASR) systems offer the most natural front-end for human-computer interface because humans naturally communicate through speech. ASR is the automatic process of transforming spoken speech into a textual representation of corresponding word sequences. It allows for applications, such as command and control, dictation, dialog systems, audio indexing, and speech translation.

However, traditional ASR is based on the acoustic representation of speech and thus comes with several challenges. First of all, it requires the user to speak audibly. This may disturb bystanders and may also jeopardize a confidential communication. For example, telephone-based service systems often require the user to provide confidential information such as passwords or credit card numbers. If the call is made in public places, this confidential information might be eavesdropped by others. At the same time, making a phone call might distract or annoy bystanders, for example if a phone call is made in a quiet environment such as in the theater or during a meeting. The second major challenge of traditional ASR is the lack of robustness in case the acoustic channel is

A. Fred, J. Filipe, and H. Gamboa (Eds.): BIOSTEC 2008, CCIS 25, pp. 305–320, 2008.
© Springer-Verlag Berlin Heidelberg 2008

disturbed by ambient noise. Since the input speech signal is transmitted over the air and usually is picked up by a standard microphone, all other air-transmitted acoustic signals are picked up by this microphone as well. In most cases it is impossible to accurately extract the relevant speech signal from the overlapping noise. Usually, such a corrupted speech signal results in a dramatic decrease of ASR performance.

We address these challenges of traditional ASR by using a transmission channel that is robust against ambient noise. Instead of relying on the acoustic signal we switch to electromyographic biosignals emitted from our body when speaking. Electromyography (EMG) is a technique for measuring the electrical potential generated by muscle cells during muscle activity. Speech is produced by the activity of the articulatory apparatus, which is moved by a large variety of articulatory muscles. By placing surface electrodes on the relevant articulatory muscles, we measure the electrical potentials during the speech production process. Our recognition system learns the typical muscle activity patterns. After this training process it can then recognize a produced sound from the corresponding electromyographic signal.

Automatic speech recognition based on electromyographic biosignals is inherently robust to ambient noise because the EMG electrodes measure the muscle activity at the skin tissue and do not rely on any air-transmitted signals. In addition, there is another major advantage: since EMG-based ASR does not rely on any air-transmitted signal, it is no longer necessary to speak audibly. Rather it could be shown that it is possible to recognize spoken speech even if it is only mouthed without making any sound. As a result, EMG-based speech recognition provides answers to all three major challenges of traditional speech recognition. It is robust to ambient noise, it allows for confidential input in public places and the input process does not disturb any bystanders or quiet environments. In summary we believe that the proposed EMG-based interface driven by non-audible speech will be of significant benefit to the community. We see three major purposes:

1. Robustness and Environment: interfaces for non-audible speech will enable people to communicate silently without disturbing bystanders or contaminating the environment with noise.
2. Privacy and Security: silent speech interfaces keep confidential spoken input safe and secure.
3. Health and Aging: recognizing speech on the bases of electromyographic signals may offer an alternative for people with speech disabilities and also for elderly who need to preserve energy and want to speak with less effort.

2   Related Work

EMG-based ASR is a very young discipline and several challenges have yet to be overcome. While capturing the eletromyographic biosignal has proved to be a very useful tool to analyze speech research since the 1960's [1], the application of surface EMG signals to automatic speech recognition happened very recently. It was first proposed by Chan et al. [2] in 2002. Their research focused on recognizing short commands and digits spoken by jet pilots during a flight mission, i.e. the speech was captured in an extremely noisy environment. Jorgensen et al. [3] proposed sub-auditory speech recognition using two pairs of EMG electrodes attached to the throat and was the first to demonstrate sub-vocal isolated word recognition using different feature extraction and classification methods [3,4,5]. Manabe et al. showed that silent speech recognition is feasible when the EMG is recorded at the skin surface using electrodes pressed to the skin [6,7]. They applied this technique to discriminate five Japanese vowels and to recognize a small vocabulary of 10 Japanese digits, carefully spoken with pauses in between each event. Maier-Hein et al. investigated non-audible and audible EMG speech recognition and focused on important aspects, such as dependencies on speakers and effects of electrode re-positioning [8].

While the described pioneering studies show some of the potential of EMG-based speech recognition, they are limited to a very small vocabulary ranging from five to forty isolated spoken words. The main motivation for this limitation is to simplify the classification task by treating the complete utterance, spoken in isolation, as one class to be identified. In contrast, the standard practice in traditional large vocabulary continuous speech recognition (LVCSR) consists of modeling a word or utterance as a sequence of phones, i.e. in terms of the smallest possible sound unit. The rationale behind this modeling scheme is that more reliable models can be trained for smaller units since they appear more often during training. Also, the complete set of phone units is usually small for most languages (around 50 units) and after training all phones of a language, each word of this language can be built by simply concatenating the corresponding phones.

Consequently, lifting the constraints of whole word models in EMG-based ASR by introducing phones as a basic recognition unit is one of the major stepping stones necessary to enable large vocabulary continuous speech recognition and thus to open up silent speech technologies to a large number of applications.

3   Experimental Setup

This paper describes our efforts in developing an EMG-based speech recognition system for a 100-word vocabulary - a size which already allows for useful applications. We achieve this by creating a phone-based EMG ASR system that is based on a novel feature extraction method tailored toward EMG signals. This recognizer is benchmarked, analyzed, and enhanced by articulatory features (AF). Furthermore, we investigate the relationship between AFs and EMG signals, present issues and current limitations in the signal capturing process, discuss the extraction of relevant features and optimize the signal processing step for the purpose of speech recognition. Finally, we integrate the novel EMG features and the AF classifiers into the phone-based EMG speech recognition system using a stream architecture, and report speech recognition performance numbers in terms of word error rates.

3.1   Data Collection

For this study we collected data from one male speaker in a single recording session. The speaker was sitting in a quiet office room and read English sentences as prompted on a computer screen. To compare acoustic and electromyographic speech signals we recorded both signals simultaneously in a parallel setup. For this comparison to be valuable we recorded audibly spoken speech, so all results reported in this paper refers to

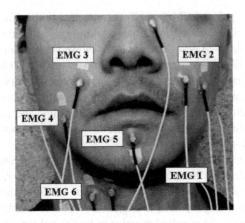

Fig. 1. EMG Electrode positioning

audible speech, not silent speech. The acoustic signal was recorded with a Sennheiser HMD 410 close-speaking microphone in 16 kHz sampling rate, 16 bit resolution, and linear PCM encoding. The electromyographic signal was recorded with six pairs of Ag/Ag-Cl surface electrodes attached to the skin (see Fig. 1 with description below) using a Varioport EMG recorder [9], sampled at 600 Hz, with the same resolution and encoding as the acoustic signal. Additionally, a common ground reference for the EMG signals is connected via a self-adhesive button electrode placed on the left wrist. Synchronization of the signals was ensured by a push-to-talk scenario, where pushing the button generated a marker signal that was fed into both, the EMG recorder and the acoustic sound card. Furthermore, care was taken such that the close-speaking microphone did not interfere with the EMG electrode attachment.

The speaker read 10 times a set of 38 phonetically-balanced sentences and 10 times a set of 12 news article sentences. The resulting 380 phonetically-balanced utterances were used for training and the 120 news article utterances were used for testing. The total duration of the training and test set are 45.9 and 10.6 minutes, respectively. We also recorded ten utterances of 5-second silence for normalization purposes (see below).

3.2   Electrode Positioning

Fig. 1 shows the positions of the six pairs of Ag/Ag-Cl surface electrodes in the face of the speaker to pick up the electromyographic signals of the most relevant articulatory muscles: the levator angulis oris (EMG2,3), the zygomaticus major (EMG2,3), the platysma (EMG4), the orbicularis oris (EMG5), the anterior belly of the digastric (EMG1), and the tongue (EMG1,6). In earlier studies [2,8] these and similar positions had shown to be the most effective ones. Two of these six channels (EMG2,6) are captured with the traditional bipolar configuration, with a 2cm center-to-center interelectrode spacing. In the other four cases, one electrode is placed directly onto the articulatory muscle, while the other electrode is used as a reference attached to either the nose (EMG1) or to both ears (EMG 3,4,5). In order to reduce the impedance at the electrode-skin junctions, a small amount of electrode gel was applied. All the electrode pairs were connected to the EMG recorder [9], in which each of the detection electrode pairs pick up the EMG signal and the ground electrode provides a common reference. EMG responses were differentially amplified, filtered by a 300 Hz low-pass and a 1Hz high-pass filter. We chose to not apply a 50Hz notch filter to avoid the loss of information in this frequency band.

As described above the material for training and testing the recognizer was taken from the same recording session. This way we controlled the impact of electrode repositioning, a challenge for state-of-the-art EMG speech recognition that we had previously studied in [8].

3.3 Modeling Units for EMG-Based Speech Recognition

In our earlier work Walliczek et al.[10] compared different sound units for EMG-based speech recognition. Three model granularities were investigated, models based on full-words, on syllables, and based on phones. In addition, these models were refined to incorporate context information similar to traditional acoustic speech recognition. Three types of context models were explored, context independent, context dependent, and context clustered models. Experiment were conducted to recognize *seen* words, i.e. those words which had been seen in training and test. With a 32-word vocabulary, the word model performs the best with a word error rate of 17.1%, while context-dependent syllable model achieved 20.7% and context-clustered phone model gave 20.2% performance. However, when experiments were carried out on *unseen* words, i.e. on those test words which had not been seen during training but were covered by the test vocabulary, the picture changes drastically. In this case the full-word model does not have the flexibility to recognize unseen words, thus this test was performed on syllable and phone models only. Experimental result showed that the phone-based model outperforms the syllable-based model with a word error rate of 37.6% and 44.9% respectively. Consequently, we will focus the remainder of this work on phone-based speech recognition systems.

3.4 Bootstrapping EMG-Based Speech Recognition

In order to initialize the phone-based EMG speech recognizer we generated a forced alignment of the audible speech data with a pre-existing Broadcast News (BN) recognizer [11] that was trained using our Janus Speech Recognition Toolkit. Due to our simultaneous recording setup these forced-aligned labels can be used to bootstrap the EMG speech recognizer. Since the training set is small, we limited the recognizer to context-independent acoustic models, applying a 3-state Hidden Markov Model scheme using a total of 3.3k Gaussians divided among 136 states. The amount of Gaussians used to model one state is decided automatically in a data-driven fashion. For decoding, the resulting acoustic model was combined with a standard trigram BN language model and a vocabulary that was restricted to the words in the test set. In total the decoding vocabulary contains 108 words of which 35 have been seen in the training. The test set was explicitly excluded from the language model corpus. We are aware that this setup ignores some aspects of large vocabulary speech recognition, such as the problems of out-of-vocabulary words. However, the aim of this study is to focus on the signal preprocessing and unit modeling aspects for electromyographic signals rather than on back-end language related challenges.

3.5 Articulatory Feature Classifier and Stream Architecture

Articulatory features are expected to be more robust than cepstral features since they represent articulatory movements, which are less affected by speech signal variation or noise. We derive the AFs from the IPA phonological features of phones as described in [12]. AFs have binary values, e.g. the values of the horizontal positions of the dorsum FRONT, CENTRAL, and BACK are either present or absent. To classify an AF as present or absent, we compare the likelihood score of the present model with that of the absent model. The models also consider priors based on the frequency of features seen in the training data. Similar to the phone units of the EMG recognizer, the AF classifiers are trained solely on the EMG signal, no acoustic signal is used. In total there are 29 AF classifiers, each is modeled by a Gaussian Mixture Model of 60 mixtures.

Finally the AF classifiers are combined with the phone-based HMMs using a stream architecture. This approach had proved to be successful for traditional ASR since AFs provide complementary information to the phone-based models. The stream architecture employs a list of parallel feature streams, each of which contains one of the phone-based or articulatory features. Information from all streams are combined with a weighting scheme to generate the final EMG acoustic score for decoding [12].

3.6 Feature Extraction for EMG

For baseline experiments we extracted traditional spectral features from the signals of each recorded EMG channel. First a 27ms hamming window with 10ms frame-shift is applied and a Short Time Fourier Transform is computed. From the resulting spectral features, 17 delta coefficients are calculated together with the mean of the time domain values in the 27ms observation window. This results in an 18-dimensional feature vector per channel. If the signals of more than one channel are used for classification, the corresponding feature vectors are concatenated to form the final vector.

Since the EMG signal differs substantially from the acoustic speech signal, we explored other features. First, we normalized the DC offset to zero by estimating the offset based on the 5-second silence utterances and subtract the value from all EMG signals. We denote the EMG signal with normalized DC as $x[n]$ and its short-time Fourier spectrum as $\mathbf{X}$. The nine-point double-averaged signal is given by $w[n]$, the high frequency signal $p[n]$, and the corresponding rectified signal $r[n]$. We define the time-domain mean features $\bar{\mathbf{x}}, \bar{\mathbf{w}}$, and $\bar{\mathbf{r}}$ of the signals $x[n], w[n]$, and $r[n]$, respectively. In addition, we use the power features $\mathbf{P_w}$ and $\mathbf{P_r}$ and we define $\mathbf{z}$ as the frame-based zero-crossing rate of $p[n]$. To improve the context modeling, we apply several contextual filters. The delta filter: $D(\mathbf{f}_j) = \mathbf{f}_j - \mathbf{f}_{j-1}$. The trend filter: $T(\mathbf{f}_j, k) = \mathbf{f}_{j+k} - \mathbf{f}_{j-k}$. The stacking filter: $S(\mathbf{f}_j, k) = [\mathbf{f}_{j-k}, \mathbf{f}_{j-k+1}, ..., \mathbf{f}_{j+k-1}, \mathbf{f}_{j+k}]$, where j is the frame index and k is the context width. After the feature extraction process we apply a linear discriminant analysis (LDA) on the final features to reduce the dimensionality to a constant value of 32 [13].

4 Articulatory Feature Classifiers

In this chapter we describe the development and benchmarking of the Articulatory Feature Classifiers. The baseline AF classifiers are first trained using the forced-alignments

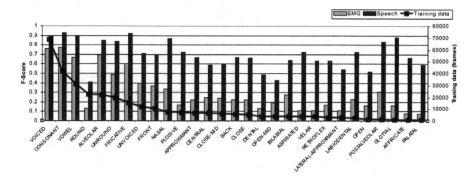

Fig. 2. AF classification performance (F-score) for acoustic and EMG signals

derived from the acoustic speech recognizer using the BN system as described in section 3.4. The AF classifiers are trained on middle frames of the phones, since the alignment to the middle frames is assumed to be more stable than to the begin and end states. To evaluate the performance, the trained AF classifiers are applied to the test data to generate frame-based hypothesis. The performance metrics are F-scores with $\alpha = 0.5$, i.e. F-score $= 2PR/(P + R)$, and precision $P = C_{tp}/(C_{tp} + C_{fp})$, recall $R = C_{tp}/(C_{tp} + C_{fn})$, C_{tp} = true positive count, C_{fp} = false positive count, and C_{fn} = false negative count.

The AF training and test procedure was applied to both, the acoustic speech and the EMG signals. For the EMG signals we concatenated all six channels (EMG1 - EMG6) into a single feature vector as described in section 3.6. The average F-score for all 29 AFs is 0.814 for the acoustic signal and 0.467 for the EMG signal. Fig. 2 gives a breakdown of F-scores for each single AF for both acoustic and EMG signal. It also shows the amount of data available to train each feature (frame counts, 10ms per frame).

First, we observe that the F-score of the AFs trained on acoustic signals significantly outperforms the F-score trained on the EMG signal. This is true for all articulatory features, however for some AFs the decrease is less severe than for others. The AFs VOICED, CONSONANT, VOWEL, and ALVEOLAR seem to be less effected than the rest. Second, we see that the amount of training data is naturally biased toward more general categories, such as VOICED-UNVOICED, VOWEL-CONSONANT.

4.1 Time Delay between Acoustic and EMG Signals

We investigated why some of the EMG-based AF classifiers show a larger decrease in performance compared to the acoustic ones than others. For the evaluation above we used the offset between the two signals as given by the marker channel of our simultaneous recording setup, i.e. all EMG channels are synchronized to the acoustic channel. However, the human articulator movements are anticipatory to the acoustic signal since the speech signal is a product of articulator movements and source excitation [2]. Consequently, the time alignment we used for bootstrapping our EMG-based system might lead to mis-alignments for the EMG signals. Therefore, we investigated the delay time of the acoustic signal to the EMG signals by applying different offset times for the

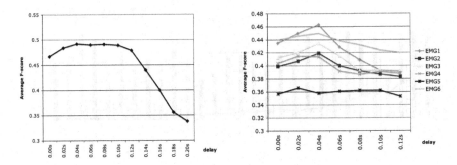

Fig. 3. Classification performance (F-score) over different time delays between EMG and acoustic signals. Six-channel concatenated EMG (left) and individual EMG channels (right).

forced-alignment labels of the acoustic signal. As shown on the left-hand side of Fig. 3, the initial time-alignment (delay = 0) does not give the best F-score. The best F-scores are achieved with time delays between 0.02 to 0.12 seconds. After 0.12 seconds the performance decreases drastically. These results suggest that EMG signals precede the acoustic speech signal.

In the next set of experiments we explore if the anticipatory effect between the articulator movements and the acoustic speech differs for particular articulatory muscles. We assumed that this would be the case since some muscle activity result in longer range movements than others and thus may be more sensitive to a matching time delay. To explore the impact of the time delay on single EMG channels we conducted the same experiments as above, this time separate for each single EMG channel. The right-hand side of Fig. 3 depicts the impact of the time delay for each of the six EMG channels. As expected, some EMG signals are more sensitive to time delay than others, e.g. EMG1 (digastric, the muscle which moves the jaw, among other functions) has a clear peak at 0.04 seconds delay, while EMG6 (tongue muscle) is more consistent over the various time delays. The time delays vary over the channels but the performance peaks range between 0.2 and 0.10 seconds. When the optimal time delay is applied to each EMG channel, the F-score increases to 0.502 compared to the baseline of 0.467. It also outperforms a uniform delay of 0.04 seconds which gave 0.492.

4.2 Impact of Time Delay on Articulatory Features

In the last set of experiments on time delay we investigated the impact of the delay from a different angle. Rather than looking at single EMG channels and averaging the F-score over all AF classifiers, we explored the impact on the single Articulatory Features and averaged over all channels. Fig. 4 shows the analysis for six AFs that represent different characteristics of performance changes with different delays. For example the F-scores for VOICED are rather stable across various delays, while BILABIAL and ALVEOLAR are rather sensitive. We were not able to find a conclusive explanation yet, but will investigate this further on a multiple speaker data set.

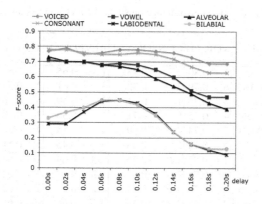

Fig. 4. Performances of six representative AFs

4.3 EMG Channel Combination

Results in [8] for EMG speech recognition based on full-word models had indicated that the concatenation of multiple EMG channels into one feature vector usually outperforms single EMG channel features. In the following experiments we wanted to explore if this finding holds for the combination of time-delayed signals and for AF classifiers. For this purpose we conducted experiments on EMG-pairs in which each EMG signal is adjusted with its optimal time offset.

The first row in Table 1 shows the F-scores averaged over all AFs for the single channel baseline when no time delay is applied. The second row gives the F-scores when the optimal time delay for the individual channel is applied. The third to last row give the F-scores for combinations of two EMG channels, i.e. the F-score 0.464 in row EMG2 + ... column EMG6 shows that the combination of channel EMG2 + EMG6 outperforms the single channel EMG2 (F-score = 0.419) and the single channel EMG6 (F-score = 0.450). This indicates that the two channels carry complementary information. Similar effects could be observed with the combination of EMG1 and EMG3.

Table 2 lists the top-4 articulators with the best F-scores. For single channels, EMG1 performs the best across these top-performance articulators, while EMG1-3, EMG1-6, and EMG2-6 are the best pairwise channel combinations. Interestingly, even though

Table 1. F-scores for single EMG Channels and Pairwise Combination

F-Scores	EMG1	EMG2	EMG3	EMG4	EMG5	EMG6
delay = 0	0.435	0.399	0.413	0.404	0.357	0.440
opt. delay	**0.463**	0.419	0.435	0.415	0.366	0.450
EMG1 + ...		0.439	**0.465**	0.443	0.417	0.458
EMG2 + ...			0.440	0.443	0.414	**0.464**
EMG3 + ...				0.421	0.414	0.449
EMG4 + ...					0.400	0.433
EMG5 + ...						0.399

Table 2. AF Classification Performance on single EMG Channels and Combination

AFs	VOICED		CONSONANT		ALVEOLAR		VOWEL	
Sorted F-Score (single EMG)	1	0.80	2	0.73	1	0.65	1	0.59
	6	0.79	3	0.72	3	0.61	2	0.59
	3	0.76	1	0.71	2	0.59	6	0.56
	4	0.75	6	0.71	6	0.56	3	0.52
	2	0.74	4	0.69	4	0.55	4	0.51
	5	0.74	5	0.63	5	0.45	5	0.51
Sorted F-Score (Paired EMGs)	1-6	0.77	1-6	0.76	1-3	0.69	2-6	0.64
	1-3	0.76	2-3	0.75	1-6	0.67	2-4	0.62
	1-2	0.76	3-6	0.74	1-2	0.66	2-5	0.62
	2-6	0.75	2-4	0.74	2-6	0.66	1-6	0.62
	3-6	0.75	2-6	0.74	2-3	0.65	1-3	0.61

EMG5 performs the worst as a single channel classifier, EMG5 can be complemented with EMG2 to improve the classification performance for VOWEL.

5   Feature Extraction for EMG Speech Recognition

In this chapter we investigate the extraction of relevant features for EMG-based speech recognition. We report performance numbers based on the Word Error Rate (WER) of the recognizer. Word error rate is given as WER $= \frac{S+D+I}{N}$, with S = word substitution count, D = word deletion count, I = word insertion count, N = number of reference words. In the following experiments the final EMG features are generated by stacking single-channel EMG features of the channels EMG1, EMG2, EMG3, EMG4, and EMG6. The channel EMG5 was not considered because it was found to be rather noisy. After stacking, an LDA was applied to reduce the dimensions to 32 throughout the experiments.

Spectral Features. In earlier work we found that spectral coefficients outperform cepstral and LPC coefficients on EMG-based speech recognition [8]. Therefore, we use the spectral features as baseline in this paper. The spectral features are denoted by $\mathbf{S0} = \mathbf{X}$, $\mathbf{SD} = [\mathbf{X}, D(\mathbf{X})]$, and $\mathbf{SS} = S(\mathbf{X}, 1)$. The left-hand side of Fig. 5 depicts the Word Error Rates for the spectral features. It shows that the contextual features improve the performance. Additionally, adding time delays for modeling the anticipatory effects also helps, which is consistent with the results of the articulatory feature analysis.

Spectral + Temporal (ST) Features. Following the results of [8] we added the following time-domain features: $\mathbf{S0M} = \mathbf{X_m}$, $\mathbf{SDM} = [\mathbf{X_m}, D(\mathbf{X_m})]$, $\mathbf{SSM} = S(\mathbf{X_m}, 1)$, and $\mathbf{SSMR} = S(\mathbf{X_{mr}}, 1)$, where $\mathbf{X_m} = [\mathbf{X}, \bar{\mathbf{x}}]$ and $\mathbf{X_{mr}} = [\mathbf{X}, \bar{\mathbf{x}}, \bar{\mathbf{r}}, \mathbf{z}]$. The performance of the resulting speech recognition system is shown on the right-hand side of Fig. 5. Enhancing the spectral features by time-domain features improves the performance quite substantially.

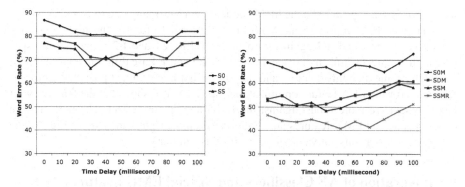

Fig. 5. Word Error Rate on Spectral (left) and Spectral+Temporal (right) Features

Specialized EMG Features. We observed that the spectral features are still noisy for the model training of EMG-based speech recognition. Therefore we designed specialized EMG features that are normalized and smoothed to extract relevant information from EMG signals more robustly. The performance of these EMG features are given on the left-hand side of Fig. 6, where the EMG features are

$$\mathbf{E0} = [\mathbf{f0}, D(\mathbf{f0}), D(D(\mathbf{f0})), T(\mathbf{f0}, 3)], \qquad \text{where } \mathbf{f0} = [\bar{\mathbf{w}}, \mathbf{P_w}]$$
$$\mathbf{E1} = [\mathbf{f1}, D(\mathbf{f1}), T(\mathbf{f1}, 3)], \qquad \text{where } \mathbf{f1} = [\bar{\mathbf{w}}, \mathbf{P_w}, \mathbf{P_r}, \mathbf{z}]$$
$$\mathbf{E2} = [\mathbf{f2}, D(\mathbf{f2}), T(\mathbf{f2}, 3)], \qquad \text{where } \mathbf{f2} = [\bar{\mathbf{w}}, \mathbf{P_w}, \mathbf{P_r}, \mathbf{z}, \bar{\mathbf{r}}]$$
$$\mathbf{E3} = S(\mathbf{E2}, 1)$$
$$\mathbf{E4} = S(\mathbf{f2}, 5)$$

The essence of the design of these specialized EMG feature extraction methods is to reduce noise while preserving the most relevant information for classification. Since the EMG spectral feature is noisy, we first extract the time-domain mean feature, which is empirically known to be useful. By adding power and contextual information to the time-domain mean, **E0** is generated and it by itself outperforms all the spectral-only features. Since the mean and power represent low-frequency components only, we added

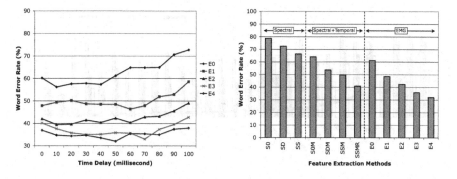

Fig. 6. Word Error Rate on EMG Features over Delay (left) and Summary of Improvements (right)

the high-frequency power and the high-frequency zero-crossing rate to form **E1**, which gives another 10% improvement. With one additional feature of the high-frequency mean, **E2** is generated. **E2** again improves the WER. **E1** and **E2** show that the specific high-frequency information can be helpful. **E3** and **E4** use different approaches to model the contextual information, and they show that large context provides useful information for the LDA feature optimization step. They also show that the features with large context are more robust against the EMG anticipatory effect. The performance of the specialized EMG features are summarized on the right-hand side of Fig. 6. The delay was set to a constant value of 50 ms for this summary.

6   Integration of AF Classifiers and Special EMG Features

In the following experiments we describe the integration of the newly developed specialized EMG features into the Articulatory Feature classifiers. Similar to the experiments described above, we bootstrapped the EMG recognizer from the forced alignments based on the acoustic data. The baseline system is created with setting the delay to zero. Different from the experiments described in section 4 we only applied five EMG channels (all but EMG5) and compared the spectral plus time domain features (EMG-ST) with the specialized EMG features (EMG-E4). The average F-scores over the five channels of all 29 AFs are 0.492 for EMG-ST, 0.686 for EMG-E4, and 0.814 for the acoustic signal (the acoustic signal remains the same, so the F-score has the same value as reported in section 4). Fig. 7 shows the classification performance for the individual AFs for both, the EMG and acoustic signal, along with the amount of training data in frames. The results indicate that EMG-E4 significantly outperforms EMG-ST. Also, performance of the EMG-based recognizer closes the large gap to the acoustic based recognizer that was initially observed in our baseline experiments (see Fig.2).

We also conducted time delay experiments similar to the ones described above to investigate the anticipatory effect of EMG signals [14]. Fig. 8 shows the F-scores of E4 over various LDA frame sizes and delays. We observe similar anticipatory effects of E4-LDA and ST with time delays around 0.02 to 0.10 seconds. Compared to the 90-dimensional ST feature, E4-LDA1 has a dimensionality of 25 but demonstrates a much higher F-score. Fig. 8 also indicates that a wider LDA context width provides a higher

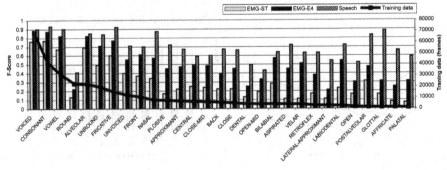

Fig. 7. AF classification performance (F-score) for acoustic and EMG signals with E4 features

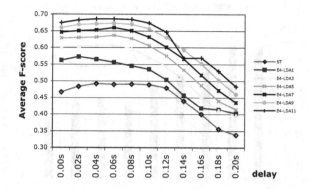

Fig. 8. AF Classification (F-scores) for five-channel EMG-ST and EMG-E4 over different LDA frame sizes and time delays

F-score and seems to be more robust for modeling the anticipatory effect. We believe this results from that fact that LDA is able to pick up useful information from the wider context.

6.1 EMG Channel Combination

In order to analyze E4 for individual EMG channels, we trained the AF classifiers on single channels and pairwise channel combinations. The F-scores are shown in Fig. 9 and prove that the E4 features outperform ST features in all configurations. Moreover, E4 on single-channel EMG 1, 2, 3, 6 are already better than the all-channel ST's best F-score 0.492. For ST, the pairwise channel combination provides only marginal improvements; in contrast, for E4, the figure shows significant improvements of pairwise

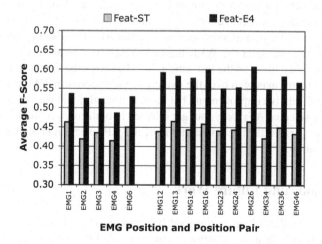

Fig. 9. AF Classification (F-scores) on single and combined EMG channels for EMG-ST and EMG-E4 features

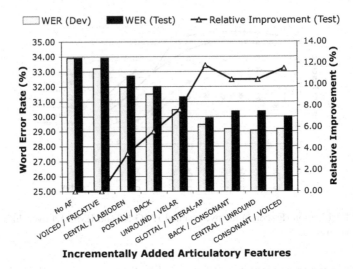

Fig. 10. EMG recognition performance (WER) and relative improvements for incrementally added AF classifiers. The two AF sequences correspond to the best AF-insertion on a two-fold cross-validation.

channels combinations compared to single channels setups. We believe this significant improvements comes from a better decorrelated feature space provided by the E4 features.

6.2 Decoding in the Stream Architecture

Finally we conducted full decoding experiments by integrating the phone-based and the AF-based information into our stream architecture. The test set was divided into two equally-sized subsets, on which the following procedure was done in two-fold cross-validation. On the development subset, we incrementally added the AF classifiers one by one into the decoder in a greedy fashion, i.e., the AF that achieves the best WER was kept in the stream. After the WER improvement was saturated, we fixed the AF sequences and applied them on the test subset. Fig. 10 shows the word error rate and its relative improvements averaged over the two cross-validation turns. With five AFs, the WER tops 11.8% relative improvement, but there is no additional gain by adding more AFs. Among the selected AFs, only four are selected in both cross-validation turns. This inconsistency suggests a further investigation of the AF selection procedure to ensure better generalization.

7   Conclusions and Future Work

We presented our studies of automatic speech recognition based on electromyographic biosignals captured from the articulatory muscles in the face using surface electrodes. We designed our data collection and experiments such that audibly spoken speech was simultaneously recorded as acoustic and as electromyographic data. This way we could study the time delay between the acoustic signals captured by a microphone and the EMG signals resulting from the corresponding articulatory muscle activity. Our experiments indicate that the EMG signal precedes the acoustic signal by about 50 to 60ms. Furthermore, we found that the time delay depends on the captured articulatory muscle.

Based on the collected dataset we develop a phone-based EMG speech recognizer and described how the performance of this recognizer improves by carefully designing and tailoring the extraction of relevant speech features toward electromyographic signals. The specialized EMG features gave significant improvements over spectral and time-domain features. Furthermore, we introduce articulatory feature classifiers, which had recently shown to improve classical speech recognition significantly. We describe that the classification accuracy of articulatory features clearly benefits from the tailored feature extraction.

Finally, we integrated the AF classifiers into the overall decoding framework by applying a stream architecture. The AF classifiers gave a 11.8% relative improvement over the phone-based recognizer. Our final EMG based speech recognition system achieves a word error rate of 29.9% on a 100-word recognition task and thus comes within performance ranges useful for practical applications.

EMG-based speech recognition is a very young research area and many aspects are waiting to get explored. In the near future we expect to study effects such as the impact of speaker dependencies, electrode re-positioning, and kinematic differences between audible and non-audible speech. In order to investigate some of these aspects we are currently collecting a database of EMG speech that targets a large number of speakers uttering audible and non-audible speech.

References

1. Fromkin, V., Ladefoged, P.: Electromyography in speech research. Phonetica 15 (1966)
2. Chan, A., Englehart, K., Hudgins, B., Lovely, D.: Hidden Markov model classification of myoelectric signals in speech. IEEE Engineering in Medicine and Biology Magazine 21(4), 143–146 (2002)
3. Jorgensen, C., Lee, D., Agabon, S.: Sub auditory speech recognition based on EMG signals. In: Proc. IJCNN, Portland, Oregon (July 2003)
4. Jorgensen, C., Binsted, K.: Web browser control using EMG based sub vocal speech recognition. In: Proc. HICSS, Hawaii (January 2005)
5. Betts, B., Jorgensen, C.: Small vocabulary communication and control using surface electromyography in an acoustically noisy environment. In: Proc. HICSS, Hawaii (January 2006)
6. Manabe, H., Hiraiwa, A., Sugimura, T.: Unvoiced speech recognition using EMG-Mime speech recognition. In: Proc. CHI, Ft. Lauderdale, Florida (April 2003)
7. Manabe, H., Zhang, Z.: Multi-stream HMM for EMG-based speech recognition. In: Proc. IEEE EMBS, San Francisco, California (September 2004)
8. Maier-Hein, L., Metze, F., Schultz, T., Waibel, A.: Session independent non-audible speech recognition using surface electromyography. In: Proc. ASRU, San Juan, Puerto Rico (November 2005)
9. Becker, K.: Varioport (2005), http://www.becker-meditec.de
10. Walliczek, M., Kraft, F., Jou, S.C., Schultz, T., Waibel, A.: Sub-word unit based non-audible speech recognition using surface electromyography. In: Proc. Interspeech, Pittsburgh, PA (September 2006)

11. Yu, H., Waibel, A.: Streaming the front-end of a speech recognizer. In: Proc. ICSLP, Beijing, China (2000)
12. Metze, F., Waibel, A.: A flexible stream architecture for ASR using articulatory features. In: Proc. ICSLP, Denver, CO (September 2002)
13. Jou, S.C., Schultz, T., Walliczek, M., Kraft, F., Waibel, A.: Towards continuous speech recognition using surface electromyography. In: Proc. Interspeech, Pittsburgh, PA (September 2006)
14. Jou, S.C., Schultz, T., Waibel, A.: Continuous electromyographic speech recognition with a multi-stream decoding architecture. In: Proc. ICASSP, Honolulu, Hawai'i (April 2007)

A Multiphase Approach to MRI Shoulder Images Classification

Gabriela Pérez[1], J.F. Garamendi[2], R. Montes Diez[3], and E. Schiavi[4]

[1] Departamento de Ciencias de la Computación
[2] Laboratorio de Imagen Médica y Biometría
[3] Departamento de Estadística e Investigación, Operativa
[4] Departamento de Matemática Aplicada
Universidad Rey Juan Carlos, Madrid, Spain
{gabriela.perez,juanfrancisco.garamendi,raquel.montes,
emanuele.schiavi,lncs}@urjc.es
http://www.urjc.es/lncs

Abstract. This paper deals with a segmentation (classification) problem which arises in the diagnostic and treatment of shoulder disorders. Classical techniques can be applied successfully to solve the binary problem but they do not provide a suitable method for the multiphase problem we consider. To this end we compare two different methods which have been applied successfully to other medical images modalities and structures. Our preliminary results suggest that a successful segmentation and classification has to be based on an hybrid method combining statistical and geometric information.

Keywords: MRI, shoulder complex, segmentation classification, multiphase Chan-Vese model.

1 Introduction

Shoulder imaging is one of the major applications in MRI and the primary diagnostic tool in the evaluation of musculoskeletal disease, [1], [2].

Accurate diagnosis and treatment of painful shoulder and others musculoskeletal complaints and disorders (such as arthritis, abnormalities, bone tumors, worn-out cartilage, torn ligaments, or infection) may prevent from functional loss, instability and disability. Recent interest is also in musculoskeletal tumor and disorders associated with HIV infection and AIDS, [3] and [4]. In order to provide a reliable method for successful clinical evaluation an increasing effort has to be done in mathematical engineering and biomedical imaging where the specific protocols of 2D segmentation, 3D reconstruction, feature extraction and 4D motion are modeled. In this approach for image guided analysis of shoulder pathologies, automatic and unsupervised segmentation and classification represent the first challenging task. In fact, practical difficulties arise due to the high resolution required for visualization of small but critical structures, to the gross inhomogeneities of field coil response, to the degree of noise present with the signal and to extreme low contrast details between some distinct anatomical structures (fat, bone regions, muscle and tendons, ligaments and cartilage). The existence of a general technique able to cope with

A. Fred, J. Filipe, and H. Gamboa (Eds.): BIOSTEC 2008, CCIS 25, pp. 321–329, 2008.
© Springer-Verlag Berlin Heidelberg 2008

all these difficulties for all 3D MRI images sequences is still an open question. A preliminary analysis of the model problem is done here, where a multiphase (2 phases, 4 classes) variational framework is considered for 2D image segmentation and classification. Notice that 2D segmentation is a fundamental step towards the 3D morpho-dynamic reconstruction problem of automatic segmentation. This in turn allows for motion tracking for 4D reconstruction and visualization of musculoskeletal structures.

2   Material and Methods

This contribution is devoted to the preliminary analysis and application of a modified multiphase segmentation and classification algorithm based on previous work of Chan and Vese [5]. This multiphase approach can manage the classification problem underlying the segmentation exercise so broadening the scope of these PDE-based segmentation models.

In order to validate our results we compare with a mixture density estimation algorithm for image classification previously presented in [6], in the context of brain MRI images.

As an application of our method we consider coronal and transverse (axial), 2D MRI shoulder images extracted from two 3D sequences. The images are courtesy of the Ruber International Hospital in Madrid.

The shoulder joint is composed of three bones: the clavicle (collarbone), the scapula (shoulder blade), and the humerus (upper arm bone). The bones of the shoulder are held in place by muscles, tendons and ligaments. Tendons are tough cords of tissue that attach the shoulder muscles to bone and assist the muscles in moving the shoulder. Ligaments attach shoulder bones to each other, providing stability. The ends bones are covered by cartilage which provides painless motion. See Figure 1.

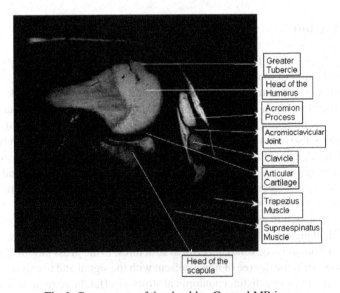

Fig. 1. Components of the shoulder, Coronal MR image

The classification problem we are about can be considered in the framework of minimal partition problems [7] and cannot be dealt with classical techniques whereas binarization of image sequence is not suitable to produce the segmentation of all the classes we are interested in. Nevertheless it is interesting to compare the binary images obtained with thresholding techniques for the two classes (1 phase) problem in order to assess the performance of our algorithms when the full classification problem is considered. To cope with the difficulties above mentioned we consider two different approaches based on density mixture estimations (see [6]) and a variational model formulated in a level set framework [5].

The proposed algorithms are described in the next sections. The results obtained with classical (global and local thresholding or the popular k-means algorithm) techniques for the 2 classes (1 phase) problem are also reported for comparison. In particular we used the original Otsu's method [8] and the Ridler-Calvard technique [9]. We show the results obtained in figures 3-4 (d) and e)).

2.1 Density Mixture Estimation

In the case of the density mixture estimations framework (see [10] for an application to functional brain images) the original magnitude images have been pre-processed in order to eliminate the high frequencies associated to noise and to increase the low contrast present in some parts of the image (see [11], [12]). We consider a low pass homomorphic filter in the frequency domain which has been successfully used in previous works [12].

The initial pre-processing step is performed with a homomorphic filter in order to correct the gray scale inhomogeneity field (bias correction). In fact, MR images are usually corrupted by a smooth, spatially varying artifact that modulates the intensity of the image (bias). These inhomogeneities are known to appear in MR images as systematic changes in the local statistical characteristics of tissues and are often quite subtle to the human eye. However, even inhomogeneities that are invisible to the human observer alter tissue characteristics enough to hamper automated and semi-automated classification.

Then, in a denoising step, the homogenized image is then filtered again with an adaptative filter to produce 2D wiener denoised sequence of the original image. The denoised slices are then normalized using a dynamical range operator in order to increase the (low) contrast present in the images. We then characterized the different soft tissues and bony structures in 4 classes (bone, muscle, cartilage, fat) partitioning the shoulder complex estimating their initial parameter statistics.

In order to show the basic steps of the algorithm we follow the Bayesian mixture parameter estimation method proposed in [6].

Let $Z = (X, Y)$ be two random fields where $Y = \{Y_s, s \in S\}$ represents the field of observations corresponding to the pixels of the MR image and $X = \{X_s, s \in S\}$ corresponds to the label field representing the segmented image. Following a Bayesian approach, the posterior distribution of (X, Y), $P(x|y)$, will result by combining the prior distribution, assumed to be stationary and Markovian,

$$P(x) = exp\{-\sum_{<s,t>} \beta(1 - \delta(x_s, x_t))\},$$

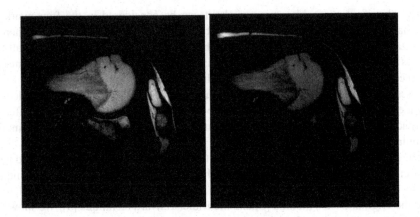

Fig. 2. On the (*left*) the original image and on the (*right*) the pre-processed, corrected image

and the site-wise likelihood $P(y_s|x_s)$, modelled as a mixture of densities

$$P(y_s|x_s, k, \Phi_k) = \sum_{k=1}^{K} \pi_k P(y_s|x_s, k, \Phi_k)$$

where π_k are the mixing proportion ($\sum_k \pi_k = 1$) and where $P(y_s|x_s, k, \Phi_k)$ define Gaussian distributions, with parameters $\Phi_k = (\mu_k, \sigma_k^2)$ in each segmented class k. Let $\Phi = (\Phi_1, \Phi_2, \ldots, \Phi_K)$ and $\pi = (\pi_1, \pi_2, \ldots, \pi_K)$. In order to develop the segmentation procedure, we perform the following algorithm:

0. Initialize parameters $(\Phi^{[0]}, \pi^{[0]})$.

Given $(\Phi^{[p]}, \pi^{[p]})$, we can calculate $(\Phi^{[p+1]}, \pi^{[p+1]})$ by

1. Using the Gibbs sampler, simulate one realization x from the posterior distribution $P(x|y)$ with parameter $(\Phi^{[p]}, \pi^{[p]})$.
2. Define $(\Phi^{[p+1]}, \pi^{[p+1]})$ as the ML estimation of the data (Y, x)
3. Repeat till convergence is achieved.

2.2   Active Contours without Edges

Since the work of Kass [13] is well known that the segmentation problem of digital images can be dealt with in the framework of variational calculus. Nevertheless in medical images there are often no sharp-gradient induced edges at all and region-based active contours driven by gradients can fail in automatic approaches. Recently a new model has been suggested by Chan-Vese which can be deduced from the Mumford-Shah minimal partition problem, [7], a basic problem in computer vision. Successful applications of this method have been reported in many papers and fields (see [5] and [14]). Our aim is to show that this active contour without edges model (or statistical feature driven model) can be used to solve the classification problem we consider here where a multiphase level set framework for image segmentation is implemented. The basic idea is that, fixed the number of classes in which we are interested in (fat, bone regions, muscle and tendons, ligaments and cartilage), it is sufficient to consider a two phase model, say ϕ_1, ϕ_2, in order to provided partition of the image in four classes (($\phi_1 > 0$ and $\phi_2 > 0$), ($\phi_1 < 0$ and $\phi_2 > 0$), ($\phi_1 > 0$ and $\phi_2 < 0$), ($\phi_1 < 0$ and $\phi_2 < 0$)).

Now, we explain the one phase (binary) and two phases models considered in the experiments. Let $\Omega \subset I\!\!R^2$ be an open, bounded domain (usually a square) where $(x, y) \in \Omega$ denotes pixel location and $I(x, y)$ is a function representing the intensity image values. Let moreover the level sets functions denoted by $\phi_1, \phi_2 : \Omega \to I\!\!R$. The union of the zero-level sets of ϕ_1 and ϕ_2 represents the edges of segmentation. Using this formalism the functions ϕ_1 and ϕ_2 can be characterized as the minimum of the following energy functional:

$$
\begin{aligned}
F(C, \Phi) = & \int_\Omega (I - c_{11})^2 H(\phi_1) H(\phi_2) dx dy \\
& + \int_\Omega (I - c_{10})^2 H(\phi_1)(1 - H(\phi_2)) dx dy \\
& + \int_\Omega (I - c_{01})^2 (1 - H(\phi_1)) H(\phi_2) dx dy \\
& + \int_\Omega (I - c_{00})^2 (1 - H(\phi_1))(1 - H(\phi_2)) dx dy \\
& + \nu \int_\Omega |\nabla H(\phi_1)| dx dy + \\
& + \nu \int_\Omega |\nabla H(\phi_2)| dx dy
\end{aligned}
\tag{1}
$$

where $C = (c_{11}, c_{10}, c_{01}, c_{00})$ is a constant vector representing the mean of each region (or class), $\Phi = (\phi_1, \phi_2)$, ν is a parameter of smoothness and $H(x)$ is the Heaviside function, $H(x) = 1$ if $x \geq 0$ and $H(x) = 0$, otherwise.

The Euler-Lagrange equations obtained by minimizing (1) with respect to C and Φ are solved with a gradient descent method leading to the coupled parabolic PDE system [14]:

$$
\begin{aligned}
\frac{\partial \phi_1}{\partial t} = & \delta_\epsilon(\phi_1) \left\{ \nu \nabla \cdot \left(\frac{\nabla \phi_1}{|\nabla \phi_1|} \right) - \left[(I - c_{11})^2 - (I - c_{01}^2) \right] H(\phi_2) + \right. \\
& \left. + \left[(I - c_{10})^2 - (I - c_{00}^2) \right] (1 - H(\phi_2)) \right\}
\end{aligned}
\tag{2}
$$

$$
\begin{aligned}
\frac{\partial \phi_2}{\partial t} = & \delta_\epsilon(\phi_2) \left\{ \nu \nabla \cdot \left(\frac{\nabla \phi_2}{|\nabla \phi_2|} \right) - \left[(I - c_{11})^2 - (I - c_{01}^2) \right] H(\phi_1) + \right. \\
& \left. + \left[(I - c_{10})^2 - (I - c_{00}^2) \right] (1 - H(\phi_1)) \right\}.
\end{aligned}
\tag{3}
$$

where δ_ϵ denotes a smooth (not compactly supported) approximation to the Dirac delta distribution with these notations the image function u can be expressed in form:

$$u = c_{11}H(\phi_1)H(\phi_2) + c_{10}H(\phi_1)(1 - H(\phi_2)) +$$
$$c_{01}(1 - H(\phi_1))H(\phi_2) + c_{00}(1 - H(\phi_1))(1 - H(\phi_2)).$$

Notice that the equations (2) and (3) are (weakly) coupled in the lower order terms. In case of two regions (binary segmentation) only one level set function ϕ is needed. The resulting one phase energy functional to minimize is as follows:

$$E_{cv}(c_1, c_0, \phi) = \nu \int_\Omega |\nabla H(\phi)| dx dy +$$
$$+ \int_\Omega H(\phi)|I(x,y) - c_1|^2 dx dy +$$
$$+ \int_\Omega (1 - H(\phi))|I(x,y) - c_0|^2 dx dy$$

$$(4)$$

and the associated gradient descent equation is :

$$\frac{\partial \phi}{\partial t} = \delta_\epsilon(\phi) \left[\nu \nabla \cdot \left(\frac{\nabla \phi}{|\nabla \phi|} \right) - |I(x,y) - c_1|^2 + |I(x,y) - c_0|^2 \right]. \qquad (5)$$

The equations (2), (3) or (5) have to be complemented with feasible (due to the non-uniqueness of the corresponding steady states) initial conditions and homogeneous boundary conditions of Neuman type (no flux). As in Chan and Vese [5] the steady states associated to system (2), (3) or the eq. (5) can be asymptotically reached by using a gradient descent method where δ_ϵ is substituted by 1 (this is possible because δ_ϵ has no compact support). Numerically, as we are concerned with the quality of the classification and not in to speed it up, we used a simple first order (in time) Euler explicit finite difference scheme and weighted, centered, second order formulas in space, with a regularization of the (degenerate) diffusion term to avoid division by zero (which occurs in homogeneous, very low gradient regions which are located far from the active contour and do not affect the final segmentation as soon as the regularizing parameter is small). The time steps have been choosen accordingly in order to preserve numerical stability and convergence.

3   Results

We present the results obtained by applying the above methods to a pair of slices extracted from a volume MRI sequence of the shoulder complex. The slices dimensions are 512x512.

Binary segmentations obtained with both methods (the bayesian density mixture estimation and the PDE-based hybrid active contours method without edges) are shown in Figures 3-4 before of the multiphase classification, see Figures 5-6.

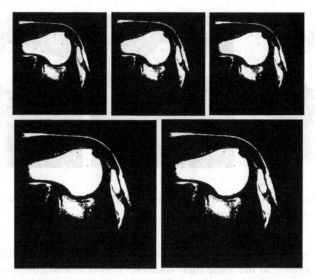

Fig. 3. Slice 1. Segmentation image for one phase (2 classes) with: a) k-means b) Density mixture c) Active contours without edges d) Otsu's and e) Ridler Calvard algorithm

For comparison and in both cases, we also include the results provided by classical methods. Binary segmentation is also used to assess the parameters involved in the model equations and to provide automatic, robust initial conditions for the evolutive problem in the multiphase case.

In Figures 3-4 we see that in both cases the bony structures (head of scapula, head of humerus, clavicle, acromion) are properly classified. Background, skin, and muscle

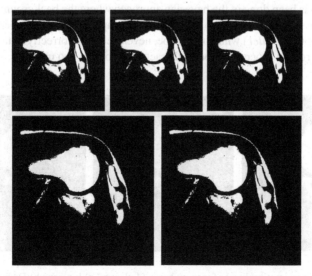

Fig. 4. Slice 2. Segmentation image for one phase (2 classes) with: a) k-means b) Density mixture c) Active contours without edges d) Otsu's and e) Ridler Calvard algorithm

328   G. Pérez et al.

 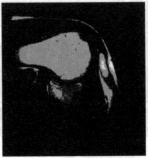 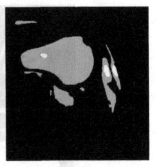

Fig. 5. Slice 1. Segmentation image for two phases (4 classes) a) k-means b) Density mixture c) Active contours without edges algorithm

are also characterized in the binary images as the soft (tissue) class. Visual inspection suggests convergence to the same limit solution. This is indeed confirmed when the differences images are computed and classical methods (*first row*) are compared.

As aspected, some more differences between the quality reconstruction of the different methods can be appreciated in the multiphase (four classes) classification problem. In Figures 5-6 we report the results obtained with the classical (k-means) algorithm (*left*), the bayesian mixture model (*center*) and the Chan-Vese model (*right*). The greater tubercle and the head of the humerus are properly classified and shaped with our methods (center and right) while the classical k-means fails in both aspects (and in both slices, see Figures 5-6, (*left*), where the bone is under-estimated and muscle is wrongly detected). Articular cartilage has been detected in (*center and right*) but not in (*left*). Muscle is properly classified with the Chan-Vese model (*right*) and the classical method (*left*) but no classification has been done in the bayesian approach where the background is assigned to the same class. At the same time the head of the scapula has been properly classified in (*right*) but not in (*center*) where the shape, nevertheless is correctly obtained. Notice also that the acromial process has been characterized by the two methods.

 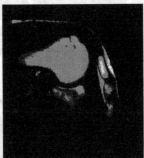 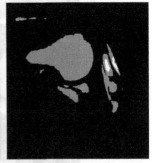

Fig. 6. Slice 2. Segmentation image for two phases (4 classes) a) k-means b) Density mixture c) Active contours without edges algorithm

4 Conclusions

We considered the problem of automatic segmentation of 2D images using an hybrid, statistical and geometrical model based on Chan-Vese work. This method provides correct classification of bony structures but soft tissues are not yet properly classified. This is also manifested in the bayesian approach. The differences between the results obtained with the two methods suggest the conclusion that hybrid methods can give better results as far as the right statistics are included in the model and this will be the aim of our future work.

Acknowledgements. The authors would like to thank financial support of the following projects: URJC-CM-2006-CET-0603, URJC-CM-2006-CET-0638, MTM2006-14961-C05-05, TSI2007-66706-C04-01, URJC-CM-2006-CET-0371 and MTM2005-03463.

References

1. Vahlensieck, M.: MRI of the shoulder. European Radiology 2, 242–249 (2000)
2. Ehman, R.L., Megibow, A.J., McCauley, T.R., Bluemke, D.A., Steinbach, L.S.: Musculoskeletal imaging. In: 24th Annual Course of the Society of Computed Body Tomography and Magnetic Resonance (SCBT/MR), Symposium, Miami, Florida, March 19 (2001)
3. Biviji, A.A., Paiement, G.D., Steinbach, L.S.: Musculoeskeletal manifestations of the human immunodeficiency virus infection. Of the American Academy of Orthopedic Surgeons 10, 312–320 (2002)
4. Johnson, R., Steinbach, L.S. (eds.): Essentials of Musculoskeletal Imaging. American Academy of Orthopedic Surgeons, Chicago (2003)
5. Chan, T.F., Vese, L.A.: Active Contours Without Edges. IEEE Transactions on Image Processing 10 (2001)
6. Mignotte, M., Meunier, J., Soucy, J.P., Janicki, C.: Classification of brain SPECT images using 3D Markov Random Field and density mixture estimations. In: 5th World Multi-Conference on Systemics, Cybernetics and Informatics. Concepts and Applications of Systemics and Informatics, Orlando, vol. 10, pp. 239–244 (2001)
7. Mumford, D., Shah, J.: Optimal Approximation by Piecewise Smooth Functions and Associated Variational Problems. Communications on Pure Applied Mathematics 42, 577–685 (1989)
8. Otsu, N.: A Threshold selection method from gray level histograms. IEEE transactions on systems, man, and cybernetics 9(1), 62–66 (1979)
9. Ridler, T., Calvard, S.: Picture Thresholding Using an Iterative Selection Method. IEEE transactions on systems, man, and cybernetics 8(8), 620–632 (1978)
10. Ashburner, J., Friston, K.J.: Unified segmentation. Neuroimage 26, 839–851 (2005)
11. Brinkmann, B., Manduca, A.: Optimized Homomorphic Unsharp Masking for MR Grayscale Inhomogeneity Correction. IEEE transactions on medical imaging 17(2), 62–66 (1998)
12. Pérez, G., Montes Diez, R., Hernández, J.A., Martín, J.S.: A New Approach to Automatic Segmentation of Bone in Medical Magnetic Resonance Imaging. In: Jose, M.B., Fernando, M.S., Victor, M.F. (eds.) ISBMDA 2004. LNCS, vol. 3337, pp. 21–26. Springer, Heidelberg (2004)
13. Kass, M., Witkin, A., Terzopoulos, D.: Snakes: Active contour models. Intl. J. Comput. Vision 1, 321–331 (1987)
14. Vese, L.A., Chan, T.F.: A Multiphase Level Set Framework for Image Segmentation Using the Mumford and Shah Model. International Journal of Computer Vision 50, 271–293 (2002)

Human-Like Rule Optimization for Continuous Domains

Fedja Hadzic and Tharam S. Dillon

DEBII, Curtin University of Technology, Perth, Australia
{f.hadzic,t.dillon}@curtin.edu.au

Abstract. When using machine learning techniques for data mining purposes one of the main requirements is that the learned rule set is represented in a comprehensible form. Simpler rules are preferred as they are expected to perform better on unseen data. At the same time the rules should be specific enough so that the misclassification rate is kept to a minimum. In this paper we present a rule optimizing technique motivated by the psychological studies of human concept learning. The technique allows for reasoning to happen at both higher levels of abstraction and lower level of detail in order to optimize the rule set. Information stored at the higher level allows for optimizing processes such as rule splitting, merging and deleting, while the information stored at the lower level allows for determining the attribute relevance for a particular rule. The attributes detected as irrelevant can be removed and the ones previously detected as irrelevant can be reintroduced if necessary. The method is evaluated on the rules extracted from publicly available real world datasets using different classifiers, and the results demonstrate the effectiveness of the presented rule optimizing technique.

Keywords: Data Mining, Rule Optimization, Feature Selection.

1 Introduction

Large amounts of data are being collected for different industrial, commercial or scientific purposes, where the aim is to discover new and useful patterns from data that gives rise to discovery of valuable domain knowledge. This process is termed knowledge discovery and the step concerned with applying programs that are capable of learning and generalizing from presented information is called data mining. The end result is a knowledge model that should be easy for human comprehension. This knowledge model is then used by the people involved in that domain for particular domain specific tasks. For example in many businesses it is used as a decision support tool, in medical domains it aids in diagnostic tasks and in more general terms it provides an organization with the basic knowledge of the concepts and their roles and relationships which occur in that particular domain.

The process of using the underlying rules of a knowledge model for classifying future unseen instances is in the data mining field known as the prediction task. The knowledge model can be evaluated based on its predictive accuracy which corresponds to the percentage of correctly classified instances from an unseen data set. In addition to predictive accuracy, simple rules are preferred since they are more

A. Fred, J. Filipe, and H. Gamboa (Eds.): BIOSTEC 2008, CCIS 25, pp. 330–343, 2008.

comprehensible and are expected to perform better on unseen data since they are more general. Taking these observations into account, a rule optimizing process needs to make a between the misclassification rate (MR), coverage rate (CR) and generalization power (GP) [1]. MR is measured as the number of incorrectly classified instances while CR is the number of instances that are captured by the rule set. MR should be minimized while the CR should be maximized. Good GP is achieved by simplifying the rules. The trade-off is especially evident when optimizing the rule set from a domain characterized by continuous attributes. An optimal constraint on the attribute range needs to be determined as in many cases increasing the attribute range usually leads to the increase in CR but at the cost of an increase in MR. In regards to obtaining good GP by simplifying the rule, there is a trade-off since if the rules are too general, the MR may increase, since the simple rules may be incapable of distinguishing some more specific cases of the domain. Rule optimization is a type of uncertain reasoning technique and a number of different techniques have been adopted in the literature [1, 2, 3, 4].

The aim of this work is to present a stand-alone rule optimizing technique that is capable of reasoning at the higher level of abstraction as well as lower level of detail. It is an extension of the rule optimizing technique discussed in [5, 6], which was an integral part of the neural network learning method for continuous domains. The reasoning only occurred at the higher level where similar rules were merged and the ones with high MR were split into more specific rules. The lower level instance information was not used and any poor choices made during the network pruning [5] stage could not be corrected. In the proposed extension the lower level information corresponds to the relationships between the values of the defining attributes and the implying class of a rule. This information allows one to determine the relevance of rule attributes at any stage of the rule optimizing (RO) process. Attributes previously found as irrelevant can be re-introduced if they become relevant at a later stage in the process, and at the same time attributes that have lost their predictive capability can be deleted. This corresponds to being able to measure an attributes sequential variation in its predictive capability. This characteristic is very useful for RO since it can often be the case that other attributes may become relevant for more specific distinguishing of a new rule which resulted from splitting of an original rule. On the other hand, some attributes can loose their relevance for predicting the class value implied by a rule, when that rule is obtained through merging of two or more specific rules. Furthermore the method from [5, 6] is only applicable for optimizing of the rules learned by the neural network. The proposed RO technique is applicable to any sets of rules, not only those extracted from a specifically developed system. As such it is capable of incorporating domain expert knowledge which can be represented as a set of rules and which can then be refined and adapted to the future cases. The effectiveness of the proposed method is demonstrated by evaluating it on the rules learned from publicly available real world datasets.

The rest of the paper is organized as follows. In Section 2 we discuss the theoretical motivations of the proposed rule optimizing technique. It also provides an overview of how the theory of concept formation will be mimicked in the proposed method which is described in detail in Section 3. Section 4 provides an experimental evaluation of the method and some general remarks. The paper is concluded in Section 5.

2 Theoretical Motivation for the Proposed RO Method

From the biological perspective of AI which studies the way humans perform intelligent tasks, a machine learning technique should resemble the way that humans learn. While neural networks resemble this greatly at the brain level of neural interaction, there is also a higher level of reasoning that occurs when a human reasons about the formed knowledge or beliefs. Hence, from the same perspective it would be useful for a rule optimizing technique to resemble the way concept formation and refinement occurs in humans.

The way that humans engage in concept or category formation has been studied extensively in the psychology area. In general terms it corresponds to the process by which a person learns to sort specific observations into general rules or classes, and thereby allows one to respond to events in terms of their class membership rather than uniqueness [7]. It is the elementary form by which humans adjust to their environment. One needs to identify the attributes of relevance for learning or applying a particular rule for formulating a concept [8]. Humans consistently seek confirming information by actively searching they environment for appropriate examples which can confirm or modify the newly discovered concepts [8, 9, 10] Hence, there exists one level at which the concepts or categories have been formed and there is another level where the observations are used for confirming or adjusting the learned concepts and their relationships [11]. When an observation appears to be contradictory to a formed belief, one may engage in thinking at the lower level of detail where the constituents of a belief and the examples that formed it are investigated. Some preconditions can be added or removed from the constituents of that particular belief so that the updated belief will not contradict the current observations. In this process of aiming for a reliable belief, it is often the case that features previously found as irrelevant are re-introduced or the ones previously thought to be important are removed from the constituents of a belief. In the context of this work, the term 'belief' corresponds to the rule a human uses to classify the examples or observations into classes or categories.

Performing this type of task is highly desirable for machine learning and our main focus is to allow for such a mechanism in the rule optimizing process. The proposed RO technique is mainly motivated by the psychological studies of human concept formation performed in [7]. In one of the experiments, the human subjects were presented with a number of instances which were classified according to a rule and their task was to discover the rule. In this process, the human subjects formed initial rules from a few observations and then would refine or update these rules when instances contradicting their currently formed rule were observed [7]. This process is simulated in our RO technique since it starts with the initial set of rules and then uses a set of class labelled instances to refine or update the rules according to the instances. The aim is to optimize the rules in such a way that there is an optimal trade-off between misclassification rate, coverage rate and the generalization power. This is particularly important when continuous attributes are in question since a slight increase in the allowed range leads to a higher coverage rate, but at the same time may increase the misclassification rate. Further, the rules are presented in a structure that is capable of adapting itself according to the future instances. Furthermore we are interested in a structure which can adapt itself to the changes in a domain. The higher level of abstraction would correspond to the rule structure while the low level of detail would correspond to the instance information stored in the structure at the attribute level rather than rule level. So at the higher level we have the rules with the attribute constraints and the predicting class values while at the lower level we have the relationships between attribute values and the occurring class values as detected from the input instances. This information provides the necessary means for determining the relevance of attributes with respect to a particular rule. In other words, we measure the importance of an attribute in predicting the class value implied by a rule. The irrelevant attributes can then be deleted (or re-introduced) at the higher level which will affect the rule coverage rate and generalization power. It is advantageous to integrate a feature selection mechanism since initial bad choices made about the attribute relevance could be corrected as learning progresses.

3   Description of the Proposed RO Technique

This section starts by providing a brief overview of the developed RO technique, and then proceeds to explain each step in more detail in each of the subsequent sections. The general steps of the proposed rule optimization method are presented in the flow chart of Fig. 1. The method takes as input a file describing the rules detected by a particular classifier and the domain dataset from which the rules were learned. As the first step, the rules need to be appropriately represented so as to enable optimization reasoning to occur. The rules are represented in a graph structure (GS) where each rule has a set of attribute constraints and a target vector that stores a set of weighted links pointing to one or more target values. Reading the rule set determines the set of attribute constraints for each rule. In order to set the target vector of each rule, the domain dataset is read in on top of the GS triggering those rules whose attribute constraints best match the attribute values in the presented instance from the dataset. Whenever a rule is triggered, its target vector is updated. The target vector of a rule is updated by updating the weight on the link to the target value that occurred in the instance that triggered that particular rule. At this stage, the GS contains the high level information about the domain at hand in the form of rules, their attribute constraints and their target vectors. This information is used for reasoning at the higher level of abstraction. During this process, the rules can undergo a process of splitting, merging and deleting as will be described later in the paper.

When two or more rules are merged, it is possible that some attributes have lost their relevance. On the other hand, when a simplified rule is split into two more specific rules, some attributes may become relevant for distinguishing more specific data object characteristics and need to be reintroduced. This is the reason for the reasoning at a lower level that determines the attribute relevance of the rules to occur after the higher level reasoning has been completed. Hence, after the reasoning at the higher level, the target vectors of all the rules in the GS are reset and the domain dataset is read in again, this time storing the low level information. The collected information corresponds to the relationships between attribute and target values as detected in the instances from the dataset. This kind of information allows for the calculation of the Symmetrical Tau [12] criterion for determining the relevance of the attributes for a

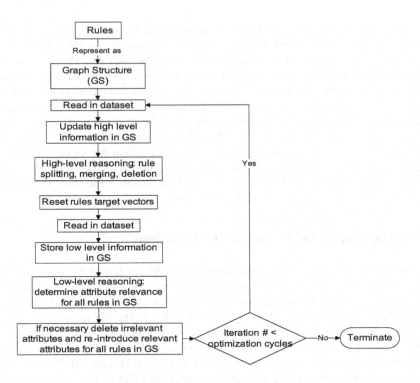

Fig. 1. General steps of the proposed rule optimization method

particular rule. Hence, irrelevant attributes can be removed from the rule and attributes previously detected as irrelevant can be re-introduced if the relevance turns out to be sufficiently high throughout the RO process. Each of the RO steps (except for initial rule representation in graph structure) explained in the following subsections repeats for a number of chosen iterations. The following section describes how the rules and related information can be described using a graph structure.

3.1   Formation of the Graph Structure

In order for the *GS* to be formed two files are read, one describing the rules detected by a classifier and the other containing the total set of instances from which the rules were learned. The rules are in form of attribute constraints while the implying class of each rule is ignored. The reason is that during the whole process of *RO*, the implying class values can change as some clusters will be merged or split. Rather the domain dataset is read according to which the weighted links between the rules and class values are set. The implying class value of a rule becomes the highest weighted link to a particular class value node. This class value has most frequently occurred in the instances which were captured by the rule. An example of the *GS* after a dataset is read in is shown in Fig. 2. The implying class of Rule1 and Rule 3 would be class value 1 while for Rule2 it is class value 2. Even though it is not shown in the figure, each rule has a set of attribute constraints associated with it, which we refer to as the weight vector (*WV*) of that rule. The set of attribute values occurring in the instance being processed are referred to as the input vector (*IV*). Hence, to classify an instance we match the *IV* against the *WV*s of the available rules. A constraint for a continuous attribute is given in terms of a lower range (lr) and an upper range (ur) indicating the set of allowed attribute values.

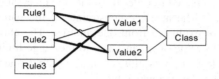

Fig. 2. Example graph structure from high level

3.2 Representing Lower Level Information

Previous sub-section has explained the *GS* formation at the top level which is used mainly for determining the implying class values of the rules. In this section we discuss how lower level instance information is stored for each rule. This low level information is necessary for the reasoning at the lower level.

As previously mentioned each rule has a set of attribute constraints associated with it, which are stored in its *WV*. For each of the attributes in the *WV* we collect the occurring attribute values in the instances that were captured by that particular rule. Hence each attribute has a value list (*VL*) associated with it which stores all the occurring attribute values. Furthermore, each of the value objects in the list has a set of weighted links to the occurring class values in the instance where that particular value occurred. This is necessary for the feature selection process which will be explained later. For a continuous attributes there could be many occurring values and values close to one another are merged into one value object when the difference between the values is less than a chosen merge value threshold. Hence the numerical values stored in a list of a continuous attribute will be ordered so that a new value is always stored in an appropriate place and the merging can occur if necessary. Fig. 3 illustrates how this low level information is stored for a rule that consists of two continuous attributes A and B. The attribute A has the lower range (lr) and the upper range (ur) in between which the values v1, v2 and v3 occur. The 'lr' of A is equal to the value of v1 or the 'lr; of v1 if v1 is a merged value object, while the 'ur' of A is equal to the value of v3 or the 'ur' of v3 if v3 is a merged value object.

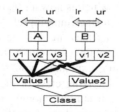

Fig. 3. Storing low level instance information

3.3 Higher Level Reasoning

Once the implying classes are set for each of the rules the dataset is read in again in order to check for any misclassifications and update the rule set accordingly. When a rule captures an instance that has a different class value than the implication of the rule, a child rule will be created in order to isolate the characteristic of the rule causing the misclassification. The attribute constraints of the parent and child rule are updated so that they are exclusive from one another. The child attribute constraint ranges from the attribute value of the instance to the range limit of the parent rule to which the input attribute value was closest to. The parent rule adopts the remaining range as the constraint for the attribute at hand.

To illustrate the process of making an attribute constraint exclusive from one another in the parent and child rule, please consider Fig. 4. At the top of the figure we show the lower (LR) and upper range (UR) of an attribute a_k occurring in the weight vector of the parent rule (r_s) at position k and an input value (IVk) that has occurred within the range. In this example, the value IVk was closer to the upper limit of the range. The newly created attribute constraint for the child rule (bottom right of Fig. 4) will have its lower range (i.e. LR') set to value of IVk and the upper range (i.e. UR') is equal to UR. The constraint in the parent rule (bottom left of Fig. 4) is updated so that the new upper range is set to be a small value (i.e. $smlVal$) away from IVk . The process for all the cases is more formally explained below.

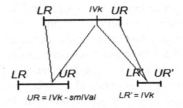

Fig. 4. Illustrating the update of range constraints for an attribute a_k of parent and child rule

After the whole dataset is read in there could be many child rules created from a parent rule. Some child rules may be merged together first but explanation of this is to come later once we discuss the process of rule similarity comparison and merging. If a child rule points to other target values with high confidence it become a new rule and this corresponds to the process of rule splitting, since the parent rule has been modified to exclude the child rule which is now a rule on its own. On the other hand if the child rule still mainly points to the implying class value of the parent rule it is merged back into the parent rule (if they are still similar enough). An example of a rule which has been modified to contain a few children due to the misclassifications is displayed in Fig. 5. The reasoning explained would merge 'Child3' back into the parent rule since it points to the implying class of the parent rule with high weight. This is assuming that they are still similar enough. On the other hand Child1 and Child2 would become new rules since they more frequently capture the instances where the class value is different to the implying class of the parent rule. Furthermore if they are similar enough they would be merged into one rule.

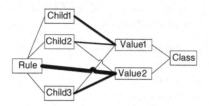

Fig. 5. Example of rule splitting

In order to measure the similarity among the rules we make use of a modified Euclidean distance (*ED*) measure. This measure is also used to determine which rule captures a presented instance. An instance is always assigned to the rule with the smallest *ED* to the *IV*. Even though one would expect the *ED* to be equal to 0 when classifying instances this may not always be the case throughout the *RO* process. The *ED* calculation is calculated according to the difference in the allowed range values of a particular attribute. The way that *ED* is calculated is what determines the similarity among rules, and therefore we first overview the *ED* calculation and then proceed onto explaining the merging of rules that may occur in the whole *RO* process.

3.3.1 Euclidean Distance Calculation

For a continuous attribute a_i occurring at the position i of *WV* of rule R, let 'a_ilr' denote the lower range, 'a_iur' the upper range, and 'a_iv' the initial value if the ranges of a_i are not set. The value from the i-th attribute of *IV* will be denotes as iva_i. The i-th term of the *ED* calculation between *IV* and *WV* of R for continuous attributes is:

- case 1: a_i ranges are not set
  - 0 iff $iva_i = a_iv$
  - $iva_i - a_iv$ if $iva_i > a_iv$
  - $a_iv - iva_i$ if $iva_i < a_iv$
- case 2: a_i ranges are set
  - 0 iff $iva_i \geq a_ilr$ and $iva_i \leq a_iur$
  - $a_ilr - iva_i$ if $iva_i < a_ilr$
  - $iva_i - a_iur$ if $iva_i > a_iur$

The input merge threshold used for continuous attribute (*MT*) also needs to be set with respect to the number of continuous attributes in the set. It corresponds to the maximum allowed sum of the range differences among the *WV* and *IV* so that the rule would capture the instance at hand.

When calculating the *ED* for the purpose of merging similar rules there are four possibilities that need to be accounted with respect to the ranges being set in the rule attributes, and the *ED* calculation is adjusted. For rule R_1 let r_1a_i denote the attribute occurring at the position i of *WV* of rule R_1, let 'r_1a_ilr' denote the lower range, 'r_1a_iur' the upper range, and 'r_1a_iv' the initial value if the ranges of r_1a_i are not set. Similarly for rule R_2 let r_2a_i denote the attribute occurring at the position i of *WV* of rule R_2, let 'r_2a_ilr' denote the lower range, 'r_2a_iur' the upper range, and 'r_2a_iv' the initial value if the ranges of r_2a_i are not set. The i-th term of the ED calculation between the WV of R_1 and WV of R_2 for continuous attributes is:

- case 1: both r_1a_i and r_2a_i ranges are not set
  - 0 iff $r_1a_iv = r_2a_iv$
  - $r_1a_iv - r_2a_iv$ if $r_1a_iv > r_2a_iv$
  - $r_2a_iv - r_1a_iv$ if $r_1a_iv < r_2a_iv$
- case 2: r_1a_i ranges are set and r_2a_i ranges are not set
  - 0 iff $r_2a_iv \geq r_1a_ilr$ and $r_2a_iv \leq r_1a_iur$
  - $r_1a_ilr - r_2a_iv$ if $r_2a_iv < r_1a_ilr$
  - $r_2a_iv - r_1a_iur$ if $r_2a_iv > r_1a_iur$
- case 3: r_1a_i ranges are not set and r_2a_i ranges are set
  - 0 iff $r_1a_iv \geq r_2a_ilr$ and $r_1a_iv \leq r_2a_iur$
  - $r_2a_ilr - r_1a_iv$ if $r_1a_iv < r_1a_ilr$
  - $r_1a_iv - r_2a_iur$ if $r_1a_iv > r_2a_iur$
- case 4: both r_1a_i and r_2a_i ranges are set
  - 0 iff $r_1a_ilr \geq r_2a_ilr$ and $r_1a_iur \leq r_2a_iur$
  - 0 iff $r_2a_ilr \geq r_1a_ilr$ and $r_2a_iur \leq r_1a_iur$
  - $\min(r_1a_ilr - r_2a_ilr, r_1a_iur - r_2a_iur)$ iff $r_1a_ilr > r_2a_ilr$ and $r_1a_iur > r_2a_iur$
  - $\min(r_2a_ilr - r_1a_ilr, r_2a_iur - r_1a_iur)$ iff $r_2a_ilr > r_1a_ilr$ and $r_2a_iur > r_1a_iur$
  - $(r_1a_ilr - r_2a_iur)$ iff $r_1a_ilr > r_2a_iur$
  - $(r_2a_ilr - r_1a_iur)$ iff $r_2a_ilr > r_1a_iur$

For a rule to capture an instance or for it to be considered sufficiently similar to another rule the ED would need to be smaller than the MT threshold.

3.3.2 Merging of Similar Rules

As mentioned at the start of Section 3.3 the child rules may be created when a particular rule captures an instance that has a different class value than the implying class value of that rule (i.e. misclassification occurs). After the whole file is read in the child rules that have the same implying class values are merged together if the ED between them is below the MT. Thereafter the child rules either become a new rule or are merged back into the parent rule, as discussed earlier. Once all the child rules have been validated the merging can occur among the new rule set. Hence if any of the rules have the same implying class value and the ED between them is below the MT the rules will be merged together and the attribute constraints updated. After this process the file is read in again and any of the rules that do not capture any instances are deleted from the rule set.

3.4 Lower Level Reasoning

Once the rules have undergone the process of splitting and merging, the relevance of rule attributes should be calculated as some attributes may have lost their relevance through merging of two or more rules. Other attributes may have become relevant as a more specific distinguishing factor of a new rule which resulted from splitting of an original rule. For this purpose we make use of the Symmetrical Tau [12] feature selection criterion whose calculation is made possible by the information stored at the lower level of the graph structure. We start this section by discussing the properties of the symmetrical tau and then proceed onto explaining how the relevance cut-off is determined and the issue of choosing the merge value threshold for the value objects in a value list.

3.4.1 Feature Selection Measure

Symmetrical Tau (τ) [12] is a statistical measure for the capability of an attribute in predicting the class of another attribute. The τ measure is calculated using a contingency table which is used in statistical area to record and analyze the relationship between two or more variables. If there are I rows and J columns in the table, the probability that an individual belongs to row category i and column category j is represented as $P(ij)$, and $P(i+)$ and $P(+j)$ are the marginal probabilities in row category i and column category j respectively, the Symmetrical Tau measure is defined as [12]:

$$\tau = \frac{\sum_{j=1}^{J}\sum_{i=1}^{I}\frac{(Pij)^2}{P(+j)} + \sum_{i=1}^{I}\sum_{j=1}^{J}\frac{P(ij)^2}{P(i+)} - \sum_{i=1}^{I}P(i+)^2 - \sum_{j=1}^{J}P(+j)^2}{2 - \sum_{i=1}^{I}P(i+)^2 - \sum_{j=1}^{J}P(+j)^2}$$

For the purpose of feature selection problem one criteria in the contingency table could be viewed as an attribute and the other as the target class that needs to be predicted. In our case the information contained in a contingency table between the rule attributes and the class attributes is stored at the lower level of the graph structure as explained in Section 3.2. The τ measure was used as a filter approach for the feature subset selection problem in [13]. In the current work its capability of measuring the sequential variation of an attribute's predictive capability is exploited.

3.4.2 Calculating Relevance Cut-Off

For each of the rules that are triggered for multiple class values we calculate the τ criterion and rank the rule attributes according to the decreasing τ value. The relevance cut-off point is determined as the point in the ranking where the τ value of an attribute is less than half of the previous attribute's τ value. All the attributes below the cut-off point are considered irrelevant for that particular rule and are removed from the rule's WV. On the other hand, if some of the attributes above the relevance cut-off point were previously excluded from the WV of the rule, they are now re-introduced since their τ value indicates their relevance for the rule at hand.

As mentioned in Section 3.2 when the occurring values stored in the value list of an attribute are close together they are merged and the new value object represents a range of values. The merge value threshold chosen determines when the difference among the value objects is sufficiently small for merging to occur. This is important for appropriate τ calculation. Ideally a good merge value threshold will be picked with respect to the value distribution of that particular attribute. However, this information is not always available and in our approach we pick a general merge threshold of around 0.02. This has some implications for the calculated τ value since when the categories of an attribute A are increased more is known about attribute A and the error in predicting attribute B may decrease. Hence, if the merge value threshold is too large many attributes will be considered as irrelevant since all the occurring values could be merged into one value object which points to many target objects and this aspect would indicate no distinguishing property of the attribute. On the other hand, if it is too small many value objects may exist which may wrongly indicate that the attribute has high relevance in predicting the class attribute.

3.5 Summary of the Method

The whole set of RO processes is usually repeated for around 10 iterations. Each time a new iteration is started, the dataset is read in so that the implying class values of the rules can be determined. The dataset is read in again during which misclassifications are detected, and the rules where the misclassifications occur are split in order to isolate the characteristic which leads to the wrong prediction of the class value. The child rules which have the same implying class value and are similar according to the *ED* are merged together. Thereafter, the child rules are either merged back into the parent rule or become a new rule if they are not similar to the parent rule with respect to the implying class value and the *ED* between their weight vectors. The whole rule set is then traversed to merge any further similar rules. The dataset is then read in to store the lower level instance information according to which the τ criterion can be calculated and irrelevant attributes deleted for a rule, and relevant ones reintroduced when necessary. After a number of iterations, the unseen test file is used to determine the predictive accuracy of the optimized rule set.

4 Method Evaluation

The proposed method was evaluated on two rule sets learned from publicly available real world datasets [14]. The rule optimizing process was run for 10 iterations for each of the tested domains. The first set of rules we consider has been learned from the 'Iris' dataset using the continuous self-organizing map [5] so that we can compare the improvement of the extension to the rule optimizing method. The merge cluster threshold *MT* was set to 0.1 and the merge value threshold *MVT* for attribute values was set to 0.02. The rules obtained using the CSOM technique [5] are displayed in Fig. 6. When the rules obtained after retraining were taken as input by our proposed rule optimization method the resulting rule set was different in only one rule. The rule 4 was further simplified to exclude the attribute constraint from sepal-width and the new attribute constraint was only that petal-width has to be between the values of 0.667 and 1.0 for the class value of Iris-virginica. Hence the process was able to detect another attribute that has become irrelevant during the *RO* process. The predictive accuracy remained the same.

With respect to using CSOM to extract rules from the 'Iris' domain we have performed another experiment. The initial rules extracted by CSOM without the network pruning and retraining of the network were optimized. When network pruning occurs the network should be re-trained for new abstractions to be properly formed. In this experiment we wanted to see how the *RO* technique performs by itself without any network pruning or retraining.

Fig. 6. Iris rule set as obtained by using the traditional rule optimizing technique

By applying the *RO* technique the rule set was reduced to four rules as displayed in Fig. 7. However, not as many attributes were removed from each of the rules and two instances were misclassified. Hence, performing network pruning and retraining prior to *RO* may achieve a more optimal rule set. However, in the cases where retraining the network may be too expensive the *RO* technique can be applied by itself. In fact compared to the initial set of rules detected by CSOM, which consisted of nine rules with three misclassified instances this is still a significant improvement.

Rules	Implying class
0.33 < PL < 0.678 0.375 < PW < 0.792	Iris-versicolor
0.208 < SW < 0.542 0.627 < PL < 0.847 0.54 < PW < 1.0	Iris-virginica
0.778 < SL < 1.0 0.25 < SW < 0.75 0.814 < PL < 1.0 0.625 < PW < 0.917	Iris-virginica
0.0 < SL < 0.417 0.41 < SW < 0.917 0.0 < PL < 0.153 0.0 < PW < 0.208	Iris-setosa

Fig. 7. Optimized initial rules extracted by CSOM Notation: SL – sepal_length, SW – sepal_width, PL – petal _length, PW – petal_width

The second set of experiments was performed on the complex 'Sonar' dataset which consists of sixty continuous attributes. The examples are classified into two groups one identified as rocks (R) and the second identified as metal cylinders (M). The learned decision tree by the C4.5 algorithm [15] consisted of 18 rules with the predictive accuracy equal to 65.1%. These rules were taken as input in our *RO* technique and the *MT* was set to 0.2 while the *MVT* was set to 0.0005. The optimized rule set consisted of only two rules i.e $0.0 < a11 \leq 0.197 \rightarrow$ R and $0.197 < a11 \leq 1.0 \rightarrow$ M. When tested on an unseen dataset the predictive accuracy was 82.2 % i.e. 11 instances were misclassified from the available 62. Hence the *RO* process has again proved useful in simplifying the rules set without the cost of increasing the number of misclassified instances.

5 Conclusions

This work has described a rule optimizing technique suitable for domains character-ized by continuous attributes. It is applicable to the optimization of rules obtained from any data mining techniques, and with such a characteristic it allows for the in-corporation of domain expert knowledge. This domain knowledge can be represented as a set of rules, which are then to be fine tuned according to the newly collected data from the domain. Hence the method as a whole is adaptable to the changes in the domain it is applied to. Being able to swap from higher level reasoning to the reason-ing at the lower instance level has indeed proven useful for determining the relevance of attributes throughout the rule optimizing process. The evaluation of the method on the rules learned from real world data by different classifier methods has shown its effectiveness in optimizing the rule set. As a future work method needs to be extended so that categorical attributes can be handled as well. Furthermore, it would be interest-ing to explore the possibilities of the rule optimizing method in becoming a stand-alone machine learning method itself.

References

1. Wang, D.H., Dillon, T.S., Chang, E.: Trading off between misclassification, recognition and generalization in data mining with continuous features. In: Hendtlass, T., Ali, M. (eds.) IEA/AIE 2002. LNCS, vol. 2358, pp. 303–313. Springer, Heidelberg (2002)
2. Abe, S., Sakaguchi, K.: Generalization Improvement of a Fuzzy Classifier with Ellipsoidal Regions. In: Proceedings of the 10th IEEE International Conference on Fuzzy Systems (FUZZ-IEEE 2001), Melbourne, pp. 207–210 (2001)
3. Chen, Z.: Data mining and uncertain reasoning: an integrated approach. John Wiley & Sons, Inc., New York (2001)
4. Engelbrecht, A.P.: Computational intelligence: an introduction. J. Wiley & Sons, Hoboken (2002)
5. Hadzic, F., Dillon, T.S.: CSOM: Self Organizing Map for Continuous Data. In: 3rd Interna-tional IEEE Conference on Industrial Informatics (INDIN 2005), Perth, August 10-12 (2005)
6. Hadzic, F., Dillon, T.S.: CSOM for Mixed Data Types. In: Fourth International Sympo-sium on Neural Networks, Nanjing, China, June 3-7 (2007)
7. Bruner, J.S., Goodnow, J.J., Austin, G.A.: A study of thinking. John Wiley & Sons, Inc., New York (1956)
8. Sestito, S., Dillon, S.T.: Automated Knowledge Acquisition. Prentice Hall of Australia Pty Ltd., Sydney (1994)
9. Kristal, L.: ABC of Psychology (ed.). Michael Joseph, London, pp. 56–57 (1981)
10. Pollio, H.R.: The psychology of Symbolic Activity. Addison-Wesley, Reading (1974)
11. Roch, E.: Classification of real-world objects: Origins and representations in cognition. In: Johnson-Laird, P.N., Wason, P.C. (eds.) Thinking: Readings in Cognitive Science, pp. 212–222. Cambridge University Press, Cambridge (1977)

12. Zhou, X., Dillon, T.S.: A statistical-heuristic feature selection criterion for decision tree induction. IEEE Transactions on Pattern Analysis and Machine Intelligence 13(8), 834–841 (1991)
13. Hadzic, F., Dillon, T.S.: Using the Symmetrical Tau (τ) Criterion for Feature Selection in Decision Tree and Neural Network Learning. In: Proceedings of the 2nd Workshop on Feature Selection for Data Mining: Interfacing Machine Learning and Statistics, in conjunction with the 2006 SIAM International Conference on Data Mining, April 22, Bethesda (2006)
14. Blake, C., Keogh, E., Merz, C.J.: UCI Repository of Machine Learning Databases. University of California, Department of Information and Computer Science, Irvine (1998), http://www.ics.uci.edu/~mlearn/MLRepository.html
15. Quinlan, J.R.: Probabilistic Decision Trees. In: Kadratoff, Y., Michalski, R. (eds.) Machine Learning: An Artificial Intelligence Approach, vol. 4, Morgan Kaufmann Publishers, Inc., San Mateo (1990)

Automated Discrimination of Pathological Regions in Tissue Images: Unsupervised Clustering vs. Supervised SVM Classification

Santa Di Cataldo, Elisa Ficarra, and Enrico Macii

Dep. of Control and Computer Engineering, Politecnico di Torino, Cso Duca degli Abruzzi 24, 10129 Torino, Italy
{santa.dicataldo,elisa.ficarra,enrico.macii}@polito.it

Abstract. Recognizing and isolating cancerous cells from non pathological tissue areas (e.g. connective stroma) is crucial for fast and objective immunohistochemical analysis of tissue images. This operation allows the further application of fully-automated techniques for quantitative evaluation of protein activity, since it avoids the necessity of a preventive manual selection of the representative pathological areas in the image, as well as of taking pictures only in the pure-cancerous portions of the tissue. In this paper we present a fully-automated method based on unsupervised clustering that performs tissue segmentations highly comparable with those provided by a skilled operator, achieving on average an accuracy of 90%. Experimental results on a heterogeneous dataset of immunohistochemical lung cancer tissue images demonstrate that our proposed unsupervised approach overcomes the accuracy of a theoretically superior supervised method such as Support Vector Machine (SVM) by 8%.

Keywords: Tissue segmentation, tissue confocal images, immunohistochemistry, K-means clustering, Support Vector Machine.

1 Introduction

Detecting tumor areas in cancer tissue images and disregarding non pathological portions such as connective tissue are critical tasks for the analysis of disease state and dynamics. In fact, by monitoring the activity of proteins involved in the genesis and the development of multi-factorial genetic pathologies we can obtain a useful diagnostic tool. It leads to classify the pathology in a more accurate way through its particular genetic alterations, and to create new opportunities for early diagnosis and personalized predictive therapies [1].

An approach for monitoring and quantifying the protein activity in pathological tissues is to analyze, for example, images of the tissue where the localization of proteins is highlighted by fluorescent marked antibodies that can detect and link the target proteins. The antibodies are marked with particular stains whose intensity is related to protein activity intensity. This procedure is called *immunohistochemistry* (IHC). The increased use of immunohistochemistry in both clinical and basic research settings has led to the development of techniques for acquiring quantitative information from

A. Fred, J. Filipe, and H. Gamboa (Eds.): BIOSTEC 2008, CCIS 25, pp. 344–356, 2008.

immunostains and automated imaging methods have been developed in an attempt to standardize IHC analysis.

Tissue segmentation for tumor areas detection is the first fundamental step of automated IHC image processing and protein activity evaluation. In fact the quantification of a target protein's activity should be performed on tumor portions of the tissue without taking into account the non pathological areas eventually present in the same IHC images. In Fig. 1 examples of IHC tissue images are reported where connective tissue (i.e. non tumoral tissue) is outlined in black (for details about these images see Section 2).

Several methods have been proposed in the last few years to perform automated segmentation of tissue images [2], [3], [4], [5], [6]. However the most accurate approaches are those that provide a well-suited framework for incorporating primary expert knowledge into the adaptation of algorithms, such as supervised learning algorithm (e.g. Neural Networks, Machine Learning, kernel-based) [6].

The most prominent algorithm among these is the *Support Vector Machine* (SVM) proposed by V.Vapnik [7] for binary classification.

SVM is a theoretically superior machine learning method which has often been shown to achieve great classification performance compared to other learning algorithms across most application fields and tasks, including image processing and tissue image processing in particular [8], [9], [10].

Moreover, the SVM method is more able to handle very high dimensional feature spaces than traditional learning approaches [11], [12]. This is in fact the case of the images targeted by our work.

However, the IHC tissue images we considered in our study present an intrinsic complexity, such as very different characteristics of staining, intensity distribution, considerable variation of tissue shape and/or size and/or orientation and, finally, considerable variation of the signal intensity within the same tissue areas due for example to superimposed staining.

Because of the heterogeneity of the representative features related to each tissue, it is very difficult for the supervised methods to obtain a satisfactory fixed classifier able to distinguish between tumor areas (i.e. epithelial tissue) and non cancerous tissue portions (such as connective tissue).

For this reason we designed a fully-automated unsupervised approach that is based only on the characteristics of the input image rather than on a fixed model of the ground truth.

In this paper we present our fully-automated unsupervised method and we compare its performance to that provided by a SVM approach applied on the same IHC tissue image target. We demonstrate that our method enables more accurate tissue segmentation compared with SVM.

In Section 2 we detail our fully-automated unsupervised method and we briefly introduce the SVM method. The implementation and the set-up are discussed in Section 3. Experimental results conducted on a large set of heterogeneous immunohistochemical lung cancer images are reported and discussed in Section 4. Finally, the Conclusions are reported in Section 5.

2 Method

The images we analyzed in this work were acquired through high-resolution confocal microscopy and show lung cancer tissue cells stained with marked antibodies (see Fig. 1). They are characterized by a blue hematoxylin stain as a background colour and a brown DAB stain in cellular regions where a receptor of the EGF-R/*erb-B* or TGF-alpha family is detected (i.e. membranes or cytoplasm, respectively). Cellular nuclei are blue-coloured and show a staining intensity darker than background. In all the images a remarkable portion of connective or other non cancerous tissue components is present, which appears as a blue-coloured mass (since brown DAB-stained cells are only in cancerous tissue) with quite well-defined borders.

Connective tissue is usually characterized by shorter inter-cellular distances and smaller nuclei than epithelial component; however, a generalization of this remark is impossible because shape and dimensions distributions of cancerous cells are often not predictable. As we outlined in the Introduction, in order to perform accurate and robust cell segmentation and protein activity quantification [13] these non cancerous tissue portions have to be identified and isolated from the representative epithelial tissue.

Here we present two different segmentation approaches to perform this critical task: i) an unsupervised procedure based on a K-means clustering of brown intensities followed by some morphological and edge-based refinement steps (see Fig. 2); ii) a supervised classification of RGB features through Support Vector Machine (see Fig. 5). Experimental results obtained with each approach on the same real-life datasets are presented and compared in Section 4.

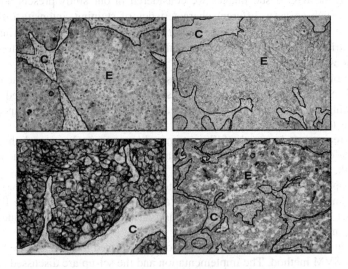

Fig. 1. IHC tissue images. First row, from the left: x400 image with EGF-R positive reactions, x200 image with EGF-R positive reactions; second row: x400 image with EGF-R positive reactions, x200 image with TGF-alpha positive reactions. Connective tissue is outlined in black and labelled with *C*; epithelial tissue is labelled with *E*.

Fig. 2. Flow-chart of the unsupervised procedure based on K-means clustering

2.1 Unsupervised Procedure

Since non cancerous cells do not show positive reactions at the EGF-R/TGF-alpha receptors, the monochromatic pure-DAB component instead of the original RGB image can be analyzed to perform tissue segmentation: in fact in this simpler color space connective components can be easily identified as wide bright regions with a quite homogeneous appearance (see Fig. 3(b)).

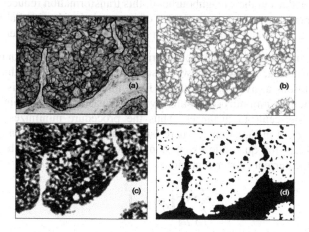

Fig. 3. Unsupervised procedure: (a) original IHC image with connective regions manually outlined (*in black*); (b) pure-DAB image (c) results after K-means clustering (pixels belonging to different clusters are mapped with grey intensity proportional to the cluster centroid); (d) cluster with highest centroid value (*in black*); as outlined in section 2.1 step 4, some small and round-shaped epithelial particles still have to be removed

An unsupervised learning algorithm (K-means, in our work) can be efficaciously applied to isolate bright regions; then areas which show morphological and edge characteristics which are typical of connective tissue can be selected to refine tissue segmentation.

Main steps of the procedure are (see Fig. 2 and 3):

Step 1: DAB-Component Separation. To separate pure-DAB from pure-hematoxylin component a color deconvolution algorithm based on stain-specific RGB absorption is applied on the original RGB image [14], [15].

Differently from classical color segmentation approaches based on transformation of RGB information to HSI or to another specific color representation [16], this method has been demonstrated to perform a good color separation even with colocalized stains.

This critical condition, due to chemical reactions of stains linking the target proteins and to the tissue superposition during the slicing of samples before image acquisition, is very common in the images targeted by our method.

For this step, the free color deconvolution plugin developed by [17] was integrated to our algorithm.

Step 2: Preprocessing. In pure-DAB images, connective tissue can be differentiated from epithelial tissue through its higher intensity (see Fig. 3(b)); anyway some pre-processing is needed in order to homogenize and separate the intensity distributions of the two tissues, thus improving K-means' performance.

First of all, a mean filter is performed: this operation replaces each pixel value with the average value in its neighbourhood, thus smoothing intensity peaks and decreasing the influence of single non-representative pixels.

Then a minimum filter is applied. The filter replaces pixels values with the minimum intensity values in their neighbourhood: this transformation reduces the intensity dynamic and performs a further separation of connective and epithelial intensity distributions, since the former shows minimum values higher than the latter.

Step 3: K-Means Clustering. To isolate bright pixels belonging to connective tissue a K-means clustering, the well-known unsupervised learning algorithm [18] which iteratively partitions a given dataset into a fixed number of clusters, is applied. This iterative partitioning minimizes the sum, over all clusters, of the within-cluster sums of point-to-cluster-centroid distances. Thus the procedure minimizes the so-called *objective function*, J in Equation 1, where k is the number of clusters, n is the number of data points and the quadratic expression is the distance measure between a data point $x_i^{(j)}$ and the current cluster centroid c_j.

$$J = \sum_{j=1}^{k} \sum_{i=1}^{n} \left\| x_i^{(j)} - c_j \right\|^2 . \tag{1}$$

The cluster with the highest centroid value is selected as representative of the connective tissue (see Fig. 3(c)). The number of clusters k was empirically set to four (see Section 3.1 for details about the parameter set-up).

Step 4: Refinement by Size and Circularity Analysis. Bright epithelial regions with low EGF-R/TGF-alpha activity have to be removed from the connective cluster to refine tissue segmentation.

As shown in Fig. 3(d), a large number of these regions are approximately round-shaped and are considerably smaller than connective mass: then a selective removal of particles with a low area and a high circularity compared to threshold values T_S and T_C is performed (parameters set-up in Section 3.1).

Equation 2 shows the proposed index for circularity evaluation (a value of 1 indicates a perfect circle, a value approaching 0 an increasingly elongated polygon).

$$Circularity = 4\pi \frac{Area}{Perimeter^2} . \tag{2}$$

Step 5: Refinement by Gradient Magnitude Analysis. Other bright epithelial regions can be removed from the connective cluster through their edge characteristics, since connective tissue usually shows a well-defined boundary w.r.t. epithelial background in terms of intensity gradient variation. On the base of this remark, in this step areas which show along their boundary a percentage of edge pixels (i.e. pixels with high gradient intensity variation w.r.t. background) lower than a threshold value T_E are selectively removed from connective cluster (parameter set-up in Section 3.1). Edge detection is performed through a Sobel detector followed by automated intensity global thresholding.

2.2 Supervised Procedure

An alternate approach for tissue segmentation is supervised learning; for this purpose a *Support Vector Machine* (SVM) classification is proposed.

The SVM [7] is a theoretically superior machine learning method which has often been shown to achieve great classification performance compared to other learning algorithms across most application fields and tasks including image processing [19].

Here we propose a procedure based on binary SVM classification, in which the input elements (in this work, small tissue regions) are associated to one of two different classes, connective or epithelial, on the base of a set of representative characteristics, the *features vector*. To perform a reliable classification, the SVM is previously trained with a set of elements whose class is well-known, the so-called *training instances*. The classification is based on the implicit mapping of data to a higher dimensional space via a *kernel function* and on the consequent solving of an optimization problem to identify the *maximum-margin hyperplane* that separates the given training instances (see Fig. 4). This hyperplane is calculated on the base of boundary training instances (i.e. elements with characteristics which are border-line between the two classes), the so-called *support vectors*; new instances are then classified according to the side of the hyperplane they fall into. In order to handle linearly nonseparable data, the optimization *cost function* includes an error minimization term to penalize the wrongly classified training instances.

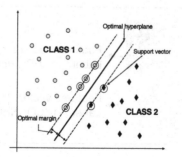

Fig. 4. Maximum-margin hyperplane in SVMs (linearly separable case). The boundary training instances (*support vectors*) are indicated by an extra circle.

See the references provided in the text for a technical description of SVMs.

Our proposed supervised procedure for tissue segmentation consists in three main steps (see Fig. 5):

Fig. 5. Flow-chart of the supervised procedure based on SVM

Step 1: Training Features Extraction. In order to obtain a good generalization of the SVM, a skilled operator was asked to select from a large number of real-life tissue images small rectangular regions wherein both connective and epithelial tissue were present. The images showed various staining levels and very different characteristics of tissue shape and intensity distribution. In each representative sample the operator manually traced the boundaries of connective and epithelial tissue. Then a NxN square sliding window was horizontally and vertically shifted over the samples (shift value s), thus covering the entire surface of the image; for each shifted window, a features vector was generated with the RGB values of 256 equally-spaced pixels (see Fig. 6, parameters set-up in Section 3.2). In this way, a features vector of 3x256 variables was created for each single shift. A +1 label was assigned to windows with a prevalence of epithelial tissue pixels, a -1 label to windows with a prevalence of connective tissue pixels.

Step 2: Training. The labelled features vectors were fed into the SVM for the training; for details about the parameters set-up see Section 3.2.

Step 3: Classification. The optimized SVM obtained in the training step is used to perform tissue classification for new images. For this purpose, the input images are processed to generate features vectors as in step 1; then input features vectors are fed into the trained SVM. At the end of the classification, the SVM automatically associates positive labels to epithelial patterns and negative labels to connective patterns. The output is then processed to reconstruct two-dimensional results as those shown in Fig. 8.

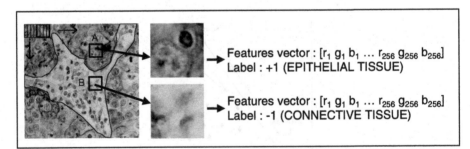

Fig. 6. Generation of the features vectors for SVM training. A NxN square window is horizontally and vertically shifted on the sample, thus covering the entire surface of the image. For each shift a features vector is generated with RGB values of 256 equally spaced pixels, as for *Window A* and *Window B* in the figure above. Epithelial instances are labelled with a +1, connective instances with a -1.

3 Implementation

The algorithm was implemented in Java as a plugin for ImageJ [20], a public domain image analysis and processing software which runs on all the standard operating systems (Windows, Mac OS, Mac OS X and Linux): therefore it is totally hardware-independent, flexible and upgradeable. We inherited the whole class hierarchy of the open-source ImageJ 1.37 API and the free plugins for color deconvolution [17] and K-means clustering [21] and we implemented our own functions and classes. A user-friendly interface enables the user to set different parameters values without modifying the source code.

For the supervised procedure we used the cSVM tool for binary classification [22], since it uses the state-of-art optimization method SMO, i.e. Sequential Minimal Optimization [23]. This cSVM tool implements the algorithm described in [24], which was successfully used to solve different real world problems. Our ImageJ plugins for features vectors generation and output reconstruction were integrated to the SVM tool.

The parameters of the proposed algorithms were empirically tuned by a skilled operator after running several experiments on a large dataset of real tissue images which showed very different characteristics of staining intensity, resolution, EGF-R/TGF-alpha activity level, tissue shape. In the following subsections, we report some details about the implementation of both the unsupervised and the supervised classification procedures and we outline the experimental set-up of the main parameters.

3.1 Unsupervised Procedure

The *number of clusters k* (see Section 2.1 step 3) was set to 4 after running the algorithm with values varying from 2 to 5 and evaluating each time K-means performance in terms of sensibility (power to detect connective components) and selectivity (power to avoid misclassification of epithelial components). For values lower than 4 we often experienced a very good sensibility but a not sufficient selectivity; for higher values the sensibility was frequently poor. A k value equal to 4 assured a good performance of K-means in all the tested images.

The *size threshold T_S* (see Section 2.1 step 4) was varied from 1000 to 5000 pixels with a step of 1000 and was finally set to 3000. Increasing values led to a progressive improvement of selectivity in the connective tissue selection; with values higher than 3000 the lack in sensibility was often not acceptable. Similarly, the *circularity threshold T_C* (see Section 2.1 step 4) was decreased from 0,9 to 0,3. A value of 0,7 assured a good selectivity enhancement without altering sensibility in any of the images.

The *edge threshold T_E* (see Section 2.1 step 5) was increased from 20% to 35% with a step of 5%, evaluating each time the parameter performance in terms of selectivity enhancement and sensibility preservation. A value of 25% assured the best improvement in selectivity without altering sensibility in any of the tested images.

3.2 Supervised Procedure

The *window size N* for features vectors generation (see Section 2.2 step 1) should grant a visible differentiation between connective and epithelial tissue; since nuclei are blue-colored and quite similar in both the tissues, the window has to be large enough to contain a whole nucleus and some surrounding tissue. On the other hand, lower-sized windows allows a better selectivity.

After running several experiments with values varying from 16 to 72 pixels, N was set to 32 for x200 images and to 64 for x400 images.

Since the optimal window size depends on image resolution, x200 and x400 images were respectively classified with SVM trained with x200 and x400 samples.

The *shift* value s (see Section 2.2 step 1) was set to $N/4$, which granted the best compromise between selectivity of classification and computational time.

After running experiments with linear, gaussian and polynomial kernels, we finally chose the *normalized polynomial kernel* shown in Equation 3, where x_1 and x_2 are feature vectors, $n=768$ is the input space dimension and $p=2$ is the kernel hyperparameter; see [24] for technical details.

$$K(x_1, x_2) = \frac{(x_1 \cdot x_2 + n)^p}{\sqrt{(x_1 \cdot x_1 + n)^p} \sqrt{(x_2 \cdot x_2 + n)^p}}. \tag{3}$$

4   Experimental Results

We tested the performance of both the algorithms on a large dataset extracted from real tissue images which presented positive reactions at the EGF-R or at the TGF-alpha receptor activation (see Fig. 1 for examples); reactions are localized in cellular membranes for EGF-R and in cytoplasm for TGF-alpha. Images were acquired from different samples with two different enlargements, x200 or x400.

A skilled operator was asked to manually draw the boundaries of connective tissue in each of the testing datasets. The manual segmentations performed by the operator were pixel-by-pixel compared to those obtained by both the unsupervised and the supervised algorithms.

Connective tissue selection was evaluated in terms of *sensibility* (i.e. power to detect connective tissue) and *selectivity* (power to avoid misclassification of non-connective tissue): for this purpose, the percentage of respectively connective and non-connective pixels which were equally classified by manual and automated segmentation was calculated.

The segmentation *accuracy* was then calculated as weighted average of sensibility and selectivity, as shown in Equation 4.

$$Accuracy = \frac{2}{3} \cdot Sensibility + \frac{1}{3} \cdot Selectivity. \tag{4}$$

Different weights were used because sensibility is more critical for automated measures of protein activity, which is the principal application targeted by our method: in fact, in order to obtain a reliable measure, it is fundamental to eliminate as much as possible non representative tissues from the range of interest; on the contrary, erroneous removal of some epithelial regions is more tolerable, since it has a lower influence on the final measure.

Results obtained for both the automated algorithms are reported in Table 1.

Table 1. Experimental results of unsupervised and supervised classifications. As outlined in Section 3.2, in supervised classification two different SVMs trained respectively with x200 and x400 samples were used (the number of training and validation instances extracted from each dataset is reported for both x200 and x400 classifiers). Training instances were removed from the validation dataset, which was considerably larger.

		UNSUPERVISED ALGORITHM			SUPERVISED ALGORITHM				
Data-set		Sensibility (%)	Selectivity (%)	Accuracy (%)	Number of training instances	Number of validation instances	Sensibility (%)	Selectivity (%)	Accuracy (%)
x200	1	81,89	90,54	84,77	1692	28308	57,91	91,38	69,07
	2	94,64	84,94	91,41	912	20263	94,05	79,20	89,10
	3	95,21	97,99	96,14	220	20192	91,09	94,75	92,31
	4	86,60	87,32	86,84	408	19142	84,41	91,18	86,66
x400	5	91,77	86,20	89,91	558	6942	67,48	82,35	72,43
	6	91,30	78,56	87,05	640	6860	66,48	90,02	74,32
	7	99,67	93,33	97,56	252	7248	93,53	87,46	91,51
	8	89,21	86,28	88,23	300	5888	87,29	85,39	86,66

The number of training instances extracted from each dataset is reported too for both x200 and x400 SVMs. The classification performance was evaluated on a large validation dataset (91137 and 28688 validation instances for x200 and x400 SVM, respectively) which did not include the patterns used for training. Some examples of tissue segmentation are shown in Fig. 8. As shown in Table 1 and Fig. 7, our unsupervised procedure achieved the best results: this method performed tissue segmentations highly comparable with those provided by the skilled operator in all the testing datasets; mean accuracy was 90,24%, with values generally around 90% and always above 85%. SVM performed worse in all the tested datasets; mean accuracy was about 7,5% lower than our unsupervised method.

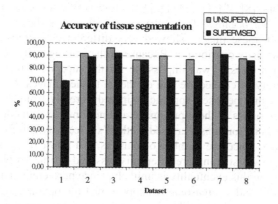

Fig. 7. Accuracy of tissue segmentation; comparison between unsupervised and supervised procedure

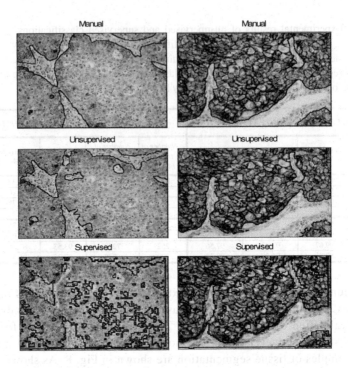

Fig. 8. Examples of tissue segmentation (*in black*). First row: manual segmentation performed by a skilled operator; second row: segmentation performed by the unsupervised procedure; third row: segmentation performed by the supervised SVM approach.

As we previously outlined, SVM is a theoretically superior machine learning method which has often been shown to achieve great   classification performance compared to other learning algorithms across most application fields and tasks including image processing [8], [9], [10], [11], [12]. However, in this case its classification performance was poor because of the intrinsic complexity of the images targeted by our method: in fact, these images showed very different characteristics of staining, tissue shape and intensity distribution. Because of the heterogeneity of the representative features of each class, it was impossible for the supervised method to obtain a satisfying separability of connective and epithelial tissue. Images' heterogeneity was less critical for the unsupervised approach, since differently from SVMs it is based only on the characteristics of the input image and not on a fixed model of the ground truth. On the other hand, our unsupervised method's selectivity is influenced by tissue composition: in fact, since the number of clusters is a-priori fixed, some epithelial regions with low brown staining are often misclassified in images without any connective tissue. Despite this eventuality is unlikely, since pure-epithelial tissue samples are very uncommon (and we reasonably suppose that the operator would escape the automated tissue segmentation in this case), we are working on the solution of the problem: in particular, the introduction of an adaptive number of clusters is in development. As regards the supervised approach, other learning methods such as neural networks and artificial neural networks (ANN) will be tested in the future.

5   Conclusions

We presented a fully-automated unsupervised tissue image segmentation method that allows to distinguish tumor areas in immunohistochemical images and disregard non pathological areas such as connective tissue. This procedure is critical for automated protein activity quantification in tumor tissues in order to analyze the pathology dynamics and development. We described the original processing steps we designed. Finally, we carried out an extensive experimental evaluation on a large set of heterogeneous images that demonstrated the high accuracy achievable by the proposed technique (90% on average) compared to a more traditional approach based on Support Vector Machines (SVM). As future work, we will compare the proposed approach to artificial neural networks (ANN), and we will eventually study the possibility of their integration.

Acknowledgments. We acknowledge the Dep. of Pathology of the S.Luigi Hospital of Orbassano in Turin, Italy, for providing IHC images and for the helpful and stimulating discussions.

References

1. Taneja, T.K., Sharma, S.K.: Markers of Small Cell Lung Cancer. World Journal of Surgical Oncology 2(10) (2004)
2. Demandolx, D., Davoust, J.: Multiparameter Image Cytometry: from Confocal Micrographs to Subcellular Fluorograms. Bioimaging 5(3), 159–169 (1997)
3. Nedzved, A., Ablameyko, S., Pitas, I.: Morphological Segmentation of Histology Cell Images. In: 15th International Conference on Pattern Recognition (ICPR 2000), vol. 1, p. 1500 (2000)
4. Malpica, N., de Solorzano, C.O., Vaquero, J.J., Santos, A., Vallcorba, I., Garcia-Sagredo, J.M., del Pozo, F.: Applying Watershed Algorithms to the Segmentation of Clustered Nuclei. Cytometry 28(4), 289–297 (1997)
5. Dybowski, R.: Neural Computation in Medicine: Perspectives and Prospects. In: Proc. of the ANNIMAB-1 Conference (Artificial Neural Networks in Medicine and Biology), pp. 26–36 (2000)
6. Nattkemper, T.W.: Automatic Segmentation of Digital Micrographs: A Survey. Medinfo. 11(Pt 2), 847–851 (2004)
7. Vapnik, V.: Statistical Learning Theory. Wiley-Interscience, New York (1998)
8. Angelini, E., Campanini, R., Iampieri, E., Lanconelli, N., Masotti, M., Roffilli, M.: Testing the Performances of Different Image Representation for Mass Classification in Digital Mammograms. Int. J. Mod. Phys. 17(1), 113–131 (2006)
9. Osuna, E., Freund, R., Girrosi, F.: Training Support Vector Machines: an Application to Face Detection. In: IEEE Computer Society Conference on Computer Vision and Pattern Recognition (CVPR 1997), p. 130 (1997)
10. Twellmann, T., Nattkemper, T.W., Schubert, W., Ritter, H.: Cell Detection in Micrographs of Tissue Sections Using Support Vector Machines. In: Proc. of the ICANN: Workshop on Kernel & Subspace Methods for Computer Vision, Vienna, Austria, pp. 79–88 (2001)
11. Muller, K.R., Mika, S., Ratsch, G., Tsuda, K., Scholkopf, B.: An Introduction to Kernel-Based Learning Algorithms. IEEE Trans. Neural Networks 12(2), 181–201 (2001)

12. Cai, C.Z., Wang, W.L., Chen, W.Z.: Support Vector Machine Classification of Physical and Biological Datasets. Int. J. Mod. Phys. 14(5), 575–585 (2003)
13. Ficarra, E., Macii, E., De Micheli, G.: Computer-aided Evaluation of Protein Expression in Pathological Tissue Images. In: Proc. of IEEE Symposium on Computer-Based Medical Systems (CBMS), pp. 413–418 (2006)
14. Ruifrok, A.C., Johnston, D.A.: Quantification of Histochemical Staining by Color Deconvolution. Anal. Quant. Cytol. Histol. 23(4), 291–299 (2001)
15. Ruifrok, A.C., Katz, R., Johnston, D.: Comparison of Quantification of Histochemical Staining by Hue-Saturation-Intensity (HSI) Transformation and Color Deconvolution. Appl. Immunohisto. M. M. 11(1), 85–91 (2004)
16. Brey, E.M., Lalani, Z., Hohnston, C., Wong, M., McIntire, L.V., Duke, P.J., Patrick, C.W.: Automated Selection of DAB-labeled Tissue for Immunohistochemical Quantification. J. Histochem. Cytochem. 51(5), 575–584 (2003)
17. Landini, G.: Software, http://www.dentistry.bham.ac.uk/landinig/software/software.html
18. Jain, A.K., Dubes, R.C.: Algorithms for clustering data. Prentice Hall, Englewood Cliffs (1988)
19. Statnikov, A., Aliferis, C.F., Tsamardinos, I., Hardin, D., Levy, S.: A Comprehensive Evaluation of Multicategory Classification Methods for Microarray Gene Expression Cancer Diagnosis. Bioinformatics 21(5), 631–643 (2005)
20. Rasband, W.S.: ImageJ. U.S. National Institutes of Health, Bethesda, Maryland, USA, http://rsb.info.nih.gov/ij/
21. Sacha, J.: K-means clustering, http://ij-plugins.sourceforge.net/plugins/clustering/
22. Anguita, D., Boni, A., Ridella, S., Rivieccio, F., Sterpi, D.: Theoretical and Practical Model Selection Methods for Support Vector Classifiers. In: Support Vector Machines: Theory and Application. Studies in Fuzziness and Soft Computing, vol. 177, pp. 159–179. Springer, Heidelberg (2005)
23. Platt, J.: Fast Training of Support Vector Machines Using Sequential Minimal Optimization. In: Advances in Kernel Methods - Support Vector Learning. MIT Press, Cambridge (1999)
24. Wang, L.: Support Vector Machines: Theory and Applications. Springer, Berlin (2005)

ECoG Based Brain Computer Interface with Subset Selection

Nuri F. Ince[1,2], Fikri Goksu[1], and Ahmed H. Tewfik[1]

[1] Department of Electrical and Computer Engineering, University of Minnesota, 200 Union St.
SE, Minneapolis, MN 55455, U.S.A.
[2] Brain Sciences Center, VA Medical Center, One Veterans Drive
Minneapolis, MN 55417 U.S.A.
{firat,goks0002,tewfik}@umn.edu

Abstract. We describe an adaptive approach for the classification of multichannel neural recordings for a brain computer interface. A dual-tree undecimated wavelet packet transform generates a structured redundant feature dictionary with different time-frequency resolutions computed on multichannel neural recordings. Rather than evaluating the individual discrimination performance of each electrode or candidate feature, the proposed approach implements a wrapper strategy combined with pruning to select a subset of features from the structured dictionary by evaluating the classification performance of their combination. The pruning stage and wrapper combination enables the algorithm to select a subset of the most informative features coming from different cortical areas and/or time frequency locations with faster speeds, while guaranteeing high generalization capability and lower error rates. We show experimental classification results on the ECoG data set of BCI competition 2005. The proposed approach achieved a classification accuracy of 93% by using only three features. This is a marked improvement over other reported approaches that use all electrodes or require manual selection of sensor subsets and feature indices and at best achieve slightly lower classification accuracies.

Keywords: Electrocorticogram, Brain Computer Interface, Time Frequency, Undecimated Wavelet Packet Transform.

1 Introduction

Recent advances in computational neuroscience show that after appropriate signal processing the electrical activity of brain can be used as a new source to help people suffering spinal cord injury, amyotrophic lateral sclerosis etc.. In this scheme, a brain-computer interface (BCI) records and processes the electrical activity of the brain by means of EEG, MEG or ECoG to be used for communication and control by handicapped people [1]. Electroencephalogram (EEG) is widely used in BCIs due to its non-invasiveness. However, the low signal to noise ratio (SNR) and spatial resolution of EEG limit its effectiveness in BCIs. On the other hand invasive methods such as single neuron recordings have higher spatial resolution and SNR. However, they have clinical risks. Furthermore, maintaining long term reliable recording with implantable

A. Fred, J. Filipe, and H. Gamboa (Eds.): BIOSTEC 2008, CCIS 25, pp. 357–374, 2008.
© Springer-Verlag Berlin Heidelberg 2008

electrodes is difficult. An electrocorticogram (ECoG) has the ability to provide long term recordings from the surface of brain. Furthermore, ECoG signals also provide information about oscillatory activities in the brain with a much higher bandwidth than EEG [2]. Therefore, existing algorithms for EEG classification are readily applicable to ECoG processing.

Various events in brain signals such as slow cortical potentials, motor imagery (MI) related sensorimotor rhythms, and visual evoked potentials were used in construction of EEG and ECoG based BCIs [1, 3]. In MI based BCIs, the subjects are asked to perform an imagined rehearsal of either hand/finger or foot movement without any muscular output. Related events in sensorimotor rhythms such as alpha (7-13Hz) and beta (16-32Hz) bands are processed to recognize the executed task using only brain waves. Several methods have been proposed to extract relevant features from rhythmic activities to be used in classification. Methods such as autoregressive modeling and subband energies in predefined windows are widely used in single trial classification [4 and 5]. When used with multi channel recordings, all of these methods need to deal with the high dimensionality of the data. Selecting the most informative electrodes and adapting to subject specific oscillatory patterns is critical for accurate classification. However, due to the lack of prior knowledge, selection of the most informative electrode locations can be difficult. Furthermore, it is well known that there exists a great deal of inter subject variability of EEG and ECoG patterns in spatial, temporal, and frequency domains [2, 3, 4, 6 and 7]. In [8], the common spatial patterns (CSP) method was proposed to classify multichannel neural recordings. The CSP method weighs each electrode location for classification and uses the correlation between channels to increase the SNR of the extracted features. Although the performance is increased with CSP, it has been shown that this method requires a large number of electrodes to improve classification accuracy and that it is very sensitive to electrode montage. Furthermore, since it uses the variance of each channel, this method does not account for the spatiotemporal differences in distinct frequency subbands. Recently, time-frequency (t-f) methods have been proposed as an alternative strategy for the extraction of MI related patterns in BCI's [6, 7 and 9]. These methods utilized the entire time-frequency plane of each channel and integrate components with different temporal and spectral characteristics. Promising results were reported on well known data sets while classifying multichannel neural activity. One of the main difficulties with these methods is once again dealing with the high dimensionality of the data collected. Furthermore, the adaptation to important patterns is implemented by only accounting for the discrimination power of individual electrode locations.

Another important problem arises with the complexity of the constructed BCI. Many of the methods proposed in the literature process a large number of simultaneous recorded sensors. Or a large number of features were extracted for real-time processing. Both cases require either high computational power or long term robustness over neural recordings which are very difficult to maintain with multiple channels.

In this study we tackle these problems by implementing a combinatorial optimization to select space-time-frequency plane features and a classification strategy. The proposed system selects a minimum number of neural patterns from multichannel recordings associated with higher classification accuracies. To our knowledge,

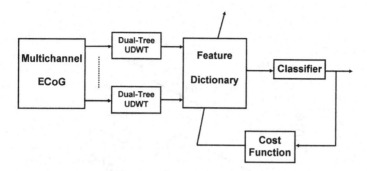

Fig. 1. The block diagram of the proposed feature extraction and feature subset selection technique

this is the first time that an approach implements a joint processing of ECoG features with different time and frequency resolutions coming from distinct cortical areas for classification purposes. The algorithm proposed in this paper requires no prior knowledge of relevant time-frequency indices and related cortical areas. In particular, as a first step, the proposed approach implements a time-frequency plane feature extraction strategy on each channel of a multichannel ECoG array by using a dual-tree undecimated wavelet packet transform (UDWT). The dual-tree undecimated wavelet packet transform forms a redundant, structured feature dictionary with different time-frequency resolutions. In the next step, this redundant dictionary is used for classification. Rather than evaluating the individual discrimination performance of each electrode or candidate feature, the proposed approach selects a subset of features from the redundant structured dictionary by evaluating the classification performance of their combination using a wrapper strategy. This enables the algorithm to optimally select the most discriminative features coming from different cortical areas and/or time-frequency locations. A block diagram summarizing the technical concept is given in Fig. 1. In order to evaluate the efficiency of the proposed method we test it on the ECoG dataset of BCI competition 2005.

The paper is organized as follows. In the next section we describe the extraction of structural time-frequency features with dual-tree undecimated wavelet transform. In section 3 we discuss several subset selection procedures and details of our proposed solutions. We describe the multichannel ECoG data in section 4. Finally we provide experimental results in section 5 and discuss our findings in section 6.

2   Feature Extraction

Let us describe our feature dictionary and explain how it is computed from the wavelet-based dual-tree structure. A schematic diagram of the dual tree is shown in Fig. 2. As indicated in the previous sections, the ECoG signal from each channel can be divided into several frequency subbands with distinct and subject depended characteristic. In order to extract information from these subject specific rhythms, we examine subbands of the ECoG signal by using an undecimated wavelet transform. In each subband, a second pyramidal tree is utilized to extract the time varying characteristics of the subband.

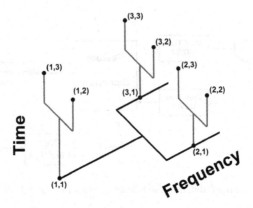

Fig. 2. This dual tree uses 1-level in both planes. Each node of the horizontal tree is a frequency subbands. Node (1,1) represents unfiltered original signal, node(2,1) represents low pass filtered signal and node (3,1) high pass filtered. Each of these subbands is segmented in time into 3 segments, as shown in the vertical tree. Segment (1,1),(2,1) and (3,1) covers the whole subband, segment. (1,2),(2,2) and (3,2) covers the first and segments with time indices three the second half of it.

2.1 Undecimated Wavelet Transform

Discrete Wavelet Transform (DWT) and its variants have been extensively used in 1D and 2D signal analysis [10]. However, the downsampling operator at the outputs of each filter produces a shift variant decomposition. In practice, a shift in the signal is reflected by abrupt changes in the extracted expansion coefficients or related features. In [11] the undecimated wavelet transform is proposed to extract subband energy features which are shift invariant. This is achieved by removing the downsampling operation. The output at any level of pyramidal filter bank is computed by using an appropriate filter which is derived by upsampling the basic filter.

A filter g(n) with a z-transform G(z) that satisfies the quadrature mirror filter condition

$$G(z)G(z^{-1}) + G(-z)G(-z^{-1}) = 1 \qquad (1)$$

is used to construct the pyramidal filter bank (Fig. 3). The high-pass filter h(n) is obtained by shifting and modulating g(n). Specifically, the z transform of h(n) is chosen as

$$H(z) = zG(-z^{-1}). \qquad (2)$$

The subsequent filters in the filter bank are then generated by increasing the width of f(n) and g(n) at every step, e.g.,

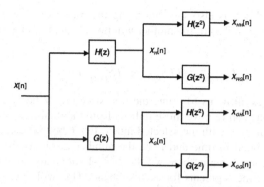

Fig. 3. The pyramidal undecimated wavelet tree

$$G_{i+1}(z) = G(z^{2^i})$$

$$H_{i+1}(z) = H(z^{2^i}), \text{(i=0,1,....., N).}$$

(3)

In the signal domain, the filter generation can be expressed as

$$g_{i+1}(k) = [g]_{\uparrow 2^i}$$

$$h_{i+1}(k) = [h]_{\uparrow 2^i}$$

(4)

where the notation $[]_{\uparrow m}$ denotes the up-sampling operation by a factor of m.

The horizontal pyramidal tree of Fig.2 provides subband decomposition of the ECoG signal. Next, we segment the signal in each subband with rectangular time windows. Such an approach will extract the temporal information in each subband. As in the frequency decomposition tree, every node of the frequency tree is segmented into time segments with a pyramidal tree structure. Each parent time window covers a space equal to the union of its children windows. In a given level, the length of a window is equal to $L/2^t$ where L is the length of signal and t denotes the level. The time segmentation explained above forms the second branch (vertical) of the double tree. After segmenting the signal in time and frequency, we retain the energy of each node of the dual-tree as a feature. By using a dual tree structure we can calculate a rich library of features describing the ECoG activities with several spectro-temporal resolutions. From now on we keep the index information of the dual tree structure to be used in the later stage for dimension reduction via pruning.

To summarize this section the reader is referred to the double tree structure in Fig. 2. Note that the dual tree structure satisfies two conditions:

- For a given node in the frequency tree, the mother band covers the same frequency band width (BW) as the union of its children

$$BW_{Mother} \supset (BW_{Child1} \cup BW_{Child2})$$

(5)

- This same condition is also satisfied along the time axis. For a given node, the number of time samples (TS) of the mother window is equal to that of the union of its children

$$TS_{Mother} \supset (TS_{Child1} \cup TS_{Child2}).$$ (6)

These two properties allow us to prune the tree structure. When a particular feature index is selected, one can remove those indices from the dual tree structure that over-lap in time and frequency with the selected index. Let T be the number of levels used to decompose the signal in time and F be the number levels used to decompose the signal in the frequency domain, there will be $2^{(F+1)}-1$ subbands (including the original signal) and $2^{(T+1)}-1$ time segments for each subband. This will make the total number of potential features $NF=(2^{(F+1)}-1)(2^{(T+1)}-1)$.

3   Subset Selection

Calculating the dual-tree features for each electrode location forms a redundant fea-ture dictionary. The redundancy comes from the dual tree structure. As explained in the previous section, the dual tree has a total of $NF=(2^{(F+1)}-1)(2^{(T+1)}-1)$ features for each channel where F is the total number of frequency levels and T is the total num-ber of time levels. In a typical case, $T=3$, $F=4$ and 64 electrodes are used. The result-ing dictionary than has around thirty thousand features. In such a high dimensional space ($NF=29760$) the classifier may easily go into over-learning and provide a lower generalization capability. Therefore, we would like to select a small number of fea-tures from the available dictionary to reduce the computational complexity, increase generalization capability and use implicitly minimal number of channels for online applications.

Here, we incorporate the structural relationship between features in the dictionary and use several feature subset selection strategies to reduce the dimensionality of the feature set. Since the features are calculated in a tree structure, efficient algorithms were proposed in the past for dimensionality reduction. In [12] a pruning approach was proposed which utilizes the relationship between the mother and children sub-spaces to decrease the dimensionality of the feature set. In particular, each tree is individually pruned from bottom to top by maximizing a distance function. The re-sulting features are sorted according to their discrimination power and the top subset is used for classification. Although such a filtering strategy with pruning will provide significant dimension reduction by keeping the most predictive features, it does not account for the interrelations between features in the final classification stage. Here, we reshape and combine the pruning procedure for feature selection with a wrapper strategy. In particular, we quantify the efficiency of each feature subset by evaluating its classification accuracy with a cost measure and we use this cost to reformulate our dictionary via pruning.

Four different types of methods are considered for feature selection in this study. The structure in Fig. 1 is general representation of each of the four methods. The left most box in Fig. 1 is the rich time-frequency feature dictionary. On the right end a linear discriminant (LDA) is used both for classification and extraction of the relationship among combinations of features. This output is fed to a cost function to measure the discrimination power for that combination of features. This measure will be used to select the best among all other feature combinations. Furthermore, depending on the selected feature index, a pruning operation will be implemented to reduce the dimensionality in the rich feature dictionary.

In this particular study, the Fisher discrimination (FD) criterion is used as a cost function. The Fisher discrimination criterion is defined as:

$$FD = \frac{\left(\mu_1 - \mu_2\right)^2}{\sigma_1^2 + \sigma_2^2}. \tag{7}$$

The four different strategies mentioned above are: (i) Sequential forward feature selection (SFFS), (ii) SFFS with pruning (SFFS-P), (iii) Cost function based pruning and feature Selection (CFS), and (iv) CFS with principal component analysis (PCA) post processing.

3.1 Sequential Forward Feature Selection: SFFS

The SFFS is a wrapper strategy which selects a subset of features one by one. A cost function is used on the classifier output to measure the efficiency of each feature. By using LDA, the feature vectors are projected on a one dimensional space. Then the FD criterion is used to estimate the efficiency of the projection. After this search is done over all feature vectors, the best feature index is selected by comparing the cost values of each feature vector. In the next step, the feature vector which will do the best in combination with the ones selected earlier is identified by searching over the remaining feature vectors. This procedure is run until a desired number of features is reached. Note that SFFS uses all the boxes and connections in Fig. 1 except for the feedback from the cost function to the dictionary. Since no dimension reduction is implemented on the entire feature space, this approach has high computational complexity.

3.2 SFFS with Pruning: SFFS-P

The SFFS-P is also a wrapper strategy with an additional pruning module for dimension reduction. Once a feature index is selected, the corresponding frequency tree and time tree indexes are calculated on the dual-tree. Then the nodes that overlap with the selected feature index in time and frequency are removed. Next, the feature which will do best in combination with the first selected feature is identified by searching the pruned dictionary. In other words, the dictionary is pruned based on the last selected feature. This procedure is run until the desired number of features is reached. Therefore, the only difference between SFFS and SFFS-P is that in the latter strategy, pruning is done on the dictionary based on the selected features. This provides a fast decrease in the number of candidate features and complexity is much smaller than SFFS.

3.3   Cost Function Based Pruning and Feature Selection (CFS)

The CFS is a filtering approach that uses the structure in the feature dictionary for pruning. After finalizing the pruning procedure for each electrode location, it uses a cost function to rank the features. In particular, it uses the FD criterion to rank the features. It does not use either the LDA or the feedback path in Fig. 1. Instead, using the FD measure, a cost value is computed for each node on the double tree individually. Then, a pruning algorithm is run on the double tree by keeping the nodes with maximum discrimination. Once a node is selected, all nodes overlapping with the selected one are removed. This procedure is iterated until no pruning can be implemented. After pruning the dual-trees for each electrode location, the resulting feature set is sorted according to their corresponding discrimination power and input to the classifier. In this way the most predictive features are entered into the classification module. Since no feedback is used from the classifier, the CFS has lower computational complexity than the other two methods.

The CFS method works as a filter on the electrodes by only keeping those indices with maximum discrimination power. However, since features are evaluated according to their discrimination power individually, such a method does not account for the correlations between features. In [6 and 7], PCA analysis is performed on a subset of top sorted features to obtain a decorrelated feature set. The PCA post processed features are sorted according to their corresponding eigenvalues in decreasing order and used in classification. Here, we will also use the PCA as a post processing step with the CFS to obtain a decorrelated feature set. We will refer this method as CFS-PCA.

4   Multichannel ECoG Data

In order to evaluate the performance of the proposed method we used the multichannel ECoG [13] dataset of BCI competition 2005 [14]. During the BCI experiment, the subject was asked to perform imagined movements of either the left small finger or the tongue. The ECoG data was recorded using an 8x8 ECoG platinum electrode grid which was placed on the contralateral (right) motor cortex as shown in Fig. 4. A sampling rate of 1000Hz was used to record the neural data. Every trial consisted of either an imagined tongue or an imagined finger movement and was recorded for 3 seconds duration. The recording intervals started 0.5 seconds after the visual cue had ended to avoid visually evoked potentials. Each channel was filtered with a low pass filter in 0-120Hz band. The filtered data was down sampled by a factor 4 to 250Hz. Each trial was expanded from 750 samples into 768 samples by symmetric extension on the right side to enable segmentation in a pyramidal tree structure. Besides monopolar data, we also consider ECoG data that is processed using a surface Laplacian derivation. More specifically, each electrode data is subtracted from the weighted average of the surrounding 6 electrodes. The electrodes on the border are eliminated from the analysis resulting in a total of 36 electrodes (See Fig. 4). For monopolar data all 64 electrodes were used for analysis. We used 278 trials for training and 100 trials for testing. The training and test data were recorded from the same subject and with the same task, but on two different days with about 1 week in between.

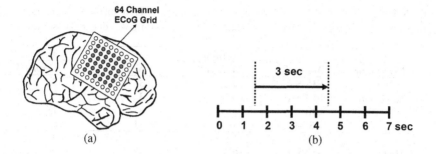

Fig. 4. (a) The 8x8 electrode grid was placed on the right hemisphere over the motor cortex. (Modified from Lal 2005). For surface Laplacian derivation only marked electrodes are used. (b) The timing diagram of the experimental paradigm. The go cue for motor imagery is given at second one. A three second time window starting after 500ms of go cue is used to classify ECoG data.

5   Results

To extract the dual tree features we select $T=3$ and $F=4$. For a 125 Hz bandwidth, the frequency tree provided around 8Hz resolution at the finest level. Along the time axis, the finest time resolution was 375ms. Several filter settings with increasing vanishing moments have been evaluated. In particular we tested Daubechies filters with different vanishing moments (db4, db5, db6, db7 and db8) in constructing the frequency tree of the UDWT. In order to learn the most discriminant time-frequency indices and the corresponding cortical areas, we utilized a 10 times 10 fold cross validation in the training dataset. The optimal number of features at which the classification error is minimal is selected from the averaged cross validation error curves. Then, the learned feature indices are used in testing the classifier on the test set. The training and test results obtained with the different methods with db6 are presented in Table.1.

5.1   Comparison among Methods

We inspected the combination of several methods and wavelet filters during our experimental studies. We note that the SFFS and SFFS-P provided the highest classification accuracy of 93% with only three features on the test set using the Laplacian derivation. Therefore in the rest of this section we provide detailed results related to different wavelet filters obtained only with this method. The training and test set accuracies for different filters for SFFS-P algorithm is given in Table 2. We note that the same classification accuracies were obtained with filters db5, db6 and db7. The db4 and db8 filters provided poor classification accuracies on the test set compared to other filter settings. The tradeoff between time and frequency localization with increasing vanishing moments is a well known problem. While higher vanishing moments provide good frequency localization they are poor in time localization and vice versa. We note that the db5, db6 and db7 filters provide a healthy balance in this perspective. In the rest of the paper we report the results related to the db6 filter unless stated otherwise.

Table 1. The cross validation (CV) and test error rates of different methods and related number of features (NoF) used for final classification. The filter setting was db6 and cost measure FD.

Derivation	Method	Training CV Error (%)	NoF	Test Error (%)
	SFFS	10.2	3	7
Laplacian	SFFS-P	10.2	3	7
	CSF	9.9	27	18
	CSF-PCA	9.6	11	8
	SFFS	12.6	3	20
Monopolar	SFFS-P	10.3	4	9
	CSF	11.7	22	12
	CSF-PCA	11.2	14	8

Although a lower error rate was achieved by CFS with the training data, interestingly, the testing error rate of the CFS was higher than those of the other methods. We also note that a large number of features were used by CFS to achieve 9.9% error rate in the training set. In contrast, the SFFS and SFFS-P algorithms used only 3 features to achieve the minimum 10.2% error rate. The cross validation error curves versus the number of features are given in Fig. 5. Since the results using Laplacian derivation outperformed those obtained with monopolar data, only the results corresponding to the former are provided.

Table 2. The cross validation and test error rates achieved with SFFS-P method for different filter settings

Wavelet Filter	Train Error	NoF	Test Error
db4	12.4	4	11
db5	12.6	3	7
db6	10.2	3	7
db7	10.1	3	7
db8	10.7	4	8

As can be seen clearly from these curves, SFFS and SFFS-P select the best combination of features and achieve the minimal error with only three features. Furthermore, using the structure of the feature dictionary, SFFS-P achieves this result with reduced complexity due to pruning. The pruning process provides a dimension reduction and feature decorrelation. The CFS approach, on the other hand, achieves the minimal error using a large number of features. The interactions among the selected features cannot be taken into account with this approach. In addition, the correlated neighbor areas may result in a duplication of information in the sorted features. In order to decorrelate the features, a Principal Component Analysis (PCA) was employed on the CFS ordered features. This post processing step provided lower error rates than those achieved by CFS alone. The test error rate was 8% for the PCA post processed features. It should be noted here that CFS-PCA produced comparable results with those of SFFS and SFFS-P. However, one should note that PCA induces an

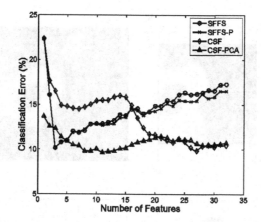

Fig. 5. The cross validation error curves for the different methods in the training data

additional complexity. This method requires all 32 features to be extracted from the ECoG signals which lead to a much higher computational complexity compared to the three features selected by SFFS and SFFS-P. In addition, since only three features are used by SFFS and SFFS-P, they are more robust to intra-subject variability in ECoG signals. Note also that the error rate in monopolar derivation is much higher than that of the Laplacian derivation. We observed large DC changes in ECoG signals in the test data set. Since the Laplacian derivation provides a differential operator, large baseline wanders affecting many electrodes are eliminated in this setup. However, for the monopolar recordings the features are very sensitive to this type of changes.

Note also that the classification accuracies of SFFS and SFFS-P in the test set are higher than that of the training set. One of the underlying reasons could be that the subject can control his/her brain patterns with a higher accuracy with the increasing number of trials. In addition, the SNR of the signals might have improved over time due to tissue electrode interaction.

Since the testing data was recorded on another date, variability in the ECoG signal is expected. The results obtained indicate that the CFS algorithm is very sensitive to this type of variability. We believe that the correlated activity across cortical areas is an important reason why CFS selects the same information repeatedly. Since the SFFS and SFFS-P have the advantage of examining the interactions between different cortical areas and t-f locations, these algorithms can form a more effective subset of features for classification.

5.2 Patterns in Space Time and Frequency

In order to support our hypothesis given above and to provide additional information about the spatial content of discrimination, we show the discriminatory cortical maps of monopolar and Laplacian derivations in Fig. 6. To generate these images we used the most discriminant feature of each electrode location and produced an interpolated matrix over the 8x8 grid to present the distribution of the most discriminative locations. Furthermore, we mark the electrode locations selected by SFFS, SFFS-P, and

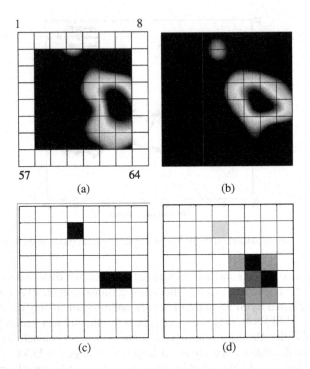

Fig. 6. Discriminant cortical areas (a) Laplacian (b) Monopolar. The number of selected features from different electrode locations in Laplacian derivation for SFFS-P (c) and for CSF (d) are given. The darker areas indicate a higher number of features are selected from these regions. Note that SFFS-P provides a balanced feature distribution. The CSF selected most of 27 features from the same region.

CFS for classification. After inspecting Fig. 6 (a) and (b) we noticed that a large number of neighbor electrode locations carry discriminant information.

The CFS method used a large number of electrodes from this region for classification. In contrast, the SFFS and SFFS-P methods selected another cortical area from the upper side of the grid. Even though this electrode location does not seem to be very discriminative, it played a key role in achieving a lower classification rate.

We recently obtained the information about the true locations of the ECoG grid[1]. The real locations of selected sensor positions are shown in Fig. 7. In order to emphasize the contrast in extracted features we weighted the selected sensor positions with the signed Fisher discrimination power. In particular, the sign reflects the difference between the class means (Tongue Imagery-Hand Imagery). We note that the sign is positive for the sensor in the upper region around the hand area and negative for the lower region around the speech area. This could be related to the differences in ERD and ERS levels in hand and tongue area during motor imageries. The hand imagery produces a higher ERD around the hand area than tongue imagery and, vice versa.

---

[1] By the time initial results of this study were presented at BIOSTEC 2008 congress the authors were unaware about the real positions of ECoG electrodes.

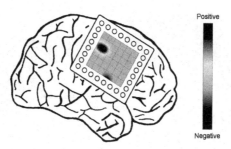

Fig. 7. The sign weighted FD criterion of selected electrodes. The sign reflects the mean difference between tongue imagery vs. hand imagery. Note that the sign is positive for the left upper electrodes around the hand are and it is negative for the electrode at the bottom around speech area. Also the discrimination power around the hand area is larger than speech area.

The time frequency locations of the selected three features are given in Fig. 8. The image order from left to right is the same as they were selected by the algorithm. We observed that all features were extracted from the second subband which corresponds to the alpha frequency range of ECoG.

Interestingly the temporal characteristics of the selected features within this band were distinct. The features F10473 and F10940 originated from two neighboring sensor positions around the hand area as visualized in Fig.6 and Fig.7. Along the time axis, the feature F10940 preceded the feature F10473 by two time segments (=750ms). Interestingly, the entire three second interval was selected in feature F1171 from the speech area. Our results indicate that there exists a dynamic structure in the temporal domain.

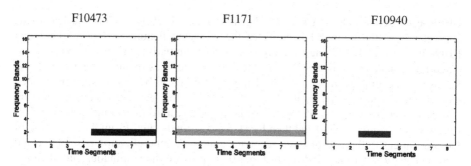

Fig. 8. The extracted time-frequency features from selected electrode positions. The features F10473 and F10940 were neighboring electrodes located around hand area and feature F1171 was located around speech area. All features were selected from alpha frequency band. The darker features indicate the more discrimination power.

5.3 Dynamic Decoding for Continuous Feedback

Although our proposed algorithm extracts features from the entire three second time frame and uses them in a single shot classification, one can also decode the dynamic

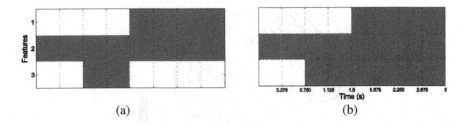

Fig. 9. The weighting functions for each feature (rows) and each time segment (columns) are visualized. The $w_P[m]$ is the mask along the time axis for computing the cumulative energy of a particular feature (a). The graph on the right represents enable function which encodes if a particular feature is available to be used in classification (b). The gray regions represent ones and white regions zero values.

structure described above and provide continuous classification feedback. In particular, since these features are well localized in time and frequency, their locations will be used as weighting functions. Furthermore, when one of the features becomes available it will be used for classification. Let us briefly summarize the dynamic decoding framework shown in Fig. 9:

Let $x_P[l]$ represent the alpha band energy of ECoG channel P in 375ms intervals of consecutive time windows, $l = 1, 2, .., 8$. Then the cumulative energy $E_P[m]$ up to time window m is defined as

$$E_P[m] = \sum_{l=1}^{m} x_P[l] w_P[l] \tag{8}$$

where $w_P[l]$ denotes the mask along the time axis (Fig. 9 (a)). Note that each row corresponds to the temporal location of the extracted features F10473, F1171 and F10940, respectively. As the time progresses the cumulative energy, $E_P[m]$, is computed for each channel using (8). These features are used as inputs to the LDA classifier as they become available in time. The graph in Fig. 9 (b) indicates the time when these features can be used by the LDA classifier. In this particular study the second feature F1171 (second row of Fig. 9 (b)) is the first feature to become available. It is followed by feature F10940 and then F10473 1.5 seconds later. From the beginning of the experiment to the end of three second segment, the classification is implemented within each 375ms time window. In Fig. 10 we provide the continuous classification results when such a dynamic decoding framework is implemented. We also considered the case where one uses only the channel and subband information as used in general BCI applications. Specifically, in such an approach in each time window features coming from the three channels are always evaluated simultaneously for classification with no regard to when they become available as in dynamic decoding. For each channel the mask along the three second time window is one. We refer to this strategy "static decoding". After inspecting the classification error curves we note that there is almost a 10% difference in overall classification accuracies between dynamic and static decoding. We also note that dynamic decoding approached lower error rates faster than the static one. After 1.85 seconds, its classification error was around 10%. These results indicate that the features selected by the proposed algorithm can be effectively used in continuous feedback BCIs with a dynamic decoding framework. Please also note that the dynamic decoding algorithm can be easily extended to those cases where the features are originating from different frequency subbands for a particular channel or with different temporal characteristic within the same subband.

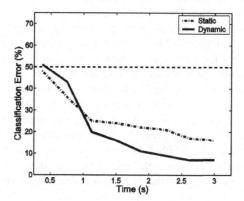

Fig. 10. The continuous classification error rate versus time. Almost 1.8 s later the system approaches around 10% error rate. Solid curve corresponds to the dynamic decoding and dashed one corresponds to the static decoding framework. Recall that the temporal structure of extracted features is not evaluated in static case.

5.4 Computational Complexity and Subset Selection

The final system utilized the selected subset of features for classification and has low computational complexity for realtime application. However, selecting three features from a redundant dictionary is quite demanding. In particular, the system faces a great deal of computational complexity while constructing the LDA classifier which includes calculation of covariance matrices and computation of inverses. Repeating these steps in the entire dictionary boosts the learning time with increasing numbers of selected featured. Therefore, the pruning stage reduces the number of iterations by removing overlapped tree indices. Since the pruning is implemented on a channel basis, this stage reduces the dimensionality of the dictionary slowly.

Here we evaluate the relationship between Fisher LDA and least squares regression (LS) to speed up the feature selection process. In [15] it has been shown that the weights of Fisher LDA are the scaled version of LS regression weights computed over features where the target variable $y \in \{-1, 1\}$ is a vector of class labels. In this perspective the sequential selection of features with Fisher LDA also corresponds to the selection of predictors with the largest correlation in magnitude during each step of an iterative solution of the orthogonal least squares (OLS) regression [16]. Selecting the features with the OLS algorithm decreases the training time dramatically compared to that of using SFFS. Furthermore, the observation of this equivalence leads us to view the feature subset selection as a sparse signal representation problem where the target variable is the signal to be approximated and each predictor is a basis element (column) of matrix A. We therefore employed the Lasso algorithm for subset selection to approximate the target signal with minimal number of features. The Lasso algorithm [17] solves for the vector x the following convex optimization problem

$$\min \|Ax\text{-}y\|_2^2 + \lambda \|x\|_1 . \tag{9}$$

The first error term measures how good the target variable y is approximated and the second term measures the sparsity of x by means of $L1$ norm. Finally, the term λ controls the trade off between these two contradicting objectives.

We select the interval of λ values on the ECoG training data empirically such that the number of nonzero elements of the corresponding solution vectors varies on average from 1 to 50. Using the same cross validation procedure as is done for the previously provided methods, the optimum λ is selected on the training data. Then, the solution corresponding to this optimal complexity is found using the whole training data and applied to test data. In this experiment we used the SparseLab toolbox of [18] to implement the Lasso algorithm. The training and test error rates obtained with Lasso were 9.3% and 12% respectively. We note that although the training error rate of Lasso was lower than the SFFS-P algorithm, the test error rate was higher. We observed that the Lasso algorithm selected 10 features from the dictionary. Higher number of features can be one of the reasons for obtaining higher error rates in the test set.

5.5   Comparison with the BCI2005 Competition Results

Finally, we compared our results with those of achieved at the BCI competition 2005 using the same ECoG data. The classification accuracies and related methods are presented in Table 3. Our method achieved the best result of 7% error and outperformed both the CSP and AR model based techniques. We note that all the methods shown in Table 3 including ours, used linear classifiers. Hence, the comparison among features extraction engines is fair. Although the classification error obtained with the CSP method is close to our result, one should note that this method requires all ECoG channels to be processed. As indicated in the introduction section, maintaining robust recordings over long periods is difficult. The other two methods involve manual selection of ECoG channels and features to be used in the classification. Within this perspective our proposed solution provides an automated way of feature extraction associated with higher classification accuracies. Since the main challenge is to evaluate the generalization performance of the algorithms on the test set, at this

Table 3. The comparison of the proposed method with the best three methods from the BCI 2005 competition

Features Used	Classifier	Test Error (%)
UDWT based subband energies	LDA	7
Common Spatial Subspace Decomposition	Linear SVM	9
ICA combined with spectral power and AR coefficients	Regularized logistic regression	13
Spectral power of manually selected channels	Logistic regression	14

point, it is difficult to assign any statistical significance between the differences of classification results including the Lasso algorithm presented in the previous section.

6 Conclusions

In this paper we proposed a new feature extraction and classification strategy for multi-channel neural recordings in a BCI task. In particular, our aim was to obtain higher generalization capability and lower error rates by selecting a minimum number of features from a redundant dictionary. A feature dictionary was obtained by decomposing the neural activity into subbands with an undecimated wavelet transform and then segmenting each subband in time successively over a pyramidal dual tree structure. Rather than using predefined frequency indices or manually selecting cortical areas, the algorithm implemented a sequential forward feature selection procedure combined with dictionary pruning to select a small subset of features for classification. The pruning step basically used the structure in the dictionary and reduced dimensionality to speed up the training process. The computational complexity of training stage can be reduced further by noting the equivalence between our proposed sequential feature subset selection and orthogonal least squares regression. We tested our approach on the multichannel ECoG dataset of BCI competition 2005, which was recorded during finger and tongue movement imagery. After applying our proposed approach we note that, using only three features, the method achieved 93% classification accuracy on the test set which was recorded a week after the training set. We observed that these three features originated from three different sensor positions located around hand and speech area. We extended the classification module for continuous feedback scenario by incorporating a dynamic decoding framework. Finally, we compared our results to existing methods used in BCI on the same data. We observed that our method outperformed the traditional techniques such as CSP and AR by using small number of features, sensors. In addition the algorithm can be seen as a powerful technique in extracting parsimonious features from multichannel neural recordings. These results show that the proposed method is a good candidate for the construction of a robust ECoG based invasive BCI system. The overall computational complexity of the proposed algorithm is low in both training and testing stages. The features are easy to compute which requires only a linear filtering process that is crucial for realtime BCI applications.

References

1. Wolpaw, J.R., et al.: Brain-Computer Interface Technology: A review of the first international meeting. IEEE Trans. On Rehab. Eng. 8, 164–173 (2000)
2. Leuthardt, E.C., Schalk, G., Wolpaw, J.R., Ojemann, J.G., Moran, D.W.: A brain–computer interface using electrocorticographic signals in humans. Journal of Neural Engineering, 63–71 (2004)
3. Pfurtscheller, G., Neuper, C.: Motor Imagery and Direct Brain-Computer Interface. Proceedings of IEEE 89, 1123–1134 (2001)
4. Schlögl, A., Flotzinger, D., Pfurtscheller, G.: Adaptive autoregressive modeling used for single trial EEG classification. Biomed. Technik 42, 162–167 (1997)

5. Prezenger, M., Pfurtscheller, G.: Frequency component selection for an EEG-based brain computer interface. IEEE Trans. on Rehabil. Eng. 7, 413–419 (1999)
6. Ince, N.F., Tewfik, A., Arica, S.: Extraction subject-specific motor imagery time-frequency patterns for single trial EEG classification. Comp. Biol. Med (2007)
7. Ince, N.F., Arica, S., Tewfik, A.: Classification of single trial motor imagery EEG recordings by using subject adapted non-dyadic arbitrary time-frequency tilings. J. Neural Eng. 3, 235–244 (2006)
8. Ramoser, H., Müller-Gerking, J., Pfurtscheller, G.: Optimal spatial filtering of single trial EEG during imagined hand movement. IEEE Trans. Rehab. Eng. 8(4), 441–446 (2000)
9. Wang, T., He, B.: Classifying EEG-based motor imagery tasks by means of time–frequency synthesized spatial patterns. Clin. Neuro. 115, 2744–2753 (2004)
10. Vetterli, M.: Wavelets, approximation, and compression. IEEE Signal Proc. Magazine, 59–73 (September 2001)
11. Unser, M.: Texture classification and segmentation using wavelet frames. IEEE Trans. Image Proc. 4(11), 1549–1560 (1995)
12. Saito, N., et al.: Discriminant feature extraction using empirical probability density and a local basis library. Pattern Recognition 35, 1842–1852 (2002)
13. Lal, T.N., et al.: Methods Towards Invasive Human Brain Computer Interfaces. In: Saul, L.K., Weiss, Y., Bottou, L. (eds.) Advances in Neural Information Processing Systems (NIPS 17), pp. 737–744. MIT Press, Cambridge (2005)
14. http://ida.first.fraunhofer.de/projects/bci/competition_iii/
15. Duda, R.O., Hart, P.E., Stork, D.G.: Pattern Classification. John Wiley & Sons, Chichester (2006)
16. Chen, S., Cowan, C.F.N., Grant, P.M.: Orthogonal least squares learning for radial basis function networks. IEEE Transactions on Neural Networks 2, 302–309 (1991)
17. Tibshirani, R.: Regression shrinkage and selection via the lasso. J. Royal. Statist. Soc B 58(1), 267–288 (1996)
18. http://sparselab.stanford.edu/

Part III

HEALTHINF

THE-MUSS: Mobile U-Health Service System

Dongsoo Han, Sungjoon Park, and Minkyu Lee

School of Engineering, Information and Communications University
119, Munjiro, Yuseong-gu, Daejeon, Korea
{dshan,sungjoon,niklaus}@icu.ac.kr

Abstract. In this paper, we introduce a mobile u-health service system, named THE-MUSS, which supports u-health service development and running, with functions, modules, and facilities that are commonly required for various mobile u-health services. Basic modules to support bio-signal capturing, processing, analysis, diagnosis, feedbacks are prepared and stacked in the system. Reusability and evolvability are elicited as the primary design goals to achieve in developing THE-MUSS after the understanding of u-health service characteristics. U-health service platform, u-health ontology incorporated u-health service design tool, Matrix based disease group identification framework, and u-health portal are the major components constructing the layered architecture of THE-MUSS. Mobile stress and weigh management services are developed on THE-MUSS to confirm and evaluate the usefulness of THE-MUSS in developing mobile u-health services. According to the evaluation, it turned out that THE-MUSS has strength in reusability and evolvability, but also in system flexibility, adaptability, interoperability, and guideline provision for developing u-health services.

Keywords: e-Health, mobile u-Health service system, ontology, prediction system, BPM, THE-MUSS, stress service.

1 Introduction

With the wide spread use of mobile phones and the announcement of new mobile bio-sensors attached to the mobile phones[10], there are many attempts for development of mobile u-health services on the mobile phones both in research and commercial areas[11]. Diverse forms and types of mobile u-health services will be available on the mobile phones in the near future. However, most u-health services on current mobile phones have been developed conformable to specific sensors or applications. That is, when new bio-sensors become available, services for the bio-sensors should be developed from the scratch, and thus the functions and modules developed for one service is hardly reused in other services.

If the services are developed on a commonly available u-health service platform, more efficient and systematic u-health service development would be possible[1, 2]. In this paper, we introduce a mobile u-health service system, named Total Health Enriching-Mobile U-health Service System (THE-MUSS), which supports u-health service development and execution, with functions, modules, and facilities that are

A. Fred, J. Filipe, and H. Gamboa (Eds.): BIOSTEC 2008, CCIS 25, pp. 377–389, 2008.

commonly required for various mobile u-health services. Basic primitive modules and services for bio-signal capturing, processing, analysis, diagnosis, feedbacks are prepared and provided in highly sharable manner in the system. Ordinary u-health services are developed by integrating modules and services equipped on the system.

Understanding u-health services are essential for the setting of design goals for a u-health service system. For example, the understanding that u-health services may look different from the service point of view, they often share many common features at various levels such as service structure, unit service, and data. That leads us to emphasize the reusability design goal. So, we analyze the characteristics of u-health services to elicit the design goals of THE-MUSS. Evolvability is also set up as another primary design goal for THE-MUSS after understanding the u-health service characteristics.

BPMS-based u-health service platform, u-health ontology incorporated u-health service design tool, matrix based disease group identification framework, mobile client and u-health portal are the major components constituting THE-MUSS's layered architecture. Each of the components has some roles and functions to achieve the reusability and evolvability design goals.

THE-MUSS has several unique features. First, it interprets and treats mobile u-health service as service process and extends Business Process Management System (BPMS) to healthcare service area. As a consequence, THE-MUSS provides general functions and facilities of BPMS. Process definition, execution, and management are the typical functions of BPMS. In regard with the u-health service process, we define a typical mobile u-health service process template and deploy the process template on THE-MUSS.

Second, THE-MUSS is equipped with a very unique matrix based patient group identification method. The method is useful for mobile u-health services where less precise bio-signals data is gathered more frequently from a large number of users. Third, the platform provides several advanced features and functions by incorporating ontology technologies in defining mobile u-health services and inferring or analyzing diseases.

We implemented THE-MUSS and then developed mobile stress and weigh management services on THE-MUSS to confirm and evaluate the usefulness of THE-MUSS in mobile u-health services. The evaluation is performed based on well-known software quality attributes[8]. According to the evaluation, THE-MUSS turned out to have strength in service flexibility, accessibility, evolvability, reusability, adaptability, interoperability and guideline provision for developing u-health services.

2   Characteristics of U-Health Services and System Design Goals

Understanding the characteristics of u-health services is an essential step to elicit meaningful design goals and core components for a u-health service system. In this section, we describe the characteristics of u-health services with the design goals, technologies, and core components adopted in designing and implementing THE-MUSS.

Like the services in other business sectors, most u-health services are process oriented as well. That is, a complete u-health service process is constructed by connecting and integrating small service units and associated data. Many u-health services can be constructed in the form of variations of a commonly sharable u-health service scenario. This strongly back ups the adoption of Business Process Management System (BPMS) as base system for a u-health service system[6].

Moreover, many units of a u-health service are reusable by other u-health services. In other words, u-health services often share many common features with each other at various levels such as service structure, unit service, and data levels. These u-health characteristics lead us to emphasize the reusability design goal. Since BPMS is one of typical approaches to accomplish reusability at various levels among the processes, THE-MUSS is built on WebVine BPMS[3].

Evolving, in other words, the quality and completeness of u-health services are improved as more user and user feed back data are accumulated and the better unit services become available. Thus a u-health service system should be able to support the evolving nature of u-health services. Ontology technology is mainly adopted to support evolving features of u-health services.

It is desirable if a u-health service can be developed not only by programmers but also by medical experts such as doctors and nurses. A u-health service system should provide a service development environment and a guideline for medical experts to develop own u-health services. THE-MUSS provides a u-health ontology incorporated u-health service design tool to cope with this requirement.

Meanwhile, the advantage of u-health service originates from the capability of obtaining user's bio data from mobile bio-sensors at anywhere at anytime. But the bio-sensors for u-health services are limited in terms of size and precision against off-line high cost biomedical sensors. As a result, less precise bio-signals are obtained more frequently from large number of users in u-health services. A u-health service system must compensate for such drawbacks of u-health services and take advantage of the advantages of u-health services. An adequate disease group prediction method for u-health services would be helpful in providing u-health services.

THE-MUSS provides a matrix-based disease group prediction method for the requirement in the above. The method is devised to deal with less precise but large amount of data with a feedback mechanism, and it accommodates incremental learning. Since newly obtained data is reflected to the method, it can be considered supporting evolving features of u-health services.

In summary, BPMS based u-health service platform, u-health ontology incorporated u-health service design tool, matrix based disease group identification framework, and mobile clients and u-health portal are the core components constituting THE-MUSS to make use of and support the characteristics of u-health services. The detailed description of each component is in the next section.

3 THE-MUSS

THE-MUSS is a total mobile u-health service system that provides developing and running environments of u-health services for users. The system integrates various components such as bio-sensors, cellular phones, associated software, and devices that are essential for u-health services. In this section, we describe the details of the core components that constitute THE-MUSS.

3.1 BPMS Based U-Health Service Platform

BPMS based u-health service platform is an extension of BPMS for u-health services. Shareable service units, modules, and functions among u-health services are placed on the platform so that they can be reused in other u-health services. In order to identify such sharable parts among u-health services, we need to understand the core activities for u-health services. In this section, we explain the core activities for u-health services and then illustrate the architecture of our u-health service platform.

3.1.1 Mobile U-Health Service Scenario

Fig. 1 illustrates the generic service scenario for mobile u-health services. In the first step of the service scenario, users fill out questionnaires to provide information about physical symptoms and their environments, which cannot be obtained from bio-sensors. The sensors embedded in cellular phones capture necessary bio-signal and relay the data to the u-health server.

After gathering bio-data from questionnaires and sensors, a mobile u-health service process is initiated by an associated BPM engine. The next activity of the scenario process is to store the relayed data to a database so that the history of bio-data of a person can be persistently kept for further analyses.

In the data analysis and decision support steps, the data ontology manager, which keeps semantically structured data for a specific set of diseases, plays a key role in identifying potential diseases based on the symptoms and bio-signals. For each symptom and bio-signal, the data ontology manager assigns a weight value that represents the degree of association between the symptom or bio-signal and a certain disease.

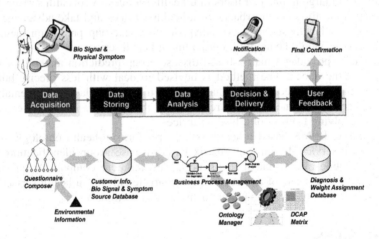

Fig. 1. Mobile u-health service scenario

The final diagnosis decision is made by some mechanisms or methods for the service. If there is an general disease group prediction method for u-health services would be helpful for this step. THE-MUSS is equipped with a unique prediction method.

3.1.2 Bio-data Gathering and Management

A mobile u-health service starts its function with periodically or randomly gathering input data i.e. capturing the bio-signal data from users. We use some bio-sensors that are wearable by the users or imbedded into cellular phones. Thus, the bio-sensor devices and cellular phones, which act as a gateway between the bio-sensors and the u-health server, are the essential components for gathering input data.

Besides, questionnaires that can be provided via cellular phones are necessary to obtain information that cannot be gathered from bio-sensors. The physical symptoms that a user experiences, the location of the user, and the weather information are examples of such information that need to be obtained directly from the users by using the questionnaires. We have developed a generic questionnaire composer to accommodate various symptoms and environmental information in making questionnaires.

Our mobile u-health service platform provides the sensing modules and questionnaire interfaces independently from the core functionality of u-health services. This improves the reusability of the sensing modules and questionnaires for different healthcare services.

Since huge amount of bio-data need to be gathered and analyzed in mobile u-health services, it is essential to have an efficient data structure that allows the system to effectively store and manage bio-data. We found that a matrix is one of the good candidates for storing and analyzing a large quantity of data. We devised a bio-signal and symptom combination matrix, in which the appearances of bio-signal and symptom combination pairs of a normal user group and those of a patient user group are registered. We call the matrix as the Disease Combination Appearance Probability (DCAP) matrix. Two DCAP matrices are created, one for a normal group, and another for a patient group. The accumulated data in DCAP matrices becomes the basis of the next step of the u-health service process, the knowledge extraction and decision support.

3.1.3 Knowledge Extraction and Decision Support

Once a large amount of bio-data is accumulated, knowledge extraction and decision support for diagnosis become possible through data mining techniques. Sometimes, we can use well established health indices to diagnose certain diseases for u-health services. However, in many cases, new health indices may need to be developed through machine learning or pattern recognition on the accumulated data.

For the accumulation and management of bio-data and information obtained from questionnaires, we have developed data ontology. Based on this data ontology, symptom, disease and bio-signal data can be archived in a structured manner. Some additional information such as a weight value to represent the degree of association between each input data and a certain disease type is assigned to the data ontology. Note that the weight values are defined not by users but by domain experts. The data ontology grows as new services and/or disease information are incorporated, and this contributes to make our platform evolvable.

One assumption that we have made in identifying health indices is that diseases can be diagnosed based on bio-signal and symptom pair information. Another assumption is that bio-signal and symptom data of normal and patient groups can be obtained in some way. If the difference between the data gathered from the normal group and the patient group is obvious, the interrelation patterns between bio-signals and symptoms provide good criteria to classify users into two groups. A probabilistic equation is devised based on the DCAP matrices to discern a normal group from a patient group. The details are in section 3.2.

3.1.4 Platform Architecture

The enactment service for u-health service processes is performed by the service platform. The mobile u-health service platform is a middleware that enables the integration of diverse services on BPMS. It serves as a hub to integrate techniques and functions circumventing mobile u-health services, and provides an environment to develop and run the u-health services. Fig. 2 shows the architecture of platform with the major components and the connections to the surrounding elements in the u-health service framework.

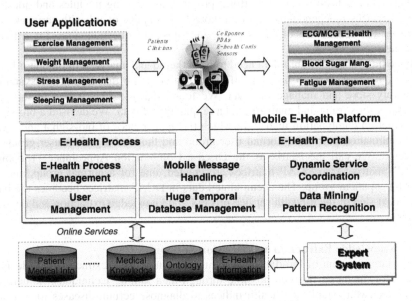

Fig. 2. Mobile u-health service platform architecture

Since cellular phones play the role of a gateway between bio-sensors and servers, mobile message handling is essential for the platform. The mobile message handling module relays all the messages from bio-sensors to the server. Not only the bio signal data but also service request messages are delivered by the message handling module. Sometimes, it contributes to filter out some noise signals from the received messages.

The bio-data delivered to a server is stored and managed by the huge temporal data management module in secure manner. The bio-data is stored in diverse forms so that various services can utilize the data to perform their functions.

Data mining and pattern recognition techniques are used to identify health index from the accumulated bio-data. Also, an external or internal expert system may refer to the data as feedback information. In order to support this, the database schema of the temporal database needs to be designed to satisfy the requirements of data mining or pattern recognition techniques and expert systems.

As explained in Section 3.1, in order to develop a mobile u-health service on the service platform, the u-health service should be defined in a form of process containing steps like bio-data gathering, storing, analysis, and result reporting. The u-health process block in Fig. 2 denotes a set of mobile u-health processes derivable from the u-health service scenario process explained in Section 3.1.

The u-health process definition tool enables developers to easily define new mobile u-health services. Mobile u-health services defined in a form of process are controlled and enacted by the u-health process management module. The u-health process management module provides not only the enactment function but also monitoring and administration functions for u-health processes.

The mobile u-health management module and u-health process definition tool play a key role in making the platform evolvable. When a new mobile u-health service process is defined, reusable process templates, steps, and data structures are identified and registered to the server for later use. The management module ties together a set of u-health services into a group by specifying execution dependencies and the data-flow between the services. There may be also some constraints that prevent a particular service to be initiated until some other services finish their functions.

The user management module is essential for providing personalized service to individual users. The user management module stores user profile information such as age, gender, and occupation. Since a user is a participant in a mobile u-health process as well, this module is closely connected to the participant information in the u-health process management module.

The dynamic service coordination module ensures reliable u-health service execution by replacing one service with an alternative one when a fault occurs in the service or the user's requirements are changed. Data mining and pattern recognition modules may need to be developed for specific u-health services. Whenever such modules become available, they are placed on the data mining/pattern recognition module. DCAP based disease group prediction system can be one of such modules. An expert system engaged with a u-health service may need to access the functions in the data mining/pattern recognition module to get a decision making support.

3.2 DCAP Matrix Based Disease Group Prediction Method

In this section, we describe DCAP matrix based disease group prediction method which is one of unique features supported only in THE-MUSS. Since the method includes feedback mechanism and the newly obtained user data is reflected to the method, it makes our system evolvable in terms of prediction.

One assumption that we made in identifying health indices is that diseases can be diagnosed based on the combination of bio-signals and symptom information. Another assumption is that bio-signals and symptom data of normal and patient groups can be obtained in some way. If the difference between the data gathered from the normal group and the patient group is obvious, the interrelation patterns between bio-signals and symptoms provide good criteria to classify users into the two groups.

3.2.1 The Big Picture of DCAP

In the learning stage, members of each group are characterized in terms of bio-signals and symptoms. The characterized results of a normal and patient group are summarized and stored in DCAP and DCAP matrices respectively. Then the two matrices are merged into a DCAP matrix.

A probabilistic equation is developed based on the DCAP matrix to distinguish a normal group from a patient group. Once a DCAP matrix and weight vectors are prepared, we can predict if a user with specific bio-signals and symptoms belongs to a disease group or not, using (Eq. 1). In (Eq. 1), CP(p) denotes bio-signal and symptom pairs of a person p, and X(p), Y(p) denote bio-signal and symptom weight vectors of a person p, respectively. We leave the details of (Eq. 1) to our previous work and we do not delve into the details in this paper[4].

$$P(p \in \text{Patient Group} \mid CP(p)) =$$
$$\sum_{i=1}^{m} [\frac{X(p)_i}{\sum_{j=1}^{m} X(p)_j} \cdot (\sum_{u=1}^{n} \frac{Y(p)_u \cdot DCAP(p)_{i,u}}{\sum_{v=1}^{n} Y(p)_v})] \qquad \text{(Eq. 1)}$$

The obtained prediction results by applying (Eq. 1) are delivered to the corresponding user, then the user confirms the results and gives feedback information to the system. The feedback information from the users are accumulated and reflected to the system. The feedback information is handled in two different ways: private feedback and group feedback. The private feedback is reflected by changing weigh vectors of a corresponding user and the group feedback is reflected to DCAP matrix. Fig. 3 shows the big picture of the method.

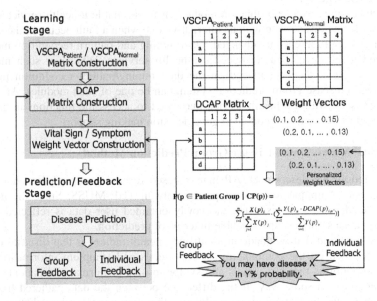

Fig. 3. The big picture of matrix-based disease group prediction method

3.2.2 Rationales of Using DCAP

DCAP based prediction method is general in the sense that it can be used for any disease as long as the disease group can be distinguished by bio-signals and symptoms. The types of bio-signals and symptoms are decided in flexible manner by medical experts according to the types of diseases. They can easily add or eliminate new bio-signals or symptoms if necessary.

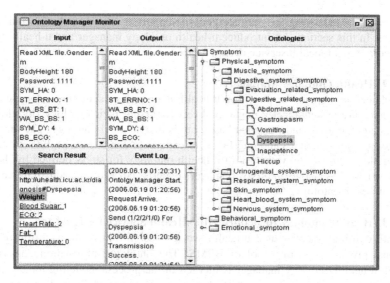

Fig. 4. Ontology manager interface

Besides, DCAP based prediction method has several benefits against conventional machine learning techniques. First, unlike conventional machine learning methods, incremental learning is possible in the method. That is, the new data obtained while running the prediction system is easily reflected back to the system by just changing the values of DCAP matrix. Continuously generated data is the source for improving prediction accuracies in u-health environment, and incremental learning is an appropriate approach for this situation.

Second, DCAP matrix based prediction method is good to analyze the prediction results by investigating decisive elements of DCAP matrix, i.e. the key factors in producing a certain prediction result. Note that, the prediction system for u-health services needs to explain the prediction results in detail to the users as long as possible.

Third, since matrix is a relatively simple data structure, it is intuitively understood by medical experts and easy to handle in the prediction system. Since medical experts may take a look on the matrix to get insights and make improvements on the method for the disease they are dealing with, the matrix is an adequate data structure for this situation.

Lastly, our DCAP matrix based probabilistic approach is not vulnerable to the erroneous learning data. The influences of error data for training are limited in our prediction method. When we consider that we usually deal with a large amount of erroneous data in u-health service environment, error resiliency of DCAP based prediction method is a desirable feature for u-health services.

3.3 U-Health Ontology Incorporated U-Health Service Design Tool

Eventually, u-health service should be designed by the medical experts without the help of programmers. Conventional process modelling tool for BPMS is not sufficient to meet such requirements. THE-MUSS provides u-health ontology incorporated u-health service design tool for this. Although the tool is not complete yet, it is the first step toward for achieving the ultimate goal of supporting experts in health and medical areas. In this section, we introduce u-health ontology manager and u-health ontology incorporated u-health service design tool.

3.3.1 U-Health Ontology Manager

As explained earlier, in our platform, u-health data such as symptoms, diseases, and bio-signals, are systematically managed by constructing their ontology. In addition, services are classified and managed based on their functional ontology. The ontology manager provides modules and interfaces for experts to construct u-health service and data ontology. Fig. 4 shows a case of u-health ontology for stress defined by the ontology manager. On the ontology manager, experts can browse, add, remove, and search u-health ontology.

3.3.2 U-Health Ontology Incorporated U-Health Service Design Tool

U-health ontology incorporated u-health service design tool is an extension of conventional process modeling tool of BPMS. The ontology information constructed through u-health ontology manager is merged in the process modeling tool with service recommendation mechanisms. With service and data ontology, the service recommendation mechanism becomes the basis for semi-automatic u-health service design for medical experts. For the complete support for medical experts to develop their own services without the help of programmers, we need to define a complete and systematic u-health service development process. Mobile u-health service scenario shown in Figure 1 is the starting point for the definition. But we do not delve into details in this paper.

THE-MUSS provides a u-health ontology incorporated u-health service design tool on which u-health service developers semi-automatically compose mobile u-health services. In the process of designing a service, the tool recommends a set of candidate services that can be used for each step of the service. The recommendation is performed by searching connectable services with the unit services that are already used for the service. If necessary, it also automatically inserts adapters and converters when it finds some syntactic and semantic mismatches between the connected service units in the service.

4   U-Health Services on THE-MUSS

Two mobile u-health services are developed on THE-MUSS to confirm and evaluate the usefulness of THE-MUSS. One is a mobile stress management service and the other one is a mobile weight management service. In this section, we focus on the mobile stress management service due to space limitations.

Stress management is known as one of most complex u-health services in terms of diagnosis and service process[9]. This means that if the development process of stress management service on THE-MUSS is lubricant, other u-health services can be developed on THE-MUSS relatively easily. In this section, we introduce DCAP matrix for stress service and extend the service scenario in Figure 1 for stress service.

4.1 DCAP Matrix for Stress Service

The construction of DCAP matrix is a one of core activities for a mobile stress management service. Prior to the construction of DCAP matrix for stress service, types of bio-signal and symptoms should be decided. Heart Rate Variability (HRV), blood sugar level, body temperature, body fat ratio are used for bio-signals, and symptoms are decided based on the attainable information from the result of Stress Response Inventory (SRI) questionnaire.

In order to fill up the values of DCAP matrix, bio-signal and symptom data are collected from 360 subjects at Seoul Paik Hospital in Korea as training data. Among the subjects, 51 subjects are classified into a stress group and the rest 318 subjects are classified into a normal group according to a Stress Response Inventory (SRI) value [7].

THE-MUSS provides functions and modules for constructing DCAP matrix from training files. Thus, once the training data is prepared, the construction of DCAP matrix is rather straightforward. Fig. 5 shows the constructed DCAP matrix (Fig. 5(b)) and distributions of normal and disease groups (Fig. 5(a)).

Once DCAP matrix is constructed, we can compute the probability of a subject with a certain bio-signal and symptom data belonging to a particular group using (Eq. 1). The corresponding elements of DCAP matrix for the subject in computing the probability is marked in DCAP (Fig. 5(b)). From this we can figure out which bio-signals and symptoms have affected how much degree to the prediction results. So the prediction system can give the information on the primary factors and associated explanations of the results to the users.

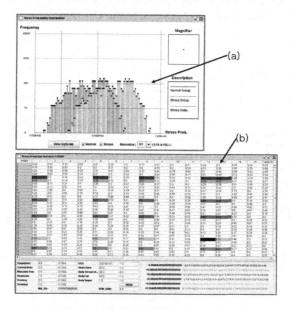

Fig. 5. Disease combination appearance probability matrix – (a) Distributions of disease and normal group; (b) Test result for a user health condition by the DCAP matrix

In the future, the more data will be accumulated from the users by running the system and then the data can be used for training data. Since DCAP based prediction system supports incremental learning, the obtained data can be reflected to the prediction system immediately. This will improve the prediction accuracy of the system.

4.2 Stress Service Scenario and Mobile User Interfaces

The service scenario in Fig. 1 is not sufficient to describe a particular u-health service. Necessary service steps for a mobile phone and a portal are missing in the scenario. We have revised and refined the service scenario in Fig. 1 for the stress service. Fig. 6 shows the service scenario for the mobile stress service.

Although the details are not specified in the scenario, we highlight the feedback mechanism for the stress service. Users can respond to the prediction results either positively or negatively. When the users' responses are negative, the service system provides means to adjust the results according to the user feedbacks. For example, the change of prediction results are illustrated according to the change of personal weigh vectors of symptoms so that a user can decide and keep own weigh vectors fit to oneself.

The mobile weight management service can be developed in a similar manner to the stress service except the diagnosis part. Baysian network [5] is used instead for diagnosis because the weight management is understood better than the stress service. Since the service structures of stress and weight managements are similar with each other, only small parts of the service are added and modified to the stress service. Moreover, most parts of the stress service such as capturing, delivering, processing bio-signals are reused in weight management service because bio-signals for weight management service are part of bio-signals for stress service.

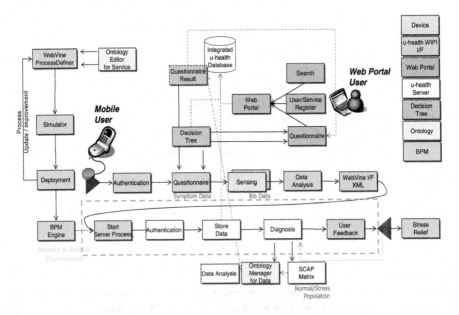

Fig. 6. Mobile stress management service: Service scenario

5 Conclusions

In this paper, we identified several core components of the u-health service system, and described the overall architecture of a mobile u-health service system, THE-MUSS. We also developed essential functions and facilities for service developers to effectively use the core components in developing various mobile u-health services with less effort.

THE-MUSS is unique in the sense that it adopts BPMS as its underlying platform, and provides DCAP matrix based prediction framework and ontology incorporated u-health service design tool, etc. Besides, the standard u-health service scenario provides a guideline to u-health service developers. Users can modify and/or extend the service scenario to develop their own u-health services. Six design goals are identified as quality attributes for the mobile u-health service system, and we have tried to achieve them in THE-MUSS.

We confirmed that THE-MUSS is useful in developing mobile u-health services by developing mobile stress and weigh management services on THE-MUSS. The more services are developed on THE-MUSS, the better services THE-MUSS can provide.

There are some pending works to make THE-MUSS more secure and reliable. We are currently pursuing supporting secure personal healthcare data management and enhancing the performance by compressing XML messages exchanged between mobile devices and the u-health server.

References

1. Han, D., Ko, I., Park, S.: An Evolving Mobile e-Health Service Platform. In: Proceedings of IEEE International Conference on Consumer Electronics – ICCE 2007, USA, pp. 337–338 (January 2007)
2. Han, D., Ko, I., Park, S., Lee, M., Jung, S.: A BPM-Based Mobile u-Health Service Framework. In: Proceedings of International Conference on Health Informatics, Portugal, vol. 1, pp. 110–117 (January 2008)
3. Han, D., Song, S., Koo, J.: WebVine suite: A web services based BPMS. In: Zhou, X., Li, J., Shen, H.T., Kitsuregawa, M., Zhang, Y. (eds.) APWeb 2006. LNCS, vol. 3841, pp. 1185–1188. Springer, Heidelberg (2006)
4. Han, D., Song, J., Matai, J., Lee, M.: A Probability-Based Prediction Framework for Stress Identification. In: 2007 9th International Conference on e-Health Networking, Application and Services – Healthcom 2007, Taipei, Taiwan, pp. 58–63 (June 2007)
5. Jensen, F.V.: Introduction to Bayesian Networks, 1st edn. Springer, New York (1996)
6. Smith, H., Fingar, P.: Business Process Management: The Third Ware. Meghan-Kiffer Press (2003)
7. Koh, K.B., Park, J.K., Kim, C.H., Cho, S.H.: Development of the Stress Response Inventory and Its Application in Clinical Practice. Psychosomatic Medicine 63, 668–678 (2001)
8. Wiegers, K.E.: Software Requirements, 2nd edn. Microsoft Press (2003)
9. Lehrer, P.M., Woolfolk, R.L., Sime, W.E., Barlow, D.: Principles and Practice of Stress Management, 3rd edn. Guilford Publications, New York (2007)
10. Quero, J.M., Tarrida, C.L., Santana, J.J., Ermolov, V., Jantunen, I., Laine, H., Eichholz, J.: Health Care Applications Based on Mobile Phone Centric Smart Sensor Network. In: Proceedings of 29th International Conference on Engineering in Medicine and Biology Society, pp. 6298–6301 (August 2007)
11. Istepanian, R.S.H., Laxminarayan, S., Pattichis, C.S.: M-Health: Emerging Mobile Health Systems. Springer Science+Business Media (2006)

Scalable Medical Image Understanding by Fusing Cross-Modal Object Recognition with Formal Domain Semantics

Manuel Möller[1], Michael Sintek[1], Paul Buitelaar[2], Saikat Mukherjee[3],
Xiang Sean Zhou[4], and Jörg Freund[5]

[1] DFKI GmbH, Knowledge Management Dept., Kaiserslautern, Germany
`manuel.moeller@dfki.de, michael.sintek@dfki.de`
[2] DFKI GmbH, Language Technology Lab, Saarbrücken, Germany
`paul.buitelaar@dfki.de`
[3] Siemens Corporate Research, 755 College Road East, Princeton NJ, U.S.A.
`saikat.mukherjee@siemens.com`
[4] Siemens Medical Solutions, 51 Valley Stream Parkway, Malvern PA, U.S.A.
`xiang.zhou@siemens.com`
[5] Siemens Medical Solutions, Hartmannstr. 16, 91052 Erlangen, Germany
`joerg.freund@siemens.com`

Abstract. Recent advances in medical imaging technology have dramatically increased the amount of clinical image data. In contrast, techniques for efficiently exploiting the rich semantic information in medical images have evolved much slower. Despite the research outcomes in image understanding, current image databases are still indexed by manually assigned subjective keywords instead of the semantics of the images. Indeed, most current content-based image search applications index image features that do not generalize well and use inflexible queries. This slow progress is due to the lack of scalable and generic information representation systems which can abstract over the high dimensional nature of medical images as well as semantically model the results of object recognition techniques. We propose a system combining medical imaging information with ontological formalized semantic knowledge that provides a basis for building universal knowledge repositories and gives clinicians fully cross-lingual and cross-modal access to biomedical information.

1 Introduction

Rapid advances in imaging technology have dramatically increased the amount of medical image data generated daily by hospitals, pharmaceutical companies, and academic medical research[1]. Technologies such as 4D 64-slice CT, whole-body MR, 4D Ultrasound, and PET/CT can provide incredible detail and a wealth of information with respect to the human body anatomy, function and disease associations. This increase in the volume of data has brought about significant advances in techniques for analyzing such data. The precision and sophistication of different image understanding

---

[1] For example, University Hospital of Erlangen, Germany, has a total of about 50 TB of medical images. Currently they have approx. 150.000 medical examinations producing 13 TB per year.

A. Fred, J. Filipe, and H. Gamboa (Eds.): BIOSTEC 2008, CCIS 25, pp. 390–401, 2008.

techniques, such as object recognition and image segmentation, have also improved to cope with the increasing complexity of the data.

However, these improvements in analysis have not resulted in more flexible or generic image understanding techniques. Instead, the analysis techniques are very object specific and not generic enough to be applied across different applications. Throughout this paper we will address this fact as "lack of scalability". Consequently, current image search techniques, whether for Web sources or for medical PACS (Picture Archiving and Communications System), are still dependent on the subjective association of keywords to images for retrieval.

One of the important reasons behind this lack of scalability in image understanding techniques has been the absence of *generic* information representation structures capable of overcoming the feature-space complexity of image data. Indeed, most current content-based image search applications are focused on indexing syntactic image features that do not generalize well across domains. As a result, current image search technology does not operate at the *semantic* level and, hence, is not scalable.

We propose to use hierarchical information representation structures, which integrate state-of-the-art object recognition algorithms with generic domain semantics, for a more scalable approach to image understanding. Such a system will be able to provide direct and seamless access to the informational content of image databases.

Our approach is based on the following main techniques:

- Integrate the state-of-the-art in semantics and image understanding to build a sound bridge between the symbolic and the subsymbolic world. This cross-layer research approach defines our road-map to quasi-generic image search.
- Integrate higher level knowledge represented by formal ontologies that will help explain different semantic views on the same medical image: structure, function, and disease. These different semantic views will be coupled to a backbone ontology of the human body.
- Exploit the intrinsic constraints of the medical imaging domain to define a rich set of queries for concepts in the human body ontology. The ontology not only provides a natural abstraction over these queries but also statistical image algorithms could be associated to semantic concepts for answering these queries.

Our focus is on filling the gap between what is current practice in image search (*i. e.*, indexing by keywords) and the needs of modern health provision and research. The overall goal is to empower the medical imaging content-stakeholders (clinicians, pharmaceutical specialists, patients, citizens, and policy makers) by providing flexible and scalable semantic access to medical image databases. Our short term goal is to develop a basic image search engine and prove its functionality in various medical applications.

In 2001 Berners-Lee and others published a visionary article on the Semantic Web [1]. The use-case they described was about the use of meta-knowledge by computers. For our goals we propose to build a system on existing Semantic Web technologies like RDF [2] and OWL [3] which were developed to lay the foundations of Berners-Lee's vision. From this point of view it is also a Semantic Web project.

Therefore we propose a system that combines medical imaging information with semantic background knowledge from formalized ontologies and provides a basis for building universal knowledge repositories, giving clinicians *cross-modal* (independent from different modalities like PET, CT, ultrasound) as well as *cross-lingual* (independent of particular languages like English and German) access to various forms of biomedical information.

2   General Idea

There are numerous advanced object recognition algorithms for the detection of particular objects on medical images: [4] at anatomical level, [5] at disease level and [6] at functional level. Their specificity is also their limitation: Existing object recognition algorithms are not at all generic. Given an arbitrary image it still needs human intelligence to select the right object recognizers to apply to an image. Aiming to gain a pseudo-general object recognition one can try to apply the whole spectrum of available object recognition algorithms. But it turns out that in generic scenarios even with state-of-the-art object recognition tools the accuracy is below 50 percent [7, 8].

In automatic image understanding there is a semantic gap between low-level image features and techniques for complex pattern recognition. Existing work aims to bridge this gap by ad-hoc and application specific knowledge. In contrast our objective is to create a formal fusion of semantic knowledge and image understanding to bridge this gap to support more flexible and scalable queries.

For instance, human anatomical knowledge tells us that it is almost impossible to find a heart valve next to a knee joint. Only in cases of very severe injuries these two objects might be found next to each other. But in most cases the anatomical intuition is correct and, hence, the background knowledge precludes the recognition of certain anatomical parts given the presence of other parts. It is in this use of formalized knowledge that ontologies[2] come into play within our framework.

In the context of medical imaging it is necessary to define image semantics for parts of human anatomy. In this domain the expert's knowledge is already formalized in comprehensive ontologies like the *Foundational Model of Anatomy* [10] for human anatomy or the *International Statistical Classification of Diseases and Related Health Problems* (ICD-10)[3] of World Health Organization for a classification of human diseases. These ontologies represent a rich medical knowledge in a standardized and machine interpretable format.

In contrast to current work which defines ad-hoc semantics, we take the novel view that within a constrained domain the semantics of a concept is defined by the queries associated with it. We will investigate which types of queries are asked by medical experts to ensure that the necessary concepts are integrated into the knowledge base. We believe that in IR applications this view will allow a number of advances which will be described in the following sections.

We chose the medical domain as our area of application. Unlike common language and many other scientific areas the medical domain has clear definitions for its technical terms. Ambiguities are rare which eases the task of finding a semantic abstraction for a

---

[2] According to Gruber [9], an ontology is a specification of a (shared) conceptualization.
[3] http://www.who.int/classifications/icd/en/

particular text or image. However, our framework is generic and can be applied to other domains with well-defined semantics.

3 Aspects of Using Ontologies

Ontologies (usually) define the semantics for a *set of objects* in the world using a *set of classes*, each of which may be identified by a particular symbol (either linguistic, as image, or otherwise). In this way, ontologies cover all three sides of the "semiotic triangle" that includes *object*, *referent*, and *sign* (see Fig. 1). *I. e.*, an *object* in the world is defined by its *referent* and represented by a *symbol* (Ogden and Richards, 1923—based on Peirce, de Saussure and others).

Currently, ontology development and the Semantic Web effort in general have been mostly directed at the *referent* side of the triangle, and much less at the *symbol* side. To allow for automatic multilingual and multimedia knowledge markup a richer representation is needed of the linguistic and image-based symbols for the object classes that are defined by the ontology [11].

From our point of view a semantic representation should not be encapsulated into a single module. Instead we think that a layered approach as shown in Fig. 2 has a number of advantages.

Once there is a representation established at the semantic level there are a number of benefits compared to conventional IR systems. For a more detailed description of the abstraction process see Sect. 4.

Cross-Modal Image Retrieval. Current systems for medical IR depend on the modality of the stored images. But in medical diagnosis very different imaging techniques

Fig. 1. Semiotic Triangle

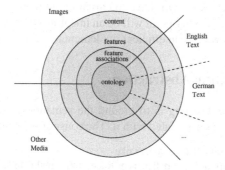

Fig. 2. Interacting Layers in Feature Extraction and Representation

are used such as PET, CT, ultrasonography, or time series data from 4D CT *etc.* Each technique produces images with characteristic appearance. For tumor detection, for example, often PET (to identify the tumor) and CT (to have a view on the anatomy) are combined, to formulate a precise diagnosis with a proper localization of the tumor. The proposed system will allow to answer queries based on semantic similarity and not only visual similarity.

Full Modality Independence. Cross-Modality even can be driven another step forward by integrating documents of any format into one single database. We plan to also include text documents like medical reports and diagnoses. On the level of semantic representation they will be treated like the images. Accordingly, the system will be able to answer queries not only with images but also with text documents including similar concepts as in the retrieved images.

Improved Relevance of Results. Current search engines retrieve documents which contain the keyword from the query. The documents in the result set are ranked by various techniques using information such as their inter-connectivity, statistical language models, or the like. For huge datasets search by keyword often returns very large result sets. Ranking by relevance is hard.

This holds for low-level image retrieval as well. Here only two similarity measures are applicable: through visual similarity which can be completely independent from the object and context and via a comparison between keywords and image annotations. With current IR systems the user is forced to use pure keyword-based search as a detour while in fact he or she is searching for documents and/or images including certain concepts.

Our notion of keyword-based querying goes beyond current search engines by mapping keywords from the query to ontological concepts. Our system provides the user with a semantic representation. That allows the user to ask for a concept or a set of concepts with particular relationships. This allows far more precise queries than a simple keyword-based retrieval mechanism and likewise better matching between query and result set.

Inferencing of Hidden Knowledge. By mapping the keywords from a text-based query to ontological concepts and the use of semantics the system is able to infer[4] implicit results. This allows us to retrieve images which are not annotated explicitly with the query concepts but with concepts related to them through the ontology.

To represent the complex knowledge of the medical domain and allow a maximum of flexibility in the queries we will have to enrich the ontology by rules and allow to use rules in the queries. Another point will be an integration of spatial representation of anatomical relations as well as an efficient implementation of spatial reasoning.

4   Levels of Semantic Abstraction

Our notion of semantic imaging is to ground the semantics of a human anatomical concept on a set of queries associated with it. The constrained domain of a human body

---

[4] We aim at using standard OWL reasoners like
http://www.sts.tuharburg.de/ r.f.moeller/racer/Racer, http://owl.man.ac.uk/factplusplus/ FaCT++ or http://pellet.owldl.com/Pellet.

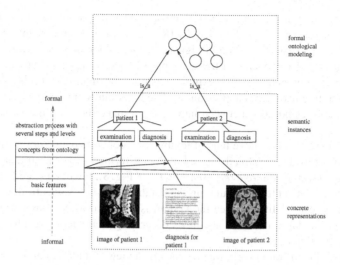

Fig. 3. Abstraction Model

enables us to have a rich coverage of these queries and, consequently, define image semantics at various levels of the hierarchy of the human anatomy.

Fig. 3 gives an overview of the different abstraction levels in the intended system.

For the proposed system we want to take a step beyond the simple dichotomy between a symbolic and subsymbolic representation of images. Instead, from our perspective there is a spectrum ranging from regarding the images as simple bit vectors over color histograms, shapes and objects to a fully semantic description of what is depicted. The most formal and generic level of representation is in form of an ontology (*formal ontological modeling*). The ontology holds information about the general structure of things. Concrete entities are to be represented as *semantic instances* according to the schema formalized in the ontology.

To emphasize the difference to the dichotomic view, we call the lower end of this spectrum *informal* and the upper end *formal* representation. From our perspective the abstraction has to be modeled as a multi-step process across several sublevels of abstraction. This makes it easier to close the gap between the symbolic and subsymbolic levels from the classic perspective. Depending on the similarity measure that is to be applied for a concrete task different levels of the abstraction process will be accessed.

If a medical expert searches for images that are *looking* similar to the one he or she recently got from an examination, the system will use low-level features like histograms or the bit vector representation. In another case the expert might search for information about a particular syndrome. In that case the system will use features from higher abstraction levels like the semantic description of images and texts in the database to be able to return results from completely different modalities.

We believe that text documents have to be understood in a similar way. Per se, a text document is just a string of characters. This is similar to regarding an image as a sheer bit vector. Starting from the string of characters, in a first step relatively simple methods can be used to identify terms. In further steps technologies from concept based Cross

Language Information Retrieval (CLIR)[5] are applied to map terms in the documents to concepts in the ontology [12,13,14,15]. CLIR currently can be divided into three different methods: approaches based on bilingual dictionary lookup or Machine Translation (MT); corpus based approaches utilizing a range of IR-specific statistical measures; and concept-driven approaches, which exploit semantic information to bridge the gap between surface linguistic form and meaning. The latter allows not only to cross borders between modalities but also between languages to make the system *cross-modal* and *cross-lingual*.

Modern hospitals often have tens of terabytes of concrete data from medical cases. In many cases this data is already stored in digital format. In our abstraction model it belongs to the level of *concrete representations* (*cf.* Fig. 3). It will be stored in a conventional database. The entities populating the database can be images from all available medical imaging modalities. Since the system is designed to be completely independent from the modality it can be also texts describing examination results, diagnoses, medical publications, *etc.*

Recall that, in our framework, we take the novel view that the semantics of a concept is defined by the queries associated with it. These queries can either be defined by domain experts, such as physicians, radiologists, nurses, or they can be mined from text corpora. Medical knowledge repositories, such as clinical books, journals, *etc.* , contain information on image-centric questions on different body parts which are of interest to physicians. For instance, typical image queries of the heart could be about detection of the ventricles, *etc.* Similar information also exists in physician reports, laboratory notes, *etc.*

We will make use of methods for extracting information from unstructured text for automatically discovering these queries. There is a significant research on information extraction [16] from natural language text as well as learning patterns of free-text questions from examples [17, 18]. We will apply these techniques for identifying relevant questions and their answers in natural language text. This will help us in collecting a broad coverage of possible questions associated with concepts.

5   Supported Query Scenarios

Iterative Retrieval Process. In most present systems the retrieval is started by a query sent by the user which is subsequently answered by a result set. We think that the retrieval should be merely understood as an iterative process. The user starts the process of information retrieval by submitting a query. In many cases this query will be either too general resulting in an result set which is too huge. The other extreme is a query which is too specific leading to an empty result set. To support the user with the navigation through the available information we aim to have a close interaction between user and system. Step by step the user can refine the search query using aid which is given proactively from the domain knowledge of the system.

We envision three primary ways in which users will query the semantic imaging platform. Users can query either through sample images, or pose structured queries using

---

[5] The research project *MUCHMORE* (http://muchmore.dfki.de/http://muchmore.dfki.de/) was focused on this aspect.

conceptual descriptors, or use natural language to describe queries. In the following, we explain each of these different methods.

Query by Sample Image. Basically there are two different approaches to image based queries. The *first approach* retrieves images from the database which are *looking* similar. Only low-level image features are used to select results for this type of query. The ability to match the image of a current patient to similar images from a database of former medical cases can be of great help in assisting the medical professional in his diagnosis (see Sect 4) we believe that image understanding has to be an abstraction over several levels. To answer queries by sample image we will make use of the more informal features extracted from the images. The support for these queries is based on state-of-the-art similarity-based image retrieval techniques [19].

Today there are various image modalities in modern medicine. Many diseases like cancer require to look at images from different modalities to formulate a reliable diagnosis (see example in Sect. 3). The *second approach* therefore takes the image from the query and extracts the semantics of what can be seen on the image. Through mapping the concrete image to concepts in the ontology, an abstract representation is generated. This representation can be used for a matching on the level of image semantics with other images in the database. Applying this method makes it possible to use a CT image of the brain to search for images from all available modalities in the database.

Query by Conceptual Descriptions. Similar to the use of SQL for querying structured relational databases, special purpose languages are also required for querying semantic metadata. Relying on well-established standards we propose using a language on top of RDF, such as SPARQL, for supporting generic structured semantic queries.

Query by Natural Language. From the point of the medical expert having a natural language interface is very important. Through a textual interface the user directly enters keywords which are mapped to ontology concepts. Current systems like the IRMA-Project (see Sect. 6) only allow to search for keywords which are extracted offline and stored as annotations. Since our system has to compose a semantic representation of each query, the ontological background knowledge can be used in an iterative process of query refinement. Additionally, it will be possible to use complete text documents as queries.

In cases where the system cannot generate a semantic representation—due to missing knowledge about a knew syndrome, therapy, drug or the like—it will fall back to a normal full text search. If the same keyword is used frequently this can be used as evidence that the foundational ontology has to be extended to cover the corresponding concept(s).

6   Related Work

Most current work in content based image retrieval models object recognition as a probabilistic inferencing problem and use various mathematical methods to cope with the problems of image understanding. These techniques use image features which are tied to particular applications and, hence, suffer from a lack of scalability.

Among extant work in fusing semantics with image understanding, [20] describes a technique for modeling the MPEG-7 standard, which is a set of structured schemas for describing multimedia content, as an ontology in RDFS. There has been some research [21, 22, 23, 24, 25, 26, 27] on semantic imaging relying primarily on associating word distributions to image features. However, these works used hierarchies of words for semantic interpretation and did not attempt to model image features themselves in levels of abstraction. Furthermore, the lack of formal modeling made these techniques susceptible to subjective interpretations of the word hierarchies and, hence, were not truly scalable. Especially in the context of medical imaging, our notion of semantics is tied to information gathered from physics, biology, anatomy, *etc.* This is in contrast to perception-based subjective semantics in these works.

The goal of the project "Image Retrieval in Medical Applications" (IRMA) [28] was an automated classification of radiographs based on global features with respect to imaging modality, direction, body region examined and the biological system under investigation. The aim was to identify image features that are relevant for medical diagnosis. These features were derived from a database of 10.000 a-priori classified and registered images. By means of prototypes in each category, identification of objects and their geometrical or temporal relationships are handled in the object and the knowledge layer, respectively. Compared to the system proposed in this paper the IRMA project only used a taxonomy without a formal representation of anatomical, functional, or pathological relations. Therefore it lacks any inference of implicit statements in the query as well as using the background knowledge for a relevance test during the automatic annotation of new images and to score results.

A number of research publications in the area of ontology-based image retrieval emphasize the necessity to fuse sub-symbolic object recognition and abstract domain knowledge. [29] proposes an integration of spatial context and semantic concepts into the feature extraction and retrieval process using a relevance feedback procedure. [30] combine sub-symbolic machine learning algorithms with spatial information from a domain ontology. [31] present a system that aims at applying a knowledge-based approach to interpret X-ray images of bones and to identify the fractured regions. [32] present a hybrid method which combines symbolic and sub-symbolic techniques for the annotation of brain Magnetic Resonance images. While it focuses only on one modality and body region, their approach shares the use of OWL DL [3], SWRL rules [33] and DL reasoning with our proposal. The BOEMIE EU project[6] also focuses on knowledge acquisition independent from modalities. But while we obtain the formal medical domain knowledge from existing large-scale ontologies they use a bootstrapping approach to evolve ontologies to also cover missing concepts [34].

7   Conclusions and Future Work

In this paper we proposed a close integration of subsymbolic pattern recognition algorithms and semantic domain knowledge represented in formal ontologies. The vision is to combine the techniques from both fields to bridge the gap between a symbolic and subsymbolic world for a generic understanding of medical images and text. We take the

---

[6] http://www.boemie.org/

novel view that within a constrained domain the semantics of a concept, as described in a physics-based ontology of human anatomy, is defined by typical queries associated with it. Thus, our framework is different from research which fuses image understanding with subjective semantics.

The use of formal ontologies, and their reasoning capabilities, forms the essence behind better information retrieval. By abstracting from the syntactic content representation, it is possible to perform semantic matching between queries and the content. Additionally, the user is provided with an extremely flexible interface which allows cross-modal as well as cross-lingual queries. By matching at the level of semantic concepts, abstracting from syntactic representations where necessary, and using low-level features where necessary, our framework enables scalable querying on images and text across different anatomical concepts.

Extensive research has been done on the extraction of semantics from text documents. Therefore this component of our system can rely on an existing state-of-the-art. The next research task in implementing the proposed system will be the integration of formalized background knowledge with low-level object recognition algorithms. The section Related Work shows that this is currently an area of intensive work. All existing approaches lack the generality which we aim at for the proposed system. Therefore developing a truly generic and scalable integration will be our next step.

Acknowledgements. This research has been supported in part by the THESEUS Program in the MEDICO Project, which is funded by the German Federal Ministry of Economics and Technology under the grant number 01MQ07016. The responsibility for this publication lies with the authors.

References

1. Berners-Lee, T., Hendler, J., Lassila, O.: The semantic web – a new form of web content that is meaningful to computers will unleash a revolution of new possibilities. Scientific American (2001)
2. Brickley, D., Guha, R.: RDF vocabulary description language 1.0: RDF Schema (2004)
3. McGuinness, D.L., van Harmelen, F.: OWL web ontology language overview (2004)
4. Hong, W., Georgescu, B., Zhou, X.S., Krishnan, S., Ma, Y., Comaniciu, D.: Database-guided simultaneous multi-slice 3D segmentation for volumetric data. In: Leonardis, A., Bischof, H., Pinz, A. (eds.) ECCV 2006. LNCS, vol. 3954, pp. 397–409. Springer, Heidelberg (2006)
5. Tu, Z., Zhou, X.S., Bogoni, L., Barbu, A., Comaniciu, D.: Probabilistic 3D polyp detection in CT images: The role of sample alignment. IEEE Computer Society Conference on Computer Vision and Pattern Recognition 2, 1544–1551 (2006)
6. Comaniciu, D., Zhou, X., Krishnan, S.: Robust real-time myocardial border tracking for echocardiography: an information fusion approach. IEEE Transactions in Medical Imaging 23(7), 849–860 (2004)
7. Chan, A.B., Moreno, P.J., Vasconcelos, N.: Using statistics to search and annotate pictures: an evaluation of semantic image annotation and retrieval on large databases. In: Proceedings of Joint Statistical Meetings (JSM) (2006)
8. Müller, H., Deselaers, T., Lehmann, T., Clough, P., Kim, E., Hersh, W.: Overview of the ImageCLEFmed 2006 medical retrieval and annotation tasks. In: Peters, C., Gey, F.C., Gonzalo, J., Müller, H., Jones, G.J.F., Kluck, M., Magnini, B., de Rijke, M., Giampiccolo, D. (eds.) CLEF 2005. LNCS, vol. 4022, Springer, Heidelberg (2006)

9. Gruber, T.R.: Toward principles for the design of ontologies used for knowledge sharing. International Journal of Human-Computer Studies 43, 907–928 (1995)

10. Rosse, C., Mejino, R.L.V.: A reference ontology for bioinformatics: The Foundational Model of Anatomy. Journal of Biomedical Informatics 36, 478–500 (2003)

11. Buitelaar, P., Sintek, M., Kiesel, M.: A lexicon model for multilingual/multimedia ontologies. In: Sure, Y., Domingue, J. (eds.) ESWC 2006. LNCS, vol. 4011. Springer, Heidelberg (2006)

12. Volk, M., Ripplinger, B., Vintar, S., Buitelaar, P., Raileanu, D., Sacaleanu, B.: Semantic annotation for concept-based cross-language medical information retrieval. International Journal of Medical Informatics (2003)

13. Vintar, S., Buitelaar, P., Volk, M.: Semantic relations in concept-based cross-language medical information retrieval. In: Proceedings of the ECML/PKDD Workshop on Adaptive Text Extraction and Mining (ATEM) (2003)

14. Carbonell, J.G., Yang, Y., Frederking, R.E., Brown, R.D., Geng, Y., Lee, D.: Translingual information retrieval: A comparative evaluation. International Journal of Computational Intelligence and Applications 1, 708–715 (1997)

15. Eichmann, D., Ruiz, M.E., Srinivasan, P.: Cross-language information retrieval with the UMLS metathesaurus. In: Proceedings of the 21st annual international ACM SIGIR conference on Research and development in information retrieval, pp. 72–80 (1998)

16. Laender, A., Ribeiro-Neto, B., Silva, A., Teixeira, J.: A brief survey of web data extraction tools. Special Interest Group on Management of Data (SIGMOD) Record 31(2) (2002)

17. Ravichandran, D., Hovy, E.: Learning surface text for a question answering system. In: Proceedings of the Association of Computational Linguistics Conference (ACL) (2002)

18. Ling, C., Gao, J., Zhang, H., Qian, W., Zhang, H.: Improving encarta search engine performance by mining user logs. International Journal of Pattern Recognition and Artificial Intelligence 16(8) (2002)

19. Deselaers, T., Weyand, T., Keysers, D., Macherey, W., Ney, H.: FIRE in ImageCLEF 2005: Combining content-based image retrieval with textual information retrieval. In: Working Notes of the CLEF Workshop (2005)

20. Hunter, J.: Adding multimedia to the semantic web - building an MPEG-7 ontology. In: International Semantic Web Working Symposium (SWWS) (2001)

21. Barnard, K., Duygulu, P., Forsyth, D., de Freitas, N., Blei, D., Jordan, M.: Matching words and pictures. Journal of Machine Learning Research 3, 1107–1135 (2003)

22. Lavrenko, V., Manmatha, R., Jeon, J.: A model for learning the semantics of pictures. In: Proceedings of the Neural Information Processing Systems Conference (2003)

23. Lim, J.H.: Learnable visual keywords for image classification. In: DL 1999: Proceedings of the fourth ACM conference on Digital libraries, pp. 139–145. ACM Press, New York (1999)

24. Carneiro, G., Vasconcelos, N.: A database centric view of semantic image annotation and retrieval. In: SIGIR 2005: Proceedings of the 28th annual international ACM SIGIR conference on Research and development in information retrieval, pp. 559–566 (2005)

25. Town, C.: Ontological inference for image and video analysis. International Journal of Machine Vision and Applications 17, 94–115 (2006)

26. Mezaris, V., Kompatsiaris, I., Strintzis, M.: An ontology approach to object-based image retrieval. In: International Conference on Image Processing (ICIP) (2003)

27. Mojsilovic, A., Gomes, J., Rogowitz, B.E.: Semantic-friendly indexing and querying of images based on the extraction of the objective semantic cues. International Journal of Computer Vision 56, 79–107 (2004)

28. Lehmann, T., Güld, M., Thies, C., Fischer, B., Spitzer, K., Keysers, D., Ney, H., Kohnen, M., Schubert, H., Wein, B.: The IRMA project. A state of the art report on content-based image retrieval in medical applications. In: Proceedings 7th Korea-Germany Joint Workshop on Advanced Medical Image Processing, pp. 161–171 (2003)

29. Vompras, J.: Towards adaptive ontology-based image retrieval. In: Stefan Brass, C.G. (ed.) 17th GI-Workshop on the Foundations of Databases, Wörlitz, Germany, Institute of Computer Science, Martin-Luther-University Halle-Wittenberg, pp. 148–152 (2005)
30. Papadopoulosa, G.T., Mezaris, V., Dasiopoulou, S., Kompatsiaris, I.: Semantic image analysis using a learning approach and spatial context. In: Proceedings of the 1st international conference on Semantics And digital Media Technologies (SAMT) (2006)
31. Su, L., Sharp, B., Chibelushi, C.: Knowledge-based image understanding: A rule-based production system for X-ray segmentation. In: Proceedings of Fourth International Conference on Enterprise Information System, Ciudad Real, Spain, vol. 1, pp. 530–533 (2002)
32. Mechouche, A., Golbreich, C., Gibaud, B.: Towards a hybrid system using an ontology enriched by rules for the semantic annotation of brain MRI images. In: Marchiori, M., Pan, J.Z., de Sainte Marie, C. (eds.) RR 2007. LNCS, vol. 4524, pp. 219–228. Springer, Heidelberg (2007)
33. Horrocks, I., Patel-Schneider, P.F., Boley, H., Tabet, S., Grosof, B., Dean, M.: SWRL: A semantic web rule language combining OWL and RuleML. Technical report, W3C Member submission, May 21 (2004)
34. Castano, S., Ferrara, A., Hess, G.: Discovery-driven ontology evolution. In: SWAP 2006: Semantic Web Applications and Perspectives - 3rd Italian Semantic Web Workshop (2006)

Human Factors Affecting the Patient's Acceptance of Wireless Biomedical Sensors

Rune Fensli[1] and Egil Boisen[2]

[1] University of Agder, Faculty of Engineering and Science, Grimstad, Norway
rune.fensli@uia.no
[2] Aalborg University, Department of Health Science and Technology, Aalborg, Denmark
eb@hast.aau.dk

Abstract. In monitoring arrhythmia, the quality of medical data from the ECG sensors may be enhanced by being based on everyday life situations. Hence, the development of wireless biomedical sensors is of growing interest, both to diagnose the heart patient, as well as to adjust the regimen. However, human factors such as emotional barriers and stigmatization, may affect the patient's behavior while wearing the equipment, which in turn may influence quality of data. The study of human factors and patient acceptance is important both in relation to the development of such equipment, as well as in evaluating the quality of data gathered from the individual patient. In this paper, we highlight some important aspects in patient acceptance by comparing results from a preliminary clinical trial with patients using a wireless ECG sensor for three days out-of-hospital service, to available published results from telehomecare projects, and discuss important aspects to be taken into account in future investigations.

Keywords: Quality of life, Patient Acceptance, Ambulatory Monitoring, Telehomecare, Wearable sensors.

1 Introduction

Several research projects have been focusing on wearable biomedical sensors and their benefits for ambient assisted living, where the patients are remotely monitored by different sensors placed on the body for vital signs recording[1]. However, little efforts have been done investigating how the patients experience and manage this new technology.

Chronically ill patients experience a greater degree of freedom and are more involved in the treatment with daily monitoring of vital information during hospitalization in their own home, than with the traditional treatment procedures at hospital. Introducing advanced medical technology in the patient's own home will influence the patient's situation as it makes empowerment and self-management possible[2,3]. Such concepts are emerging due to anticipations of reducing the cost of caring for the aging population, and at the same time they are intended to improve the quality of the services[4].

Furthermore, coordinated follow-up and new workflow procedures for the healthcare services need to be implemented in order to give the patient satisfactory support by virtual visits in his home[5]. However, this support also must be integrated in the

A. Fred, J. Filipe, and H. Gamboa (Eds.): BIOSTEC 2008, CCIS 25, pp. 402–412, 2008.
© Springer-Verlag Berlin Heidelberg 2008

self-monitoring of vital signs information performed by the patients, with understandable interpretations of the results.

The primary goal of developing wireless ECG sensors is to find ways of monitoring the everyday life of the patient as closely as possible. Compared to conventional monitoring systems, where the patient is hospitalized so as to be monitored for a few days, mobile solutions with wearable sensors attached to the chest for three days of continuous monitoring do not prohibit the patient from carrying out normal daily activities, while still being under continuous medical monitoring. Since arrhythmias may occur during the patient's physical activities, such solutions allowing the patient's behavior to be closer to his normal routines may thus enhance the quality of data gathered. In other words, the over-all rationale is to increase the quality of data being collected by decreasing the impact on the patient's everyday life when he or she is being monitored. In consequence, the patient's acceptance and the integration of the monitoring systems into his or her everyday life routines plays a vital role in enhancing the quality of medical data from the ECG sensors. It is therefore necessary to study human factors in patient acceptance of such equipment, both in developing equipment with less impact on everyday life, as well as in evaluating the data being collected by the individual patient.

1.1 Usability and Telehomecare

Telehomecare is a growing field of patient treatment and follow-up, but most research studies have focused on the technology, and so questions of human factors and patient satisfaction need to be addressed. A systematic review of studies of patient satisfaction with telemedicine was done by Mair and Whitten, arguing that "available research fails to provide satisfactory explanations of the underlying reasons for patient satisfaction or dissatisfaction with telemedicine". They found that the studies concerning patient satisfaction mainly used simple survey instruments and that many of the studies had only a few participants[6]. Williams et al. found in their review surveying 93 studies within telemedicine and telehomecare that questionnaires "tend to be brief, quantitative, and generally lack validation and standardization"[7]. Maglaveras et al. also found that the user acceptance and user friendliness are unresolved, during their project focusing on technologies for home-care[8].

According to Friedman & Wyatt, usability studies can be useful in the evaluation of new biomedical equipment, with the aim of observing speed of use, user comments and completion of simple tasks[9]. They describe field function studies as useful in the validation of prototypes or released versions of new biomedical equipments, but the trials should be conducted in situations with real use of the equipment, and with the aim of observing speed and quality of data collected and accuracy of advice provided by the devices.

The principles of user-centered design can be useful when designing new telehomecare devices. Adlam et al. describe how user evaluation can be implemented in the design process[10]. They start with discovering the "real problem" and the users' requirements, which can be accomplished with a simple prototype demonstrating the actual functionality of a device. However, interacting with real users in their own environments will be a challenging task, and prototypes with limited functionality can give restrictions on the use of the new solution to be developed, resulting in tests, which do not reflect a more complicated use situation, i.e. the daily activities of the patients.

Kaufmann et al. developed a system design for both usability inspection and usability testing in the patient's home for a diabetes telemedicine system. First, they used a cognitive walkthrough identifying goals and sequences of actions to anticipate potential user problems. Second, field usability testing was performed as a series of tasks to be accomplished by the subjects in their home, followed by immediate semi-structured interviews to reveal problems and barriers to efficient and safe use of the system[11]. A similar approach has been suggested by Kushniruk and Patel, who have developed a low-cost portable usability testing solution intended to be used by patients in their homes[12].

In a study of the patient's perspectives on high-tech homecare technology, Lehoux found that user-acceptance was dependent on both technical and human dimensions, where the technology is integrated into the patients' private and social lives[13]. One aspect was found to be concerned with technical competence, where lay people are supposed to use high-tech medical devices. Also, Lehoux found that user-acceptance was dependent on different types of anxiety, mainly being related to the actual equipment and the procedures of using it. Furthermore, some patients complained of reductions in daily activities, as well as of feelings of stigmatization, leading to withdrawal from social activities in order to hide the medical equipment from the eyes of visitors. As found by Lehoux, for a patient wearing a permanent catheter, this will alter the patient's body image.

If the use of wireless sensors after a time can be accepted by the patients, this can be similar to what Haggard and Wolpert describes as being integrated into the person's 'body scheme', as a dynamic, and typically unaware, representation of the positions of one's body parts. They argue that a patient's body scheme can change on a short timescale, to incorporate additional objects as new segments of the body representation[14].

Hopp et al. measured the outcome for patients receiving telehealth home-care and used a questionnaire at baseline and after six months, where they used a modified version of the SF-36 to measure Health Related Quality Of Life (HRQOL)[15]. In addition they used separate questions to ask about satisfaction with the telehealth equipment for the intervention group, with questions from the National Ambulatory Care Survey and modified for use in evaluating telehealth services. They found a high degree of satisfaction with the telemedicine equipment, but few patients reported that their family members had been taught how to use the equipment installed in their homes.

2   Objectives

In this paper we will focus on how user acceptance of wireless biomedical sensors can be monitored in order to highlight obstacles and possible improvements both in relation to developing the wireless solutions, as well as in relation to evaluating the quality of data gathered by the individual patient. Based on preliminary results from a clinical study where patients have used wearable sensors for a three day period of out-of-hospital service, we focus on the impact on everyday life during the patients' use of the wireless ECG-sensor, and their experiences with daily behavior as well as general patient satisfaction. Furthermore, in order to observe the patients' influence on the data collected via the new telehomecare services, we discuss observations concerning future research methods into user acceptance of wireless sensors.

3 Methods

3.1 Study Design

Patients referred to long-term ambulatory "Holter" arrhythmia procedures at the out-patient clinic at Sørlandet Hospital HF in Arendal, Norway, were asked to participate in a study wearing our newly developed wireless ECG-sensor[16]. By signing the informed consent form, they participated in the study during their ordinary arrhythmia investigation. The inclusion criteria was patients with suspected arrhythmias, and the exclusion criteria was patients with dementia who were anticipated not being able to handle the equipment and contribute in filling out the required questionnaires.

After signing the informed consent form, the patients were given information of the research project and they received several questionnaires to be filled in during the time they used the wireless ECG-system. Because influences from participating in a three day trial of the new recording solutions can be a bias in the evaluation of patient acceptance, a reference group with patients undergoing a "normal" routine investigation at the hospital using conventional "Holter" monitoring equipment (Huntleigh Healthcare)[17] was established.

The questionnaires focused on several topics, based on our description of a Sensor Acceptance Model[18]. It is anticipated that the patient's acceptance of wearable sensors is influenced by several dimensions as shown in Fig. 1. We have defined 23 questions or items which are used to quantify the 5 dimensions: Hygienic Aspects (3

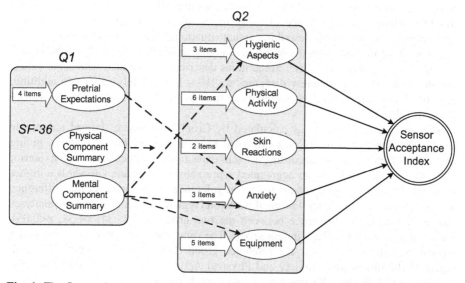

Fig. 1. The Sensor Acceptance Model shows the actual dimensions influencing the patient's perceived acceptance using wearable sensors, defined as Sensor Acceptance Index. Two different questionnaires are used; Q1 is to be filled in at startup of the sensor use, while Q2 is to be filled in after the actual use of the sensor to be evaluated. Q1 contains both items defining the dimension Pretrial Expectations and the SF-36v2 Acute Form defining the Physical Component Summary and the Mental Component Summary. Dotted lines indicate significant influence on the dimensions[18].

items), Physical Activity (6 items), Skin Reactions (2 items), Anxiety (3 items), Equipment (5 items). In addition, the dimension Pretrial Expectations (4 items) is included in the questionnaire Q1 together with the SF-36 which is a standardized questionnaire to be used as a generic measure of Health Related Quality of Life[19].

During the period from November 2006 to May 2007, 11 patients were enrolled in the study, and 25 patients were included in the reference group. After a three day period of arrhythmia investigation with the use of the wireless sensor, 4 patients in the intervention group were followed-up with qualitative interviews, in order to discover more general experience with regard to the use of a wearable sensor.

3.2 Ethical Considerations

The study has been accepted by the Regional Ethical Committee as well as the hospital's Ethical Committee. In addition, permission to use the new technological equipment was obtained from The Directorate for Health and Social Affairs in Norway. In order to perform the empirical work, permission with respect to data privacy requirements was given by the Norwegian Social Science Data Services on behalf of The Norwegian Data Inspectorate.

4 Exploratory Results

In the questionnaire, the patients are asked about their experiences with the use of the equipment, where we have used an 11-point semantic differential scale. The patients should evaluate their experience of using the wireless equipment in terms of their agreement to the statements describing actual situations of the equipment usage.

For some of the questions the scale was ranging from "0 – Extremely problematic" to "10 – Without any problems". Some items described a statement with scale values ranging from "0 – I completely disagree" to "10 – I completely agree". In addition, they filled in some general characteristics, such as gender and age. The results are given in Table 1.

The internal consistency, as calculated by Cronbach's alpha, showed acceptable levels. The construct validity was evaluated by confirmatory factor analysis, giving reasonable factor loading according to our expectations. Calculations of a Sensor Acceptance Index (SAI) as an aggregated score showed a tendency towards a higher score for the wireless group compared to the reference group; however this difference was not at a significant level. For the dimension Hygienic Aspects regression analyses showed a significant difference between the two groups ($F_{(1, 34)} = 4.51$; $p < 0.05$), with a higher score for the wireless group. The dimension Skin Reactions showed a significantly higher score for the reference group ($F_{(1, 31)} = 5.95$; $p < 0.05$). With regard to the dimensions Anxiety and Physical Activity, the wireless sensor showed higher scores; however, this was not a significant difference.

In the interviews with four of the patients in the intervention group, we tried to discover some general experiences from the patients' use of the wireless recording equipment. All patients reported some anxiety because of what they found to be a degree of uncertainty, as they did not receive any feedback from the recording system

Table 1. Patients' general characteristics and results from the calculated Sensor Acceptance Index (SAI) and the defined dimensions based on the questionnaires, Mean (SD). Calculations of Cronbach's alpha for the dimension Hygienic Aspects was 0.83, for Physical Activity 0.88, for Skin Reactions 0.65, for Anxiety 0.80 and for Equipment 0.86.

General characteristics and Dimensions	Wireless sensor (n=11)	Reference group (n=25)	
Gender: Man/woman	6 / 5	7 / 18	
Age	40.2 (19.4)	56.4 (13.2)	
SAI	8.2 (1.0)	7.6 (1.9)	
Hygienic Aspects	8.6 (1.6)	6.6 (2.9)	*
Physical Activity	9.2 (0.8)	8.0 (2.8)	
Skin Reactions	6.3 (2.5)	8.6 (2.4)	*
Anxiety	9.0 (1.3)	7.3 (3.5)	
Equipment	8.0 (1.7)	7.5 (2.8)	

* significant difference p<.05.

of the measurements made. They expected a quick feedback from the hospital, and two of the patients expressed the need for patient influence, while one of the respondents said that he was not concerned with influence.

The hygienic factor focused on actual tasks related to the patient's ability to perform body wash and use of the equipment during the night while asleep, in order to obtain information about any problems relating to the wearable recording sensor. The survey showed a significant difference for the hygienic aspects, and the wireless solution obviously was preferable, since it is easy to wear and does not involve any hindrances like cables. This was confirmed both by responses to the open questions in the questionnaire and by statements in the interviews, as the patients generally expressed high satisfaction with the wireless solution compared to the existing "Holter" system. They felt free to carry out daily activities without any hindrance.

With respect to the equipment used, one patient complained about the "Holter" recorder, and said that he had "a feeling of being a living medical instrument"; because of all the cables he had to wear. With regard to the wireless sensor, he said: "The wireless sensor was comfortable to wear, and most of the time I just forgot I was wearing this system". He said that the sensor after a while became "a part of me".

Another patient said the wireless system made it possible for her to participate in physical exercise. It was much easier to wear, especially during the night, and it did not prevent her from being able to take a shower. The Holter equipment was troublesome with all the cables, and made the hygienic activities more problematic. In her view, the differences in use between those two systems were huge, and they can not be compared at all.

One of the patients expressed some dissatisfaction with wearing this equipment, and she wanted to hide the equipment from other people. Similar expressions of embarrassment were also found in responses to the open questions in the questionnaire, and even if the calculated difference was not at a significant level, it showed a trend toward more anxiety regarding the use of the Holter equipment.

5 Discussion

The results presented above are of interest in the evaluation of patient experience with wearable sensors attached to the chest for three days of continuous monitoring. Even though this clinical trial was limited both in terms of time and the number of patients, some interesting aspects have been discovered in relation to integration into the patients' everyday life. By comparing results and experiences from the clinical trial with available published results from other telehomecare projects, it has been possible to discover some general aspects and point out ideas for future investigations.

First, our findings point to the issue of stigmatization, as some patients wish to hide the wireless recorder from the eyes of other people. This was similar to the findings by Myers and colleagues, who studied the impact of home-based monitoring on patients with congestive heart failure, and followed up patients for a 2 month period. They found it necessary to train patients in telemonitoring procedures on an ongoing basis[20]. During their study, 13.5% of patients withdrew due to anxiety or because they did not "like" the telemonitoring procedures or equipment. Their experiences of patient perception and ability to learn how to use the equipment indicate that the emotional barriers and stigmatization were a challenging factor and time was needed to allow the patients to adapt to this new situation. Our findings, however, also showed that when patients feel like the sensor is becoming "a part of me", as expressed by one of the patients, the stigmatization factor does not seem to represent any problems for the patient during daily activities, and can be adopted within his or her body scheme.

Second, our findings point to a need of constant feedback from the system or the health professionals to the patients. In this study, we did not implement a daily reporting schema between the patient and the hospital, which was probably the reason why the patients expressed some worries about what the technology was measuring in terms of irregularities in their heart beats. Even though they trusted the equipment, they expressed that they would like to see the results and the doctor's evaluation of the results when they felt irregular heart beats. These findings were in line with our experiences from an earlier study where patients underwent a 24-hour Holter recording procedure[21], and quick feedback from the doctor was evaluated to be of utmost importance. However, although the patients appreciated good information during the research project, they expressed some uncertainty with respect to from whom they would receive an answer concerning arrhythmia findings. Their misperception of the health care sector as a "clear and strictly coordinated service", capable of giving them the desired follow-up, shows that organizational issues will be of utmost importance when introducing new telemedical solutions. If the co-ordination within the health care sector is not clearly defined, questions from the patients will not be correctly addressed and there will easily be situations where patients will suffer from not having received the required feedback to the situation at hand.

Third, during the interviews the patients expressed overall good confidence with using the wearable sensor, mostly because of the ease of use during daily activities, which was confirmed by the significant difference calculated for the Hygienic Aspects. As a general measure of satisfaction (SAI), we found a relatively high score at 8.2 for the intervention group and 7.6 for the reference group. According to the intentions of home-based health care as expressed by Barlow and colleagues, those expressions of satisfaction were somewhat as expected[2]. This has also been confirmed by Whitten and Mickus in their study of patients with congestive heart failure in addition to chronic pulmonary disease, finding that the patients were satisfied with the technology[22]. Wootton and Kvedar have also pointed to the importance of changes in the health care services[5], and their findings are also in line with what was reported by Scalvini et al. in their study of chronic heart patients and the effects of home-based telecardiololgy[23]. In our study, the scores for Anxiety were relatively high indicating a low degree of anxiety, with 9.0 in the intervention group and 7.3 in the reference group. The age of patients in the reference group was slightly higher and consisted of more female patients, which may represent a bias. However, being confident with using the wearable equipment combined with a feeling of safety is important to the patients.

Patient acceptance of home hospitalization equipment on a long-term basis does not seem to have been given the necessary attention in previous studies of telehomecare. Following a systematic study of observed effects in home telemonitoring of patients with diabetes, Jaana and Pare found that studies should be extended in time and involve larger samples of patients in order to generalize the findings and obtain sufficient validity[24]. Long-term evaluation may probably also discover some underlying reasons for the feelings of anxiety as reported by Lehoux[13].

We therefore propose future studies to follow the patients' use of wearable sensors and telehomecare equipment for a relatively long time in order for the patient to adopt the technology into his/her daily activities and body scheme. Attention should be paid to the patients' ability to carry out hygienic activities such as body wash or even taking a shower, and it should be possible to participate in physical sports activities while using wearable sensor recorders. In a future scenario, it can be possible for the patient to wear a wireless ECG sensor communicating to a hand held device, which detects arrhythmias and transmits cardiac events to the cardiology specialist at the hospital, as shown in Fig. 2. Compared to conventional monitoring systems, such mobile solutions can enhance the patient's ability to carry out normal daily activities, while still being under continuous medical monitoring. Since arrhythmias may occur during the patient's physical activities, such solutions can allow the patient's behavior to be closer to his normal routines, and may thus enhance the quality of data measured.

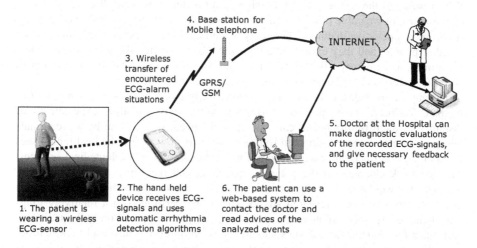

Fig. 2. A future scenario for wearable sensors communicating to the doctor at the hospital

Also, not only the patient but also his/her spouse or partner should be taken into account and given the necessary information about how the treatment should be performed and how to interact with the health care services in a coordinated manner using tele-medical equipment. This is in accordance with observations made by Dinesen et al., who in a home hospitalisation project found that the impact of the spouse or partner "included elements of increased responsibility, nervous tension and invasion of privacy"[3]. The increased anxiety of the patient's spouse or partner may affect the patient's well being, and as the telehomecare interferes with the private sphere those aspects should be carefully observed and analyzed in future studies.

As proposed by Kushniruk and Patel, multi-method evaluations can be important, and even the use of video recording can be performed in the patients' home[12]. Hence, a multi-method approach can be employed, where questionnaires are supplemented with qualitative methods to provide information about the patients, his or her behaviour, and acceptance of the technology, where interviews in the patient's home can be combined with video recordings and, later on, analysis of the process data to obtain a more thorough understanding of obstacles and barriers to the use of such solutions. Evaluating the communication between the patient and health care services will also be of special importance, and new e-health-based forms of communication should be investigated. As quick responses are required by the patients, quality factors in the communication between the patient and the health care service should be observed.

6 Conclusions

Implementation of wireless sensors for vital signs recording for the use in home hospitalization can be a great benefit for the patient, as this gives a feeling of freedom compared to ordinary hospital stay. During the use of wireless ECG-sensors, the patients in this study were satisfied with the ease of use in a daily life situation, and gave a higher score in the factor Hygienic Aspects than patients in the reference group. With respect to the factor Anxiety no significant differences were discovered. However, the use of wearable medical equipment can also affect the patients' everyday life situation in a negative manner, where they tended to hide the equipment from the eyes of other people, and they expressed anxiousness for not using the telemedical equipment in a correct manner. But at the same time, the patients also expressed confidence with the system and tended to adopt its use into their daily life.

During the interviews, the patients, however, expressed worries of not having immediate feedback and responses to irregular heart beats, and their expectations of instant follow-up by the health care sector were revealed. This can be a great challenge where necessary coordinated routines and workflow within the health care sector should be defined and established before implementing new telemedical services. The feedback channels to the patients' questions and recordings should be timely defined and validated. In order to investigate the use of telemedical technology in everyday life situations, research studies should be performed in a relatively long-term manner. Also, by employing a multi-method approach based on both qualitative and quantitative data, such studies should focus on observing how and to which degree the patient adopts this new technology into his or her everyday life as well as body scheme or body image. Additionally, such studies should take into account the interaction between the patient and the system, as well as the interaction between the patient and the health care professionals. Finally, it seems necessary to also investigate how the patient's partner or spouse experiences the technology in daily use.

Acknowledgements. The study is supported by the Research Council of Norway. The clinical trials are done in close co-operation with Sørlandet Hospital HF, Arendal and the company WPR Medical AS. The authors thank Torstein Gundersen, Ellen Ytrehus and Åse Løsnesløkken at Sørlandet Hospital, and Eirik Aanonsen at WPR Medical for good assistance with the patients during clinical trials. We also thank Kjersti Nag, Hjørdis Løvdal Gulseth and Hedda Løvland for their assistance with interviewing the patients.

References

1. Jones, V., van Haltern, A., Dokovsky, N., Koprinkov, G., Peuscher, J., Bults, R., et al.: Mobihealth: Mobile Health Services based on Body Area Networks. In: Istepanian, R.S.H., Laxminarayan, S., Pattichis, C.S. (eds.) M-Health: Emerging Mobile Health Systems, pp. 219–236. Springer, Heidelberg (2006)
2. Barlow, J., Wright, C., Sheasby, J., Turner, A., Hainsworth, J.: Self-management approaches for people with chronic conditions: a review. Patient Educ. Couns. 48(2), 177–187 (2002)
3. Dinesen, B., Nøhr, C., Andersen, S.K., Sejersen, H., Toft, E.: Under surveillance, yet looked after: telehomecare as viewed by patients and their spouse/partners. Eur. J. Cardiovasc. Nurs. 7(3), 239–246 (2008)
4. Lukowicz, P., Kirstein, T., Tröster, G.: Wearable systems for health care applications. Methods Inf. Med. 43(3), 232–238 (2004)
5. Wootton, R., Kvedar, J.C.: Home Telehealth: Connecting Care Within the Community. RSM Press (2006)
6. Mair, F., Whitten, P.: Systematic review of studies of patient satisfaction with telemedicine. BMJ 320(7248), 1517–1520 (2000)
7. Williams, T.L., May, C.R., Esmail, A.: Limitations of Patient Satisfaction Studies in Telehealthcare: A Systematic Review of the Litterature. Telemed. J. E Health 7(4), 293–316 (2001)
8. Maglaveras, N., Chouvarda, I., Koutkias, V., Meletiadis, S., Haris, K., Balas, E.A.: Information Technology Can Enhance Quality in Regional Health Delivery. Methods Inf. Med. 41(5), 393–400 (2002)
9. Friedman, C.P., Wyatt, J.C.: Evaluation Methods in Biomedical Informatics, 2nd edn. Springer, New York (2006)
10. Adlam, T., Orpwood, R., Dunn, T.: User evaluation in pervasive healthcare. In: Bardram, J.E., Mihailidis, A., Wan, D. (eds.) Pervasive Computing in Healthcare, USA, pp. 243–274. CRC Press, Inc., Boca Raton (2006)
11. Kaufmann, D.R., Patel, V.L., Hilliman, C., Morin, P.C., Pevzner, J., Weinstock, R.S., et al.: Usability in the real world: assessing medical information technologies in patients. homes. J. Biomed. Inform. 36, 45–60 (2003)
12. Kushniruk, A.W., Patel, V.L.: Cognitive and usability engineering methods for the evaluation of clinical information systems. J. Biomed. Inform. 37(1), 56–76 (2004)
13. Lehoux, P.: Patients' Perspectives on High-Tech Home Care: A Qualitative Inquiry into the User-Friendliness of Four Interventions. BMC Health Services Research 4, 28 (2004)

14. Haggard, P., Wolpert, D.M.: Disorders of body schema. In: Freund, H.-J. (ed.) Higher-order Motor Disorders: From Neuroanatomy and Neurobiology to Clinical Neurology, ch. 14, pp. 261–272. Oxford University Press, Oxford (2005)

15. Hopp, F., Woodbridge, P., Subramanian, U., Copeland, L., Smith, D., Lowery, J.: Outcomes Associated with a Home Care Telehealth Intervention. Telemedicine and e-Health 12(3), 297–307 (2006)

16. Fensli, R., Gunnarson, E., Gundersen, T.: A wearable ECG-recording System for Continuous Arrhythmia Monitoring in a Wireless Tele-Home-Care Situation. In: Proceedings 18th IEEE Symposium on Computer-Based Medical Systems, pp. 407–412 (2005)

17. Huntleigh Healthcare. Medilog AR4 Digital Holter Recorder,
    http://www.medilogdarwin.com/Medilog_AR4.html

18. Fensli, R., Pedersen, P.E., Gundersen, T., Hejlesen, O.: Sensor Acceptance Model - Measuring Patient Acceptance of Wearable Sensors. Methods Inf. Med. 47(1), 89–95 (2008)

19. Ware, J.E., Kosinski, M., Dewey, J.E. How to Score Version 2 of the SF-36® Health Survey. Lincoln, RI. QualityMetric Incorporated (2000)

20. Myers, S., Grant, R.W., Lugn, N.E., Holbert, B., Kvedar, J.C.: Impact of Home-Based Monitoring on the Care of Patients with Congestive Heart Failure. Home Health Care Management & Practice 18(6), 444 (2006)

21. Fensli, R., Gundersen, T., Gunnarson, E.: Design Requirements for Long-Time ECG recordings in a Tele-Home-Care Situation, A Survey Study. In: Scandinavian Conference in Health Informatics, pp. 14–18 (2004)

22. Whitten, P., Mickus, M.: Home telecare for COPD/CHF patients: outcomes and perceptions. J. Telemed. Telecare 13(2), 69–73 (2007)

23. Scalvini, S., Capomolla, S., Zanelli, E., Benigno, M., Domenighini, D., Paletta, L., et al.: Effect of home-based telecardiology on chronic heart failure: costs and outcomes. J. Telemed. Telecare 11(S1), 16–18 (2005)

24. Jaana, M., Pare, G.: Home telemonitoring of patients with diabetes: a systematic assessment of observed effects. J. Eval. Clin Pract. 13(2), 242–253 (2007)

The Policy Debate on Pseudonymous Health Registers in Norway

Herbjørn Andresen

Norwegian Research Centre for Computers and Law, University of Oslo, Norway
herbjorn.andresen@jus.uio.no

Abstract. Patient health data has a valuable potential for secondary use, such as decision support on a national level, reimbursement settlements, and research on public health or on the effects of various treatment methods. Unfortunately, extensive secondary use of data has disproportionate negative impact on the patients' privacy. The Norwegian health data processing regulation prescribes four different ways of organizing health registers (anonymous, de-identified, pseudonymous or fully identified data subjects). Pseudonymity is the most innovative of these methods, and it has been available as a legitimate means to achieve extensive secondary use of accurate and detailed data since 2001. Up to now, two different national health registers have been organized this way. The evidence from these experiences should be encouraging: Pseudonymity works as intended. Yet, there is still discernible reluctance against extending the pseudonymity principle to encompass other national health registers as well.

Keywords: Pseudonymity, Public Health Policy, Privacy Regulation, Confidentiality, Data Security.

1 Introduction

Any patient must accept some processing of his personal data, within the confines of a medical treatment. Some data are collected from the patient himself, and some data may be generated along the course of the treatment. The confidentiality will inevitably be at risk; the risks need to be monitored and handled. Privacy regulations and data security measures, along with professional ethics, safeguard against unwarranted processing and filing practices, and against deliberate misuse.

Privacy is also at stake when the patient's health data is used beyond the appointed medical treatment. Such use could be named secondary purposes for processing data. The subject matter of this paper is a particular variety of secondary purposes, namely a group of national health data registers. A register is a service that comprises a database, an operating organization, and a legal authority defining responsibilities, duties to report to the register, restrictions on use and so on. In colloquial language the term register is normally taken to mean the database itself. The organization and the legal authority are implied.

The registers have two important features, from a privacy point of view. Firstly, they are centralized systems, containing aggregated data. Personal health data are collected from different hospitals or other treatment entities. As Norway is a small

A. Fred, J. Filipe, and H. Gamboa (Eds.): BIOSTEC 2008, CCIS 25, pp. 413–424, 2008.
© Springer-Verlag Berlin Heidelberg 2008

country, centralized registers usually cover the entire nation. The procedure of collecting data could either be by electronic exchange or by printed reports which are re-typed into the centralized system.

Secondly, the data is collected and processed for secondary purposes, somewhat remote from the patient's immediate needs and interests. Roughly stated, these secondary purposes are governmental administration and medical research. Governmental administration includes both macro-level decision support and reimbursement control procedures. The demand for data is, at least in principle, limited and foreseeable. Medical research will also in most cases demand a stable amount of foreseeable data, yet in some cases it could be beneficial to use excess data or to perform inventive couplings involving different data sources. The future value of ingenious data mining is by definition unknown.

Regulations, security measures and ethics are of course at least as required for the registers as they are for data in the immediate care systems. In addition, the registers are vulnerable to expansions of their stated and legitimate purposes. Proponents of strict privacy regulations warn against "the slippery slope". It can be increasingly difficult on each particular occasion to turn down a proposition for extended use of a register. Such propositions often serve legitimate purposes, which is to achieve new and even benign goals more easily. Consequently, the patients' privacy is in danger of being scooped out in the long run.

1.1 The Adage of Norway's Favorable Conditions for Health Registers

Norway introduced a national identity number quite early. Starting in 1964, Statistics Norway assigned a unique 11-digit identification number to every individual. The primary purpose of the national identity number was to produce accurate statistics. Large public agencies, such as the Tax Administration and the National Insurance Administration, soon adopted the new unique identification number. No one imagined the vast future use of this new identity number. There was no explicit legal support for it, and hence there were no expressed restrictions on its use either [1].

Due to the lack of restrictions on the use of population register data in the early years, the identity number is now the key to personal data in thousands of public as well as private IT systems throughout the country, including primary health care systems and hospital systems. Most Norwegians will have to type or pronounce his unique personal identity number to some electronic apparatus (or to its human gatekeeper) several times a week. It is "open sesame" to enter both caves of treasures and caves of dung.

Due to the widely used national identity number, Norway may have favorable conditions for national health registers. Because the identity numbers are used in systems that are vital to virtually everyone, the data quality of the primary systems reporting to the registers almost takes care of itself. The registers could technically be suitable for collecting perfectly linkable data from the cradle to the grave.

However, the early years of vastly expanding the use of national identity numbers was succeeded by almost three decades of efforts to impart restrictions on their use. Lawful use of the national identity number is subject to a "norm of necessity". Roughly, it goes like this: When an exact identification is necessary, then the controller ought to use a unique identifier, while using the same unique identifier would be unlawful when unique identification is not necessary (this is a crude simplification, on my own behalf, of the Personal Data Act [2] section 12). The result from these efforts to limit the lawful use of unique identifiers has not been any actual decrease in the use of the national identity numbers. Yet, many researchers perceive some uncertainty on whether they can use national identity numbers lawfully in their projects. A health register may not impart non-anonymous patient health data to research projects unless the recipient can provide sufficient justification for it.

No matter how favorable the conditions for health registers may be; legal restrictions prevent them from being used to their full potential. This situation induces two parallel debates: The first one is about the balance between privacy and legitimate uses of a health register. The second debate is about the possibility to circumvent patient identification without sacrificing the benefits of a health register.

1.2 The Origins of Digital Pseudonyms

A pseudonym is, literally, a "false name". For ages, pseudonyms have been used by authors and artists, or even in the rare event of modest researchers, to disguise their identity. The notion of a digital pseudonym first appeared in a paper by David Chaum. He invented digital pseudonyms as a means to conceal an individual's real identity in electronic transactions [3]. The intended field of application in Chaum's paper was banking and electronic commerce. A pseudonym concealed the identity of the person who actually paid the goods.

In a few consecutive papers, he developed both the methods and the rationale further. The public key distribution system provided a secure cryptographic pseudonym. For the holder of a pseudonym to be able to communicate or inspect his own personal data, a trusted third party could manage the pseudonyms. The rationale was to introduce a new paradigm for data protection; using technological means to put the individual in control of his own data [4]. Organizations would not be able to share data about the individual without the data subject "acting out" his consent, so to speak. No one could collect the complete history of your transactions, debts or savings. The holder of the pseudonym would also hold the key to reverse it.

As for the proposed new paradigm of privacy in banking and electronic commerce, it seems to have lost completely to the old paradigm of widespread use of fully identified data subjects. Meanwhile, the fields of health administration and medical research have revived the idea of digital pseudonyms.

1.3 The Pseudonymization Process

There are different ways to carry out the process of generating a digital pseudonym to conceal the data subject's real identity.

The simplest form of pseudonyms, used for decades in research projects based on samples, is to assign a sequence number to each respondent. To enhance the respondents' trust, the researcher could hire a third party to perform the assigning process. This method works well for one-time surveys. For a panel study over time, managing sequence numbers becomes increasingly more difficult. Coupling data with relevant data from other sources would require an overt process of reversing the sequence numbers. The pseudonyms would be illusory. To exchange a "real identity" with an unrelated sequence number is only trustworthy when the researcher grants the respondent permanent anonymity, without adding to the data later. It is not a viable method for a long-term and multi-purpose health register.

A digital pseudonym in a health register involves advanced cryptography. The input to the algorithm that generates the pseudonym will have to be a stable identifying number, which does not change over time for the same patient. In Norway, the national identity number provides a convenient unique input. The health register will not need to store the national identity number, the algorithm secures that the same pseudonym is assigned to the same patient when more data is added to the register.

With a reliable and stable identification, there are, conceptually, two different ways to generate a pseudonym. One way is to use an asymmetric hash function. The encryption algorithm then generates a digest that is unique to the input, but there is no way to reverse from the encrypted digest back to the input identifier. Because the same input identifier always transforms to the same digest, it is possible to add data about the same patient in the same health register. It is, however, not possible to generate data couplings between individual-level data from two different health registers. This method provides a very high degree of confidentiality, but is on the other hand inflexible. Two health registers cannot be merged, and it would not be possible to address any registered patient, for instance if a new treatment method vital to his particular decease is developed.

The alternative way to generate a pseudonym resembles the "public key" encryption technology, and is basically the same as Chaum invented (see section 1.2 above). The input to the algorithm is the same stable and reliable patient identity number. An encryption algorithm, using the "public key" of a key pair, generates the pseudonym. The same input, and the same public key, will make it possible to add data about the same patient to the same health register. In addition, a decryption algorithm can reverse the pseudonym back to the "real identity", by using the "private key" of the same key pair that was used for encryption. A trusted third party, which is an independent pseudonym manager, carry out the encryption, and if requested, the decryption. The health register will never see the real identity of the patient. The trusted third party, who is able to decrypt the pseudonym, does not have access to any sensitive information about the patients. This process provides more flexibility, at the cost of more fragile pseudonyms. The confidentiality of the patient is to a higher degree based on trust. Violating the pseudonyms will be somewhat easier from a technical point of view.

The two existing pseudonymous health registers in Norway use a combination of these methods. The pseudonyms data key in the health registers are non-reversible hash values computed from the national identification number. The communication between the entities reporting to the register and the pseudonym manager, and then between the pseudonym manager and the health register, is based on the public key infrastructure. However, the computed hash values can not actually guarantee non-reversible pseudonyms: Due to the weak and well-known structure of the national identification numbers in Norway, any person who knows the algorithm could compute a complete list of pseudonyms matching each possible identification number.

2 Legislative Support for Pseudonyms

Recent technological innovations often seem to be far ahead of developments in legislation. Society's toolbox for protecting values and for distributing rights and obligations usually adapts slowly, to fit technological changes that have already taken place.

The introduction of pseudonyms in Norwegian health registers differs from this typical path of history. The first Norwegian national register based on pseudonyms was established in 2004. By that time participants in various legislation processes had already advocated this method for more than a decade. Technologists and professional users of the registers remained skeptic. Pseudonymous health registers have not at any rate been "technology-driven" in Norway, it would be far more correct to call it a "legislation-driven" development.

Norway has had registers for specific diseases, such as The Cancer Register, for decades. They started out as paper files, and were later converted to computer databases. The specific health registers had proved to be useful over time, and the health authorities started to nourish a desire to establish a General National Health Register, not to be limited to any particular diagnosis.

2.1 An Early, Avant-Garde Proposal

Though the advantages of a General National Health Register were convincing, The Parliament was also much concerned about the impact on the patients' privacy. In 1989, they urged The Government to appoint a committee with a mandate to examine ways and methods to establish such register "without threshing individuals' privacy" [5].

The appointed committee issued a report in 1993 [6]. An "Official Norwegian Report" is in most cases the product of an appointed drafting committee, at an early stage of the legislation process. After an official hearing among relevant stakeholders, both The Government and eventually The Parliament may make changes to the original draft, or even turn down the entire proposition all together.

The drafting committee proposed, in their report, a new act to provide legal authority to the desired General National Health Register. The proposed act was very much ahead of its time. It contained regulations on cryptographic pseudonyms generated and managed by trusted third parties, along with a profound set of rules on ensuring legitimate use of the register, data quality, and the patients' right to access and so on.

However, neither the health authorities nor the medical research community was in favor of this avant-garde way to organize their much-desired new health register. As the main stakeholders did not support the proposition, The Government put it on hold, and it remained so for about eight years.

Instead of either a fully identified register (which was what the health authorities wanted) or a pseudonymous health register (the proposition they turned down), the health authorities established the Norwegian Patient Register (see section 3.3 below) in 1997. The Norwegian Patient register was originally established as a de-identified register (see section 2.3 below). This was acceptable under the Personal Data Filing Systems Act of that time, and it did not require The Parliament to pass any new legislation.

2.2 Specific Privacy Regulations for Health Data and Health Registers

The Parliament passed a new general Personal Data Act [2] on April 14, 2000. The primary motivation for replacing the old act of 1978 was to comply with the European Union Directive on protection of personal data [7].

The Personal Data Act regulates all processing of personal data, for any legitimate purpose. Therefore, the rules are quite flexible, leaving most assessments and decisions to the discretion of the controller. For the processing of health data, The Parliament did not consider the general act sufficient. On May 18, 2001, they passed the Personal Health Data Filing Systems Act [8], containing rules that are somewhat more specific. The Personal Health Data Filing Systems Act too complies with the European Union Directive, and it has many important features in common with the general Personal Data Act. For instance, the information security requirements are essentially the same.

The primary guiding rule for processing health data is a requirement to obtain the patients' consent. However, the act also recognizes a need in some situations to process data without consent. A typical exception to requiring consent would be the kind of health registers where complete coverage is vital to fulfill the purpose of the register.

2.3 Four Different Levels of Patient Identification in a Health Register

The key to the regulation of health registers is section 8 of the Personal Health Data Filing Systems Act. The initial position is simply that central health registers are prohibited, unless authorized by this act or by another statute.

The remainder of section 8 spans the possibilities and preconditions for establishing health registers, providing they have an adequate legal authority. The purpose of a register shall be "to perform functions pursuant to," specified health services (the relevant acts are listed in section 8). Those functions include "the general management and planning of services, quality improvement, research and statistics". In addition, the Government shall prescribe subordinate legislation for each health register, defining specific rules, responsibilities and organization.

An interesting feature of section 8 is the way it categorizes health registers into four distinct levels of patient identification. Every health register has to conform to one of these four levels. The *choice* of a level of identification encapsulates the privacy balancing process for each register.

Table 1. Outline of levels of patient identification (adapted from [9])

Personal data (being subject to privacy regulation)		Not personal data	
Data refers to unambiguous individuals		Data may refer to ambiguous data subjects	
Fully identified	Pseudonyms	De-identified	Anonymous

The bottom row of the table above shows the four different levels of patient identification. Their order, from the left to the right, reflects an order from more to less strain on the patients' privacy.

The middle row of the table shows the main division of whether the data refer to unique patients or not. Fully identified patients and pseudonymous patients both have the same granularity. They will provide the same level of statistical data accuracy.

The top row of the table merely shows that only three out of the four levels of patient identification are strictly within the definition of personal data.

Generally, fully identified health registers shall only process data about patients who consent. The only exceptions are a moderate number of health registers particularly named in section 8. By the end of 2007 there are exactly nine fully identified health registers not requiring the patients to consent (the number of such registers was six by the time the act was originally passed, in 2001). The Parliament has to pass a formal change to section 8, specifically naming the new register, before anyone can establish a new fully identified central health register with a complete coverage (i.e. not requiring consent). That is the beauty of this construct in the Personal Health Data Filing Systems Act; it ensures an overt and highly democratic legislation process to be carried out before establishing a new register.

By using pseudonyms instead of fully identified patients, the health authorities can establish a new health register by issuing subordinate legislation. This means a Parliament decision is not necessary to establish the register. It also means the register can omit the patients' consent, if it needs complete coverage of the data. The option of pseudonymous health registers thus entered Norwegian legislation in 2001, eight years after it was first proposed.

A de-identified and a pseudonymous register have the same legal status according to section 8. The health authorities may establish a de-identified register by issuing subordinate legislation. A de-identified register means that any clear and manifest identifying information is removed. The advantage of a de-identified over a pseudonymous register is that the de-identified register is technically easier and less expensive to operate. The paramount disadvantage of a de-identified register is that the data is not on a strictly individual level. If a hospital carries out the same surgical procedure, say four times, a de-identified register cannot tell whether it involved four different individual patients or if the same patient was involved four times.

Pseudonymous and de-identified registers share the same risk of unlawful re-identification through computational analysis of the stored data elements [10]. The uniqueness of each registered individual becomes more transparent as the number of detailed variables increase. Coping with the risk of re-identification first requires the register owner to keep his sobriety on what data is stored and processed. Second, there is still an indispensable need for rigid access control and other conventional information security measures with pseudonymous and de-identified registers.

In an anonymous register, all information that can possibly identify individual patients is removed. In addition to removing the manifest identifiers, the register also removes, or reclassifies into categories that are more general, any data suitable for re-identification by analysis. An anonymous register takes the granularity of the data into account. Making the data anonymous often means to take deliberate action to sacrifice their accuracy.

Anonymous data may be published, and they will not require extensive data protection. The downside to anonymous data, which is why they are unapt for health registers in most cases, is that it is virtually impossible to add meaningfully to the data.

2.4 The Professionals' Responsibilities, and a Democratic Safety Valve

To summarize, the Norwegian legislation allows four different methods for storing and processing personal health registers. A method granting more privacy is less effective for achieving administration and research goals. This inverse ratio is at the heart of any privacy regulation. All the four levels of identification are in use in some existing health register, and they have all proved to work as intended. Apart from the likes and the dislikes of different stakeholders: The four different nominal levels of identification themselves provide relatively objective aids for an informed policy debate. We know what the options are, and how they work. We also know how each of these options influence the privacy of the patients, versus the accuracy of governmental decision support data and the possibilities for providing valuable research data.

The health authorities remain chiefly responsible for all aspects of the health registers. However, the very strict preconditions for establishing a fully identified register (unless requiring patients to consent) constitute a striking democratic safety valve. Any such register require The Parliament to pass a formal change to section 8 of the Health Data Filing Systems Act. Professional agenda owners and stakeholders need not, and may not, decide alone on such privacy invasive registers. Though the process may be cumbersome and time-consuming, it also secures a highly democratic participation in the balancing between privacy and the well-grounded benefits of a health register.

3 A Current Status and Recent Developments

Up to the end of 2007, the Ministry of Health has seriously considered and deliberated on the option to make a health register pseudonymous on four different occasions. On two out of these four occasions, they actually decided to establish the proposed register with pseudonyms as its level of patient identification. For the other two registers, one of them was "promoted" to be a fully identified register, while the other one was "demoted" to be a de-identified register.

3.1 The Norwegian Prescription Database

The first pseudonymous national health register in Norway is called *The Norwegian Prescription Database* ("Reseptregisteret", in Norwegian). In October 2003 The Ministry of Health issued the subordinate legislation providing legal authority for the register, as required by section 8 of the Health Data Filing Systems Act. The register was actually established in the beginning of 2004.

Before the Prescription Database was established, the medicine statistics were based on sales figures reported from wholesale dealers. Unquestionably, the data was insufficient both for straightforward knowledge about use of medicine, and for research on effects thereof. Various stakeholders demanded statistics based on prescriptions and actual dispatch from pharmacies to individual patients. The intended purposes were neither to control any patient's catch at the pharmacy nor to supervise how named doctors carried out the business of prescribing. Pseudonyms both ensure the demanded capacities of the register, and safeguard against undesirable infringements of privacy.

All pharmacies report the prescription data electronically every month. A central data collecting point transfers the data to a trusted third party. The trusted pseudonym manager is in this case Statistics Norway. They transfer the pseudonymous data to the register owner, which is the Norwegian Institute of Public Health. Both the patient's identity (his national identity number) and the doctor's identity (his authorized license identifier) are replaced with pseudonyms. The pharmacies are fully identified, on an enterprise level; their license identifier is not pseudonymized [11].

The Prescription Database was in many ways a "quiet reform". The changes have been virtually invisible to the patients. They still pick up their prescriptions and carry them to the pharmacies the same way they did before. They are not asked for consent. The existence of the Prescription Database is not a secret in any way, but neither is it of much concern to the patients. They only know about the pseudonymous data if they take a particular interest in detailed level health politics.

3.2 National Statistics Linked to Individual Needs for Care (IPLOS)

The second pseudonym-based health register is named the *National statistics linked to individual needs for care* (its Norwegian acronym is "IPLOS", which is derived from "Individbasert pleie- og omsorgsstatistikk"). The Ministry of Health issued the subordinate legislation providing legal authority for the register in February 2006. The first mandatory reporting term to the register was February 2007, collecting data from health care services throughout all Norwegian municipalities. The register owner in this case is the Directorate for Health. The Tax Administration is the trusted pseudonym manager, which illustrates the point that the main feature of a pseudonym manager is its institutional independence.

Contrary to the Prescription Database, the IPLOS has not been a "quiet reform". The information about individual needs for care was not readily available from any existing process. Even though the patients can trust the confidentiality of the central register, they had to answer a new set of questions. Someone would type their answers into a local database before they were sent electronically to the pseudonymous register. The crucial question was not anymore whether the pseudonym provided sufficient privacy. Many patients felt offended by some of the most invasive questions in the form. The forms were changed as a result from complaints about some of the questions, such as whether a handicapped patient needs help after going to the toilet or needs help with handling the menstrual period.

Pseudonyms only remedy privacy issues that become present after the patients have left off their participation. An important lesson is that health registers mainly deal with data that the patients hardly are aware of. In many cases, the limits to a health register are with the processes of eliciting and collecting data, and not with the confidentiality of the register itself.

3.3 The Norwegian Patient Register

The Norwegian Patient Register is a hospital and outpatient clinic discharge register. Data on each patient is collected from every hospital in Norway. The acronym NPR is used both in English and in Norwegian when this register is referred to.

The history of the NPR is complicated, and truly interesting from a privacy point of view. NPR is the actual instantiation of the "General National Health Register" which initiated the committee back in 1989, who proposed a pseudonym-based solution register in their 1993 report.

After the proposed pseudonymous register was put on hold, the NPR was revived in 1997 as a de-identified register. The NPR was established in March 1997. It receives reports on operative procedures extracted from the patient administrative systems at all hospitals. Age, sex, place of residence, hospital and department, diagnosis, surgical procedure, and dates of admission and discharge are included in the register [12]. The name and national ID number of the patients are not included.

The NPR has proved to be a valuable register, providing much demanded data for both research and administration purposes. Yet, as a de-identified register it does have obvious shortcomings. The data do not refer to strictly unambiguous individuals.

Over the last few years, the health authorities have made efforts to "promote" the NPR into a fully identified register. Proponents of privacy argued that promoting it into a pseudonymous register would be sufficient for all purposes of the register. The health authorities and the research community argued that a pseudonymous register might not provide adequate data quality. After a heated debate, The Parliament finally passed the necessary change to section 8 of the Personal Health Data Filing Systems Act on February 1, 2007, and included the NPR to be a fully identified health register. Subordinate legislation, specifically regulating "the new" NPR was issued on December 12, 2007.

3.4 The Abortion Register

The latest example so far, of the Ministry of Health having seriously considered pseudonyms, is The Abortion Register. They made a proposition, intent to establish this as a pseudonymous register. The proposition went to a formal hearing; the closing date for the hearing was January 13, 2006. A large number of the bodies entitled to comment on the hearing were skeptic to a register containing as sensitive information as abortions.

The proposal was met with reluctance from different sides. Many answers to the hearing pointed out the particular strain on some of the women who decide to go through an abortion. Induced abortions are legal in Norway, yet there is a risk of social or religious condemnation from parts of the society, making the burden heavier. The Data Inspectorate, for instance, argued that the knowledge of an abortion register might influence on actual decisions on whether to have an abortion or not. Thus the register could affect, and not merely reflect, the health care activities. On June 21, 2007, The Government decided to make The Abortion Register de-identified, and not pseudonymous.

The policy debate on The Abortion Register revealed an interesting limit to people's trust in a pseudonymous register. The confidentiality is based on trust in society as we know it today. The possibility of reversing a pseudonym could be exploited sometime in the future, when privacy values may be worse off.

4 Conclusions

4.1 Pseudonymous Identities Work

Pseudonyms are a legal and a viable means for protecting personal data in Norwegian Registers. It has been one out of four lawful levels of patient identification since 2001. There are only two health registers based on pseudonyms so far. The numbers of both fully identified registers and de-identified registers are much higher.

A case study on the Prescription Database shows that the vital functions of a pseudonymous register work as intended [9]. Neither the trusted pseudonym manager nor the register owner – nor anybody else for that matter – gets to see both the "real" identification numbers and unencrypted health data relating to the individuals at the same time. There were initial problems with the data quality for some time, but more recently, the rate of errors have been approximately the same as they are for fully identified registers.

4.2 Pseudonyms Are Still Controversial

This brief report on pseudonymous health registers in Norway reveals some broad categories of arguments for and against pseudonyms.

The pro arguments are primarily a lower risk of disclosing information about patients, to people who do not need it, and for purposes where identification is unnecessary. A pseudonym increases privacy, while maintaining the statistical accuracy of the data.

The contra arguments are increased expenses due to the third party process, a risk of re-identification by analysis of the non-identifying data, and a danger of unforeseeable decrease in privacy if privacy policies change in the future. Finally, the argument most often raised against pseudonyms, is the data quality issue. The register owners will have reduced opportunities for discovering and fixing errors on their own. However, the third party may assist in structured "data laundering"-procedures.

Looking at the pros and cons of pseudonymous health registers, the amplitude of the controversies seems somewhat exaggerated. The pseudonymous registers are plainly an in-between solution. Moving either one step to the left or one step to the right, referring to the outline of levels of identifications in table 1, is merely a change in the balance of the arguments for or against pseudonyms. Choosing another level of patient identification will neither release all the advantages nor solve all the problems that may occur to a pseudonymous register. A proposal to make a particular health register pseudonymous can expect attacks from both sides; some will say identifying unambiguous individuals pseudonymously infringes privacy anyway, others will say pseudonyms place too heavy restrictions on the register.

In my opinion, though, the fact that the health authorities have only established two pseudonymous registers since that option was legislated for in 2001, while The Parliament has accepted three new fully identified registers during the same period, sadly indicates that proponents of pseudonymous registers in Norway are perhaps fighting a loosing battle.

To the credit of the Personal Health Data Filing Systems Act, the construct of choosing between only four lawful levels of identification is in my opinion both clever and successful. It ensures a broad and overt democratic process, calling the attention of all stakeholders to voice their opinion.

References

1. Selmer, K.: Hvem er du: Om systemer for registrering og identifikasjon av personer (Title translated: Who are you: On systems for registering and for identifying individuals). Lov. og Rett., 311–334 (1992)
2. Act of 14 April 2000 No. 31 relating to the processing of personal data (Norwegian statute), An English translation, http://www.ub.uio.no/ujur/ulovdata/lov-20000414-031-eng.pdf
3. Chaum, D.L.: Untraceable electronic mail, return addresses, and digital pseudonyms. CACM 24(2), 84–90 (1981)
4. Chaum, D.L.: A New Paradigm for Individuals in the Information Age. In: 1984 IEEE Symposium on Security and Privacy, pp. 99–103. IEEE Computer Society Press, Washington (1984)
5. Boe, E.: Pseudo-identities in Health Registers: Information technology as a vehicle for privacy protection. The International Privacy Bulletin 2(3), 8–13 (1994)
6. ONR 1993:22, Official Norwegian Report: Pseudonyme helseregistre [Title translated: Pseudonymous Health Registers] (1993)
7. Directive 95/46/EC of the European Parliament and of the Council of 24 October 1995 on the protection of individuals with regard to the processing of personal data and on the free movement of such data (1995)
8. Act of 18 May 2001 No. 24 on Personal Health Data Filing Systems and the Processing of Personal Health Data (Norwegian statute), An English translation, http://www.ub.uio.no/ujur/ulovdata/lov-20010518-024-eng.pdf
9. L'Abée-Lund, Å.: Pseudonymisering av personopplysninger i sentrale helseregistre [Title translated: Pseudonymizing personal data in central health registers]. Master thesis, University of Oslo (2006)
10. Malin, B.: Betrayed by My Shadow: Learning Data Identity via Trail Matching. Journal of Privacy Technology. Paper number 20050609001. Pittsburg, PA (2005), http://www.jopt.org/publications/20050609001_malin.pdf
11. Strøm, H.: Reseptbasert legemiddelregister [Title translated: A prescription-based medicine register]. Norsk Farmaceutisk Tidsskrift 2004(1), 7–9 (2004)
12. Bakken, I.J., Nyland, K., Halsteinli, V., Kvam, U.H., Skjeldestad, F.E.: The Norwegian Patient Registry. Nor. J. Epidem. 14(1), 65–69 (2004)

Application of Process Mining in Healthcare – A Case Study in a Dutch Hospital

R.S. Mans[1], M.H. Schonenberg[1], M. Song[1], W.M.P. van der Aalst[1], and P.J.M. Bakker[2]

[1] Department of Information Systems Eindhoven University of Technology, P.O. Box 513
NL-5600 MB, Eindhoven, The Netherlands
{r.s.mans,m.h.schonenberg,m.s.song,w.m.p.v.d.aalst}@tue.nl
[2] Academic Medical Center, University of Amsterdam, Department of Innovation and Process
Management, Amsterdam, The Netherlands
p.j.bakker@amc.uva.nl

Abstract. To gain competitive advantage, hospitals try to streamline their processes. In order to do so, it is essential to have an accurate view of the "careflows" under consideration. In this paper, we apply process mining techniques to obtain meaningful knowledge about these flows, e.g., to discover typical paths followed by particular groups of patients. This is a non-trivial task given the dynamic nature of healthcare processes. The paper demonstrates the applicability of process mining using a real case of a gynecological oncology process in a Dutch hospital. Using a variety of process mining techniques, we analyzed the healthcare process from three different perspectives: (1) the control flow perspective, (2) the organizational perspective and (3) the performance perspective. In order to do so we extracted relevant event logs from the hospitals information system and analyzed these logs using the ProM framework. The results show that process mining can be used to provide new insights that facilitate the improvement of existing careflows.

1 Introduction

In a competitive health-care market, hospitals have to focus on ways to streamline their processes in order to deliver high quality care while at the same time reducing costs [1]. Furthermore, also on the governmental side and on the side of the health insurance companies, more and more pressure is put on hospitals to work in the most efficient way as possible, whereas in the future, an increase in the demand for care is expected.

A complicating factor is that healthcare is characterized by highly *complex* and extremely *flexible* patient care processes, also referred to as "careflows". Moreover, many disciplines are involved for which it is found that they are working in isolation and hardly have any idea about what happens within other disciplines. Another issue is that within healthcare many autonomous, independently developed applications are found [5]. A consequence of this all is that *it is not known what happens in a healthcare process for a group of patients with the same diagnosis.*

The concept of process mining provides an interesting opportunity for providing a solution to this problem. Process mining [9] aims at extracting process knowledge from

A. Fred, J. Filipe, and H. Gamboa (Eds.): BIOSTEC 2008, CCIS 25, pp. 425–438, 2008.

so-called "event logs" which may originate from all kinds of systems, like enterprise information systems or hospital information systems. Typically, these event logs contain information about the start/completion of process steps together with related context data (e.g. actors and resources). Furthermore, process mining is a very broad area both in terms of (1) applications (from banks to embedded systems) and (2) techniques.

This paper focusses on the *applicability* of process mining in the healthcare domain. Process mining has already been successfully applied in the service industry [7]. In this paper, we demonstrate the applicability of process mining to the healthcare domain. We will show how process mining can be used for obtaining insights related to careflows, e.g., the identification of care paths and (strong) collaboration between departments. To this end, in Section 3, we will use several mining techniques which will also show the diversity of process mining techniques available, like control flow discovery but also the discovery of organizational aspects.

In this paper, we present a case study where we use raw data of the AMC hospital in Amsterdam, a large academic hospital in the Netherlands. This raw data contains data about a group of 627 gynecological oncology patients treated in 2005 and 2006 and for which all diagnostic and treatment activities have been recorded for financial purposes. Note that we did not use any a-priori knowledge about the care process of this group of patients and that we also did not have any process model at hand.

Today's Business Intelligence (BI) tools used in the healthcare domain, like Cognos, Business Objects, or SAP BI, typically look at aggregate data seen from an external perspective (frequencies, averages, utilization, service levels, etc.). These BI tools focus on performance indicators such as the number of knee operations, the length of waiting lists, and the success rate of surgery. Process mining looks "inside the process" at different abstraction levels. So, in the context of a hospital, unlike BI tools, we are more concerned with the care paths followed by individual patients and whether certain procedures are followed or not.

This paper is structured as follows: Section 2 provides an overview of process mining. In Section 3 we will show the applicability of process mining in the healthcare domain using data obtained for a group of 627 gynecological oncology patients. Section 4 concludes the paper.

2   Process Mining

Process mining is applicable to a wide range of systems. These systems may be pure information systems (e.g., ERP systems) or systems where the hardware plays a more prominent role (e.g., embedded systems). The only requirement is that the system produces *event logs*, thus recording (parts of) the actual behavior.

An interesting class of information systems that produce event logs are the so-called *Process-Aware Information Systems* (PAISs) [2]. Examples are classical workflow management systems (e.g. Staffware), ERP systems (e.g. SAP), case handling systems (e.g. FLOWer), PDM systems (e.g. Windchill), CRM systems (e.g. Microsoft Dynamics CRM), middleware (e.g., IBM's WebSphere), hospital information systems (e.g., Chipsoft), etc. These systems provide very detailed information about the activities that have been executed.

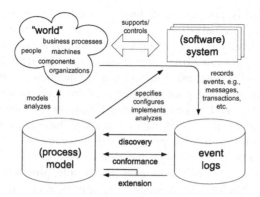

Fig. 1. Three types of process mining: (1) Discovery, (2) Conformance, and (3) Extension

However, not only PAISs are recording events. Also, in a typical hospital there is a wide variety of systems that record events. For example, in an intensive care unit, a system can record which examinations or treatments a patient undergoes and also it can record occurring complications for a patient. For a radiology department the whole process of admittance of a patient till the archival of the photograph can be recorded. However, frequently these systems are limited to one department only. However, systems used for billing purposes have to ensure that all services delivered to the patient will be paid. In order for these systems to work properly, information from different systems needs to be collected so that it is clear which activities have been performed in the care process of a patient. In this way, these systems within the hospital can contain information about processes *within* one department but also *across* departments. This information can be used for improving processes within departments itself or improving the services offered to patients.

The goal of process mining is to extract information (e.g., process models) from these logs, i.e., process mining describes a family of *a-posteriori* analysis techniques exploiting the information recorded in the event logs. Typically, these approaches assume that it is possible to sequentially record events such that each event refers to an activity (i.e., a well-defined step in the process) and is related to a particular case (i.e., a process instance). Furthermore, some mining techniques use additional information such as the performer or originator of the event (i.e., the person/resource executing or initiating the activity), the timestamp of the event, or data elements recorded with the event (e.g., the size of an order).

Process mining addresses the problem that most "process/system owners" have limited information about what is actually happening. In practice, there is often a significant gap between what is prescribed or supposed to happen, and what *actually* happens. Only a concise assessment of reality, which process mining strives to deliver, can help in verifying process models, and ultimately be used in system or process redesign efforts. *The idea of process mining is to discover, monitor and improve real processes* (i.e., not assumed processes) *by extracting knowledge from event logs.* We consider three basic types of process mining (Figure 1): (1) *discovery*, (2) *conformance*, and (3) *extension*.

Discovery. Traditionally, process mining has been focusing on *discovery*, i.e., deriving information about the original process model, the organizational context, and execution properties from enactment logs. An example of a technique addressing the control flow perspective is the α-algorithm [10], which constructs a Petri net model describing the behavior observed in the event log. It is important to mention that there is no a-priori model, i.e., based on an event log some model is constructed. However, process mining is not limited to process models (i.e., control flow) and recent process mining techniques are more and more focusing on other perspectives, e.g., the organizational perspective, performance perspective or the data perspective. For example, there are approaches to extract social networks from event logs and analyze them using social network analysis [6]. This allows organizations to monitor how people, groups, or software/system components are working together. Also, there are approaches to visualize performance related information, e.g. there are approaches which graphically shows the bottlenecks and all kinds of performance indicators, e.g., average/variance of the total flow time or the time spent between two activities.

Conformance. There is an a-priori model. This model is used to check if reality conforms to the model. For example, there may be a process model indicating that purchase orders of more than one million Euro require two checks. Another example is the checking of the so-called "four-eyes" principle. Conformance checking may be used to detect deviations, to locate and explain these deviations, and to measure the severity of these deviations.

Extension. There is an a-priori model. This model is extended with a new aspect or perspective, i.e., the goal is not to check conformance but to enrich the model with the data in the event log. An example is the extension of a process model with performance data, i.e., some a-priori process model is used on which bottlenecks are projected.

At this point in time there are mature tools such as the ProM framework [8], featuring an extensive set of analysis techniques which can be applied to real-life logs while supporting the whole spectrum depicted in Figure 1.

3   Healthcare Process

In this section, we want to show the *applicability* of process mining in healthcare. However, as healthcare processes are characterized by the fact that *several organizational units* can be involved in the treatment process of patients and that these organizational units often have their own specific IT applications, it becomes clear that getting data, which is related to healthcare processes, is not an easy task. In spite of this, systems used in hospitals need to provide an integrated view on all these IT applications as it needs to be guaranteed that the hospital gets paid for every service delivered to a patient. Consequently, these kind of systems contain process-related information about healthcare processes and are therefore an interesting candidate for providing the data needed for process mining.

To this end, as case study for showing the applicability of process mining in health care, we use raw data collected by the billing system of the AMC hospital. This raw data contains information about a group of 627 gynecological oncology patients treated in

2005 and 2006 and for which all diagnostic and treatment activities have been recorded. The process for gynecological oncology patients is supported by several different departments, e.g. gynecology, radiology and several labs.

For this data set, we have extracted event logs from the AMC's databases where each event refers to a service delivered to a patient. As the data is coming from a billing system, we have to face the interesting problem that for each service delivered for a patient it is only known on which *day* the service has been delivered. In other words, we do not have any information about the actual timestamps of the start and completion of the service delivered. Consequently, the ordering of events which happen on the same day do not necessarily conform with the order in which events of that day were executed.

Nevertheless, as the log contains *real* data about the services delivered to gynecological oncology patients it is still an interesting and representative data set for showing the applicability of process mining in healthcare as still many techniques can be applied. Note that the log contains 376 different event names which indicates that we are dealing with a non-trivial careflow process.

In the remainder of this section we will focus on obtaining, in an explorative way, *insights into the gynecological oncology healthcare process*. So, we will only focus on the *discovery* part of process mining, instead of the *conformance* and *extension* part. Furthermore, obtaining these insights should not be limited to one perspective only. Therefore, in section 3.2, we focus on the discovery of *care paths followed by patients*, the discovery of *organizational aspects* and the discovery of *performance related information*, respectively. This also demonstrates the diversity of process mining techniques available. However, as will be discussed in Section 3.1, we first need to perform some preprocessing before being able to present information on the right level of detail.

3.1   Preprocessing of Logs

The log of the AMC hospital contains a huge amount of distinct activities, of which many are rather low level activities, i.e., events at a low abstraction level. For example, for our purpose, the logged lab activities are at a too low abstraction level, e.g. determination of chloride, lactic acid and erythrocyte sedimentation rate (ESR). We would like to consider all these low level lab tests as a single lab test. Mining a log that contains many distinct activities would result in a too detailed spaghetti-like model, that is difficult to understand. Hence, we first apply some preprocessing on the logs to obtain interpretable results during mining. During preprocessing we want to "simplify" the log by removing the excess of low level activities. In addition, our goal is to consider only events at the department level. In this way, we can, for example, focus on care paths and interactions between departments. We applied two different approaches to do this.

Our first approach is to detect a *representative* for the lower level activities. In our logs, this approach can be applied to the before mentioned lab activities. In the logs we can find an activity that can serve as representative for the lab activities, namely the activity that is always executed when samples are offered to the lab. All other (low level) lab activities in the log are simply discarded.

The log also contains groups of low level activities for which there is no representative. For instance at the radiology department many activities can occur (e.g., echo

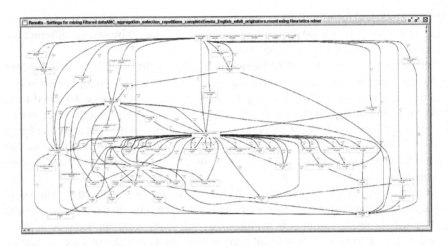

Fig. 2. Derived process model for all cases

abdomen, thorax and CT brain), but the logs do not contain a single event that occurs for every visit to this department, like a registration event for example. We apply *aggregation* for low level activities in groups without a representative by (1) defining a representative, (2) mapping all activities from the group to this representative and (3) removing repetitions of events from the log. For example, for the radiology department we define "radiology" as representative. A log that originally contains "..., ultrasound scan abdomen, chest X-ray, CT scan brain,...", would contain "..., radiology,...", after mapping low level radiology activities to this representative and removing any duplicates.

3.2 Mining

In this section, we present some results obtained through a detailed analysis of the AMC's event log for the gynecological oncology process. We concentrate on the discovery part to show actual situations (e.g. control flows, organizational interactions) in the healthcare process. More specifically, we elaborate on mining results based on three major perspectives (i.e. control flow, organizational, performance perspectives) in process mining.

Control Flow Perspective. One of the most promising mining techniques is control flow mining which automatically derives process models from process logs. The generated process model reflects the actual process as observed through real process executions. If we generate process models from healthcare process logs, they give insight into care paths for patients. Until now, there are several process mining algorithms such as the α-mining algorithm, heuristic mining algorithm, region mining algorithm, etc [10,12,11]. In this paper, we use the heuristic mining algorithm, since it can deal with noise and exceptions, and enables users to focus on the main process flow instead of on every detail of the behavior appearing in the process log [12]. Figure 2 shows the process model for all cases obtained using the Heuristics Miner. Despite its ability to

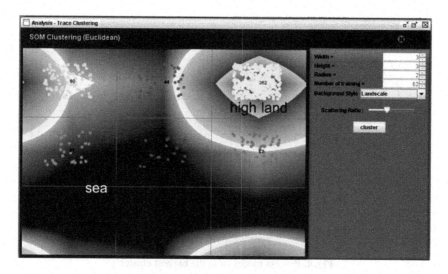

Fig. 3. Log clustering result

focus on the most frequent paths, the process, depicted in Figure 2, is still spaghetti-like and too complex to understand easily.

Since processes in the healthcare domain do not have a single kind of flow but a lot of variants based on patients and diseases, it is not surprising that the derived process model is spaghetti-like and convoluted.

One of the methods for handling this problem is breaking down a log into two or more sub-logs until these become simple enough to be analyzed clearly. We apply clustering techniques to divide a process log into several groups (i.e. clusters), where the cases in the same cluster have similar properties. Clustering is a very useful technique for logs which contain many cases following different procedures, as is the usual case in healthcare systems. Depending on the interest (e.g., exceptional or frequent procedures), a cluster can be selected. There are several clustering techniques available. Among these, we use the SOM (Self Organizing Map) algorithm to cluster the log because of its performance (i.e., speed). Figure 3 shows the clustering result obtained by applying the Trace Clustering plug-in. Nine clusters are obtained from the log. In the figure, the instances in the same cell belong to the same cluster. The figure also shows a contour map based on the number of instances in each cell. It is very useful to take a quick glance at the clusters – are there clusters with many similarities (high land), or are there many clusters with exceptional cases (sea).

By using this approach, we obtained several clusters of reasonable size. In this paper we show only the result for the biggest cluster, containing 352 cases all with similar properties. Figure 4 shows the heuristic net derived from the biggest cluster. The result is much simpler than the model in Figure 2. Furthermore, the fitness of this model is "good". The model represents the procedure for most cases in the cluster, i.e., these cases "fit" in the generated process model. A closer inspection of this main cluster by domain experts confirmed that this is indeed main stream followed by most gynecological oncology patients.

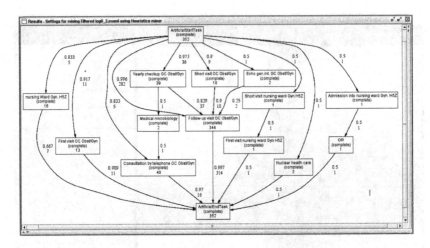

Fig. 4. Process model from the biggest cluster

When discussing the result with the people involved in the process, it was noted that patients, referred to the AMC by another hospital, only visit the outpatient clinic once or twice. These patients are already diagnosed, and afterwards they are referred to another department, like radiotherapy, for treatment and which is then responsible for the treatment process. Also, very ill patients are immediately referred to another department for treatment after their first visit.

Another approach for dealing with unstructured processes is the Fuzzy Miner. The Fuzzy Miner [4] addresses the issue of mining unstructured processes by using a mixture of abstraction and clustering techniques and attempts to make a representation of the (unstructured) process that is understandable for analysts. The miner provides a high-level view on the process by abstraction from undesired details, limiting the amount of information by aggregation of interesting details and emphasizing the most important details. The Fuzzy Miner provides an interface where these settings can be easily configured and the resulting model can directly be observed (see Figure 5a).

In addition, the Fuzzy Miner offers a dynamic view of the process by replaying the log in the model. The animation shows cases flowing through the model (depicted as white dots in Figure 5b). In the animation, frequently taken paths are highlighted, which prevents them from being overlooked. If necessary, clusters from the model can be analyzed in more detail before, or after animation (Figure 5c).

Organizational Perspective. There are several process mining techniques that address the organizational perspective, e.g., organizational mining, social network mining, mining staff assignment rules, etc. [6]. In this paper, we elaborate on social network mining to provide insights into the collaboration between departments in the hospital. The Social Network Miner allows for the discovery of social networks from process logs. Since there are several social network analysis techniques and research results available, the generated social network allows for analysis of social relations between originators involving process executions. Figure 6 shows the derived social network. To derive the

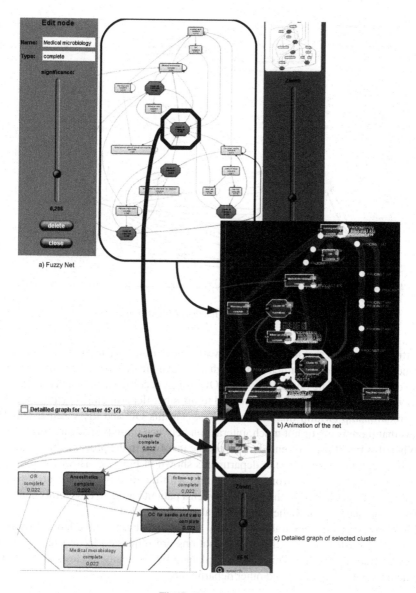

Fig. 5. Fuzzy Miner

network, we used the *Handover of Work* metric [6] that measures the frequency of transfers of work among departments.

The network shows the relationships between originators above a certain threshold. Originators, for which all relationships are below the specific threshold, appear as isolated circles. The originators that were highly involved in the process appear as larger dots in the figure. These results are useful to detect whether there are frequent interactions between originators (departments, in our case). In hospitals there are many

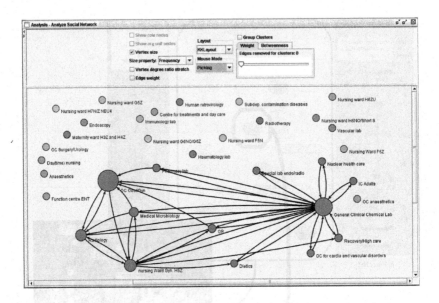

Fig. 6. Social network (handover of work metrics)

departments that interact and hand over work to each other. The mining result shows that the general clinical chemical lab is highly involved in the process and interacts with many departments. The outpatient clinic (OC) for gynecology and obstetrics is also often involved, but is not directly connected to all other departments. For instance there is no relationship (within this threshold) between this OC and the vascular lab. This means that there is no, or not much, interaction between these two departments.

When this result was presented to the people involved in the process, they confirmed the strong collaboration with the departments shown in Figure 6. However, they were surprised about the rather strong collaboration with the dietics department. Nevertheless, this can be explained by the fact that, when a patient has to go to several chemotherapy sessions, then a visit to the dietician is also often needed.

Moreover, they also noted that the many interactions between the lab and other departments is misleading as all the examinations are requested by gynecological oncology and not by the lab. This can be explained by the many lab tests and resulting interactions between the lab and other departments.

Performance Perspective. Process mining provides several performance analysis techniques. Among these, the dotted chart is a method suitable for case handling processes which are flexible and knowledge intensive business processes and focus not on the routing of work or the activities but on the case (e.g. careflows). In this paper, we use the dotted chart to show overall events and performance information of the log. Figure 7 shows the dotted chart. In the chart, events are displayed as dots, and the time is measured along the horizontal axis of the chart. The vertical axis represents case IDs and events are colored according to their task IDs. It supports several time options such as actual, relative, logical, etc. In the diagram, we use relative time which shows the

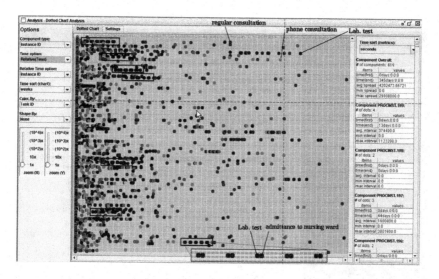

Fig. 7. Dotted Chart

duration from the beginning of an instance to a certain event. Thus it indicates the case duration of each instance. It also provides performance metrics such as the time of the first and of the last events, case durations, the number of events in an instance, etc. For example, in the figure (top right, average spread in seconds), the average case duration is about 49 days.

Users can obtain useful insights from the chart, e.g., it is easy to find interesting patterns by looking at the dotted chart. In Figure 7, the density of events on the left side of the diagram is higher than the density of those on the right side. This shows that initially patients have more diagnosis and treatment events than in the later parts of the process. When we focus on the long duration instances (i.e. the instances having events in the right side of the diagram), it can be observed that they mainly consist of regular consultation (red dot), consultation by phone (red dot), and lab test (violet dot) activities. It reflects the situation that patients have regular consultation by visiting or being phoned by the hospital and sometimes have a test after or before the consultation. It is also easy to discover patterns in the occurrences of activities. For example, seven instances have the pattern that consists of a lab test and an admittance to the nursing ward activities.

When the results were presented to the people involved in the process, they confirmed the patterns that we found. Furthermore, for the last pattern they indicated that the pattern deals about patients who get a chemotherapy regularly. The day before, they come for a lab test and when the result is good, they get the next chemotherapy.

Where the dotted chart focusses on visualizing the performance information using dots, the Basic performance analysis plug-in aims at calculation of performance values from a log and presenting this is in several ways. It allows users to draw several charts such as bar chart, pie chart, box-and-whisker chart, meter chart, etc. with performance information.

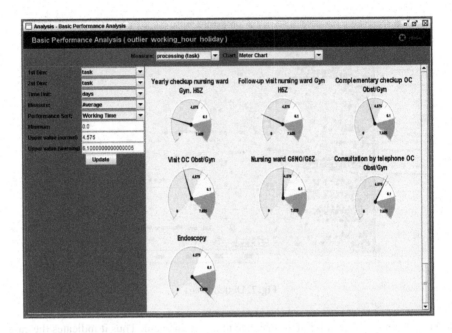

Fig. 8. Basic Performance Analysis Plug-in (meter chart)

Figure 8 shows one of the results that can be obtained with the plug-in. It depicts a meter chart for the average sojourn times of the tasks (i.e. the time between the preceding task and a certain task (the time between two dots in the dotted chart analysis)). A meter chart provides two kinds of information. They are the average sojourn time for a task and the relative position of this value amongst the average sojourn times for all tasks.

To draw the meter chart, the minimum (i.e. 0) and the maximum (i.e. 7.625 days) value among all the average sojourn times are used. The green part ranges from 0 to 60% of the maximum value, the yellow part covers 60% till 80%, and the red part starts from 80% of the maximum value. The needle in each chart shows the average sojourn time of the task. For example, in the Endoscopy chart, the average sojourn time is 7.625 days. As the needle is located in the red part, the users can easily recognize that this task is one of the tasks which have a long duration.

4   Conclusions

In this paper, we have focussed on the applicability of process mining in the healthcare domain. For our case study, we have used data coming from a non-trivial care process of the AMC hospital. We focussed on obtaining insights into the careflow by looking at the control-flow, organizational and performance perspective. For these three perspectives, we presented some initial results. We have shown that it is possible to mine complex hospital processes giving insights into the process. In addition, with existing techniques we were able to derive *understandable* models for large groups of patients. This was also confirmed by people of the AMC hospital.

Furthermore, we compared our results with a flowchart for the diagnostic trajectory of the gynaecological oncology healthcare process, and where a top-down approach had been used for creating the flowchart and obtaining the logistical data [3]. With regard to the flowchart, comparable results have been obtained. However, a lot of effort was needed for creating the flowchart and obtaining the logistical data, where with process mining there is the opportunity to obtain these kind of data in a semi-automatic way.

Unfortunately, traditional process mining approaches have problems dealing with unstructured processes as, for example, can be found in a hospital environment. Future work will focus on both developing *new* mining techniques and on using *existing* techniques in an innovative way to obtain understandable, high-level information instead of "spaghetti-like" models showing all details. Obviously, we plan to evaluate these results in healthcare organizations such as the AMC.

Acknowledgements. This research is supported by EIT, NWO-EW, the Technology Foundation STW, and the SUPER project (FP6). Moreover, we would like to thank the many people involved in the development of ProM.

References

1. Anyanwu, K., Sheth, A., Cardoso, J., Miller, J., Kochut, K.: Healthcare Enterprise Process Development and Integration. Journal of Research and Practice in Information Technology 35(2), 83–98 (2003)
2. Dumas, M., van der Aalst, W.M.P., ter Hofstede, A.H.M.: Process-Aware Information Systems: Bridging People and Software through Process Technology. Wiley & Sons, Chichester (2005)
3. Elhuizen, S.G., Burger, M.P.M., Jonkers, R.E., Limburg, M., Klazinga, N., Bakker, P.J.M.: Using Business Process Redesign to Reduce Wait Times at a University Hospital in the Netherlands. The Joint Commission Journal on Quality and Patient Safety 33(6), 332–341 (2007)
4. Günther, C.W., van der Aalst, W.M.P.: Finding structure in unstructured processes: The case for process mining. Technical report
5. Lenz, R., Elstner, T., Siegele, H., Kuhn, K.: A Practical Approach to Process Support in Health Information Systems. Journal of the American Medical Informatics Association 9(6), 571–585 (2002)
6. van der Aalst, W.M.P., Reijers, H.A., Song, M.: Discovering Social Networks from Event Logs. Computer Supported Cooperative Work 14(6), 549–593 (2005)
7. van der Aalst, W.M.P., Reijers, H.A., Weijters, A.J.M.M., van Dongen, B.F., Alves de Medeiros, A.K., Song, M., Verbeek, H.M.W.: Business process mining: an industrial application. Information Systems 32(5), 713–732 (2007)
8. van der Aalst, W.M.P., van Dongen, B.F., Günther, C.W., Mans, R.S., Alves de Medeiros, A.K., Rozinat, A., Rubin, V., Song, M., Verbeek, H.M.W., Weijters, A.J.M.M.: ProM 4.0: Comprehensive Support for Real Process Analysis. In: Kleijn, J., Yakovlev, A. (eds.) ICATPN 2007. LNCS, vol. 4546, pp. 484–494. Springer, Heidelberg (2007)
9. van der Aalst, W.M.P., van Dongen, B.F., Herbst, J., Maruster, L., Schimm, G., Weijters, A.J.M.M.: Workflow Mining: A survey of Issues and Approaches. Data and Knowledge Engineering 47(2), 237–267 (2003)

10. van der Aalst, W.M.P., Weijters, A.J.M.M., Maruster, L.: Workflow Mining: Discovering Process Models from Event Logs. IEEE Transactions on Knowledge and Data Engineering 16(9), 1128–1142 (2004)
11. van Dongen, B.F., Busi, N., Pinnaand, G.M., van der Aalst, W.M.P.: An Iterative Algorithm for Applying the Theory of Regions in Process Mining. BETA Working Paper Series, WP 195, Eindhoven University of Technology, Eindhoven (2007)
12. Weijters, A.J.M.M., van der Aalst, W.M.P.: Rediscovering Workflow Models from Event-Based Data using Little Thumb. Integrated Computer-Aided Engineering 10(2), 151–162 (2003)

Breast Contour Detection with Stable Paths

Jaime S. Cardoso, Ricardo Sousa, Luís F. Teixeira, and M.J. Cardoso

INESC Porto, Faculdade de Engenharia, Universidade do Porto, Portugal
{jaime.cardoso,rsousa,luis.f.teixeira}@inescporto.pt,
mjcard@med.up.pt

Abstract. Breast cancer conservative treatment (BCCT), due to its proven on-cological safety, is considered, when feasible, the gold standard of breast cancer treatment. However, aesthetic results are heterogeneous and difficult to evaluate in a standardized way, due to the lack of reproducibility of the subjective meth-ods usually applied. The objective assessment methods, considered in the past as being less capable of evaluating all aspects of BCCT, are nowadays being pre-ferred to overcome the drawbacks of the subjective evaluation. A computer-aided medical system was recently developed to objectively and automatically evaluate the aesthetic result of BCCT. In this system, the detection of the breast contour on the patient's digital photograph is a necessary step to extract the features sub-sequently used in the evaluation process. In this paper an algorithm based on the shortest path on a graph is proposed to detect automatically the breast contour. The proposed method extends an existing semi-automatic algorithm for the same purpose. A comprehensive comparison with manually-drawn contours reveals the strength of the proposed method.

1 Introduction

Breast cancer is the most common cancer to affect women in Europe and as 10-year sur-vival from the disease now exceeds 80%, many women are expected to live a long time with the aesthetic consequences of their treatment. The importance of good aesthetic outcome is well recognised by experts in this field although it is well known that this is often not achieved. In breast-conserving surgery for example, approximately 30% of women will have a suboptimal or poor aesthetic outcome.

A significant obstacle in auditing this problem and evaluating techniques for im-proving it has been the absence of a standard method for measuring aesthetic outcome. Most commonly used methods involve subjective assessment by an expert panel. All these methods are subject to significant intra-observer and inter-observer variability. So, there is a need to replace or enhance human expert evaluation of the aesthetic re-sults of breast surgery with a validated objective tool. This needs to be easy to employ, completely reproducible and acceptable to those who would be evaluated.

Initially, objective methods consisted of the comparison between the two breasts with simple measurements marked directly in patients or in photographs [1,2]. The correla-tion of objective measurements with subjective overall evaluation has been reported by several authors [2]. Nevertheless, objective methods usually assessed just one aspect of cosmesis such as nipple symmetry. Trying to overcome the sense that objective asym-metry measurements were insufficient, other groups proposed the sum of the individual

A. Fred, J. Filipe, and H. Gamboa (Eds.): BIOSTEC 2008, CCIS 25, pp. 439–452, 2008.

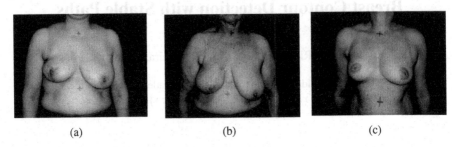

(a)   (b)   (c)

Fig. 1. Typical photographs

scores of subjective and objective individual indices [3]. More recently, a computer-aided medical system was developed to objectively and automatically perform the aesthetic evaluation of BCCT [4]. The development of this system entailed the automatic extraction of several features from the patient's photographs (Fig. 1), capturing some of the factors considered to have impact on the overall cosmetic result: breast asymmetry, skin colour changes due to the radiotherapy treatment and surgical scar appearance. In a second phase, a support vector machine classifier was trained to predict the overall cosmetic result from the recorded features [4].

In order to extract the identified relevant features from the image, the detection of the breast contour is necessary. In [5,6] we describe a *semi-automatic* method for the detection of the breast contour. The user has to *manually* identify the two endpoints of the breast contour. Subsequently, the algorithm automatically finds the contour in-between. The algorithm has been implemented in the system described above: the software automatically finds the contours, extracts relevant features and outputs a predicted overall cosmetic assessment (*excellent, good, fair,* or *poor*).

Here, we improve on the work of [5,6] in two different directions. First, we present an algorithm for the automatic detection of the endpoints of the breast contour, thus eliminating any user input from the process. Therefore a totally automatic breast contour detection is achieved. Next, we provide a thorough evaluation of the performance of the proposed method against manually-drawn breast contours. Standard metrics are employed to compare two contours.

Before presenting the proposed approach, and for completeness, we recover the framework for breast contour detection between two *known endpoints* of [5,6]. Then, in Section 3 we detail how to automatically find the endpoints of a breast contour. Examples are provided and a performance analysis is conducted in Section 4. Finally, in Section 5, we conclude the paper and present possible directions of future work.

2 A Shortest Path Approach to Contour Detection

When the two endpoints of the breast contour are known, we are left with the problem of finding the path between both endpoints going through the breast contour. Since the interior of the breast itself is essentially free of edges, the path we are looking for is the shortest path between the two endpoints, **if** paths (almost) entirely through edge pixels are favoured. More formally, let s and t be two pixels of the image and $\mathcal{P}_{s,t}$ a path over the image connecting them. We are interested in finding the path $\mathcal{P}$ that optimizes some predefined distance $d(s,t)$. This criterion should embed the need to favour edge pixels.

In the work to be detailed, the image grid is considered as a graph with pixels as nodes and edges connecting neighbouring pixels. Therefore, some graph concepts are in order.

2.1   Background Knowledge

A *graph* $G = (V, A)$ is composed of two sets V and A. V is the set of nodes, and A the set of arcs (p, q), $p, q \in V$. The graph is *weighted* if a weight $w(p, q)$ is associated to each arc. The weight of each arc, $w(p, q)$, is a function of pixels values and pixels relative positions. A path from vertex (pixel) v_1 to vertex (pixel) v_n is a list of unique vertices $v_1, v_2, \ldots, v_n$, with v_{i-1} and v_i corresponding to neighbour pixels. The total cost of a path is the sum of each arc weight in the path $\sum_{i=2}^{n} w(v_{i-1}, v_i)$.

A path from a source vertex v to a target vertex u is said to be the *shortest path* if its total cost is minimum among all v-to-u paths. The distance between a source vertex v and a target vertex u on a graph, $d(v, u)$, is the total cost of the shortest path between v and u.

A path from a source vertex v to a sub-graph Ω is said to be the shortest path between v and Ω if its total cost is minimum among all v-to-$u \in \Omega$ paths. The distance from a node v to a sub-graph Ω, $d(v, \Omega)$, is the total cost of the shortest path between v and Ω:

$$d(v, \Omega) = \min_{u \in \Omega} d(v, u). \tag{1}$$

A path from a sub-graph Ω_1 to a sub-graph Ω_2 is said to be the shortest path between Ω_1 and Ω_2 if its total cost is minimum among all $v \in \Omega_1$-to-$u \in \Omega_2$ paths. The distance from a sub-graph Ω_1 to a sub-graph Ω_2, $d(\Omega_1, \Omega_2)$, is the total cost of the shortest path between Ω_1 and Ω_2:

$$d(\Omega_1, \Omega_2) = \min_{v \in \Omega_1, u \in \Omega_2} d(v, u). \tag{2}$$

In graph theory, the shortest-path problem seeks the shortest path connecting two nodes; efficient algorithms are available to solve this problem, such as the well-known Dijkstra algorithm [7].

2.2   Base Algorithm for Breast Contour Detection

If the weight assigned to an arc in the graph captures the edge strength of the incident pixels, finding the best contour translates into computing the minimum accumulated weight along all possible curves connecting s and t:

$$d(s, t) = \min_{\mathcal{P}_{s,t}} \sum w(p, q). \tag{3}$$

Note that, if we ignore the weight component, we are simply computing the regular Euclidian distance between s and t along the path $\mathcal{P}_{s,t}$ (which will be a straight line for the shortest path). Therefore, to detect the breast contour the framework in [5] encompasses:

1. A gradient computation of the original image. In a broader view, this can be replaced by any feature extraction process that emphasizes the pixels we are seeking for.
2. Consider the gradient image as a weighted graph with pixels as nodes and arcs connecting neighbouring pixels. Assign to an arc an weight w determined by the gradient values of the two incident pixels.

The weight of the arc connecting 4-neighbour pixels p and q was expressed as an exponential law

$$\widehat{f}(g) = \alpha \, \exp(\beta \, (255 - g)) + \gamma, \tag{4}$$

with $\alpha, \beta, \gamma \in \mathbb{R}$ and g is the minimum of the gradient computed on the two incident pixels. For 8-neighbour pixels the weight was set to $\sqrt{2}$ times that value. The parameters α, β and γ were experimentally tuned using a grid search method, yielding $\alpha = 0.15$, $\beta = 0.0208$ and $\gamma = 1.85$.

2.3 Breast Contour Detection with Shape Priors

To enforce shape constraints, deformable templates are often used. Prototype-based deformable templates are usually constructed from a set of training examples. Then principal component analysis is applied to define an average template shape and modes of variation [8,9,10]. Unfortunately, when graph algorithms are used, the cost function is influenced by only two consecutive pixels at a time, whereas more pixels would be needed to encode shape information. Nevertheless, graph algorithms have the advantage over gradient descent—typical on active contours—in that they recover the optimal solution and do not get trapped in local minima. Therefore, in [6], we proposed different approaches to encode the a priori shape knowledge in the graph design. The weight function was modified by a term reflecting our knowledge of the shape probability. The shortest path problem is then formulated on this modified weighted graph. The key advantage of this approach is that the shortest path computation is left unchanged and hence the procedure remains efficient and discovers the optimal solution.

Although parametric models may seem appealing, they did not achieve a performance as good as non-parametric models for this specific task [6]. Therefore, we only summarize here the latter.

Non-parametric Mask Prior. Supposing that we have a training set $C = \{C_i | i = 1, 2, \ldots, M\}$ of M *registered* breast contours, manually outlined, we propose to construct a nonparametric breast shape prior by defining an admissible breast contour region. Considering the breast contours as closed curves (by connecting the start and ending points), we defined the difference between the reunion and the intersection of all breast contour regions as

$$\mathcal{R} = \cup_{i=1}^{M} \mathcal{C}_i - \cap_{i=1}^{M} \mathcal{C}_i.$$

The region $\mathcal{R}$ is the smallest simply connected region that contains all breast contours on the training set. If the training set is representative of the population—it was drawn from the underlying breast shape and size distribution—, one expects that unseen breast contours will be found (almost) entirely inside the region $\mathcal{R}$.

To incorporate this prior information into the breast contour detection process, the image is represented by its Euclidean distance map $\Phi(x_i, y_i)$, where the distance from each pixel is computed to the nearest pixel belonging to the region $\mathcal{R}$. The expression of the weight function is generalised to include also a term penalising pixels with a high Euclidean distance value:

$$\widehat{f}(g, d) = \alpha \, \exp\left(\beta_1 \, (255 - g) + \beta_2 \, d\right) + \gamma, \tag{5}$$

with $\alpha, \beta_1, \beta_2, \gamma \in \mathbb{R}$ and d the average of the Euclidean distance values computed on the two incident pixels. The rationale supporting this decision is that the contour points will, with high probability, be inside region $\mathcal{R}$ and the probability of belonging to the breast contour decays with the increase of the Euclidean distance. The parameters α, β_1, β_2 and γ where experimentally tuned with the same method as before, yielding $\alpha = 0.27$, $\beta_1 = 0.020$, $\beta_2 = 0.07$ and $\gamma = 2$.

Non-parametric Unimodal Model. Intuitively, the breast contour is monotonically decreasing from one endpoint down to a minimum point, located approximately in the middle (of the breast) position, and from there it monotonically increases up to the other endpoint.

This simple perception about the shape of the breast can be enforced by running the shortest path algorithm in a phased process, instead of directly finding the path between the two given endpoints. To favour a first descending path, we find the shortest path between the external endpoint and a point located at the same column as the internal endpoint, but yGap rows below. We set yGap to twice the x-difference between the two endpoints (and therefore relatively robust to the size of the patient in the image). In the second step of the process, we compute a second shortest path between the point of the first path at the x-middle position and the internal endpoint. Fig. 2 illustrates the technique.

Note that, in spite of the name, the unimodal model does capture any fluctuation or irregularity in the evolution of contour, as it retains the flexibility of the shortest path process. Secondly, although the auxiliary point is positioned only twice width of the breast bellow the endpoint, the contour can, naturally, go below that point—the actual shape is again captured by the shortest path. Finally, observe that the unimodal model relies only in Equation (4) to construct the weighted graph.

Non-parametric Template Matching. We now introduce a third non-parametric process. The template matching strategy tries to find the contour in the training set that best fits in the image. So, instead of constructing a region with all training contours, a single contour is selected as the prior. The steps of getting the most similar contour encompass creating an edge map from the image under evaluation and applying the distance transform on the extracted edge map, obtaining the Euclidean distance map $\Phi(x_i, y_i)$. The contour selected is the one minimizing the overall distance (normalized by its length):

$$\min_{\mathcal{C}_j} \sum_{i|(x_i, y_i)\in\mathcal{C}_j} \Phi(x_i, y_i)/\mathrm{length}(\mathcal{C}_j)$$

The underlying assumption is that the training set is representative of the population of possible breast contours and by that a 'sufficiently' similar to the contour to be detected is present on the training set. The template matching procedure proceeds now

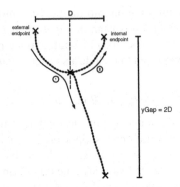

Fig. 2. Steps of the unimodal technique when applied on the left breast

exactly as the mask prior model, with the Euclidean distance computed in reference to the selected template contour. The template model relies on Equation 5 to construct the weight graph where the parameters α, β_1, β_2 and γ where experimentally tuned, yielding $\alpha = 0.17, \beta_1 = 0.0208, \beta_2 = 0.055$ and $\gamma = 2$.

3   Automatic Detection of Endpoints

The challenge now is to automatically extract the endpoints. In order to address this key problem we will assume, very reasonably, that a photo contains only the torso of the patient.

The position of the *external* endpoint of the breast contour can be assumed at the point of the body where the arm contour intersects the trunk contour. However, because patients are in the arms-down position, the arm's contour is almost indistinguishable from the trunk's contour. Therefore, we define the external endpoint of the breast contour as the highest point of the trunk contour.

The approach just delineated requires first the detection of the trunk contour. That may be searched among the strongest lines of gradient with approximate vertical direction.

3.1   Trunk Contour Detection

A strong vertical path corresponds to a path between two rows of the image, (almost) always through pixels with strong gradient values. Strong vertical paths are best modelled as paths between two regions Ω_1 and Ω_2, i.e. two rows of the image.

One may assume the simplifying assumption that the vertical paths do not zigzag back and forth, up and down. Therefore, the search may be restricted among connected paths containing one, and only one, pixel in each row between the two end-rows. Formally, let I be an $N_1 \times N_2$ image with N_1 columns and N_2 rows; define a vertical path to be

$$\mathbf{s} = \{(x(y), y)\}_{y=\Omega_1}^{\Omega_2}, \text{ s.t. } \forall y \, |x(y) - x(y-1)| \le 1,$$

where x is a mapping $x : [\Omega_1, \cdots, \Omega_2] \to [1, \cdots, N_1]$. That is, a vertical path is an 8-connected path of pixels in the image from Ω_1 to Ω_2, containing one, and only one, pixel in each row of the image.

The optimal vertical path that minimizes this cost can be found using dynamic programming (instead of resorting to a more complex algorithm for the generic shortest path problem). The first step is to traverse the image from the first to the last row and compute the cumulative minimum cost C for all possible connected paths lines for each entry (i, j):

$$C(i,j) = \min \begin{cases} C(i-1, j-1) + w(p_{i-1,j-1}; p_{i,j}) \\ \quad C(i, j-1) + w(p_{i,j-1}; p_{i,j}) \\ C(i+1, j-1) + w(p_{i+1,j-1}; p_{i,j}) \end{cases},$$

where $w(p_{i,j}; p_{l,m})$ represents the weight of the arc incident with pixels at positions (i, j) and (l, m). At the end of this process,

$$\min_{i \in \{1, \cdots, N_1\}} C(i, N_2)$$

indicates the end of the minimal connected path.

Stable Paths on a Graph. The procedure just delineated allows the computation of the single shortest path between two regions. However, we are interested in more than one strong path, possibly not exactly the shortest path (as the silhouette may constitute a shorter path). The concept of stable path allows the computation of a sort of local optimal paths.

Definition. A path $P_{s,t}$ is a stable path between regions Ω_1 and Ω_2 if $P_{s,t}$ is the shortest path between $s \in \Omega_1$ and the whole region Ω_2, and $P_{s,t}$ is the shortest path between $t \in \Omega_2$ and the whole region Ω_1.

The naming of stable path has its roots in dynamical systems, as it resembles stable fixed points. If one considers the function $\mathcal{F}_{\Omega_1 \rightarrow \Omega_2}\{\}$, mapping a node $s \in \Omega_1$ to a node $t \in \Omega_2$ by finding the shortest path $P_{s,t}$ between s and Ω_2, with $t = \mathcal{F}_{\Omega_1 \rightarrow \Omega_2}\{s\}$ as the end node of such shortest path, then

$$\mathcal{G}_{\Omega_1 \rightarrow \Omega_1}\{s\} = \mathcal{F}_{\Omega_2 \rightarrow \Omega_1}\{\mathcal{F}_{\Omega_1 \rightarrow \Omega_2}\{s\}\} = \mathcal{F}_{\Omega_1 \rightarrow \Omega_2}\{\mathcal{F}_{\Omega_2 \rightarrow \Omega_1}\{s\}\} = s$$

if and only if $P_{s,t}$ is a stable path. Note that the concept of stable path is valid for any graph and any two sub-graphs in general.

The computation of all stable paths in the graph derived from the image has only roughly twice the complexity of the shortest path computation. The first step corresponds verbatim to the computation of the shortest path presented at the beginning of this Section. In a second step one repeats the same procedure, traversing now the graph from the last to the first row. At the end of this process, if the two endpoints of a direct and reverse path coincide, we are in the presence of a stable point.

Phase 1 of the Endpoint Detection. We propose to apply the stable path concept to the bottom half of our photographs to find the trunk contour:

1. Compute the gradient of the image (see Fig. 3(a)).
2. Compute the shortest path between each point in the bottom row and the *whole* middle row of the gradient image (see Fig. 3(b)).

(a) Gradient

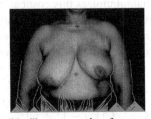

(b) Shortest paths from a pixel t in the bottom line and the whole middle row Ω_1, superimposed on the original image

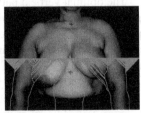

(c) Shortest paths from a pixel s in the middle row and the whole bottom row Ω_2, superimposed on the original image.

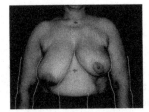

(d) Stable paths between the middle and bottom rows

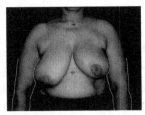

(e) Selected paths, superimposed on the original image

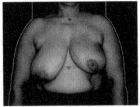

(f) Stable paths between the top and bottom rows

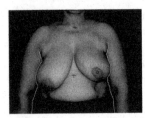

(g) The external endpoint of the breast contour is the highest point of the shortest path

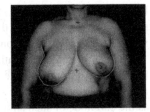

(h) Totally automated breast contour

Fig. 3. Results for a real photograph

3. Compute the shortest path between each point in the middle row and the *whole* bottom row of the gradient image (see Fig. 3(c)).
4. Discard all paths except for those common to steps 2 and 3 (see Fig. 3(d)). These first steps correspond to the computation of the stable paths between the middle and bottom rows.
5. Discard paths with a cost superior to half of the maximum possible cost (see Fig. 3(e)). Finally, the trunk contour is defined as the two contours closest to the middle of the photograph.

At the end of this phase we have already the position of the two trunk contours, but we have stopped the process at the middle of the image. It is important to stress that if the process was conducted between the bottom and top rows, the trunk contours would be lost, as the only strong paths between the top and bottom rows would be the external silhouette of the patient (see Fig. 3(f)) .

Phase 2 of the Endpoint Detection. To determine the top of the trunk contour, we need to continue the path produced in phase 1 until a certain condition is met. Towards that end, we propose to find the shortest path between the ending point of the stable contour found in phase 1 and row R_i, $R_i = middle\_row, \cdots, top\_row$. We select the highest row for which the shortest path does not contain a long sequence (length_threshold) of consecutive pixels with low gradient (gradient_threshold). Fig. 3(g) illustrates the results obtained for the exemplificative photograph.
(length_threshold and gradient_threshold were set to 12 and 48, respectively).

Before applying the algorithm presented in Section 2 to compute the breast contour we need also the internal endpoint of the breast contour. This was estimated simply as the middle point between the two external endpoints. Finally, the computation of the breast contour yields the result presented in Fig. 3(h).

4   Results

The methodology proposed in this paper was assessed on a set of photographs from 120 patients. The photographs were collected in three different institutions in Portugal. All patients were treated with conservative breast surgery, with or without auxiliary surgery, and whole breast radiotherapy, with treatment completed at least one year before the onset of the study. Breast images were acquired employing a 4M pixel digital camera.

A mark was made on the skin at the suprasternal notch and at the midline 25 cm below the first mark (see Fig. 1). These two marks create a correspondence between pixels measured on the digital photograph and the length in centimetres on the patient. In order to investigate the possibility of defining an automated method of detecting the breast contour, a set of patients with known breast contour was required. Since, ideally, the automated method should correlate coherently with human assessment, eight different observers were asked to manually draw the contours. A software tool was developed specifically to assist on this job. The user defines the contour by positioning seventeen control points of cubic splines, see Fig. 4.

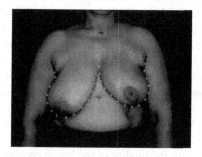

Fig. 4. Software for manual breast contour definition

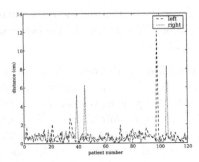

Fig. 5. Evolution of error (cm) in the position of the external endpoints of the breast contour over 120 photographs

Before applying the proposed algorithm, each image was downsized to a constant width of 768 pixels, while keeping the aspect ratio. This improves the computational performance of the implementation of the software, without degrading the quality of the final result.

Fig. 5 shows the evolution of the error when estimating the external endpoints' position of the breast contour. The error in pixels was scaled to centimetres with the help of the marks made on the skin of the patient.

Table 1 summarizes the results.

Table 1. Mean, standard deviation and maximum value of errors in the position of the endpoints

Error (cm)	mean	std dev	max
left endpoint	0.7	1.1	12.1
right endpoint	0.7	1.0	8.3
total	0.7	1.1	12.1

It can be observed that the proposed algorithm has a very interesting performance. The average error is quite low, less than 1 centimetre. Fig. 6 shows some of the photographs for each the algorithm worked satisfactorily. It represents the result after the two phases of the algorithm. The highest point of the trunk contour provides the detected external endpoint of the breast contour. It is visible in patient #35 that the algorithm is robust against cluttered background. It is also visible in Fig. 6 that, although the stable paths detected in phase 1 do not always correspond exactly to the trunk contour, the algorithm is still able to successfully detect the endpoints.

Nevertheless, four endpoints were clearly misplaced. These results, displayed in Fig. 7, bring to light some of the limitations of the current state of the proposed approach. In patients #39, #45 and #105 the shortest path followed a 'wrong' contour, misplacing the endpoint. With patient #98 the long hair created a false 'path' till the top of the image.

In a last set of experiments, the quality of the breast contour tracking algorithm was assessed. Instead of a simple subjective evaluation as provided in [5], we conducted a complete objective evaluation, based on the hausdorff and the average distances to

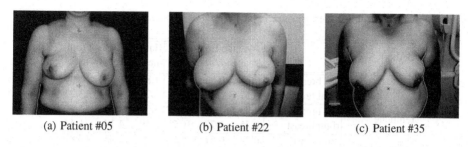

(a) Patient #05   (b) Patient #22   (c) Patient #35

Fig. 6. Selected successful results

compare two contours. The hausdorff distance is defined as the "maximum distance of a set to the nearest point in the other set". Roughly speaking, it captures the maximum separation between the manual and the automatic contours.

Comparing the different models to detect the breast contour one can conclude from Table 2 that the simplest model, the unimodal prior models, and although its simplicity, exhibits a very competitive performance. As such, the experimental analysis is continued only with this model.

As observed in Fig. 8(a) and Table 2 the experimental values obtained for the hausdorff distance correlate well with the error on the endpoints. This fact means that, most of the times, the major error on the automatic contour is located on the endpoint, with

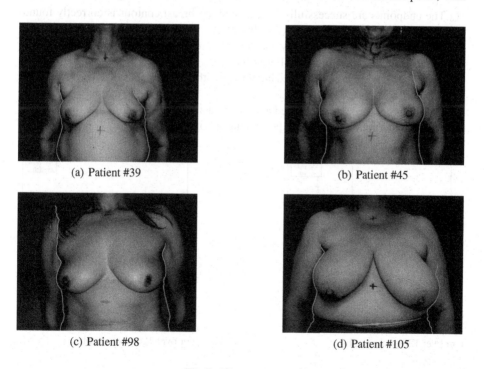

(a) Patient #39   (b) Patient #45

(c) Patient #98   (d) Patient #105

Fig. 7. All poor results

Table 2. Performance in the position of the breast contour

Method		Average Error (cm)			Hausdorff Error (cm)		
		mean	std. dev.	max.	mean	std. dev.	max.
Mask	left breast	0.25	0.61	5.52	1.81	1.78	14.04
	right breast	0.27	0.68	4.91	1.78	1.87	13.22
Template	left breast	0.75	2.08	15.87	3.30	4.63	26.90
	right breast	0.75	1.77	10.24	3.19	4.14	21.44
Unimodal	left breast	0.24	0.61	6.23	1.90	2.05	17.82
	right breast	0.30	0.88	8.81	1.98	2.45	21.44

the shortest path algorithm recovering the true contour rapidly. A clear exception is patient #73, for which the error in the endpoints is negligible but the hausdorff distance between contours is very high. Here, the contour tracking algorithm missed to follow the contour, although it received proper endpoints.

The average distance between two contours captures better the perceived quality of the automatic breast contour. Here, the distance is averaged over the whole contour. Fig. 8(b) and Table 2 summarize these results.

The application of the breast contour detection after the automatic positioning of the ending points can result in different scenarios:

1. The endpoints are successfully located and the breast contour is correctly found. This desirable result is illustrated in Fig. 9(a).
2. The endpoints are successfully located but the algorithm misses to follow adequately the breast contour (see Fig. 9(b)).
3. The endpoints are poorly located but the algorithm rapidly finds and tracks the right breast contour (see Fig. 9(c)).
4. The endpoints are poorly located and the breast contour is incorrectly tracked (this scenario did not occur in the experimental set of photographs).

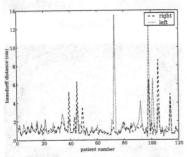

(a) Evolution of hausdorff distance (cm) in the position of the breast contour over 120 photographs

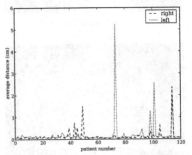

(b) Evolution of the average distance (cm) in the position of the breast contour over 120 photographs

Fig. 8. Breast contour results

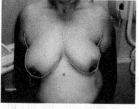
(a) Patient #35

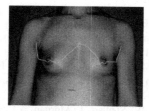
(b) Patient #114

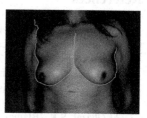
(c) Patient #98

Fig. 9. Breast contour results

As expected by visual inspection patient of #98 in Fig. 9(c), and although the error on the endpoint's location was high, the breast contour algorithm recovered rapidly the correct contour, translating into a final small average distance (but a high hausdorff distance).

The consequences of the diverse errors to the computer-aided medical system are different and have to be further studied. For example, one of the features used for the objective aesthetic evaluation of the BCCT is the difference between the levels of inferior breast contour points. This measure is quite robust over strong errors on the endpoints' position, as long as the breast contour is correctly tracked. Other features, such as the difference in the area of the breast are much more sensible to the positioning of the endpoints. One line of future investigation is the selection of features robust to the common errors of the automatic detection of notable points (contour endpoints, breast contour, nipples, etc) but still capturing adequately the aesthetic result, leading to a good classification performance.

5 Conclusions

A method has been described for applying graph concepts to the task of automatically extracting the breast contour in digital photographs of the torso of a patient, after being submitted to a breast cancer conservative treatment. In the proposed framework the problem of finding the endpoints of the breast contour is formulated as a problem of finding strong contours between two regions, a concept introduced here for the first time. The breast contour is found as the solution to the shortest problem on a graph, after conveniently modelling the image as a weighted graph. Preliminary results indicate a good performance in the tasks of finding the external endpoints of the contour and on detecting the breast contour. Future work will focus on improvements to the algorithm including generalizing the solution to other typical patient positions in these studies.

Acknowledgements. This work was partially funded by Fundação para a Ciência e a Tecnologia (FCT) - Portugal through project PTDC/EIA/64914/2006.

References

1. Limbergen, E.V., Schueren, E.V., Tongelen, K.V.: Cosmetic evaluation of breast conserving treatment for mammary cancer. 1. proposal of a quantitative scoring system. Radiotherapy and oncology 16, 159–167 (1989)
2. Christie, D.R.H., O'Brien, M.Y., Christie, J.A., Kron, T., Ferguson, S.A., Hamilton, C.S., Denham, J.W.: A comparison of methods of cosmetic assessment in breast conservation treatment. Breast 5, 358–367 (1996)
3. Al-Ghazal, S.K., Blamey, R.W., Stewart, J., Morgan, A.L.: The cosmetic outcome in early breast cancer treated with breast conservation. European journal of surgical oncology 25, 566–570 (1999)
4. Cardoso, J.S., Cardoso, M.J.: Towards an intelligent medical system for the aesthetic evaluation of breast cancer conservative treatment. Artificial Intelligence in Medicine 40, 115–126 (2007)
5. Cardoso, J.S., Cardoso, M.J.: Breast contour detection for the aesthetic evaluation of breast cancer conservative treatment. In: International Conference on Computer Recognition Systems (CORES 2007) (2007)
6. Sousa, R., Cardoso, J.S., da Costa, J.F.P., Cardoso, M.J.: Breast contour detection with shape priors. In: International Conference on Image Processing (ICIP 2008) (2008)
7. Dijkstra, E.W.: A note on two problems in connexion with graphs. Numerische Mathematik 1, 269–271 (1959)
8. Tsai, A., Yezzi, A., Wells, W., Tempany, C., Tucker, D., Fan, A., Grimson, W.E., Willsky, A.: A shape-based approach to the segmentation. of medical imagery using level sets. IEEE Transactions on Medical Imaging 22, 137–154 (2003)
9. Dydenko, I., Jamal, F., Bernard, O., D'Hooge, J., Magnin, I.E., Friboulet, D.: A level set framework with a shape and motion prior for segmentation and region tracking in echocardiography. Medical Image Analysis 10, 162–177 (2006)
10. Yan, P., Kassim, A.A.: Medical image segmentation using minimal path deformable models with implicit shape priors. IEEE Transactions on Information Technology in Biomedicine 10, 677–684 (2006)

Formal Verification of an Agent-Based Support System for Medicine Intake

Mark Hoogendoorn, Michel C.A. Klein, Zulfiqar A. Memon, and Jan Treur

Vrije Universiteit Amsterdam, Department of Artificial Intelligence
De Boelelaan 1081a, 1081 HV Amsterdam, The Netherlands
{mhoogen,michel.klein,za.memon,treur}@few.vu.nl

Abstract. In this paper we present the design of an agent system for medicine usage management. Part of it is an intelligent ambient agent which incorporates an explicit representation of a dynamical system model to estimate the medicine level in the patient's body. By simulation the ambient agent is able to analyse whether the patient intends to take the medicine too early or too late, and can take measures to prevent this.

1 Introduction

A challenge for medicine usage management is to achieve in a non-intrusive manner that patients for whom it is crucial that they take medicine regularly, indeed do so. Examples of specific relevant groups include independently living elderly people, psychiatric patients or HIV-infected persons. One of the earlier solutions reported in the literature provides the sending of automatically generated SMS reminder messages to a patient's cell phone at the relevant times; e.g. [1]. A disadvantage of this approach is that patients are disturbed often, even if they do take the medicine at the right times themselves, and that due to this after some time a number of patients start to ignore the messages.

A more sophisticated approach can be based on a recently developed automated medicine box that has a sensor that detects whether a medicine is taken from the box, and can communicate this to a server; cf. SIMpill [2]. This enables to send SMS messages only when at a relevant point in time no medicine intake is detected. A next step is to let a computing device find out more precisely what relevant times for medicine intake are. One way is to base this on prespecified prescription schemes that indicate at what time points medicine should be taken. However, this may be inflexible in cases that a patient did not follow the scheme precisely. To obtain a more robust and flexible approach, this paper explores and analyses possibilities to use an automated medicine box in combination with model-based intelligent agents to dynamically determine the (estimated) medicine level over time.

The agent-based model for medicine usage management discussed was formally specified in an executable manner and formally analysed using dedicated tools. The system incorporates a model-based intelligent agent that includes an explicitly represented dynamic system model to estimate the medicine level in the patient's body by simulation. Based on this it is able to dynamically determine at what point in time the

A. Fred, J. Filipe, and H. Gamboa (Eds.): BIOSTEC 2008, CCIS 25, pp. 453–466, 2008.

patient should take medicine, and given that, to analyse whether the patient intends to take medicine too early or too late, and to take measures to prevent this.

In this paper, Section 2 describes the multi-agent system introduced, whereas Section 3 present detailed information about the specific agents. Furthermore, Section 4 presents simulation results, and Section 5 formal analysis of these results. Finally, Section 6 is a discussion.

2 Overview of the System

Figure 1 presents an overview of the entire system as considered. The top right corner shows the patient, who interacts with the medicine box, and communicates with the patient cell phone. The Medicine Box detects whether medicine is taken out of the medicine box. The Medicine Box Agent (MBA) observes this medicine box. In case, for example, the patient intends to take the medicine too soon after the previous dose, it finds out that the medicine should not be taken at the moment (i.e., the sum of the estimated current medicine level plus a new dose is too high), and communicates a warning to the patient by a beep. Furthermore, all information obtained by this agent is passed on to the Usage Support Agent (USA). All information about medicine usage is stored in the patient database by this agent. If the patient tries to take the medicine too early, a warning SMS with a short explanation is communicated to the cell phone of the patient, in addition to the beep sound already communicated by the Medicine Box Agent.

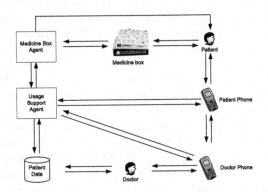

Fig. 1. Multi-Agent System: Overview

On the other hand, in case the Usage Support Agent finds out that the medicine is not taken early enough (i.e., the medicine concentration is estimated too low for the patient and no medicine was taken yet), it can take measures as well. First of all, it can warn the patient by communicating an SMS to the patient cell phone. This is done soon after the patient should have taken the medicine. In case that after some time the patient still does not take medicine, the agent can communicate an SMS to cell phone of the appropriate doctor. The doctor can look into the patient database to see the medicine usage, and in case the doctor feels it is necessary to discuss the state of affairs with the patient, he or she can contact the patient via a call using the doctor cell phone to the patient cell phone.

3 Agent Properties

The model used for the Usage Support Agent (USA) makes (re)use of elements of the Generic Agent Model GAM described in [3]. In addition, it makes use of an explicitly represented dynamical model to for the medicine level over time within the patient. Moreover, the model for the Usage Support Agent includes a reasoning method (based on simulation) to estimate the current medicine level based on the dynamical model and information on medicine taking in the past.

To express the agent's internal states and processes, a state ontology partly shown in Table 1 was specified. An example of an expression that can be formed by combining elements from this ontology is

belief(leads_to_after(I:INFO_EL, J:INFO_EL, D:REAL))

which expresses that the agent has the knowledge that state property I leads to state property J with a certain time delay specified by D. This type of expression is used to represent the agent's knowledge of a dynamical model of a process. Using the ontology, the functionality of the agent has been specified by generic and domain-specific temporal rules.

Table 1. Ontology for the Usage Support Agent Model

Formalisation	Description
belief(I:INFO_EL)	information I is believed
world_fact(I:INFO_EL)	I is a world fact
has_effect(A:ACTION, I:INFO_EL)	action A has effect I
leads_to_after(I:INFO_EL, J:INFO_EL, D:REAL)	state property I leads to state property J after duration D
at(I:INFO_EL, T:TIME)	state property I holds at time T

Note the formal form of the agent properties has been omitted below for the sake of brevity. Generic rules specify that incoming information (by observation or communication) from a source that is believed to be reliable is internally stored in the form of beliefs. When the sources are assumed always reliable, the conditions on reliability can be left out:

IB(X) From Input to Beliefs
If agent X observes some world fact, then it will believe this.
If X gets information communicated, then it will believe this.

Execution of a dynamical model is specified by:

SE(X) Simulation Execution
If it is believed that I holds at time T, and it is believed that I leads to J after time duration D, then it is believed that J holds at time T+D.

This temporal rule specifies how a dynamic model that is represented as part of the agent's knowledge can be used by the agent to extend its beliefs about the world at different points in time.

Domain-specific rules for the Usage Support Agent are shown below. The Usage Support Agent's specific functionality is described by three sets of temporal rules. First, the agent maintains a dynamic model for the concentration of medicine in the patient over time in the form of a belief about a *leads to after* relation.

USA1: Maintain Dynamic Model
The Usage Support Agent believes that if the medicine level for medicine M is C, and the usage effect of the medicine is E, then after duration D the medicine level of medicine M is C+E minus G*(C+E)*D with G the decay value. Formally:

In order to reason about the usage information, this information is interpreted (USA2), and stored in the database (USA3).

USA2: Prepare Storage Usage
If the agent has a belief concerning usage of medicine M and the current time is T, then a belief is generated that this is the last usage of medicine M, and the intention is generated to store this in the patient database.

USA3: Store Usage in Database
If the agent has the intention to store the medicine usage in the patient database, then the agent performs this action.

Finally, temporal rules were specified for taking the appropriate measures. Three types of measures are possible. First, in case of early intake, a warning SMS is communicated (USA4). Second, in case the patient is too late with taking medicine, a different SMS is communicated, suggesting to take the medicine (USA5). Finally, when the patient does not respond to such SMSs, the doctor is informed by SMS (USA6).

USA4: Send Early Warning SMS
If the agent has the belief that an intention was shown by the patient to take medicine too early, then an SMS is communicated to the patient cell phone that the medicine should be put back in the box, and the patient should wait for a new SMS before taking more medicine.

USA5: SMS to Patient when Medicine not Taken on Time
If the agent has the belief that the level of medicine M is C at the current time point, and the level is considered to be too low, and the last message has been communicated before the last usage, and at the current time point no more medicine will be absorbed by the patient due to previous intake, then an SMS is sent to the patient cell phone to take the medicine M.

USA6: SMS to Doctor when no Patient Response to SMS
If the agent has the belief that the last SMS to the patient has been communicated at time T, and the last SMS to the doctor was communicated before this time point, and furthermore, the last recorded usage is before the time point at which the SMS has been sent to the patient, and finally, the current time is later than time T plus a certain delay parameter for informing the doctor, then an SMS is communicated to the cell phone of the doctor that the patient has not taken medicine M.

USA7: Communicate Current Concentration
If the agent has the belief that the level of medicine M is C at the current time point then the agent informs the medicine box agent about this level.

The Medicine Box Agent has functionality concerning communication to both the patient and the Usage Support Agent. Generic temporal rules are included as for the Usage Support Agent (see above). Domain-specific temporal rules are both shown below. First of all, the observed usage of medicine is communicated to the Usage Support Agent in case the medicine is not taken too early, as specified in MBA1.

MBA1: Medicine Usage Communication
If the Medicine Box Agent has a belief that the patient has taken medicine from a certain position in the box, and that the particular position contains a certain type of medicine M, and taking the medicine does not result in a too high medicine concentration of medicine M within the patient, then the usage of this type of medicine is communicated to the USA.

In case medicine is taken out of the box too early, a warning is communicated by a beep and the information is forwarded to the Usage Support Agent (MBA2 and MBA3).

MBA2: Too Early Medicine Usage Prevention
If the Medicine Box Agent has the belief that the patient has taken medicine from a certain position in the box, that this position contains a certain type of medicine M, and taking the medicine results in a too high medicine concentration of medicine M within the patient, then a warning beep is communicated to the patient.

MBA3: Early Medicine Usage Communication
If the Medicine Box Agent has a belief that the patient was taking medicine from a certain position in the box, and that the particular position contains a certain type of medicine M, and taking the medicine would result in a too high concentration of medicine M within the patient, then this is communicated to the Usage Support Agent.

4   Simulation

In order to show how the above presented system functions, the executable properties have been implemented in a dedicated software environment that can execute such specifications [4]. To enable creation of simulations, a patient model is used that simulates the behaviour of the patient in a stochastic manner. The model specifies four possible behaviours of the patient, each with its own probability: (1) too early intake, (2) correct intake (on time), (3) responding to an SMS warning that medicine should be taken, and (4) responding to a doctor request by phone. Based upon such probabilities, the entire behaviour of the patient regarding medicine usage can be simulated. In the following simulations, values of respectively 0.1, 0.8, 0.9 and 1.0 have been used.

Figure 2 shows an example of a simulation trace whereby the medicine support system is active. The figure indicates the medicine level over time as estimated by the agent based on its dynamic model. Here the x-axis represents time whereas the y-axis represents the medicine level.

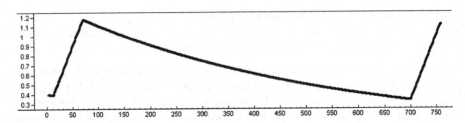

Fig. 2. Medicine level over time

Note that in this case, the minimum level of medicine within the patient is set to 0.35 whereas the maximum level is 1.5. These numbers are based on the medicine half-life value, that can vary per type of medicine. For more details on the formal properties behind the simulation, and a more elaborate discussion of the results, see [5].

5 Formal Analysis

When a model such as the one described above, has been specified, it is easy to produce various simulations based on different settings, initial conditions and external events offered. Moreover, it is possible to incorporate nondeterministic behaviours by temporal rules that involve probabilistic effects [4]. Thus large sets of traces can be generated. When such a set is given, it is more convenient to check them on relevant properties automatically, than going through them by hand. Furthermore, it may also be useful when insight is provided how dynamic properties of the multi-agent system as a whole depend on dynamic properties of the agents within the system, and further on, how these relate to properties of specific components within the agents. This section shows how this can be achieved. To this end a number of *dynamic properties* have been specified for different aggregation levels of the multi-agent system, cf. [7] and [8]. The main property considered for the system as a whole is: will the medicine level in the patient be maintained between the required minimum and maximum level? This desired situation is called 'S'. That a value V of a variable P should be within a specific range between the lower threshold TL and the upper threshold TU, is specified as follows:

$$\text{has\_value}(P,V) \wedge (V > TU \vee V < TL) \tag{S}$$

This has been applied to the variable 'medicine_level'.

GP1
At any point in time the medicine level is between TL and TU.

$$\forall T\text{:TIMEPOINT}, V\text{:REAL: state}(M, T) \models \text{has\_value}(P,V) \Rightarrow (V \leq TU \wedge V \geq TL)$$

Here M is a trace, and state(M, T) denotes the state in this trace at time T. Moreover, state(M, T) ⊨ p denotes that state property p holds in state state(M, T).

Related to this, global properties can be defined that specify that the total number of violations of the threshold values is smaller than some maximum number, or that the total duration of the violation is smaller than some maximum period. In these definitions Σ case(p, 1, 0) is a special construct in the language that calculates the sum of timepoints for which a state holds.

GP2

The total number of times that the medicine level falls below TL or raises above TU is smaller than MAX_OCCURANCES.

∀M:TRACE: ∀T1, T2:TIMEPOINT: Σ case(T1≤T & T ≤T2 & state(M, T) |= S &
state(M, T+1) |= ¬S, 1, 0) < MAX_OCCURANCES

GP3

The total time period that the medicine level is not between TL and TU is smaller than MAX_DURATION.

∀M:TRACE ∀T1, T2:TIMEPOINT: Σ case(T1≤T & T ≤T2 & state(M, T) |= ¬S, 1, 0) < MAX_DURATION

Evaluation of Traces

The formal properties have been used to evaluate the usefulness of the medicine usage management system. For this, 60 simulation traces of the medicine level within a patient have been generated, with a length of 36 hours. In half of the traces the medicine usage management system was supporting the patient, in the other half the system did not take any action, but was still monitoring the medicine usage. As a consequence of the stochastic patient model (the probabilities used are the ones mentioned in Section 5), this resulted in 60 different simulation traces.

For all traces it has been checked whether, how often and how long, the medicine level is between the required values of 0.5 and 1.35. It has also been checked whether this is the case for *preferred* values. It can be assumed that, in addition to the required range for the medicine level, there is also an optimal or preferred range. Table 2 lists the total number of violations in the 30 traces with support of the system and the 30 traces without support for different maximum and minimum levels. Table 3 shows the duration of the violations time in minutes for the same set of traces.

Table 2. Total number of violations of the threshold values (used in property GP2)

number of violations	with support	without
above 1.5 (required)	2	8
below 0.35 (required)	5	18
total required	**7**	**26**

Table 3. Total duration of violations of the threshold values (used in property GP3)

duration (in minutes)	with support	without
above 1.5 (required)	0.8	25.36
below 0.35 (required)	3.53	179.13
total required	**4.33**	**204.49**
above 1.2 (preferred)	161.6	328.20
below 0.5 (preferred)	218.1	327.63
total preferred	**379.7**	**655.83**

From the figures in tables it is immediately apparent that the medicine level is much more often and for a much longer time between the required or preferred values. However, it is also clear that even with support of the system the medicine level is sometimes outside the required range. In fact, property GP1 did not hold in 5 out of the 30 simulation traces in which the system was active. An analysis of these traces revealed that this is a side-effect of the specification of the simulation: as every communication between agents and between components within agents costs at least one time-step, it takes a number of time-steps before a message of the system has reached the patient (in fact 24 minutes). In between sending and receiving a message, it could happen that the medicine level has dropped below the minimum value, or that the patient has taken a pill already. For violations of the minimum level it is possible to compensate for this artificial delay by allowing an additional decrease that is equivalent to the decay of the medicine during the time of the delay. This means that medicine level should not drop below the 0,3335 if the delay is taken into account. Table 4 shows the duration of the violations for the corrected minimum level. In this case, there are no violations of the lower threshold in traces where the system is active.

Unfortunately, a similar correction for violations of the maximum level is not possible, as these violations are caused by taking two pills within a short period, which is a non-monotonic effect.

Table 4. Corrected values for duration the violations of the minimum level

duration (in minutes)	with support	without
below 0.3335	0	164.77

Relating Global Properties to Executable Properties

Besides the verification of properties against simulation traces, the correctness of the entire model can also be proven (given certain conditions). This proves that for all possible outcomes of the model the global properties are indeed achieved under these specific conditions. Such correctness of a model can be proved using the SMV model checker [9]. In order to come to such a proof, an additional property level is introduced, namely the external behavioural properties for the components within the system. Thereafter, the relationship between the executable properties of the Medicine Box Agent, and the Usage Support Agent are related to these behavioural properties, and furthermore, these behavioural properties are shown to entail the top-level properties.

External Behavioural Properties

First a number of external properties for the Usage Support Agent (USA) are introduced. The first property specifies that the patient should be warned that medicine should be taken in case the medicine level is close to being too low (EUSA1). Secondly, property EUSA2 specifies that the USA should warn the doctor in case such a warning has been sent to the patient, but there has been no response. Moreover, the storage of the usage history is specified in EUSA3, and the sending of an early warning message is addressed in EUSA4. Finally, EUSA5 describes that the USA should communicate the current medicine level within the patient. Note that in all properties a parameter e is used, which specifies the maximum delay after which these communications or actions should occur. Such a parameter can vary per property, and is used throughout this section for almost all non-executable properties.

EUSA1: Usage Support Agent Behaviour External View Patient Warning Take Medicine

If the Usage Support Agent received communicated medicine intake, and based on these, the estimated accumulated concentration is C, and C<TL, then it communicates an SMS to the Patient Cell Phone that medicine should be taken.

∀t:TIME, γ:TRACE, M:MEDICINE, C:REAL
history_implied_value(γ, input(USA), t, M, C) & C < TL ⇒ ∃t' t≤ t'≤ t+e & state(γ, t', output(USA)) |=
        communication_from_to(sms_take_medicine(M), usage_support_agent, patient_cell_phone)

EUSA2: Usage Support Agent Behaviour External View Doctor Warning

If the Usage Support Agent sent out a warning message to the patient, and the patient did not take medicine within X time steps after the warning, the Usage Support Agent sends a message to the Doctor Cell Phone.

∀t:TIME, γ:TRACE state(γ, t, output(USA)) |=
        communication_from_to(sms_take_medicine(M), usage_support_agent, patient_cell_phone) &
¬∃t2:TIME [t2 ≥ t & t2 < t + X & state(γ, t2, input(USA)) |= communicated_from_to(medicine_used(M),
        medicine_box_agent, usage_support_agent)]
⇒ ∃t':TIME t + X ≤ t'≤ (t + X) +e state(γ, t', output(USA)) |=
        communication_from_to(sms_not_taken_medicine(M), usage_support_agent, doctor_cell_phone)

EUSA3: Usage Support Agent Behaviour External View Store Information in Database

If the Usage Support Agent receives a communication that medicine has been taken, then the agent stores this information in the patient database.

∀t:TIME, γ:TRACE, M:MEDICINE   state(γ, t, input(USA)) |=
        communicated_from_to(medicine_used(M), medicine_box_agent, usage_support_agent) &
⇒ ∃t':TIME t ≤ t'≤ t + e
        [state(γ, t', output(USA)) |= performing_in(store_usage(M, t), patient_database)]

EUSA4: Usage Support Agent Behaviour External View Store Send Early Warning Message to Patient

If the Usage Support Agent receives a communication that the patient attempted to take medicine too early, then the agent sends an SMS to the patient cell phone.

∀t:TIME, γ:TRACE, M:MEDICINE   state(γ, t, input(USA)) |=
        communicated_from_to(too_early_intake_intention, medicine_box_agent, usage_support_agent) &
⇒ ∃t':TIME t ≤ t'≤ t + e
        [state(γ, t', output(USA)) |= communication_from_to(put_medicine_back_and_wait_for_signal,
        usage_support_agent, patient_cell_phone)]

EUSA5: Usage Support Agent Behaviour External View Send Approximated Concentration to Medicine Box Agent

If the history of medicine usage implies a certain medicine level, then the Usage Support Agent communicates this value to the Medicine Box Agent.

∀t:TIME, γ:TRACE, M:MEDICINE, C:REAL
history_implied_value(γ, input(USA), t, M, C) ⇒
∃t' t≤ t'≤ t+e & state(γ, t', output(USA)) |= communication_from_to(medicine_level(M, C),
        usage_support_agent, medicine_box_agent)

Besides the Usage Support Agent, the Medicine Box Agent (MBA) plays an important role within the system as well. From an external perspective, the MBA has three behavioural properties. The first (EMBA1) expresses that the MBA should communicate to the Usage Support Agent that medicine has been taken by the patient. This only occurs when the medicine is not taken too early. Properties EMBA2 and EMBA3 concern the communication in case of an early intake. First of all, the MBA should sound a beep (EMBA2), and furthermore, the MBA should communicate this information to the USA (EMBA3). Again, a parameter *e* is used to specify the maximum delay for these properties. Note that these properties are later referred to as EMBA, which is the conjunction of the three properties specified below.

EMBA1: Medicine Box Agent Behaviour External View Communicate Usage Non-Early Intake

When the medicine box agent observes medicine is taken from position X,Y in the box and the medicine is of type M, and the medicine level of M communicated to the agent is C, and furthermore, the medicine level plus a dose does not exceed the overall maximum, then the medicine box agent outputs medicine has been taken to the Usage Support Agent.

```
∀γ:TRACE, t:TIME, X, Y:INTEGER, C:REAL, M:MEDICINE
state(γ, t, input(medicine_box_agent) |=
            observed_result_from(medicine_taken_from_position(x_y_coordinate(X, Y)),
            medicine_box) &
state(γ, t, input(medicine_box_agent) |=
            communicated_from_to(medicine_level(M, C),
            usage_support_agent, medicine_box_agent) & C+DOSE <= MAX_MEDICINE_LEVEL &
            medicine_at_location(X, Y, M)
⇒ ∃t':TIME  t ≤ t'≤ t + e
            [ state(γ, t', output(medicine_box_agent) |= communication_from_to(medicine_used(M),
            medicine_box_agent, usage_support_agent) ]
```

EMBA2: Medicine Box Agent Behaviour External View Communicate Beep when Early Intake

When the medicine box agent observes medicine is taken from position X,Y in the box and the medicine is of type M, and the medicine level of M communicated to the agent is C, and furthermore, the medicine level plus a dose exceeds the overall maximum, then the medicine box agent outputs a beep to the Patient.

```
∀γ:TRACE, t:TIME, X, Y:INTEGER, C:REAL, M:MEDICINE
state(γ, t, input(medicine_box_agent) |=
            observed_result_from(medicine_taken_from_position(x_y_coordinate(X, Y)), medicine_box) &
state(γ, t, input(medicine_box_agent) |=
            communicated_from_to(medicine_level(M, C), usage_support_agent, medicine_box_agent) &
            C+DOSE > MAX_MEDICINE_LEVEL & medicine_at_location(X, Y, hiv_slowers)
⇒ ∃t':TIME  t ≤ t'≤ t + e
            [state(γ, t', output(medicine_box_agent) |= communication_from_to(sound_beep,
            medicine_box_agent, patient) ]
```

EMBA3: Medicine Box Agent Behaviour External View Communicate Early Intake to Usage Support Agent

When the medicine box agent observes medicine is taken from position X,Y in the box and the medicine is of type M, and the medicine level of M communicated to the agent is C, and furthermore, the medicine level plus a dose exceeds the overall maximum, then the medicine box agent outputs a communication concerning this early intake intention to the Usage Support Agent.

```
∀γ:TRACE, t:TIME, X, Y:INTEGER, C:REAL, M:MEDICINE
state(γ, t, input(medicine_box_agent) |=
            observed_result_from(medicine_taken_from_position(x_y_coordinate(X, Y)), medicine_box) &
state(γ, t, input(medicine_box_agent) |=
            communicated_from_to(medicine_level(M, C), usage_support_agent, medicine_box_agent) &
```

C+DOSE > MAX_MEDICINE_LEVEL & medicine_at_location(X, Y, hiv_slowers)
⇒ ∃t':TIME t ≤ t'≤ t + e
[state(γ, t', output(medicine_box_agent) |= communication_from_to(too_early_intake_intention, medicine_box_agent, usage_support_agent)]

Furthermore, a number of properties are specified for the external behavior of the other components. These properties include basic forwarding of information by the patient cell phone (PCP), the doctor cell phone (DCP), communication between various communicating components (CP). Furthermore, it includes the specification of the observation results of performing actions in the medicine box (EMD), the storage of information in the database (PDP), and the transfer of those actions and observations (WP).

Finally, in order for such external behavioral properties to establish the global property, certain assumptions need to be made concerning the behavior of the doctor and of the patient. For the proof, the minimal behavior of the patient is used. This minimal behavior is specified by stating that the patient should at least respond to the doctor communication. Of course, most likely would be that the patient already responds to the SMS being sent, but using the minimal behavior it can already be proven that this is not even required, as long as the patient responds to the doctor communication. This ensures, that even if the patient does not respond on an SMS his medicine level will still remain within the boundaries set. The behavior of the patient is expressed in property PB. Besides the patient, also the doctor needs to respond to the SMS of the system in a particular way, namely that he immediately contacts the patient, as expressed in property DB.

PB: Respond to Doctor Request

When a patient receives a doctor warning that the patient should take medicine M, then the patient takes the medicine from the appropriate place in the box.

∀γ:TRACE, t :TIME, M:MEDICINE, X, Y:INTEGER
state(γ, t, input(patient) |=
    communicated_from_to(doctor_request_take_medicine(M), patient_cell_phone, patient) medicine_at_location(X, Y, M)
⇒ ∃t':TIME t ≤ t'≤ t + e
    state(γ, t', output(patient) |= performing_in(take_medicine_from_position(x_y_coordinate(X, Y)), medicine_box)

DB: Warn Patient after SMS

When the doctor has received an SMS concerning a patient that has not used medicine, then the doctor uses its cell phone communicating that the patient to take medicine.

∀γ:TRACE, t :TIME, M:MEDICINE
state(γ, t, input(doctor) |=
    communicated_from_to(sms_not_taken_medicine(M), doctor_cell_phone, doctor)
⇒ ∃t':TIME t ≤ t'≤ t + e
    [state(γ, t', output(doctor) |= communication_from_to(doctor_request_take_medicine(M), doctor, doctor_cell_phone)]

Relating Properties

The relation between the external behavioral properties introduced in Section 6.2.1 and the top-level global property is shown in Figure 4. The figure also shows the relationship between the executable properties of the Usage Support Agent, and the Medicine Box Agent and the external behavioral properties thereof. Note that the external level of both the Usage Support Agent as well as the Medicine Box Agent has been represented in the figure as EUSA and UMBA respectively. This is simply the conjunction of the external properties of these agents. Furthermore, the external behavioral properties of the other components directly translate to executable properties in the specification. The following relations hold from the local to the external level.

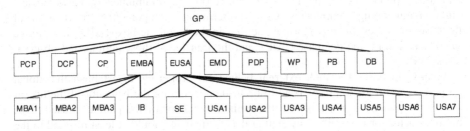

Fig. 3. Property hierarchy in the form of an AND tree

Usage Support Agent External Behavior
IB & SE & USA1 & USA5 ⇒ EUSA1
IB & USA2 & USA5 & USA6 ⇒ EUSA2
IB & USA2 & USA3 ⇒ EUSA3
IB & USA4 ⇒ EUSA4
IB & SE & USA1 & USA7 ⇒ EUSA5

Medicine Box Agent External Behavior
IB & MBA1 & MBA2 & MBA3 ⇒ EMBA

Furthermore, the global property GP1 has the following relationship with the external properties of the various components.

Global Behavior
EMBA & EUSA & PCP & DCP & CP & EMD & PDP & WP & PB & DB ⇒ GP1

In order to prove the above specified relationships, the relations have been checked within the SMV model checker. Using the following parameters, the inter-level relations within the tree expressed in Figure 4 are indeed shown to hold using SMV: The initial medicine level is set to 60, the reduction level per step is set to 98 (i.e. 0.98), the maximum medicine level is set to 150 and the minimum level to 30. Furthermore, the warning level is set to 60 (after which an SMS is sent to the patient). Finally, the delay for a doctor message is set to 10 time steps. Note that in the case of the SMV checks no communication delay is specified. Therefore, this formal verification shows that without this delay the model indeed entails the global property.

6 Discussion

In this paper, possible support in the domain of medicine usage management was addressed. A multi-agent system model that supports the users of medicine in taking their medicine at the appropriate time was discussed and formally analysed by simulation and verification. The system has been specified using a formal modeling approach which enables the specification of both quantitative as well as qualitative aspects [4] and [7]. To specify the model, both generic and domain specific temporal rules have been used, enabling reuse of the presented model. The analysis of the model has been conducted by means of (1) generation of a variety of simulation runs using a stochastic model for patients; (2) specification of dynamic properties at different aggregation levels; (3) specification of interlevel relations between these properties; (4) automated verification of properties specified in (2) against traces generated in (1), and (5) automated verification of the interlevel relations specified in (3). The simulation execution in (1) has been achieved making use of the LEADSTO software environment [4]. Specification of properties in (2) and interlevel relations in (3) have been performed using the TTL software environment [8], as has automated verification of such properties against set of simulation traces in (4). Verification of interlevel relations in (5) has been performed using the SMV model checking environment [9]. Evaluation of the system with actual users is part of future work.

The presented analysis fits well in the recent developments in Ambient Intelligence [10] and [11] as well as [12]. Furthermore, it also shows that multi-agent system technology can be of great benefit in health care applications, as also acknowledged in [13]. More approaches to support medicine usage of patients have been developed. Both in [2] as well as [14] models are presented that do not simply always send an SMS that medicine should be taken such as proposed by [1]. Both approaches only send SMS messages in case the patient does not adhere to the prescribed usage. The model presented in this paper however adds an additional dimension to such a support system, namely the explicit representation and simulation of the estimated medicine level inside the patient. Having such an explicit model enables the support agent to optimally support the patient; see also [6] for the use of such a model for mental processes of another agent.

References

1. Safren, S.A., Hendriksen, E.S., Desousa, N., Boswell, S.L., Mayer, K.H.: Use of an on-line pager system to increase adherence to antiretroviral medications. In: AIDS CARE, vol. 15, pp. 787–793 (2003)
2. Green, D.J.: Realtime Compliance Management Using a Wireless Realtime Pillbottle – A Report on the Pilot Study of SIMPILL. In: Proc. of the International Conference for eHealth, Telemedicine and Health, Med-e-Tel 2005, Luxemburg (2005)
3. Brazier, F.M.T., Jonker, C.M., Treur, J.: Compositional Design and Reuse of a Generic Agent Model. Applied Artificial Intelligence Journal 14, 491–538 (2000)
4. Bosse, T., Jonker, C.M., van der Meij, L., Treur, J.: A Language and Environment for Analysis of Dynamics by Simulation. Int. Journal of Artificial Intelligence Tools 16, 435–464 (2007)
5. Hoogendoorn, M., Klein, M., Treur, J.: Formal Design and Simulation of an Ambient Multi-Agent System Model for Medicine Usage Management. In: Proceedings of the International Workshop on Ambient Assisted Living (AAL 2007) (to appear, 2007)
6. Bosse, T., Memon, Z.A., Treur, J.: A Two-level BDI-Agent Model for Theory of Mind and its Use in Social Manipulation. In: Olivier, P., Kray, C. (eds.) Proceedings of AISB 2007, pp. 335–342. AISB Publications (2007)

7. Jonker, C.M., Treur, J.: Compositional Verification of Multi-Agent Systems: a Formal Analysis of Pro-activeness and Reactiveness. Intl. J. of Coop. Inf. Systems 11, 51–92 (2002)
8. Bosse, T., Jonker, C.M., van der Meij, L., Sharpanskykh, A., Treur, J.: Specification and Verification of Dynamics in Cognitive Agent Models. In: Nishida, T., et al. (eds.) Proceedings of IAT 2006, pp. 247–254. IEEE Computer Society Press, Los Alamitos (2006)
9. McMillan, K.L.: Symbolic Model Checking: An Approach to the State Explosion Problem. PhD thesis, School of Computer Science, CMU, Pittsburgh (1992). Kluwer Academic Publishers, Dordrecht (1993)
10. Aarts, E., Collier, R.W., van Loenen, E., de Ruyter, B. (eds.): EUSAI 2003. LNCS, vol. 2875, p. 432. Springer, Heidelberg (2003)
11. Aarts, E., Harwig, R., Schuurmans, M.: Ambient Intelligence. In: Denning, P. (ed.) The Invisible Future, pp. 235–250. McGraw Hill, New York (2001)
12. Riva, G., Vatalaro, F., Davide, F., Alcañiz, M. (eds.): Ambient Intelligence. IOS Press, Amsterdam (2005)
13. Moreno, A., Nealon, J.L. (eds.): Applications of Software Agent Technology in Health Care Domain. Birkhäuser, Basel (2004)
14. Floerkemeier, C., Siegemund, F.: Improving the Effectiveness of Medical Treatment with Pervasive Computing Technologies. In: Dey, A.K., Schmidt, A., McCarthy, J.F. (eds.) UbiComp 2003. LNCS, vol. 2864. Springer, Heidelberg (2003)

Efficient Privacy-Enhancing Techniques
for Medical Databases

Peter Schartner and Martin Schaffer

Klagenfurt University, Austria
Institute of Applied Informatics · System Security Group
{p.schartner,m.schaffer}@syssec.at

Abstract. In this paper, we introduce an alternative for using linkable unique health identifiers: locally generated system-wide unique digital pseudonyms. The presented techniques are based on a novel technique called *collision-free number generation* which is discussed in the introductory part of the article. Afterwards, attention is payed onto two specific variants of collision-free number generation: one based on the RSA-Problem and the other one based on the Elliptic Curve Discrete Logarithm Problem. Finally, two applications are sketched: centralized medical records and anonymous medical databases.

1 Introduction

In this paper, two applications of digital pseudonyms in the scope of Unique Health Identifiers (Health IDs) and digital medical records are discussed. So far, unique Health IDs are issued by some sort of centralized agency, in order to guarantee the system-wide uniqueness of the employed identifiers. This enables linking between the Health ID (and indirectly the patient's name) and the medical record. Hence, there are severe privacy concerns, which are documented by the presence of "Health IDs & Privacy" in the media [1,5] or even some web-sites which are protesting against such identifiers [7].

It is obvious that in some scenarios (e.g. electronic prescription) a link between the Health ID and the individual's name must exist, but there are several scenarios, where such a link is neither necessary nor wanted. Consider centralized medical records or anonymized databases for certain diseases (e.g. cancer registers), for instance. In these scenarios, the only requirement is that all data concerning a specific person is added to the right medical record. Thus, the real identity (i.e. the Health ID) can be replaced by a digital pseudonym, which must be unique by definition [9].

A critical problem is, the efficient generation of provably unique pseudonyms. The straightforward solution is to have a (completely) trusted centralized instance, which issues these pseudonyms. This, however, is an unrealistic assumption since no single trustworthy instance exists. The ideal solution would be each individual generating the pseudonym on his own. Without cross-checking, however, there is a risk that some pseudonyms are accidentally chosen identical. This drawback can be overcome by using the technique of collision-free number generation for the establishment of pseudonyms. Pseudonyms are then

A. Fred, J. Filipe, and H. Gamboa (Eds.): BIOSTEC 2008, CCIS 25, pp. 467–478, 2008.
© Springer-Verlag Berlin Heidelberg 2008

1. *generated locally*, i.e. derived from a system-wide unique identifier, and are
2. *system-wide unique* without being linkable to the original identifier, from which they are derived.

The remainder of this paper is organized as follows. After going through some preliminaries, we briefly describe a scheme that allows the individual to locally generate almost random numbers, which are globally (or system-wide) unique [14,15,12]. These numbers can be used as pseudonyms and hence they replace the original Health ID. The individual's name and the corresponding ID (thus the medical record) are then computationally unlinkable. After describing the original scheme with two practical approaches, we present two application scenarios. In the first, centralized databases are discussed which store a medical record for each individual to improve the quality and security of ongoing and later treatments. In the second, databases for medical records of specific diseases are discussed which are used to statistically investigate them.

2   Preliminaries

Henceforth, l_x denotes the bit-length of an integer x, i.e. $l_x = \lceil \log_2 x \rceil$.

AES Encryption [8]. Let k be a random secret AES-key. The AES-encryption of a binary message m is given through

$$\mathrm{AES}_k(m) = c,$$

where c is the computed ciphertext. Decryption is denoted straightforwardly as

$$\mathrm{AES}_k^{-1}(c) = m.$$

For large messages, the encryption will be carried out by using an appropriate mode of operation [17] with a standardized padding scheme.

RSA Encryption [11]. Let n be the product of two safe (or strong) primes p and q. The public key e is chosen at random (or system-wide static), such that $\gcd(e, \varphi(n)) = 1$. The corresponding private key d is computed such that $ed \equiv 1 \pmod{\varphi(n)}$ holds. Given (e, n), a message $m \in \mathbb{Z}_n$ is encrypted through the function $\mathrm{RSA}_{(e,n)}$, defined as

$$\mathrm{RSA}_{(e,n)}(m) = m^e \text{ MOD } n.$$

A ciphertext $c \in \mathbb{Z}_n$ can be decrypted through

$$\mathrm{RSA}_{(d,n)}^{-1}(c) = c^d \text{ MOD } n.$$

The security of the scheme relies on the problem of factoring n and the problem of computing e-th roots modulo n. The problem of finding m, given c and (e, n) is sometimes called the RSA-Problem. The latter we define as follows:

Definition 1. *Let* $n = pq$, *where p and q are primes,* $e \in \mathbb{Z}_{\varphi(n)}^*$, $m \in \mathbb{Z}_n$ *and* $c = m^e$ *MOD n. The* RSA-Problem *is the following: given c and (e, n), find m.*

It is currently believed that using $l_p \approx l_q \approx 640$ bit is sufficient for today's security needs [10]. Like all state-of-the-art implementations of RSA we employ Optimal Asymmetric Encryption Padding OAEP [6], which is a special form of padding, securing RSA against chosen ciphertext attacks.

Elliptic Curve Cryptography (ECC). To speed up computations, discrete logarithm-based cryptosystems are often run over an elliptic curve group, or in particular on a subgroup of prime order. A good introduction to ECC can be found in [2]. We use scalar multiplication in such groups as a one-way function. The corresponding intractable problem is given in the following definition.

Definition 2. *Let $E(\mathbb{Z}_p)$ be an elliptic curve group, where p is an odd prime. Let $P \in E(\mathbb{Z}_p)$ be a point of prime order q, where $q|\#E(\mathbb{Z}_p)$. The* Elliptic Curve Discrete Logarithm Problem (ECDLP) *is the following: given a (random) point $Q \in \langle P \rangle$ and P, find $k \in \mathbb{Z}_q$ such that $Q = kP$.*

To avoid confusion with ordinary multiplication we henceforth write $\mathrm{SM}(n, P)$ to express the scalar multiplication nP in $E(\mathbb{Z}_p)$. It is currently believed that using $l_p \approx 192$ and $l_q \approx 180$ bit is sufficient for today's security needs [10].

ECC Point Compression [4]. A point on an elliptic curve consists of two coordinates and thus requires $2l_p$ bits of space. It is a fact that for every x-value at most two possible y-values exist. Since these only differ in the algebraic sign, it suffices to store only one bit instead of the whole y-value. We therefore define the point compression function $\mathrm{CP} : E(\mathbb{Z}_p) \to \mathbb{Z}_p \times \{0,1\}$ as

$$\mathrm{CP}((x,y)) = (x, y \bmod 2).$$

Storing $\mathrm{CP}((x,y)) = (x, \tilde{y})$ instead of (x,y) requires only $l_p + 1$ bits of space. To uniquely recover (x, y) from $(x, \tilde{y})$, one needs some decompression function $\mathrm{DP} : \mathbb{Z}_p \times \{0,1\} \to E(\mathbb{Z}_p)$, such that

$$\mathrm{DP}((x, \tilde{y})) = (x, y).$$

How y is uniquely recovered depends on the kind of elliptic curve that is used and on how p is chosen. For instance, consider an elliptic curve defined through the equation $y^2 \equiv x^3 + ax + b \pmod{p}$, where a, b are some appropriate public domain parameters. If $p \equiv 3 \pmod 4$, one can recover y from $(x, \tilde{y})$ easily through one modular exponentiation

$$y_{0,1} \equiv \pm(x^3 + ax + b)^{(p+1)/4} \pmod{p},$$

where $y = y_{\tilde{y}}$. This is an efficient way to compute square roots modulo a prime.

ElGamal Encryption [3]. In this paper, we use a particular variant of the ElGamal encryption scheme. Let $E(\mathbb{Z}_p)$ be the elliptic curve group as described above and P a point of order q. Then $d \in_R \mathbb{Z}_q$ denotes the private key whereas $Q = \mathrm{SM}(d, P)$ denotes the corresponding public key. If done straightforwardly according to [3], a message $M \in E(\mathbb{Z}_p)$ can be encrypted by first choosing a random $r \in \mathbb{Z}_q$ and then computing the ciphertext-pair $(A, B) = (\mathrm{SM}(r, P), M + \mathrm{SM}(r, Q))$. Given (A, B) and d, the message M can be obtained through $B - \mathrm{SM}(d, A)$. However, it is not trivial to map a binary message m to a point M on an elliptic curve. An efficient method to overcome this drawback is to choose a cryptographic hash-function $H : E(\mathbb{Z}_p) \rightarrow \{0, 1\}^{l_m}$ and define the ElGamal encryption function ElG_Q as

$$ElG_Q(m) = (SM(r, P), m \oplus H(SM(r, Q))), \quad r \in_R \mathbb{Z}_q.$$

Given the ciphertext-pair $(A, B) \in E(\mathbb{Z}_p) \times \{0, 1\}^{l_m}$ and d, the plaintext m can be obtained through ElG_d^{-1}, defined as

$$ElG_d^{-1}((A, B)) = B \oplus H(SM(d, A)).$$

3   Collision-Free Number Generation

In [14] we proposed a general method for generating system-wide unique numbers in a local environment, whilst preserving the privacy of the generating instance. The described generator, called collision-free number generator (CFNG), fulfils the following requirements:

R1 (Uniqueness). A locally generated number is system-wide unique for a certain time-interval.

R2 (Efficiency). The generation process is efficient regarding communication, time and space.

R3 (Privacy). Here we distinguish three cases:

1. *Hiding*: Given a generated number, a poly-bounded algorithm is not able to efficiently identify the corresponding generator.
2. *Unlinkability*: Given a set of generated numbers, a poly-bounded algorithm is not able to efficiently decide, which of them have been generated by the same generator.
3. *Independency[1]:* Breaking the hiding-property of one particular generator does not lead to an efficient breaking of *Hiding* or *Unlinkability* of any other generator of the system.

3.1   General Construction

For efficiency reasons, an identifier-based approach is used. Every generator is (once) initialized with a system-wide unique identifier, which we denote by UI. The idea is to derive several unique numbers from UI, such that none of them is linkable to UI.

Uniqueness Generation. As a first step a routine is needed, which derives a unique number u from UI in every run of the generation process. We call such a routine *uniqueness generator* and denote it by UG. A simple definition is

$$UG() = u, \quad u := UI\|cnt\|pad,$$

where pad is a suitable padding for later use of u and cnt is an l_{cnt}-bit counter (with a random starting value), incremented modulo $2^{l_{cnt}}$ in every round. So, the output of UG is unique for at least $2^{l_{cnt}} - 1$ rounds. So far, privacy is not preserved, since UI is accessible through u.

---

[1] This further requirement was added later on in [12].

Uniqueness Randomization. A first step to provide privacy is to transform u such that the resulting block looks random. Hereby, we use an injective function f_r, where r is chosen from a set R. The idea is that u is randomized by r and hence we call f_r the *uniqueness randomization function*. One can either use an injective mixing-transformation for f_r (e.g. symmetric encryption) or an injective one-way function based on an intractable problem (e.g. the Discrete Logarithm Problem). In both cases, r is chosen at random. The output of f_r is obviously not guaranteed to be unique: let u, u', with $u \neq u'$ and r, r' random, with $r \neq r'$. Then $f_r(u) = f_{r'}(u')$ may hold since two different injective functions f_r and $f_{r'}$ map on the same output space. On the other hand, this problem cannot happen if $r = r'$, since f_r is injective. To generate a unique block o, sufficient information about the chosen function f_r has to be attached to its output. Hence, o can be defined as $o = f_r(u)\|r$. The concatenation of two blocks by writing them in a row is an unnecessary restriction, since the bits of the two blocks can be concatenated in any way by a function π. This yields (cf. Figure 1, **CFNG1**)

$$o = \pi(f_r(u), r), \quad \text{UG}() = u, \quad \text{Pre}() = r,$$

where π is a (static) bit-permutation function (or bit-permuted expansion function) over the block $f_r(u)\|r$ and the generation of r is done by a routine called *pre-processor*, denoted by Pre. The main-task of Pre is the correct selection of r. This may include a key generation process if f_r is an encryption function.

Theorem 1. *Let u be a system-wide unique number, f_r be an injective function, with $r \in R$, and π be a static bit-permutation function or bit-permuted expansion function. Then $o = \pi(f_r(u), r)$ is system-wide unique, for all $r \in R$.*

Proof. Let $o' = \pi(f_{r'}(u'), r')$ and $u \neq u'$. The case where $r \neq r'$ obviously guarantees that the pairs $(f_r(u), r)$ and $(f_{r'}(u'), r')$ are distinct. Now consider the case where $r = r'$. Since $u \neq u'$ holds per assumption $f_r(u) \neq f_{r'}(u')$ holds due to the injectivity of f_r and the fact that $r = r'$. Thus, we have $(f_r(u), r) \neq (f_{r'}(u'), r')$ for all $r, r' \in R$. It remains to show that $o \neq o'$. This is obviously the case, because π is static and injective.

Privacy Protection. Given $o = \pi(f_r(u), r)$, computing $\pi^{-1}(o) = (a, b)$ is easy since π is public. For f, two cases might occur regarding the obtainable information about u and hence about UI:

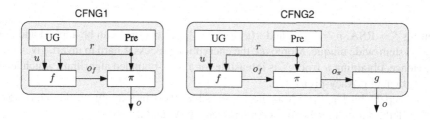

Fig. 1. Two Variants of Collision-Free Number Generation

1. Computing $f_b^{-1}(a) = u$ is computationally hard. Then we are done, since obtaining u is hard.
2. Computing $f_b^{-1}(a) = u$ is easy. Then an extension of the basic construction is necessary, which is sketched in the following.

To keep finding u hard, we suggest using an injective one-way function g to hide $\pi(f_r(u), r)$. We call g the *privacy protection function*. The new output o (cf. Figure 1, CFNG2) is defined as

$$o = g(\pi(f_r(u), r)), \quad \mathrm{UG}() = u, \quad \mathrm{Pre}() = r.$$

The extended construction still outputs unique numbers, which is shown by the following corollary.

Corollary 1. *Let u be a system-wide unique number, f_r be an injective function and $r \in R$. Furthermore, let π be a static bit-permutation function or bit-permuted expansion function and g an injective one-way function. Then $o = g(\pi(f_r(u), r))$ is system-wide unique, for all $r \in R$.*

Proof. Let $o' = g(\pi(f_{r'}(u'), r'))$ with $u \neq u'$. By Theorem 1 $\pi(f_r(u), r) \neq \pi(f_{r'}(u'), r')$. Since g is injective, $o \neq o'$ for all $r, r' \in R$.

3.2 Practical Approaches

In the following, two examples are given how CFNG1 and CFNG2 can be implemented for the generation of system-wide unique pseudonyms. The first one is based on the RSA-Problem and the second one is based on AES and the ECDLP, respectively. For both variants, we restrict ourselves to discussing the principle operation, and omit the details (e.g. bit lengths).

In order to guarantee the proper generation of the pseudonyms, we assume that each individual has been provided with a system-wide unique identifier UI (which could be the original Health ID or the ICCSN [16] of a smartcard). Henceforth, we assume that all computations are done within a smartcard.

RSA-based CFNG1. To generate a system-wide unique pseudonym N, the smartcard generates an RSA key pair consisting of the public exponent e, the random private exponent d and the random modulus n (consisting of two appropriately chosen large prime factors). Then N is generated as

$$N = \mathrm{RSA}_{(e,n)}(u)||(e, n).$$

Setting $f = \mathrm{RSA}$, $r = (e, n)$ and $\pi(x, y) := x||y$, Theorem 1 can be applied and thus N is system-wide unique. Moreover, the encryption process is hard to invert, given only N. Hence, obtaining u from N is infeasible for a poly-bounded algorithm, which leads to a sufficient privacy-protection.

To keep the bit-length of N as short as possible, every user might use the same public key e. In this case, there is no need to embed e in N, and hence

$$N = \mathrm{RSA}_{(e,n)}(u)||n.$$

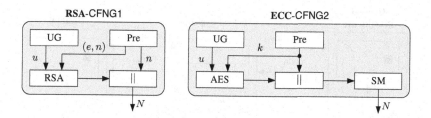

Fig. 2. RSA-based and ECC-based Pseudonym Generation

Figure 2 shows the principle design of the RSA-based approach. Pre denotes a key generator, which generates the key (e, n) and padding material for the encryption process. The system parameters includes the bit-length of n and of the padding. Note that padding in u needs to be done according to OAEP.

ECC-based CFNG2. In this variant, we employ a symmetric encryption algorithm to conceal the identity of the patient. As above, the randomly chosen key used for this encryption process (this time k) has to be concatenated to the output (the ciphertext). However, this construction is not sufficient for privacy protection, since one can easily invert the encryption process (it is symmetric). According to the design of CFNG2, one has to choose a privacy protection function g. For efficiency, we use scalar multiplication in an elliptic curve group as a one-way function. The pseudonym N is defined as

$$N = \text{SM}(\text{AES}_k(u)||k, P).$$

Figure 2 shows the principle design of the ECC-based generation process. The system parameters include the system-wide constant point P, and the bit lengths of the primes p and q, and of the key k.

Setting $f = \text{AES}$, $r = k$, $\pi(x, y) := x||y$ and $g(x) := \text{SM}(x, P)$, Corollary 1 can be applied and system-wide uniqueness is achieved. Notice that $\text{AES}_k(u)||k$ is highly random and hence inverting the ECC point multiplication process is hard due to the ECDLP. Thus, the requirement for privacy is satisfied sufficiently for practical use.

For more information about various design principles, formal proofs and applications regarding collision-free number generators we recommend to read [12].

4   Unique Pseudonyms in Medical Registers

4.1   Centralized Medical Records

In the first scenario, we employ a centralized medical database which stores a medical record for each individual. For the individual, it is essential that

- there is no link between the medical record and the real identity,
- no unauthorized person gets read or write access to the medical record, and
- additional data which is stored in the course of time is accumulated in the correct medical record.

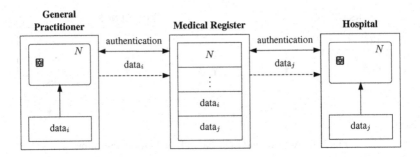

Fig. 3. Centralized Medical Registers

Initialization. In this application we employ smartcards, which are equipped with an ECC-based CFNG2 for efficiency reasons. Hence, the digital pseudonym is of the form $N = \text{CP}(\text{SM}(\text{AES}_k(u)||k, P))$, where $u = UI||cnt||pad$ (cf. Section 3.1) and UI the patient's identifier. Notice that we apply the point compression function CP to make N as short as possible. Whenever additional medical data has to be stored in the medical record, this pseudonym unambiguously identifies the record belonging to the holder of the pseudonym.

Data Storage. Prior to adding some data to his medical record, the patient has to prove that he owns the necessary rights. To protect the privacy of the patient, the authentication process (cf. Figure 3) must not include secret information on server side (this excludes symmetric encryption based authentication). An idea is to consider $Q = (x, y)$, where $N = \text{CP}(Q)$, as the public key of the ElGamal encryption scheme described in Section 2. The corresponding private key is $d = \text{AES}_k(u)||k$, since $N = \text{CP}(Q)$ and $Q = \text{SM}(\text{AES}_k(u)||k, P)$. This enables an indirect asymmetric authentication and key-exchange protocol, that provides anonymity of the patient, confidentiality of the sent medical data and freshness of the sent messages (cf. Figure 4):

1. To start the upload of medical data, the patient's smartcard contacts the server that stores the medical records, and sends his pseudonym N.
2. The server generates a random challenge c of length l_c and a session key K of length l_K, such that $l_c + l_K = l_m$ and l_K with respect to the involved symmetric encryption algorithm (here AES). Then it encrypts $c||K$ with Q using ElG_Q as described in Section 2. The resulting ciphertext-pair (A, B) is returned to the smartcard.
3. The smartcard decrypts (A, B) by using ElG_d^{-1} with the private key d and retrieves c and K. The medical data m (extended by the server's challenge) is encrypted to $C = \text{AES}_K(m||c)$ and sent to the server.
4. The server decrypts the encrypted message, compares the received challenge to the sent one and updates the medical record if they are identical. Otherwise it rejects the update request.

The authentication process only succeeds, if the patient knows d, such that $Q = \text{SM}(d, P)$. The probability of computing a correct C (that contains the same challenge as generated by the server) without knowing d is negligible.

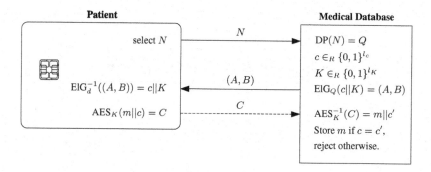

Fig. 4. Authentication and Data Storage

Notice that integrity-protection for the transferred data can be achieved easily by extending the proposed authentication protocol accordingly.

Data Retrieval. In principle, the data retrieval process is similar to the data storage process. At the beginning, the patient needs to authenticate himself to the server. Thereby, the server generates a session key. The data retrieval message is encrypted with the session key and provides the server with some data to identify the requested section of the medical record. The server selects this section, encrypts it with the session key and returns it to the patient's smartcard. Here, the message is decrypted and the medical data can be processed by the specific application.

An Extension. So far, N is of the form $CP(SM(AES_k(u)||k, P))$, and provides full access to the medical record. In order to split the medical record into several (unlinkable) parts, we append SID_i (section identifier i) to UI and generate

$$N_i = CP(SM(AES_{k_i}(u_i)||k_i, P)),$$

where u_i contains $UI||SID_i$. A pseudonym N_i will now identify and grant access to a specific section of the medical record, which enables access control and unlinkability on a finer level.

Security and Efficiency. The pseudonym is generated by a collision-free number generator that protects the privacy of the generating instance. In the current context, this means that computing UI from a pseudonym is computationally hard, i.e. requires solving the ECDLP. The generation process is very efficient, since only one symmetric encryption and one scalar multiplication is necessary.

The authentication process reveals no information about UI. The user can only respond correctly (apart from some negligible cheating probability) to the challenge, if he knows the pre-image of the pseudonym. Authentication only requires the exchanging of three messages between server and smartcard. The smartcard only performs one scalar multiplication, one run of a hash algorithm and some symmetric encryption. The server performs one run of the point decompression function DP, which equals to one modular exponentiation if $p \equiv 3 \pmod 4$. Furthermore, two scalar multiplications are done and

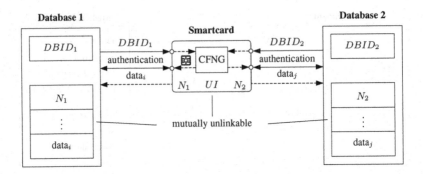

Fig. 5. Anonymous Medical Databases

some symmetric decryption is done. All computations only require minimal temporary space.

4.2 Anonymous Medical Databases

Centralized medical databases are quite commonly used to provide some anonymized data for statistical investigations of certain diseases like cancer, influenza, or tuberculosis. Information, like date of diagnosis, applied treatment, medication, chronology, and mortality is of special interest. The privacy problem with these medical databases is how to provide the anonymity of the patient. Of course, the server storing the database may be trusted, but especially in the scope of medical records, the sensitivity for privacy-endangering technologies is very high. So the best method is to remove the identifying information as early as possible. Unfortunately, the medical data concerning the disease of a specific patient is most commonly generated over a larger period of time and hence we need some identifier to link the separate data records. Here again, we can use collision-free number generators. The authentication of the data storing party (cf. Section 4.1) is moved to the application level. The smartcard of the patient provides only the unique identifier of the medical record, which now depends on the patient's identifier (UI) and the identifier of the database ($DBID$).

Initialization. The smartcard issuer has to generate appropriate system parameters, which are stored in the smartcard. Every time a new medical database is connected (cf. Figure 5), the smartcard receives the database identifier $DBID$ and generates the corresponding pseudonym. This pseudonym is stored in the smartcard for later usage.

Data Storage. During the authentication, the smartcard retrieves the session key from the database server. In order to store data, all personal data of the current medical record is removed and the remaining data is encrypted using the session key (established during the authentication process). The resulting ciphertext is sent to the database server. The message is decrypted and the medical data is stored in the record identified by the pseudonym.

Data Retrieval. This application scenario does not provide data retrieval for the individual patient. Only medical institutions may retrieve data from the centralized medical register. The process to achieve this in an authentic and confidential way is beyond the scope of this paper.

Security and Efficiency. Security and efficiency can be considered analogously to the previous application scenario. Unlinkability is again achieved through the privacy protection that holds through the design of the used ECC-based collision-free number generator.

5   Conclusions

In this paper, we introduced an alternative for using linkable unique Health IDs: locally generated but nevertheless system-wide unique digital pseudonyms. To achieve this replacement, we used so-called collision-free number generators, which have been briefly discussed in the introductory part of this article. We presented two variants: one based on the RSA-Problem, the other one based on the Elliptic Curve Discrete Logarithm Problem. Both variants fulfill the requirements uniqueness of the generated pseudonyms, privacy in terms of hiding the originator's identifier and unlinkability of individual data sets. Since the second variant is more efficient in terms of computational costs and space, it has been suggested for the use within smartcards.

Beside proposing two practical generators, a protocol has been given through which a smartcard can efficiently authenticate anonymously to a medical database (with respect to a pseudonym). Based on this protocol, medical registers can be extended and updated anonymously by the patient. Furthermore, such an authentication protocol can be used to make entries in several databases. For each database, a fresh pseudonym is used so that entries of different databases are mutually computationally unlinkable.

Regarding collision-free number generation further information on implementation issues and extended constructions can be found in [13] and [12].

References

1. CNN: National Health Identifier: Big Help or Big Brother? (2000),
   http://www.cnn.com/HEALTH/bioethics/9807/natl.medical.id
2. Hankerson, D., Menezes, A., Vanstone, S.: Guide to Elliptic Curve Cryptography. Springer, Heidelberg (2004)
3. ElGamal, T.: A Public Key Cryptosystem and a Signature Scheme Based on Discrete Logarithms. In: Blakely, G.R., Chaum, D. (eds.) CRYPTO 1984. LNCS, vol. 196, pp. 10–18. Springer, Heidelberg (1985)
4. IEEE: IEEE 1363-2000: IEEE Standard Specifications for Public-Key Cryptography. IEEE (2000)
5. Institute for Health Freedom: What's Happening with the Unique Health Identifier Plan? (2000), http://www.forhealthfreedom.org/Publications/privacy/UniqueId.html
6. Jonsson, J. Kaliski, B.: Public-Key Cryptography Standards (PKCS) #1: RSA Cryptography Specification (2002), http://www.rsa.com/rsalabs/node.asp?id=2125
7. Medical Privacy Coalition: Eliminate Unique Health Identifier (2007),
   http://www.medicalprivacycoalition.org/
   unique-health-identifier

8. NIST: FIPS PUB 197: Specification of the Advanced Encryption Standard (National Institute of Standards and Technology) (2001),
   http://csrc.nist.gov/publications/fips/fips197/fips-197.pdf
9. Pfitzmann, A., Köhntopp, M.: Anonymity, Unobservability, and Pseudonymity – A Proposal for Terminology. In: Federrath, H. (ed.) Designing Privacy Enhancing Technologies. LNCS, vol. 2009, pp. 1–9. Springer, Heidelberg (2001)
10. Post und Telekom Regulierungsbehörde: Bekanntmachung zur elektronischen Signatur nach dem Signaturgesetz und der Signaturverordnung, Bundesanzeiger Nr. 59, pp. 4695–4696 (2005)
11. Rivest, R.L., Shamir, A., Adleman, L.M.: A Method for Obtaining Digital Signatures and Public-Key Cryptosystems. Commun. ACM 21(2), 120–126 (1978)
12. Schaffer, M.: Key Aspects of Random Number Generation. VDM Verlag (2008)
13. Schaffer, M., Schartner, P.: Implementing Collision-Free Number Generators on JavaCards. Technical Report TR-syssec-07-03, University of Klagenfurt (2007)
14. Schaffer, M., Schartner, P., Rass, S.: Universally Unique Identifiers: How to ensure Uniqueness while Protecting the Issuer's Privacy. In: Alissi, S., Arabnia, H.R. (eds.) Proceedings of the 2007 International Conference on Security & Management – SAM 2007, pp. 198–204. CSREA Press (2007)
15. Schartner, P., Schaffer, M.: Unique User-Generated Digital Pseudonyms. In: Gorodetsky, V., Kotenko, I., Skormin, V.A. (eds.) MMM-ACNS 2005. LNCS, vol. 3685, pp. 194–205. Springer, Heidelberg (2005)
16. International Organization for Standardization: ISO/IEC 7812-1:2006 Identification cards – Identification of issuers – Part 1: Numbering system (2006)
17. National Institute of Standards and Technology: NIST Special Publication 800-38A: Recommendation for Block Cipher Modes of Operation – Methods and Techniques (2001)

Authentication Architecture for Region-Wide e-Health System with Smartcards and a PKI

André Zúquete[1], Helder Gomes[2], and João Paulo Silva Cunha[1]

[1] IEETA / Dep. of Electronics, Telecommunications and Informatics / Univ. of Aveiro, Portugal
[2] IEETA / ESTGA / Univ. of Aveiro, Portugal

Abstract. This paper describes the design and implementation of an e-Health authentication architecture using smartcards and a PKI. This architecture was developed to authenticate e-Health Professionals accessing the RTS (Rede Telemática da Saúde), a regional platform for sharing clinical data among a set of affiliated health institutions. The architecture had to accommodate specific RTS requirements, namely the security of Professionals' credentials, the mobility of Professionals, and the scalability to accommodate new health institutions. The adopted solution uses short-lived certificates and cross-certification agreements between RTS and e-Health institutions for authenticating Professionals accessing the RTS. These certificates carry as well the Professional's role at their home institution for role-based authorization. Trust agreements between e-Health institutions and RTS are necessary in order to make the certificates recognized by the RTS. As a proof of concept, a prototype was implemented with Windows technology. The presented authentication architecture is intended to be applied to other medical telematic systems.

1 Introduction

RTS (Rede Telemática da Saúde [1,2]) is a regional health information network (RHIN) providing an aggregated view of clinical records provided by a set of affiliated health institutions (HIs). Each HI uses its own system to produce and manage clinical records, which can be browsed and presented in different ways by RTS. The goal of RTS is not to replace the systems used by the affiliated HIs, but to provide a mediated, global view of patient's clinical records independently of the HIs holding their records.

RTS provides Portals for accessing clinical records. Two Portals were foreseen: the Patients Portal and the Professionals Portal. The first is to be used by Patients to communicate with their family doctor and to manage personal health issues, such as renovation of prescription and schedule appointments. The second is to be used by healthcare Professionals for accessing clinical records required for their normal, daily work.

The RTS Professionals' Portal is a Web server accessible through RIS (Rede Informática da Saúde[1]), a nation-wide, private network, interconnecting all HIs, including the ones affiliated with RTS. Professionals access data provided by RTS using a normal Web browser running on a computer connected to the RIS.

This paper describes an authentication architecture providing strong authentication for Professionals accessing the Professionals' Portal. Strong authentication is provided

---

[1] Health Computer Network.

A. Fred, J. Filipe, and H. Gamboa (Eds.): BIOSTEC 2008, CCIS 25, pp. 479–492, 2008.
© Springer-Verlag Berlin Heidelberg 2008

by using a two-factor approach: possession of a security token and knowledge of a secret. For the security token, we chose a smartcard. Smartcards are tamperproof devices with security-related computing capabilities that are very convenient for running computations using private keys of asymmetric key pairs.

2   Overview

European Commission recommendations [3] and individual countries' legislation reenforce the need for privacy and security for e-Health systems. Pursuing this goal, this paper describes an authentication architecture providing strong authentication for Professionals accessing the RTS Professionals' Portal. Strong authentication is highly recommended in this case, as Professionals can access sensitive data — the patients' health records. The architecture allows Professionals to roam between computers of their HI or other HIs.

Our authentication architecture had also to deal with authorization issues. In fact, the RTS Portal uses a role-based access control (RBAC) policy for deriving the Professionals' authorizations to access clinical data. Therefore, each time a Professional accesses the RTS Portal, the later must learn a role that the former may legitimately play for deriving authorisations.

The proposed architecture uses public key cryptography as the basis for its operation. Each Professional is given a smartcard for storing and using personal credentials for accessing the RTS Portal. The Professionals' authentication process uses a facility provided by Web browsers, the SSL/TLS client authentication with asymmetric keys and X.509 public key certificates (PKCs), to prove the authenticity of the Professional to the RTS Portal [4,5].

Furthermore, the PKCs used by Professionals in the SSL authentication process provide extra information to the RTS Portal, besides the identity of the Professionals, such as the HIs they are affiliated to and the role they are playing. As a Professional may play several roles simultaneously (e.g. Doctor and Chief Doctor), the PKC must contain all the roles he can play, being up to the RTS Portal to chose the role to play, from the possible ones, in each session.

Since a Professional's PKC carries roles the owner can play, a mechanism must be provided to deal with role changes. A possibility was to use certification revocation for outdating given roles. However, revocation validation requires online communication between the PKC validator and the PKC issuer, which may not be possible or convenient. Furthermore, some roles are very short in time, for example, vacation substitutions, and these dynamics can be more easily managed by short-lived certificates than by Certificate Revocation Lists (CRL).

Alternatively, we chose to use short-term validity periods for Professionals' PKCs, as in [6]. This way, Professionals' PKC get automatically invalid after a short period after their issuing and Professionals must apply for new ones. A simple and secure enrolment process was also conceived for getting new PKCs.

The public key infrastructure (PKI) for managing Professionals' credentials for accessing RTS uses a flexible, scalable grassroot approach. Each HI and the RTS have their own PKI, including root and issuing certification authorities (CA). The issuing

CA of each HI is responsible for issuing RTS credentials for local Professionals. The issuing CA of RTS is responsible for issuing credentials for the RTS Portal. The validation of certificates issued by separate PKIs is enabled by cross-certification agreements. This means that the RTS Portal will only be able to validate Professionals' credentials issued by HI CAs cross-certified by RTS; other people, including Professionals from other HIs, cannot be authenticated by the RTS Portal, therefore cannot access protected clinical data.

In this paper we mainly describe our architecture for using smartcards for authenticating Professionals and the RTS Portal when interacting with each other and for providing Professionals' roles to RTS. However, the architecture was designed taking into consideration future enhancements and synergies, such as:

- Enable Professionals to use the same smartcard for producing signed data as input for health information systems.
- Enable Professionals to give signed consents regarding accesses to the clinical data.
- Adoption of a similar authentication model for authenticating Patients, possibly using the new, smartcard-enabled Citizens Card.
- Usage of PKIs deployed for managing smartcards to generate credentials for mutual authentication within secure communications between hosts or servers used in the RTS and in HIs (e.g. with IPSec or SSL/TLS [7,5]).
- Usage of PKI deployed in each HI for managing the local authentication of Professionals accessing local services (e.g. secure wireless network access).

3   Design Goals

A set of design goals were defined at start. Those goals derived both from RTS requirements and from earlier experiences with other healthcare informatic systems.

The first goal was Professionals' mobility. The authentication architecture should not restrict the mobility of Professionals; at the end, it could be possible to use any computer, belonging to the RIS, to access RTS services. Naturally, this goal depends on software and hardware installed in client computers accessing Professionals' authentication tokens. Nevertheless, we tried to facilitate the widespread use of those tokens by using common hardware (e.g. USB ports) and free software packages (e.g. software packages already provided by operating system vendors).

The second goal was to be pragmatic regarding the implementation of a PKI for managing asymmetric keys and PKCs. Nation-wide PKIs do not exist for this purpose. Moreover, though they could be advantageous, they are difficult to deploy and to manage. Thus, we chose to start from a sort of minimalist, ad-hoc scenario, with no global PKI encompassing the RTS and all the HIs, but instead with isolated, standalone PKIs on each entity, RTS and HI.

The third goal was RTS independence regarding the management of personnel in affiliated HIs. Each HI is an independent organization, with its own Professionals, human resources management department and some kind of directory service to store the Professionals' information. It thus makes sense to reuse HI Professionals information and let each HI to manage the access of its own Professionals to the RTS. This way, we avoid replication of information and a centralized enrolment of Professionals in RTS.

The fourth goal was to minimize communication overheads related to the authentication of Professionals and fetching/validation of role membership. Namely, we tried not to use on RTS any online services from HIs to deal with details regarding the identification, authentication and role membership of Professionals. Since Professionals' information is managed solely by their home HI (our previous goal), this means that Professionals' identification and authentication credentials should convey RTS all required information.

The fifth goal was browser compatibility. To avoid the requirement of using a specific browser, no client-side active code is used in RTS. Therefore, we could not use any special code for managing the authentication of Professionals using a browser to access RTS. In other words, the authentication mechanism using a two-factor approach should be already available within the basic functionality of all browsers.

4   Authentication Architecture

The authentication architecture is resumed in Fig. 1. The Professional uses a Web browser to access the RTS Portal Web server and uses an SSL secure channel for protecting the communication from eavesdropping.

Furthermore, mutual authentication is required in the establishment of the SSL secure session, thus the browser authenticates the RTS Portal and the RTS Portal authenticates the Professional using the browser. Similarly, the Professional uses a Web browser and a mutually authenticated SSL session to access the HI Issuing CA Web server for requesting fresh RTS credentials.

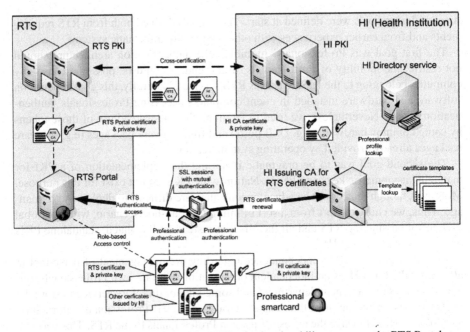

Fig. 1. Authentication architecture for HI Professionals whiling to access the RTS Portal

4.1   The Professionals' Smartcard

A Professional's smartcard carries two types of asymmetric key pair and corresponding PKCs. One type we call **RTS credentials**, which are to be used to authenticate himself when accessing the RTS Portal. The other type we call **HI credentials**, which are to be used to authenticate himself when accessing his HI issuing CA for getting new RTS credentials. Smartcards are initialised and provided by HIs to their own Professionals. At start, they only carry the HI credentials. When required, the owner uses them for requiring RTS credentials. These credentials can then be used to access the RTS Portal.

RTS credentials are short lived, lasting for one or a few days. The RTS Portal does not use remote HI services for checking for their validity. Instead, it assumes that a Professional's role revocation will naturally be enforced by not being able to get a new RTS credential containing the revoked role. On the contrary, HI credentials are long lived, because they are used for long periods for getting new RTS credentials.

4.2   Professional Authentication

The Professional authentication is requested by the SSL server-side of Web servers and conducted by the SSL client-side running on browsers. The SSL client-side authentication uses the Professional's smartcard for his authentication. The browser is configured to use smartcard services and when client-side authentication is required it will prompt the Professional for the right credentials, including the ones inside the smartcard, he intends to use. The Professional chose the right pair of asymmetric keys from the smartcard, and the PKC of the public key, and the browser uses them to provide client-side authentication.

This client-side approach is the same for accessing the RTS Portal or the HI Issuing CA. It is up to the Professional to choose the right set of credentials, from the smartcard, to be authenticated. In all cases, it needs to introduce a PIN to unblock the smartcard for producing digital signatures required by the SSL authentication protocol.

The Web servers used by the RTS Portal and the HI Issuing CA perform the following actions: (i) validate the PKC chose and presented by the Professional, (ii) use the certified public key to validate the SSL secure channel establishment and (iii) enable the service, RTS or CA, to access the Professional's PKC. The RTS learns from the PKC the Professional's identity, his home HI and his roles; the CA learns only the Professional's identity.

4.3   Role Assignment and Selection

The roles of each Professional are embedded in the PKC of his RTS credentials. These roles are stored in extension fields, namely the Extended Key Usage (EKU) field. Each role was given a numerical tag, an ASN.1 Object IDentifier, reserved at IANA[2] for RTS.

Each time a Professional requests RTS credentials, he gets, after proper authentication at the HI Issuing CA, a new PKC with the current roles he can play. This PKC is communicated to the RTS Web server during SSL authentication and, if successful, the PKC is made available for consulting by the RTS Portal during the SSL session. This

---

[2] http://www.iana.org

way, when a Professional initiates an SSL-protected session with the RTS Portal, it can easily learn the set of roles the Professional can play and prompt the Professional for selecting the right role for the current session.

4.4 Trust Relationships

Each entity, RTS and HI, uses an independent PKI for managing RTS and HI authentication credentials used by Professionals. The RTS is not meant to serve as a CA for all HIs; it only deploys a PKI mainly for managing its own certificates. HI certification hierarchies may be isolated or integrated in wider hierarchies providing large-scale validation of certificates. For the RTS that is irrelevant, all it requires is an Issuing CA for issuing RTS credentials for local HI Professionals.

Since RTS and HI certification hierarchies are isolated from each other at the beginning, some mechanism is required to enable the RTS Portal to validate Professionals' RTS credentials, issued by HIs. Similarly, some mechanism is required to enable professionals to validate the credentials of the RTS Portal, issued by RTS. This mechanism is cross-certification.

When an HI becomes affiliated to the RTS, the RTS Issuing CA issues a certificate for the public key of that HI Issuing CA. With this certificate, the RTS is able to validate all the PKCs of RTS credentials issued for the Professionals of that HI. Similarly, the HI Issuing CA issues a certificate for the RTS Issuing CA, enabling local Professionals to validate the credentials of the RTS Portal.

4.5 Validation of Certificates

With this cross-certification in place, the validating of certificates' certification chains works as follows. The RTS Portal trusts only on the (self-signed) certificate of the RTS Root CA. Similarly, the Professional trusts only on the (self-signed) certificate of his HI Root CA. Since certificate chain validations progress recursively until finding an error or a trusted certificate, the validation chains are the following:

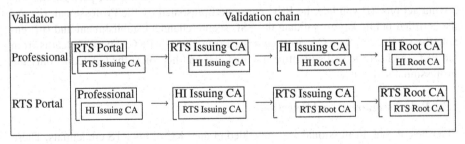

where ⌐X/Y⌐ represents the PKC of X issued by Y.

Besides cross-certification for certificate chain validation, trust relationships between the RTS, HIs and their Professionals must be complemented by common certification policies. Namely, all HIs affiliated to the RTS should follow similar policies for issuing RTS credentials. For instance, smartcards with HI credentials should be initialised by HIs and personally delivered to Professionals.

4.6   Validity of Authentication Credentials

The authentication credentials stored inside a Professional's smartcard are the HI credentials and the RTS credentials. The first ones are used to establish an authenticated session to get the second ones.

HI credentials are to be used frequently, for instance, once per day or once for a couple of days, to fetch new RTS credentials. Therefore, they should have long validity periods and CRLs must be published to prevent unwanted use of them after a given instant. For instance, if a Professional moves from one HI to another one, his smartcard from the former HI must be returned and a CRL should be issued to invalidate the public key of the HI credentials inside the smartcard. Note, however, that CRL issuing and validation are all executed within the same HI, and not by external clients.

RTS credentials are valid only during short periods, one or two days. Therefore, no CRLs are used to validate them, since the error window is too narrow to allow a Professional to play a role he is no longer allowed to. Consequently, by default the HI Issuing CA does not publish CRLs for RTS certificates.

In special cases, such as disciplinary processes and legal inquiries, it should be possible for the HIs to provide to the RTS Portal, just in time, a list of RTS certificates that should no longer be accepted while in validity period. However, since such cases should be rare, it is preferable to deal them as the exception to the general rule above stated: no CRLs exist and are checked for RTS certificates.

All the certificates used in both HI and RTS credentials do not need to be published by Issuing CAs. In fact, these certificates are used solely in the context of SSL mutual authentication, and are communicated to the interacting peers within the SSL protocol. Therefore, they need not be published in some public directory System, as other certificates do, because no one needs them for other purposes.

5   Related Work

The following e-Health Systems were analysed: HYGEIAnet and MedCom/ Sundhed.dk.

5.1   HYGEIAnet

HYGEIAnet is the RHIN of Crete, Greece. It was developed by the Institute of Computer Science (ICS) of Foundation for Research and Technology – Hellas (FORTH) to provide an integrated environment for delivery of health care services in Crete Island [8,9].

Similarly to RTS, HYGEIAnet is formed by several HIs, namely Hospitals and Primary Care Units, each with its own health data, information services and human resources. HYGEIAnet operates above these independent healthcare units, providing an infrastructure for sharing clinical information. In addition, the Integrated Electronic Health Record (I-EHR) is a key element as it aggregates the patient health information in all participating healthcare units.

The trust and security frameworks are implemented in HYGEIAnet with VPNs, SSL, smartcards, PKI, security certificates and digital signatures. A Regional certification authority issues the certificates for users and applications. These certificates can be used to authentication and digital signing of documents and in case of user certificates, they can be stored in smartcards [8].

Authentication is centralized in HYGEIAnet. All applications and services are registered in the Health Resource Service (HRS) and issued a unique ID. Each HYGEIAnet user also must register in HRS to be able to use HYGEIAnet services, and a unique user name and password is provided. The username and password are communicated to an authentication server (AS) and a certificate is issued from the regional CA.

In terms of authorization, it follows a decentralized approach, where each individual service maintains and manages roles (groups) and role based permissions. The user must be assigned to the proper role in each service he is to have access. When accessing a service, the user is authenticated through the Authentication Service and gets his individual access rights validated through the individual service.

The RTS and HYGEIAnet approaches for authentication differ: HYGEIAnet has a centralised management of resources (users and services), with certificates issued by a regional CA, and requires an online AS for user authentication. On the contrary, RTS has a decentralized management of resources, reusing the management services belonging to the affiliated HIs, with certificates issued by HI CAs for their own users and services and not requiring any online user authentication service to be used by RTS.

5.2 The Health Portal (Sundhed.dk) and the Health Data Network of Denmark (MedCom)

MedCom is the National Health Data Network of Denmark. It is working since 1994, and it connects more than 2000 Hospitals, Pharmacies, General Practitioners (GPs) and Specialists. It started has a VAN network exchanging EDIFACT messages [10]. In 2004 started the process of migration to the Internet and EDIFACT messages were translated into XML messages. Today, both message formats are used [11].

Network security is implemented at three levels [12,13]. At the first level are VPN connections connecting healthcare networks to a central hub in a star topology. This allows the reuse of Internet connections that all the health care units already have. At a second level, an agreement system controls the data flows from and to any of the local healthcare networks. When a connection between two healthcare networks is needed, a previous access to the agreement system is required to establish the connection between the two networks. The third level of security is user authentication, made locally through his username and password, or his asymmetric key pair and PKC.

The Health Portal started in December 2003. It works on top of the Health Data Network and reuses its infrastructure and services. Unlike the Health Data Network, that only provides services for Professionals, the Health Portal provides services for both Citizens and Professionals [14].

User authentication, for both Citizens and Professionals, is made using OCES certificates[3] issued by the national PKI that can be used in several national public services. Professionals can use several OCES certificates: (i) Administrative digital signature for region, hospital or GP, (ii) health professional's digital signature based on personal identifier and (iii) authorization for treating patients [15].

---

[3] OCES certificate: Public Certificate for Electronic Service.

Comparing with RTS, the Danish system extensively uses asymmetric keys and PKCs, benefiting from a nation-wide PKI. However, many of the Danish system security requirements, such as Professionals' digital signatures, are currently not required by RTS, since it is not used for entering signed data into health information systems. Nevertheless, we believe that our architecture can evolve, while keeping its basic structure, for provide security services similar to the ones provided by Danish system. Furthermore, our PKI may coexist with a national-wide PKI encompassing all HIs, though not necessarily using it.

6   Prototype Implementation

As a proof of concept, a prototype of the authentication architecture was implemented. The prototype extensively used available products for Windows operating systems, because of its dominance in the computer desktops of the HI currently affiliated to RTS. A more detailed description of the prototype can be found in [16].

The prototype included an RTS service, with a two-level PKI and a Web Server (Professionals' Portal), one HI instance, with a two level PKI, an Active Directory (AD) Server and one registered Professional (one smartcard). CAs were implemented with Windows Certification Services available in Windows 2003 Server Enterprise Edition. When installed in Enterprise mode, this CA interacts with the AD to obtain user information for certificate issuing, and uses certificate templates, stored in AD and subject to AD access control rules, for certificate issuance management.

The key aspects to test in the prototype were (i) the impact of different middleware software in smartcard deployment, (ii) the deployment of an HI PKI for the management of RTS credentials for local Professionals, and (iii) the use of short lived RTS credentials to access RTS services.

6.1   Smartcard Deployment

Since smartcards are portable devices, in theory they may be used to authenticate Professionals accessing the RTS from different computers. However, this requires some software installed in those computers: (i) the card reader driver and (ii) middleware to fill the gap between applications and smartcard services.

There are different trends in this specific middleware area. Windows applications, such as the Internet Explorer browser, use the CryptoAPI (CAPI), which can use several Cryptographic Service Providers (CSP) for interacting with different smartcards. Another approach is to use PKCS#11 [17], a standard interface for cryptographic tokens. This interface is used by Netscape and Firefox browsers.

Since middleware modules are usually specific for smartcard manufacturers and some manufacturers impose limits on the number of computers were they can be installed or do not provide similar modules for all operating systems, the following approaches were foreseen: (i) the use of smartcards with native support from the operating system, (ii) the use of open source software or free binaries and (iii) the use of non-free software providing support for multiple smartcards.

In our prototype, we used only Windows XP systems for the Professionals' computers and two smartcard tokens: Rainbow iKey 3000 and Schlumberger Cyberflex e-gate

32k. None of them was natively supported by Windows and there was not an open source solution (openSC/CSP#11) working reliably with them. For the non-free solution, we used SafeSign Standard 2.0.3 software; both smartcards worked properly after their first initialization with SafeSign. If this initialization is not made by SafeSign, chances are that smartcards are not recognized, has happened with the Cyberflex card.

After low-level initialization (e.g. personalization), smartcards were incepted with the Professional HI credentials, allowing the owner to enrolment for RTS certificates. The HI credentials cannot be renewed and the smartcard becomes useless when the HI certificate validation period expires.

6.2 Healthcare Institution

The prototype HI implements a hierarchical PKI with two levels, the higher level with the HI Root CA, the lower level with an Issuer CA. Both CAs were implemented using the Certificate Service of the Windows 2003 Server Enterprise Edition.

The Root CA was configured in Stand Alone mode for issuing only Issuer CA certificates. After that, it should be turned off (power-off). The Issuer CA was configured in Enterprise mode because it interacts with HI AD to store certificates and certificate templates, and access profiles of local Professionals. The HI also hosts an IIS 6.0 Web server providing the Professionals' RTS certificate enrolment service.

Cross-Certification with RTS. Certificates issued by the RTS, namely for the RTS Portal, need to be validated by Professionals accessing RTS. This requires the HI Issuer CA to issue a cross-certificate to the public key of the RTS Issuer CA and to provide that certificate to HI Professionals. Similarly, the RTS must certify the HI Issuer CA public key for enabling the RTS Portal to validate Professional's RTS certificates.

Trust constraints may be applied to cross-certification: name constraints, certificate policy constraints and basic constraints [5]. Microsoft further allows application policy constraints. However, only the first one can be used by browsers, as the others require some application context. Therefore, we used only name constraints in cross-certificates issued by both RTS and HI Issuer CAs.

The name constraints within the cross-certificate issued by HI impose that Professionals can only validate certificates issued by RTS for subjects which name includes the RTS organization — tags O=RTS and C=PT. Similarly, cross-certificates issued by RTS impose a similar constraint regarding HI subject names and country.

Active Directory (AD). Some new AD groups, one for each Professional role recognized by RTS, were defined for supporting the correct management of RTS certificates. These RTS role groups provide automatic access control to specific certificate enrolment, i.e., only a Professional belonging to the Doctors group can request an RTS certificate containing a Doctor role on it. All the Professionals who will access RTS must be added to one or more role groups, according to their roles.

Certificate Templates for Issuing RTS Certificates. Certificates issued by the HI Issuer CA are tailored using certificate templates. These templates allow the definition of the certificate characteristics and access control rules. To control accesses to certificate templates, each template is bound to a role group, defined in AD, according with the template role. This group is added enrolment permission to the certificate.

Specific certificate templates were created for RTS certificates, one for each role. RTS certificate templates for different roles differ in the specification of certificate extensions and certificate security. An application policy was defined for each Professional role, to be included in the EKU (Extended Key Usage) field of the RTS certificates issued for the role. Application policies are simple ASN.1 OIDs (Object IDentifiers), defined by RTS after reservation at IANA.

A certificate template was also created for HI certificates. This template includes an application policy, in the EKU field, required to sign RTS certificate renewals and to access the Web server for certificate enrolment.

The customization of certificate templates has some limitations. First, only a few certificate fields can be parameterized; many certificate extensions cannot be parameterized this way. Second, even configurable fields cannot take any value. Namely, certificate templates do not allow for validity periods shorter than two days. This may be problematic if two days is considered a large risk window for short-lived certificates without CRL validation. Nevertheless, in our opinion two days is perfectly reasonable.

Smartcard Initialization. Smartcards must be personally delivered to Professionals after proper initialization by enrolment agents of their HI. An initialized smartcard contains: (i) HI certificate and correspondent private key, (ii) HI Root CA certificate, (iii) HI Issuer CA certificate, and (iv) cross-certificate issued by the HI Issuer CA to the public key of the RTS Issuer CA. Professionals are not allowed to request or renew HI credentials; when they expire only an enrolment agent can request its renewal.

Enrolment of Role-Specific RTS Certificates. The HI provides a Web server for enrolment of local Professionals for role-specific RTS certificates. Its Web pages were adapted from the Microsoft Certificate Services Web pages, namely to require SSL client-side authentication. Therefore, to access this Web server a Professional must authenticate himself using his HI certificate and corresponding private key, for which he is requested to introduce his smartcard PIN. This PIN is required to sign data with the private key corresponding to the HI certificate for the SSL client-side authentication.

The enrolment Web page contains links to request RTS certificates for each existing role. The Web server, an IIS, was configured to apply certificate mapping, i.e., it acquires the Professional access permissions in order to limit the Professional enrolment to the certificate templates corresponding to his roles in HI. Thus, the Professional's enrolment only succeeds if he chooses a role he can play. Certificates are immediately issued and can be installed by the Professionals in their smartcards. Both CAPI and PKCS#11 enabled browsers can be used to enrol for RTS certificates, and, in both cases, the new certificate is installed in the smartcard.

Two issues exist in the RTS certificate enrolment. The first issue is that the renewal process depends on the Professional initiative, which can raise acceptance issues in the medical community. In Microsoft environments, this can be solved using certificate auto-enrolment. The second issue is that old certificates are not automatically removed from the smartcard; they must be removed manually by the Professional using some application, like SafeSign. Another solution is the development of active code to remove automatically old certificates .

6.3   Usage of RTS Credentials for Accessing the RTS Portal

The validation of Professionals' RTS credentials by the RTS Portal, an IIS 6.0 Web server, was performed at two different levels. At the IIS level, validation follows SSL rules and certification chains. At the application level, validation includes checking RTS OID values placed in EKU field of the received RTS certificate. The Portal only initiates a session with a Professional if his certificate is considered valid at both levels.

Finally, Professional can use both Internet Explorer and Mozilla Firefox to access the RTS Portal. Tests were made in order to determine if the number PKCs from the HI PKI hierarchy in the smartcard could be reduced, but due to different approaches between browsers for building and validation of certificate chain, we conclude that all certificates must be present in order to allow both browsers to be used.

7   Evaluation

In this section, we evaluate the architecture and implementation of our authentication system taking into consideration the design goals presented in Section 3.

Concerning the first goal, a pragmatic PKI implementation, it was achieved, since no specific, large-scale PKI is required. On the contrary, the PKI is build on top of independent PKIs and cross-certification agreements. Trust relationships between RTS and affiliated HIs are reflected in such cross-certification and on common policies for issuing RTS certificates for Professionals.

Concerning the second goal, Professionals' mobility, smartcards embedded in USB tokens are the most promising solution nowadays but (still) cannot be used with PDAs and smartphones. Furthermore, and more problematic, the usage of smartcards in computers still raises the problem of software installation for dealing with them. As we saw in Section 6, it is not simple to find a ubiquitous, free solution for the middleware required by different applications (browsers) to interact with many smartcards.

Concerning the third goal, leaving RTS out of the management of Professionals working at the HIs, it was fully attained. The RTS Portal only requires Professionals to have a valid certificate issued by their HI and containing a set of role on it. HIs have full control on the management of local Professionals and their role, enabling RTS access by issuing RTS certificates with the proper contents, namely Professional identity, HI affiliation and possible roles.

The fourth goal, to minimize communication overheads between RTS and HIs for authenticating Professionals and getting their role, was also fully attained. The RTS Portal is capable of authenticating Professionals just by validating their certificate, without checking CRLs remotely, and capable of learning their role also from the certificate. No online communication between RTS and HIs is required in this process.

The fifth and final goal was browser compatibility. In this case, we must say that it may be difficult to provide the same set of functionalities with all the browsers, because of the differences between the existing middleware for bridging the gap between applications and smartcards (CAPI, PKCS#11, etc.). Furthermore, some smartcard management activities, such as garbage collection of useless credentials inside the smartcard, may require the deployment of active code for running within Professionals' browsers.

8   Conclusions

In this paper we described the design and implementation of an authentication archi-tecture for Professionals working within the RTS e-Health environment. Since Profes-sionals access RTS services using a browser and an RTS Portal, the authentication of Professionals was mapped on top of SSL client-side authentication. The credentials used in this authentication are provided by their HIs and formed by a private key and a short-lived X.509 PKC, both stored inside a smartcard. The short lifetime of these certificates allows issuing CAs to simplify their PKI: they are not published and they are not listed in CRLs.

The key characteristics of the authentication architecture are (i) the use of smart-cards for strong authentication (ii) the use of short-lived RTS certificates carrying Pro-fessional identification and roles for authentication on the RTS Portal and authorization of operations required to the RTS, (iii) the use of "normal"-lived HI certificates for Professional enrolment for RTS certificates, (iv) a PKI where the RTS and each HI run their own, private PKI with (v) cross-certification for the establishment of trust relations required to validate Professionals credentials and RTS credentials within SSL sessions. This authentication architecture is highly scalable and is prepared to be applied to other medical telematic projects such as the Brain Imaging Network Grid (BING) [18] and the Grid-Enabled REpoSitories for medicine (GERESmed) [19], two medical networks now under development an IEETA/University of Aveiro.

A prototype was implemented as proof of concept and based exclusively in technol-ogy provided by Windows systems or developed for Windows systems. Regarding the browsers used by Professionals, we tested two: Internet Explorer and Mozilla Firefox. The major source of problems that we found for implementing the prototype was the use and management of smartcards by Professionals' systems and browsers. The variety of middleware for managing smartcards and different approaches followed by different browsers regarding the middleware make it very hard to provide a clean, ubiquitous interface for Professionals. Furthermore, this is a critical issue in the deployment of this authentication architecture along many different systems and computers.

Acknowledgements. This work was partially supported by the Aveiro Digital Pro-gramme 2003-2006 of the Portugal Digital Initiative, through the POSI programme of the Portuguese Government, and by the FCT (Portuguese R&D agency) through the programs INGrid 2007 (grants GRID/GRI/81819/2006 and GRID/GRI/81833/2006) and FEDER.

References

1. Cunha, J.P.S., Cruz, I., Oliveira, I., Pereira, A.S., Costa, C.T., Oliveira, A.M., Pereira, A.: The RTS Project: Promoting secure and effective clinical telematic communication within the Aveiro region. In: eHealth 2006 High Level Conf., Malaga, Spain (2006)
2. Cunha, J.P.: RTS Network: Improving Regional Health Services through Clinical Telematic Web-based Communication System. In: eHealth Conf. 2007, Berlin, Germany (2007)
3. European Commission Information Society and Media: ICT for Health and i2010: Trans-forming the European healthcare landscape (June 2006) ISBN 92-894-7060-7

4. Housley, R., Ford, W., Polk, W., Solo, D.: Internet X.509 Public Key Infrastructure Certificate and CRL Profile. RFC 2459, IETF (January 1999)
5. Dierks, T., Rescorla, E.: The TLS Protocol Version 1.1. RFC 4346, IETF (April 2006)
6. Ribeiro, C., Silva, F., Zúquete, A.: A Roaming Authentication Solution for WiFi using IPSec VPNs with Client Certificates. In: TERENA Networking Conf. 2004, Rhodes, Greece (June 2004)
7. Kent, S., Seo, K.: Security Architecture for the Internet Protocol. RFC 4301, IETF (December 2005)
8. Katehakis, D.G., Sfakianakis, S.G., Anthoulakis, D., Kavlentakis, G., Tzelepis, T.Z., Orphanoudakis, S.C., Tsiknakis, M.: A Holistic Approach for the Delivery of the Integrated Electronic Health Record within a Regional Health Information Network. Technical Report 350 (FORTH-ICS/ TR-350), Foundation for Research and Technology - Hellas, Institute of Computer Science, Heraklion, Crete, Greece (February 2005)
9. Tsiknakis, M., Katehakis, D.G., Sfakianakis, S., Kavlentakis, G., Orphanoudakis, S.C.: An Architecture for Regional Health Information Networks Addressing Issues of Modularity and Interoperability. Journal of Telecommunications and Information Technology (JTIT) 4, 26–39 (2005)
10. ISO 9735: Electronic data interchange for administration, commerce and transport (EDIFACT) (1988), http://www.iso.org
11. MedCom IV: MedCom – the Danish Healthcare Data Network. MedCom IV, Status Plans and Projects (December 2003), http://www.medcom.dk/dwn396
12. Pedersen, C.D.: An baltic healthcare network and interoperability challenges. Cisco eHealth think tank meeting (2005)
13. Voss, H., Heimly, V., Sjögren, L.H.: The Baltic ehealth Network – taking secure, Internet-based healthcare networks to the next level. Norwegian Centre for Informatics in Health and Social Care (May 2005)
14. Sundhed.dk: The Danish eHealth experience: One Portal for Citizens and Professionals (December 2006), http://dialog.sundhed.dk
15. Rossing, N.: The Health Portal and the Health Data Network of Denmark. Executive Summary of Presentaion in eHealth Athens 2005 (2005), www.sundhed.dk
16. Gomes, H., Cunha, J.P., Zúquete, A.: Authentication architecture for ehealth professionals. In: Meersman, R., Tari, Z. (eds.) OTM 2007, Part II. LNCS, vol. 4804, pp. 1583–1600. Springer, Heidelberg (2007)
17. PKCS#11: Cryptographic Token Interface Standard, v2.20. RSALaboratories (2004)
18. Cunha, J.P.S., Oliveira, I., Fernandes, J.M., Campilho, A., Castelo-Branco, M., Sousa, N., Pereira, A.S.: BING: The Portuguese Brain Imaging Network GRID. In: IberGRID 2007, Santiago de Compostela, Spain, pp. 268–276 (2007)
19. Oliveira, I.C., Fernandes, J.M., Alves, L., Pereira, A.S., Cunha, J.P.S.: GERES-med: An Architecture for Grid-Enabled scientific RepositorieS for medical applications. In: 2nd Iberian Grid Infrastructure Conf. (IBERGRID 2008), Porto, Portugal (2008)

Understanding the Effects of Sampling on Healthcare Risk Modeling for the Prediction of Future High-Cost Patients

Sai T. Moturu[1], Huan Liu[1], and William G. Johnson[2]

[1] Department of Computer Science and Engineering
School of Computing and Informatics, Arizona State University, Tempe, AZ 85287
[2] Center for Health Information & Research (CHiR), Department of Biomedical Informatics
School of Computing and Informatics, Arizona State University, Tempe, AZ 85287
{smoturu,hliu,william.g.johnson}@asu.edu

Abstract. Rapidly rising healthcare costs represent one of the major issues plaguing the healthcare system. Data from the Arizona Health Care Cost Containment System, Arizona's Medicaid program provide a unique opportunity to exploit state-of-the-art machine learning and data mining algorithms to analyze data and provide actionable findings that can aid cost containment. Our work addresses specific challenges in this real-life healthcare application with respect to data imbalance in the process of building predictive risk models for forecasting high-cost patients. We survey the literature and propose novel data mining approaches customized for this compelling application with specific focus on non-random sampling. Our empirical study indicates that the proposed approach is highly effective and can benefit further research on cost containment in the healthcare industry.

Keywords: Predictive risk modeling, health care expenditures, Medicaid, future high-cost patients, data mining, non-random sampling, risk adjustment, skewed data, imbalanced data classification.

1 Introduction

The Center for Health Information and Research (CHiR) at Arizona State University houses a community health data system called Arizona HealthQuery (AZHQ). AZHQ contains comprehensive health records of patients from the state of Arizona linked across systems and time. The data, which include more than six million persons, offer the opportunity for research that can affect the health of the community by delivering actionable results for healthcare researchers and policy makers.

One of the primary issues plaguing the healthcare system is the problem of rapidly rising costs. Many reasons have been put forward for the consistent growth in health care expenditures ranging from the lack of a free market and the development of innovative technologies to external factors like economy and population growth [1]. A first step to tackle these issues is to devise effective cost containment measures. One

A. Fred, J. Filipe, and H. Gamboa (Eds.): BIOSTEC 2008, CCIS 25, pp. 493–506, 2008.

efficient approach to cost containment is to focus on high-cost patients responsible for these expenditures and undertake measures to reduce these costs. Predictive risk modeling is a relatively recent attempt at proactively identifying prospective high-cost patients to reduce costs. We embark on the challenging task of building predictive risk models using real-life data from the Arizona Health Care Cost Containment System (AHCCCS), Arizona's Medicaid program, available in AZHQ. The AHCCCS data was selected because it contains a large number of patients who can be tracked over multiple years and it contains many features needed for the analysis in this study.

Apart from data analysis challenges due to the voluminous amount of patient records and the considerable amount of variation for similarly grouped patients, such cost data provides a bigger challenge. It has been commonly observed that a small proportion of the patients are responsible for a large share of the total healthcare expenditures. This skewed pattern has remained constant over many decades. Previous studies show that more than two-thirds of the health costs are from the top ten percent of the population [2]. Similar patterns are observed in our empirical study.

Since a tiny percentage of patients create a large portion of the impact, identifying these patients beforehand would allow for designing better cost containment measures. Early identification could help design targeted interventions for the higher risk patients who could then be part of more effective, specially designed disease or case management programs. Early identification could help defer or mitigate negative outcomes.

This approach also ensures that the different players shaping the healthcare market be satisfied. Insurers and employers who pay for the healthcare costs would stand to gain considerably from reduced costs. Employers in particular have an added incentive, as this would reduce other "indirect costs" incurred due to the time taken by the patient to return to work and the resulting loss of productivity. Additional benefits for these players include better return on investment due to an improvement in the allocation of available resources and a basis for the establishment of capitation reimbursements. On the other hand, such an approach does not directly affect providers and suppliers who provide services to the patients. However, before achieving such gains, the imbalanced nature of the data provides a considerable challenge for accurate prediction.

As a part of this study, we propose a predictive risk modeling approach to identify high-risk patients. We use data mining and machine learning techniques to design such an approach as they are known to work well with large data and in particular when the data collection has been automated and performance takes precedence over interpretability [3]. Data mining has been successfully used in the past for financial applications like credit card fraud detection, stock market prediction, and bankruptcy prediction [4].

Healthcare data provide a unique opportunity for knowledge discovery using data mining while also presenting considerable challenges. Despite the success of data mining in various areas, it has not been regularly used to tackle these challenges though limited examples exist [5][6][7]. We study the possibility of applying data mining techniques to aid in healthcare risk modeling, where we aim to forecast whether a patient would be high costing for the next year based on data from the current year.

2   Related Work

2.1   Learning from Imbalanced Data

Due to the existence of high-risk, high-cost patients, healthcare expenditure data is highly skewed. As a result, it is essential to pay attention to the data imbalance when dealing with such data. This is not uncommon and has been observed in applications like credit card fraud detection, network intrusion detection, insurance risk management, text classification, and medical diagnosis. The problems of dealing with imbalanced data for classification have been widely studied by the data mining and machine learning community [8]. Most classification algorithms assume that the class distribution in the data is uniform. Since the metric of classification accuracy is based on this assumption, the algorithms often try to improve this faulty metric while learning.

The two most common solutions to this problem include non-random sampling (under-sampling or down-sampling, over-sampling or up-sampling and a combination of both) and cost-sensitive learning. Both solutions have a few drawbacks (most importantly, under-sampling might neglect few key instances while over-sampling might cause overfitting) but they have shown improvement over conventional techniques [9][10].

Various studies have compared over-sampling, under-sampling and cost-sensitive learning. While some found that there was little difference in the results from these methods, others found one among them to be the best. Results from different studies are inconclusive in selecting the best among them [9][11][12][13]. The use of a combination of under-sampling and over-sampling has also been found to provide improved results over the individual use of these techniques. Additionally, it has been found using varying ratios of the minority and majority classes that the best results were generally obtained when the minority class was overrepresented in the training data [10][14]. The use of synthetically generated instances for the minority class has also been proposed [15] but the prudence of using this technique for highly varied instances in healthcare data needs to be evaluated.

Despite the reported success of these techniques in other domains, none has been applied with respect to healthcare expenditure data in the past. In this study, we explore the possibility of using non-random sampling as a key element in creating predictive models for identifying high-risk patients. Preliminary work has confirmed the usefulness of this approach [16].

2.2   Techniques and Predictors

Healthcare data sets have been used in the past to predict future healthcare utilization of patients where the goal varied from being able to predict individual expenditures to the prediction of total healthcare expenditures. Typically, various regression techniques have been employed in the past with varying success for these tasks but the assumptions of independence, normality and homoscedasticity are not satisfied by the skewed distribution of the costs. Regression techniques generally tend to predict the average cost for a group of patients satisfactorily but on an individual basis, the predictions are not very accurate. Other approaches include the transformation of the distribution to match the assumptions of the analysis technique and the use of the Cox proportional hazards model [17].

Apart from these statistical methods, multiple risk-adjustment models that can forecast individual annual healthcare expenses are available. These can be used to predict high-cost patients by setting a cost threshold. Popular models like Adjusted Clinical Groups (ACG), Diagnostic Cost Groups (DCG), Global Risk-Adjustment Model (GRAM), RxRisk, and Prior Expense show comparable performance [18].

The performance of predictive modeling techniques is highly dependent on the data and features used. Different sources have provided data for the prediction of future utilization. Self-reported health status information gathered from patients using surveys has been used to predict medical expenditures [19] and group patients into cost categories [5]. Unlike these studies, our work employs administrative claims-based data. For such data both demographic and disease-related features have proven to be useful in the past. Demographic variables like age have been known to work well as predictors for expenditure. Disease-related information in the form of comorbidity indices has been used in the past as predictors of healthcare costs and the use of both inpatient and outpatient information was found to be useful [20]. However, simple count measures like number of prescriptions and number of claims were found to be better predictors of healthcare costs than comorbidity indices [21]. Though the performance of comorbidity indices might vary, disease-related information is still a key predictor. Such information from various utilization classes such as inpatient, outpatient and pharmacy information has been used in the past, either separately or together to predict cost outcomes. Combining information from different utilization classes has been found to be useful [22]. In this study, we use a set of features similar to those that have proven useful in the past together with data mining techniques that have not been explored with respect to this area.

3 Predictive Risk Modeling

3.1 Data and Features

The substantially large amount of data in AZHQ necessitates the selection of a specific subset for analysis. The requirement for a multi-year claims-based data set representing patients of varied demographics and containing disease-related information from various utilization classes, AHCCCS data is well suited for risk modeling. Despite being only a small part of AZHQ, AHCCCS data provides a large sample size of 139039 patients.

Four hundred and thirty seven demographic and disease-related features, either categorical or binary, were extracted from the original AHCCCS data. The patients were categorized into the minority or rare class (high-cost) and the majority class based on the paid amount. Fig. 1 depicts the structure of the data and its division into training and test data. Since the goal is to predict future healthcare costs, features from one year and class from the following year have been used together. Training data was constructed with features from 2002 and class from 2003 while test data was constructed with features from 2003 and class from 2004.

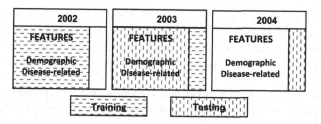

Fig. 1. Illustration of training and testing data sets

The demographic variables employed include age category (ages in groups of five), gender, race (Asian, Black, Hispanic, Native American, White and Other), marital status (single, married, divorced, separated and widowed) and county. Age and gender have been included due to previous success while race, location and marital status have been added as they could impact both financial and health aspects.

We avoid comorbidity or multimorbidity indices due to lack of flexibility. To allow the inclusion of inpatient, outpatient and emergency department information, International Classification of Diseases (ICD) procedure codes have been further grouped into twenty major diagnostic categories (MDC). For pharmacy data, the classification has been derived from the National Drug Code (NDC) classification with 136 categories. For each of these categories, information is available as the number of visits. Alternatively, a simple binary indicator can be used for each category to indicate the presence or absence of a visit.

The practice of discounting billed charges in the healthcare industry requires that the amounts paid for the services are used as measures of costs rather than the amounts charged. Payments are used in this study and we select two different thresholds for the separation of high-cost patients. These thresholds of $50,000 (954 or 0.69 % high-cost patients) and $25,000 (3028 or 2.18% high-cost patients) ensure that the resultant data is sufficiently highly skewed.

3.2 Analysis

Knowledge discovery using data mining requires clear understanding of the problem domain and the nuances of the data. These are achieved in the previous sections. Further, the analysis consists of three major steps. The first step is data preprocessing and is considered one of the most important parts of data mining. This is followed by the application of data mining techniques on training data to learn an appropriate model. Finally, this model is evaluated on test data using suitable evaluation metrics.

Training and test data are created in the data preprocessing step with required features being extracted from the data. The creation of a training data set provides a major challenge. The large size of the data makes the learning task tedious and necessitates the sampling of instances to reduce size. The nature of imbalanced data sets, which invariably result in poor performance while using conventional analysis techniques, needs to be taken into consideration for the selection of appropriate training instances. To address this challenge, non-random sampling has been employed as a combination of over-sampling the minority class and under-sampling the majority class to create a training sample. This approach is reasonable, as it has been employed successfully with such data in the past. Though the use of an equal number of training instances from both classes seems intuitive, it has been suggested that a higher number of instances from the minority class might improve sensitivity [10]. We evaluate this suggestion using multiple training samples with varying proportions of the two classes.

The next step is the creation of predictive models. We have preliminarily tested a variety of popular classification algorithms to focus on the challenge of learning from the training data. Out of the algorithms tested, five have worked considerably better. These include AdaBoost (with 250 iterations of a Decision Stump classifier), Logit-Boost (also with 250 iterations of a Decision Stump classifier), Logistic Regression, Logistic Model Trees, and the Support Vector Machine (SVM) classifier.

Performance evaluation provides the final challenge in our analysis. Since the data is highly skewed, traditional measures like accuracy are not particularly useful. We propose the following four evaluation metrics to gauge performance:

- Sensitivity: Sensitivity corresponds to the proportion of correctly predicted instances of the minority class with respect to all such instances of that class. It is equal to the number of true positives over the sum of true positives and false negatives.

$$S_T = \frac{N_{TP}}{N_{TP} + N_{FN}}$$

- Specificity: Specificity corresponds to the proportion of correctly predicted instances of the majority class with respect to all such instances of that class. It is equal to the number of true negatives over the sum of true negative and false positives.

$$S_P = \frac{N_{TN}}{N_{TN} + N_{FP}}$$

- F-measure: F-measure is typically used as a single performance measure that combines precision and recall and is defined as the harmonic mean of the two. Here we use it as a combination of sensitivity and specificity.

$$F_M = \frac{2 * S_T * S_P}{S_T + S_P}$$

- G-mean: G-mean typically refers to geometric mean and here, it is the geometric mean of sensitivity and specificity.

$$G_M = \sqrt{S_T * S_P}$$

To evaluate the performance of predictive risk models, it is necessary to understand the relevance of their predictions. The identification of high-cost patients allows for targeted interventions and better case management. Therefore, identifying most of these patients would prove useful. Such high sensitivity is achieved with a corresponding decrease in specificity, which is acceptable due to the cost benefits from identifying a large percentage of the high-cost patients. Consider the following example of two predictive models created using non-random and random sampling whose predictions are depicted through a confusion matrix in table 1. Identifying a limited number of high-cost patients (32 as opposed to 675) with greater prediction accuracy means that a large percentage of high-cost patients are unidentified and therefore a considerable portion of the health and cost benefits is unattainable. Alternatively holding targeted interventions and providing effective disease management for 22487 patients (675 correct and 21812 incorrect) could result in health benefits for the actual high-risk patients and cost benefits for the employers and insurers. This example indicates the need for high sensitivity along with an acceptable trade-off between specificity and sensitivity.

Table 1. Example: random vs. non-random sampling

	Non-random Sample		Random Sample	
	Positive	Negative	Positive	Negative
Predicted Positive	675	21812	32	82
Predicted Negative	279	116273	922	138003

3.3 Predictive Modeling

Recall that our preliminary results indicate the usefulness of non-random sampling for predictive modeling. Further, we have identified five classification algorithms that show promise and delineated four measures for performance evaluation considering the imbalance of data. These elements set the stage for an empirical study designed to indicate the usefulness of non-random sampling to our approach for predictive modeling. Further, this sampling technique is applied on suitably varied training data samples. Additionally, two class thresholds are used to check for the robustness of our approach to differently skewed data sets. These experiments help to provide a comparative outlook of our approach and indicate its benefits and flexibility.

4   Empirical Study

We first provide details of our experimental design along with the software environment, algorithms and then discuss experimental results.

4.1 Experimental Design

Employing the AHCCCS data as described earlier, we evaluate the predictions across an extensive range of experiments. All experiments have been performed using the Weka software [23]. Training data is created from the data set with features from 2002 and class from 2003. The model learned from this training data is used to predict on the test data set with features from 2003 and class from 2004. Non-random sampling was used to create training data as a default. The default class threshold used was $50,000. For each experiment, the five algorithms listed previously have been used to create predictive models with a goal of identifying the best one. The following dimensions were used for comparison.

Random Versus Non-Random Sampling. Experiments across this dimension were designed to depict the differences in performance between the sampling techniques. One set of experiments used random sampling where 50% of the data was randomly selected for training. Another set of experiments used non-random sampling where the minority class was over-sampled and the majority class was under-sampled. Twenty different random samples were obtained for both classes, with every sample containing 1,000 instances. The resulting training data sample contained 40,000 instances.

Varying Proportions of the Minority Class Instances in the Training Data. These experiments were designed to evaluate the differences of learning using non-randomly sampled data with varied proportions of rare class instances. Multiple training data sets were created with proportions of instances from the minority class being 10%, 25%, 40%, 60%, 75% and 90%. Random samples of 1000 instances each were drawn both classes according to the appropriate proportion for that training data set. However, the total number of instances was maintained at 40,000. For example, the training set with 40% rare class instances had 16 random samples from that class resulting in 16,000 instances. Six different non-randomly sampled training data sets were obtained in addition to the existing one with equal instances from both classes.

Visit Counts Versus Binary Indicators. Two different types of data were available for the various categories in each utilization class. Visit counts represent the actual visits by the patient while binary indicators represent whether the patient had a visit in a category or not. These experiments use both these in training samples similar to those described above to compare their usefulness in the creation of predictive models.

Varying the Class Threshold. Two different thresholds ($50,000 and $25,000) for the differentiation of high-cost patients have been used for the training samples described earlier to assess whether the approach is robust to variations along this boundary.

4.2 Results and Discussion

Importance of Non-Random Sampling. Both random and non-random samples are drawn from the same data set to form training data in order to build predictive models. The purpose of this experiment is twofold: (1) to verify whether non-random sampling is indeed necessary as suggested in our preliminary analysis, and (2) to use a baseline to compare predictions from the two techniques. It is apparent from Table 2 that random sampling provides very poor sensitivity with less than ten percent of the high-cost patients identified correctly. We can also consider a baseline model where patients are predicted to be in the same class as they were in the previous year. Such a model performs better with a sensitivity of 0.276 and a specificity of 0.993 for this data set resulting in an F-measure of 0.432 and a G-mean of 0.524. The low sensitivity indicates that not many high-cost patients remain in that category the following year making predictive modeling more difficult. Non-random sampling shows a marked improvement but as one would expect, this comes with a loss in specificity. Nevertheless, the F-measure and G-mean are much higher indicating that the trade-off between sensitivity and specificity is better than the baseline. These results clearly indicate the effectiveness of non-random sampling for predictive modeling.

Table 2. Random vs. non-random sampling

Algorithm	Comparison	S_T	S_P	F_M	G_M
AdaBoost	Random	0.019	1	0.037	0.138
	Non-Random	0.668	0.85	0.748	0.754
LogitBoost	Random	0.063	0.999	0.118	0.251
	Non-Random	0.646	0.894	0.75	0.760
Logistic Re-gression	Random	0.058	0.999	0.109	0.241
	Non-Random	0.646	0.899	0.752	0.762
Logistic Model Trees	Random	0	1	0	0.000
	Non-Random	0.632	0.902	0.743	0.755
SVM	Random	0.004	1	0.008	0.063
	Non-Random	0.594	0.919	0.722	0.739

Classification Algorithm Performance. Five different classification algorithms were used to learn predictive models across the experiments with the purpose of identifying the best among them. Recall that these algorithms were selected over many other algorithms based on our preliminary analysis. Results from Table 2 (and similar comparisons in the next section as shown in Fig. 2a) clearly indicate that these five algorithms perform consistently well with very similar sensitivity and specificity making it difficult to select the best one. One can only conclude that any of these algorithms could be used to learn a suitable predictive model from a non-randomly sampled training data set. Combining results from the previous section, we conclude that all classification models perform similarly poorly or well with random or non-random sampling. Hence, non-random sampling plays an instrumental role in significantly boosting performance.

Using Varied Class Proportions. Using a higher proportion of minority class instances in the training data sample is expected to improve results [10]. Experiments were designed to evaluate this expectation and this trend is observed with our data as well. Table 3 depicts the results for this comparison using the LogitBoost algorithm.

Using a higher proportion of minority class instances in the sample (60% and 75%) performs better than an equal proportion as indicated by both the F-measure and the G-mean. A receiver operating characteristics (ROC) curve can be generated from these different proportions. Fig. 2a depicts such a curve that provides a better visual representation of the improvement in results. It has to be noted that the two cases with improved results (60% and 75%) show a very different trade-off between sensitivity and specificity despite similar values for the F-measure and G-mean. Such an observation indicates a unique opportunity to deal with differences across the differently proportioned samples. It is difficult to identify a suitable trade-off without the availability of data that can establish the cost benefits to be gained from a particular trade-off. In such a scenario, such experiments can be invaluable as they provide multiple trade-offs to choose from. Upon the availability of information about the cost benefits, the suitably proportioned training data sample can be selected for analysis.

Table 3. Varying class proportions in training data

Rare Class Percentage	Comparison	S_T	S_P	F_M	G_M
10	Threshold: $25000	0.324	0.981	0.487	0.564
	Threshold: $50000	**0.289**	**0.985**	**0.447**	**0.534**
25	Threshold: $25000	0.534	0.945	0.682	0.710
	Threshold: $50000	**0.463**	**0.958**	**0.625**	**0.666**
40	Threshold: $25000	0.637	0.898	0.746	0.757
	Threshold: $50000	**0.602**	**0.919**	**0.727**	**0.744**
50	Threshold: $25000	0.637	0.863	0.733	0.742
	Threshold: $50000	**0.646**	**0.894**	**0.750**	**0.760**
60	Threshold: $25000	0.764	0.799	0.781	0.782
	Threshold: $50000	**0.731**	**0.843**	**0.783**	**0.785**
75	Threshold: $25000	0.895	0.682	0.774	0.782
	Threshold: $50000	**0.847**	**0.742**	**0.791**	**0.792**
90	Threshold: $25000	0.979	0.475	0.640	0.682
	Threshold: $50000	**0.945**	**0.553**	**0.698**	**0.723**

The Effect of Visit Counts. In the results discussed so far, we have used visit counts for the various categories in each utilization class. Alternatively, as discussed earlier, binary indicators can be used instead of visit counts. The use of limited information in

Table 4. Visit counts vs. binary indicators

Rare Class Percentage	Comparison	S_T	S_P	F_M	G_M
10	Binary	0.281	0.981	0.437	0.525
	Counts	**0.289**	**0.985**	**0.447**	**0.534**
25	Binary	0.492	0.942	0.646	0.680
	Counts	**0.463**	**0.958**	**0.625**	**0.666**
40	Binary	0.636	0.886	0.741	0.751
	Counts	**0.602**	**0.919**	**0.727**	**0.744**
50	Binary	0.708	0.842	0.769	0.772
	Counts	**0.646**	**0.894**	**0.750**	**0.760**
60	Binary	0.800	0.779	0.789	0.789
	Counts	**0.731**	**0.843**	**0.783**	**0.785**
75	Binary	0.906	0.662	0.765	0.775
	Counts	**0.847**	**0.742**	**0.791**	**0.792**
90	Binary	0.964	0.501	0.659	0.695
	Counts	**0.945**	**0.553**	**0.698**	**0.723**

the form of binary indicators is compared against the use of visit counts using the LogitBoost algorithm in Table 4. Throughout this set of experiments, the use of binary indicators provides a higher sensitivity and lower specificity compared to the use of visit counts for the same experiment.

For a balanced sample, a higher F-measure and G-mean is observed with the use of binary indicators. As the proportion of the minority class is increased in the sample, the F-measure and G-mean between the two cases grow closer with the use of visit counts proving more useful in terms of these measures at a proportion of 75%. Fig 2 shows the ROC curves in these two cases and there is little to choose between them.

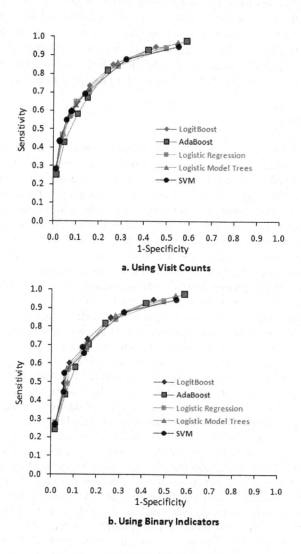

Fig. 2. ROC Curves

We can infer that the use of visit counts does not provide information that is much more useful than the binary indicators. It is also possible that our approach does not make the best use of visit counts and improvements could be made. Either way, it is clear that even limited information can provide useful results and the two situations provide a different trade-off. A closer look at the results could help decide which option is better in a particular situation.

Varying the Class Threshold. Two thresholds for the differentiation of cost categories have been used to indicate the robustness of our approach to changes in class threshold. We observe from Table 3 that results for both the thresholds are comparable with the higher threshold proving slightly better as indicated by F-measure and G-mean. Since the training data is balanced by non-random sampling, the slight underperformance from the data with lower threshold could be due to the fact that there are more patients closer to the lower threshold, increasing the chance of an error in prediction. This particular comparison serves to indicate the adaptability of our approach while using differently skewed data sets for predictive modeling.

5 Conclusions

Predictive risk modeling for forecasting high-cost patients is an important area of research and this study provides a look at a beneficial new technique using a real-world data set. Results indicate that creating training data using non-random sampling helps balance the challenges resulting from the skewed nature of healthcare cost data sets. Further, over-representing the minority class in the training data helps improve performance. Our study manifests the significance of sampling in building predictive risk models. However, it is hard to judge the best trade-off between specificity and sensitivity when there is no available data on the cost benefits. In this sense, using varied proportions of instances from the two classes in the training data can work as a boon in disguise. When data on cost benefits is available, one can test the use of different proportions of instances from the two classes to select the case with the best cost benefit. This makes our approach for predictive modeling much more adaptable.

Our comparison of classification algorithms for this task indicates that all of the selected ones work almost equally well. Though we find that it is hard to choose between these algorithms, results indicate to future users a handful of appropriate classification techniques to be used along with non-random sampling for predictive modeling. Our proposed approach creates a model by learning from the data and is therefore not restricted to the use of a specific type of data or features. Further, the threshold for high-cost patients is tunable and can be varied depending on the goals of a particular study. The approach also proves useful in the absence of actual visit counts. All these taken together signify the flexibility of predictive risk modeling for future high-cost patients using classification techniques to learn from non-randomly sampled training data and the benefits that can be obtained from such analyses.

Considering the variation in data, predictors and evaluation metrics, comparison with previous studies is improper, Nevertheless, the ROC curves in Fig. 2 is similar (the performance of the best model is comparable) to that obtained for existing risk-adjustment models [18]. The numbers are also better (our results double the sensitivity at about the same level of specificity) than a decision-tree based predictive modeling technique [5]. This validates the usefulness of this technique that is further enhanced by its flexibility. As can be observed, sampling is the most important component of this technique and is very beneficial for predictive modeling.

Predictive risk modeling is a useful technique with practical application for numerous employers and insurers in the goal to contain costs. We provide a promising approach that is valuable, flexible and proven to be successful on real-world data. Nevertheless, there is further scope to improve the interpretation of these results. It is commonly observed that a considerable percentage of high-cost patients do not remain that way every year. In addition, two patients could share very similar profiles with only one of them being high-cost. Studying these seemingly anomalous patients could provide a better understanding of how a high-cost patient is different from other patients. In addition, the current sampling approach and available classification techniques could be further tuned to improve results.

Apart from these possibilities, the most promising future direction is in working with key data partners. This avenue provides the opportunity to obtain information on the cost containment methods used and their efficiency as well as real data on the cost benefits obtained from previous predictive models. Working with such partners, we endeavor to provide a reasonable, patient-specific answer to this question that would significantly impact cost containment in the healthcare industry.

References

1. Bodenheimer, T.: High and Rising Health Care Costs. Part 1: Seeking an Explanation. Ann. Intern. Med. 142, 847–854 (2005)
2. Berk, M.L., Monheit, A.C.: The Concentration of Health Care Expenditures, Revisited. Health Affairs 20(2), 9–18 (2001)
3. Scheffer, J.: Data Mining in the Survey Setting: Why do Children go off the Rails? Res. Lett. Inf. Math. Sci. 3, 161–189 (2002)
4. Zhang, D., Zhou, L.: Discovering Golden Nuggets: Data Mining in Financial Application. IEEE Trans. Sys. Man Cybernet. 34(4), 513–522 (2004)
5. Anderson, R.T., Balkrishnan, R., Camacho, F.: Risk Classification of Medicare HMO Enrollee Cost Levels using a Decision-Tree Approach. Am. J. Managed Care 10(2), 89–98 (2004)
6. Cios, K.J., Moore, G.W.: Uniqueness of Medical Data Mining. Artificial Intelligence in Medicine 26(1-2), 1–24 (2002)
7. Li, J., Fu, A.W., He, H., Chen, J., Jin, H., McAullay, D., et al.: Mining Risk Patterns in Medical Data. In: Proc 11th ACM SIGKDD Int'l Conf. Knowledge Discovery in Data Mining (KDD 2005), pp. 770–775 (2005)
8. Chawla, N.V., Japkowicz, N., Kolcz, A.: Editorial: Special Issue on Learning from Imbalanced Data Sets. ACM SIGKDD Explorations Newsletter 6(1), 1–6 (2004)
9. McCarthy, K., Zabar, B., Weiss, G.: Does cost-sensitive learning beat sampling for classifying rare classes? In: Proc. 1st Int'l Workshop on Utility-based data mining (UBDM 2005), pp. 69–77 (2005)
10. Weiss, G.M., Provost, F.: The Effect of Class Distribution on Classifier Learning: An Empirical Study (Dept. Computer Science, Rutgers University, tech. report ML-TR-44 (2001)
11. Batista, G.E.A.P.A., Prati, R.C., Monard, M.C.: A Study of the Behavior of Several Methods for Balancing Machine Learning Training Data. ACM SIGKDD Explorations Newsletter 6(1), 20–29 (2004)

12. Drummond, C., Holte, R.C.: C4.5, Class Imbalance, and Cost Sensitivity: Why Under-Sampling beats Over-Sampling. In: ICML Workshop Learning From Imbalanced Datasets II (2003)
13. Maloof, M.: Learning When Data Sets are Imbalanced and When Costs are Unequal and Unknown. In: ICML Workshop Learning From Imbalanced Datasets II (2003)
14. Estabrooks, A., Jo, T., Japkowicz, N.: A Multiple Resampling Method For Learning From Imbalanced Data Sets. Computational Intelligence 20(1), 18–36 (2004)
15. Chawla, N.V., Bowyer, K.W., Hall, L.O., Kegelmeyer, W.P.: SMOTE: Synthetic Minority Over-sampling Technique. Journal of Artificial Intelligence Research 16, 321–357 (2002)
16. Moturu, S.T., Johnson, W.G., Liu, H.: Predicting Future High-Cost Patients: A Real-World Risk Modeling Application. In: Proc. IEEE International Conference on Bioinformatics and Biomedicine (2007)
17. Diehr, P., Yanez, D., Ash, A., Hornbrook, M., Lin, D.Y.: Methods For Analysing Health Care Utilization and Costs. Ann. Rev. Public Health 20, 125–144 (1999)
18. Meenan, R.T., Goodman, M.J., Fishman, P.A., Hornbrook, M.C., O'Keeffe-Rosetti, M.C., Bachman, D.J.: Using Risk-Adjustment Models to Identify High-Cost Risks. Med. Care 41(11), 1301–1312 (2003)
19. Fleishman, J.A., Cohen, J.W., Manning, W.G., Kosinski, M.: Using the SF-12 Health Status Measure to Improve Predictions of Medical Expenditures. Med. Care 44(5S), I-54-I-66 (2006)
20. Perkins, A.J., Kroenke, K., Unutzer, J., Katon, W., Williams Jr., J.W., Hope, C., et al.: Common comorbidity scales were similar in their ability to predict health care costs and mortality. J. Clin. Epidemiology 57, 1040–1048 (2004)
21. Farley, J.F., Harrdley, C.R., Devine, J.W.: A Comparison of Comorbidity Measurements to Predict Health care Expenditures. Am. J. Manag. Care 12, 110–117 (2006)
22. Zhao, Y., Ash, A.S., Ellis, R.P., Ayanian, J.Z., Pope, G.C., Bowen, B., et al.: Predicting Pharmacy Costs and Other Medical Costs Using Diagnoses and Drug Claims. Med. Care 43(1), 34–43 (2005)
23. Witten, I.H., Frank, E.: Data Mining: Practical machine learning tools and techniques, 2nd edn. Morgan Kaufmann, San Francisco (2005)

MEDLINE Abstracts Classification
Based on Noun Phrases Extraction

Fernando Ruiz-Rico, José-Luis Vicedo, and María-Consuelo Rubio-Sánchez

University of Alicante, Spain
frr@alu.ua.es, vicedo@dlsi.ua.es, mcrs7@alu.ua.es

Abstract. Many algorithms have come up in the last years to tackle automated text categorization. They have been exhaustively studied, leading to several variants and combinations not only in the particular procedures but also in the treatment of the input data. A widely used approach is representing documents as Bag-Of-Words (BOW) and weighting tokens with the TFIDF schema. Many researchers have thrown into precision and recall improvements and classification time reduction enriching BOW with stemming, n-grams, feature selection, noun phrases, metadata, weight normalization, etc. We contribute to this field with a novel combination of these techniques. For evaluation purposes, we provide comparisons to previous works with SVM against the simple BOW. The well known OHSUMED corpus is exploited and different sets of categories are selected, as previously done in the literature. The conclusion is that the proposed method can be successfully applied to existing binary classifiers such as SVM outperforming the mixture of BOW and TFIDF approaches.

Keywords: Text classification, SVM, MEDLINE, OHSUMED, Medical Subject Headings.

1 Introduction

In order to arrange all data in MEDLINE database, each time a new document is added, it must be assigned to one or several MESH[1] terms. More than 100,000 citations are inserted every year, leading to a tedious task, hard to be completed. During the last decades, an important effort has been focused on developing systems to automate the categorization process. In this context, several statistical and machine learning techniques have been extensively studied. We can emphasize Rocchio's based approaches, Bayesian classifiers, Support Vector Machines, Decision Trees and k-Nearest Neighbors among others [1,2,3]. Most of them treat the classified items as feature vectors, where documents are transformed into vectors using the Bag-Of-Words (BOW) representation, where commonly each feature corresponds to a single word or token.

At a first sight, some problems may arise from using the simple BOW. First, a lot of linguistic information is lost, such as word sequence. Also different terms have different importance in a text, so we should think about how to quantify the relevance of a feature so that we have a valid indicative of the degree of the information represented. From an intuitive point of view, a simple consideration of phrases as features

---

[1] Medical Subject Headings. More information in *www.nlm.nih.gov/mesh/meshhome.html*

A. Fred, J. Filipe, and H. Gamboa (Eds.): BIOSTEC 2008, CCIS 25, pp. 507–519, 2008.
© Springer-Verlag Berlin Heidelberg 2008

may increment the quality and quantity of information contained by feature vectors. For example, the expression "heart diseases" loses its meaning if both words are treated separately. Moreover, we can associate to each phrase sophisticated weights containing some statistical information such as the number of occurrences in a document, or within the whole training set or even how the phrase is distributed among different categories.

The paper is organized as follows. First, we have a look at previous efforts on the same matter by reviewing the literature and pointing out some relevant techniques for feature selection and weighting. Second, we try to remark the most important characteristics of our algorithm by explaining the intuitions that took us to carry out our experiments. Third, the details of the investigation are given by providing a full description of the algorithm and the evaluation procedure. Finally, several results and comparisons are presented and discussed.

2   Related Work

The above observations have led numerous researchers to focus on enriching the BOW model for many years. Most of them have experimented with n-grams (n consequent words) [4,5,6] and others with itemsets (n words occurring together in a document) [7,8,6]. In some cases, a significant increment in the performance was reported, but many times only marginal improvement or even a certain decrease was given.

This work proposes a new automatic feature selection and weighting schema. Some characteristics of this approach are based on ideas (noun phrases, meta-information, PoS tagging, stopwords, dimensionality reduction, etc.) that have been successfully tried out in the past [9,10,11]. However, they have never been combined altogether in the way we propose.

For concept detection and isolation we use especial n-grams as features, also known as noun phrases [4]. The ones we propose are exclusively made of nouns that may or may not be preceded by other nouns or adjectives.

For selecting the relevant expressions of each category, a lot of approaches use only the TF (*term frequency*) and DF (*document frequency*) measures calculated over the whole corpus. This way, most valuable terms occur frequently and have a discriminative nature for their occurrence in only a few documents. We propose using this concept along with the use of TF and DF as individual category membership indicatives, since the greater they are for a particular category corpus, the more related category and term are. Additionally, we define and use *category frequency* (CF) as discriminator among categories. Each of these measures (TF, DF, CF, ...) isolates a set of relevant expressions for a category using its average as discrimination threshold (*average-based discrimination*). The final representative expressions for a category will be obtained by intersecting the sets of relevant expressions for each of the proposed measures.

Finally, relevant expressions are weighted according to a new schema that aggregates all these measures into a single weight.

Normalization over the feature weights has also been proved to be effective [12]. For example, it solves the problem of differences in document sizes: long documents usually use the same terms repeatedly and they have also numerous different terms, increasing the average contribution of their features towards the query document similarity in preference over shorter documents [13].

Once each document in the corpus is represented as features, an algorithm must be provided to get the final classification. On a previous work [14], we successfully applied similar concepts for category ranking, where we obtained a ranked list of topics identifying the contents of each document. However, that approach is not suitable for MEDLINE abstracts indexation, where a document must be classified as relevant or not relevant to every particular MESH entry, by taking binary decisions over each of them. For this purpose, SVM is known to be a very accurate binary classifier. Since its complexity grows considerably with the number of features, it is often used together with some techniques for dimensionality reduction.

3   Specific Considerations

This section highlights some concepts whose analysis is considered important before describing in detail the process of feature selection and weighting.

Parts of Speech and Roots. There are several tools to identify the part of speech of each word and to get its root (word stemming). To achieve the best performance, this paper proposes using both a PoS tagger[2] and a dictionary[3] working together.

Category Descriptors. Training document collections used for classification purposes are usually built in a manual way. That is, human beings assign documents to one or more categories depending on the classification they are dealing with. To help this process, and to be sure that different people use a similar criteria, each class is represented by a set of keywords (*category descriptors*) which identifies the subject of the documents that belong to that category. A document containing some of these keywords should reinforce its relation with particular categories.

Nouns and Adjectives. There are types of words whose contribution is not important for classification tasks (e.g. articles or prepositions) because concepts are typically symbolized in texts within noun phrases [4]. Also, if we have a look at the category descriptors, we can observe that almost all the words are nouns and adjectives. So, it makes sense to think that the word types used to describe the subject of each category should be also the word types to be extracted from the training documents to identify the category they belong to.

We must assume that it is almost impossible to detect every noun phrase. Moreover, technical corpus are continuously being updated with new words and abbreviations. We propose considering these unknown terms as nouns because they are implicitly uncommon and discriminative.

Words and Expressions. When a word along with its adjoining words (a phrase) is considered towards building a category profile, it could be a good discriminator. This tight packaging of words could bring in some semantic value, and it could also filter out words occurring frequently in isolation that do not bear much weight towards

---

[2] SVMTool [15].

[3] *www-formal.stanford.edu/jsierra/cs1931-project/morphological-db.lisp*

characterizing that category [16]. Moreover, it may be useful to group the words so that the number of terms in an expression (TL or text length) can be taken as a new relevance measure.

Non-descriptive expressions. The presence of neutral or void expressions can be avoided by using a fixed list of stopwords [10]. However, if we only have a general list, some terms may be left out of it. We show that building this list automatically is not only possible but convenient.

Document's Title. The documents to be categorized have a title which briefly summarizes the contents of the full document in only one sentence. Some algorithms would discard an expression that only appears once or twice in a couple of titles because its relevance cannot be confirmed[4]. This paper proposes not only not discarding it, but giving it more importance.

4   Feature Selection and Weighting Schema

There are two main processes involved in the task of building the category prototype vector for each category. First, the training data is analyzed in order to detect and extract the most relevant expressions (*expression selection*). These expressions will be used as dimensions of the category prototype vectors. Second, the category prototypes are weighted according to the training set (*expression weighting*).

4.1   Average-Based Discrimination

An average or central tendency of a set (list) of values refers to a measure of the "middle value" of the data set. In our case, having a set E of n expressions where each expression is weighted according to a measure W ($\{w_1 \ldots w_n\}$), the average of the set E for the measure W (that we denote as $\overline{W}$) is defined as the arithmetic mean for the values w_i as follows:

$$\overline{W} = \frac{\sum_{i=1}^{n} w_i}{n}$$

Average discrimination uses the average of a measure W over a set E as threshold for discarding those elements of E whose weight w_i is higher (*H-Average discrimination*) or lower (*L-Average discrimination*) than $\overline{W}$ depending on the selected criteria. In the context of this work, this technique will be applied on different measures for selecting the most representative or discriminative expressions from the training data.

4.2   Expression Selection

The most relevant expressions are selected from the training data by using the *average discrimination measure* of different characteristics as cutting threshold.

Selecting Valid Terms. This process detects and extracts relevant expressions from each document as follows:

---

[4] A threshold of 3 is usually chosen [10], which means that terms not occurring at least within 3 documents are discarded before learning.

1. The words are reduced to their roots.
2. Only nouns and adjectives are taken into consideration. Any other part of speech (verbs, prepositions, conjunctions, etc.) is discarded. For this purpose, a PoS tagger and a dictionary are used. The words which are not found in the dictionary are considered to be nouns.
3. Sentences are divided into expressions: sequences of nouns or adjectives terminating in a noun. In regular expression form this is represented as "{Adjective, Noun}* Noun". For instance, the expressions extracted from "Ultrasound examinations detect cardiac abnormalities" are:

ultrasound   cardiac abnormality

ultrasound examination   abnormality

examination

This process will give us the set of valid terms in the whole collection. From now on the words 'term' and 'expression' are used interchangeably.

Computing Term, Document and Category Frequencies. Our starting point is m training collections, each one containing the training documents belonging to each category. Every subset is processed separately to compute the frequencies for each expression in all the categories.

We can easily organize all frequencies in a matrix, where columns correspond to expressions and rows correspond to categories.:

	e_1	e_2	...	e_j	...	e_n	
c_1	TF_{11}, DF_{11}	TF_{12}, DF_{12}	...	TF_{1j}, DF_{1j}	...	TF_{1n}, DF_{1n}	N_1
c_2	TF_{21}, DF_{21}	TF_{22}, DF_{22}	...	TF_{2j}, DF_{2j}	...	TF_{2n}, DF_{2n}	N_2
...	...	...	...	...	...	...	...
c_i	TF_{i1}, DF_{i1}	TF_{i2}, DF_{i2}	...	TF_{ij}, DF_{ij}	...	TF_{in}, DF_{in}	N_i
...	...	...	...	...	...	...	...
c_m	TF_{m1}, DF_{m1}	TF_{m2}, DF_{m2}	...	TF_{mj}, DF_{mj}	...	TF_{mn}, DF_{mn}	N_m
	CF_1	CF_2	...	CF_j	...	CF_n	

where:

- c_i = category i.
- n = number of expressions extracted from all the training documents.
- m = number of categories.
- TF_{ij} = Term Frequency of the expression e_j, that is, number of times that the expression e_j appears in all the training documents for the category c_i.
- DF_{ij} = Document Frequency of the expression e_j, that is, number of training documents for the category c_i in which the expression e_j appears.
- CF_j = Category Frequency of the expression e_j, that is, number of categories in which the expression e_j appears.
- N_i = number of expressions extracted from the training documents of the category c_i.

Every CF_j and N_i can be easily calculated from TF_{ij} by:

$$CF_j = \sum_{i=1}^{m} x_{ij} \; ; \; N_i = \sum_{j=1}^{n} x_{ij} \; ; \; x_{ij} = \begin{cases} 1 \text{ if } TF_{ij} \neq 0 \\ 0 \text{ otherwise} \end{cases}$$

Across this paper, some examples are shown with the expressions obtained from the training process. The TF and DF corresponding to each expression are put together between brackets, i.e. (TF, DF).

We also have to pay special attention to the expressions in the titles. Some experiments have been performed to get an appropriate factor which increases the weight of the expressions that appear in document's titles [14]. Doubling frequencies (TF and DF) is proved to be a consideration which optimizes the performance.

Getting the Most Representative Terms (MRT). The expressions obtained from documents are associated to the categories each document belongs to in the training collection. As a result, we will get m sets of expressions, each one representing a specific category.

For example, after analysing every document associated to the category "Carcinoid Heart Disease", some of the representative expressions are:

carcinoid disease (1,1)   tricuspid stenosis (1,1)
carcinoid heart (26,8)   ventricular enlargement (1,1)
carcinoid heart disease (26,8)   ventricular failure (4,1)
carcinoid syndrome (8,3)   ventricular volume (1,1)
carcinoid tumour (3,2)   ventricular volume overload (1,1)

For each category, we have to select the terms which best identify each category. Three criteria are used to carry out this selection:

- *Predominance inside the whole corpus.* The more times a term occurs inside the full training collection, the more important it is. L-Average discrimination using TF and DF over all the expressions in the corpus $(\overline{TF}, \overline{DF})$ is used to identify and select the best terms (BT) across the whole corpus.
- *Discrimination among categories.* The more categories a term represents, the less discriminative it is. Expressions appearing in more than half of the categories are not considered discriminative enough. Some authors use fixed stopword lists [10] for discarding expressions during the learning and classification processes. Our approach produces this list automatically so that it is adjusted to the number of categories, documents and vocabulary of the training collection.

In this case, the set of category discriminative terms (CDT) for a category is obtained by removing expressions that are representative in more than half of the categories. That is, for every category, an expression e_j will be removed if:

$$CF_j > (m/2 + 1)$$

where m stands for the number of categories.

– *Predominance inside a specific category.* The more times a term occurs inside a category, the more representative it is for that particular category. L-Average discrimination using TF and DF values over all the expressions in each category i ($\overline{TF_i}$, $\overline{DF_i}$) are used to identify the best terms in a category (BTC_i).

So, we propose using these TF, DF and CF measures for dimensionality reduction as follows. For each category i:

1. Select the set of terms that are predominant inside the corpus (BT).
2. Select the set of terms that are discriminant among categories (CDT).
3. Select the set of terms that are predominant into this category (BTC_i).

The most representative terms of the category ($MRTC_i$) are obtained from the intersection of the three enumerated sets of terms:

$$\{MRTC\}_i = \{BT\} \cap \{CDT\} \cap \{BTC\}_i$$

As a result, we will get a subset of expressions for each category.

For example, the category "Carcinoid Heart Disease" is identified by the following expressions:

carcinoid (44,8)   heart disease (40,9)

carcinoid heart (26,8)   tricuspid valve (11,5)

carcinoid heart disease (26,8)

4.3 Expression Weighting

At this point, we have the most relevant terms for each of the m categories in the training set. These expressions are now weighted in order to measure their respective importance in a category. This process is accomplished as follows.

Normalization. The corpus of each category has its own characteristics (e.g. different number of training documents, longer or shorter expressions). So, we should not use the TF, DF and TL values directly obtained from the corpus. They can be normalized so that the final weights do not depend on the size of each category's training set neither on the differences on the averaged length over the representative expressions.

As also stated in [13], we consider that expressions whose frequencies and lengths are very close to the average, are the most appropriate, and their weights should remain unchanged, i.e. they should get unit or no normalization. By selecting an average normalization factor as the pivot, normalized values for TF, DF and TL (TFn, DFn and TLn) are calculated in terms of proportion between the total values and the average over all the expressions in the category:

$$TFn_{ij} = \frac{TF_{ij}}{\overline{TF_i}} \;\; ; \;\; DFn_{ij} = \frac{DF_{ij}}{\overline{DF_i}} \;\; ; \;\; TLn_{ij} = \frac{TL_j}{\overline{TL_i}}$$

where i stands for the category c_i and j for the expression e_j respectively.

Normalized values higher than 1 indicate relevance higher than the average, therefore they point to quite significant expressions.

Expressions Matching Category Descriptors. Normalized values measure how much a term stands out over the average. Since category descriptors are special expressions which can be considered more important than the average, if an expression e_j contains some of the category descriptors of c_i, its normalized frequencies and length should be 1 or higher. To assure this, TFn, DFn and TLn are set to 1 for category descriptors with normalized weights lower than 1.

Weighting. All the proposed values are put together to get a single relevance measure (weight). The proposed weighting schema contains much more information than the common TFIDF approach. Usually, a single set of features is extracted from the training data, and each feature is assigned a single weight. We extract an individual set of features per category, obtaining also different weights for the same expression in different categories.

Every expression e_j is weighted for each of the categories c_i according the following formula:

$$w_{ij} = \frac{(TFn_{ij} + DFn_{ij}) \cdot TLn_{ij} \cdot TFnew_j}{CF_j}$$

where $TFnew_j$ stands for the single number of times that the expression e_j appears in the current document which is being represented as a vector. The greater w_{ij} becomes, the more representative e_j is for c_i. By following the intuitions explained in section 3, this equation makes the weight grow proportionally to the term length and frequencies and makes it lower when the term is more distributed among the different categories.

Since the goal is building a binary classifier, we must have a class c_p representing the positive samples in the training set. Intuitively, to get an even more separable case, the weights of the expressions representing c_p should be calculated differently from the ones representing other categories. For the latter case, we propose accumulating the weight of the negative classes. More formally, we obtain the weight w_j of the expression e_j as following:

$$w_j = \begin{cases} w_{pj} & \text{if } e_j \in c_p \\ \sum_{i=1}^{m} w_{ij} \ \forall i \neq p & \text{otherwise} \end{cases}$$

where w_{pj} is the weight calculated from the positive samples and $\sum_{i=1}^{m} w_{ij} \ \forall i \neq p$ stands for the weight calculated from the negative samples.

5 Evaluation

Comparison to previous works is proposed using SVM against the simple BOW. We have represented the input data as feature vectors under the proposed schema (noun phrases) to make comparisons using a well-known training corpus such as the OHSU-MED collection. Results will show that our method increases substantially the classification performance.

5.1   Classification Algorithm

For more accurate comparisons against previous works [10], SVM^{light} software [17] has been used for evaluation purposes as a baseline classifier. All default parameters are selected except the cost-factor ("-j") [18], which controls the relative weighting of positive to negative examples, and thus provides a way to compensate for unbalanced classes. Leave-one-out cross-validation (LOO) (turned on by "-x 1" parameter) is used to compute a training set contingency table corresponding to each setting of "-j". SVM^{light} is run multiple times for each category, once for each of the resulting values from 0.25 to 4.0 with 0.25 increments (e.g. 0.25, 0.5, 0.75 ... 3.75, 4.0).

Since SVM yields better error bounds by using euclidean norm [19], all feature vectors (both in the training and test set) are normalized to euclidean length 1.

5.2   Data Sets

The OHSUMED collection consists of 348,566 citations from medical journals published from 1987 to 1991. Only 233,445 documents contain a title and an abstract. Each document was manually assigned to one or several topics, selected from a list of 14,321 MESH terms. Since automating this process leads to a quite difficult classification problem, most of the authors use smaller data sets. We have chosen the diseases [20] and heart diseases sub-trees [10].

For the diseases hierarchy, MESH terms below the same root node are grouped, leading to 23 categories. The first 10,000 documents in 1991 which have abstracts are used for training, and the second 10,000 are used for testing.

For the heart diseases sub-tree, the categories which have no training documents are discarded, leaving only 16,592 documents and 102 possible categories. The documents from 1987 to 1990 are used as the training set, and the 1991 ones are used as the test set.

5.3   Evaluation Measures

The algorithm performance has been evaluated through the standard BEP and F1 measures [21]:

$$F_1 = \frac{2 \cdot Recall \cdot Precision}{Recall + Precision}$$

where recall is defined to be the ratio of correct assignments by the system divided by the total number of correct assignments, and precision is the ratio of correct assignments by the system divided by the total number of the system's assignments. The precision and recall are necessarily equal (BEP) when the number of test examples predicted to be in the positive class equals the true number of positive test examples.

The relevant list of topics for each category is evaluated first, and the average performance score is calculated for all documents (micro-averaged) and for all categories (macro-averaged).

5.4   Relevance of the Parameters for the Classification Task

To compute the relevance of the parameters, the micro-averaged F1 performance is obtained from the original algorithm. After removing each parameter individually, the evaluation is performed again and the percentage of deterioration from the original algorithm is calculated.

Figure 1 reflects the influence of the main characteristics for the final categorization results. The following points describe the conditions applied for the different evaluations along with their associated labels in this figure:

– **Phrases:** Expressions are made of single words.
– **Phr.PoS:** Expressions are made of one single word of any part of speech (no dictionary nor PoS tagger are used).
– **Titles:** Expressions in the titles have the same weight as the other ones.
– **Cat.Des.:** Category descriptors do not have any influence for weighting the expressions.
– **TF-DF, TFnew, CF, TL:** TFn_{ij}, DFn_{ij}, $TFnew_j$, CF_j and TLn_j respectively do not have any effect during the weighting process. This is achieved by modifying the equation given in section 4.3 to omit in each case the indicated value:

$$TF - DF \Rightarrow w_{ij} = \frac{TLn_{ij} \cdot TFnew_j}{CF_j}$$

$$TFnew \Rightarrow w_{ij} = \frac{(TFn_{ij} + DFn_{ij}) \cdot TLn_{ij}}{CF_j}$$

$$CF \Rightarrow w_{ij} = (TFn_{ij} + DFn_{ij}) \cdot TLn_{ij} \cdot TFnew_j$$

$$TL \Rightarrow w_{ij} = \frac{(TFn_{ij} + DFn_{ij}) \cdot TFnew_j}{CF_j}$$

Figure 1 indicates how relevant each parameter is for the whole categorization process. It shows the percentage of deterioration in the performance when each parameter is removed from the algorithm.

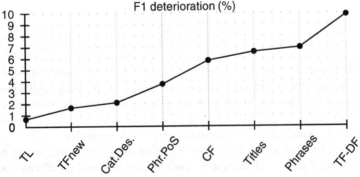

Fig. 1. Deterioration in the performance after removing each parameter. Parameters are put from left to right in increasing order, from the least to the most relevant one. The tests have been performed over the diseases sub-tree data set.

Table 1. Number of features in relation with micro-averaged F1 performance. Results obtained over the 23 diseases categories.

	# features	F1
Noun words	2055	63.9
Any words	2221	66.0
Noun phrases	24823	68.6

The TF-DF measures lead the graph, meaning that term frequencies have a crucial significance as known from many other previous works. The use of phrases or noun phrases instead of single words of any type is the second most important parameter. The increment of the weights for those expressions in titles also improves significantly the performance. Category Frequency is the fourth most important parameter, which confirms that the more categories an expression represents, the less discriminative it is.

Documents from other corpora may be better represented by taking only single words as features and using simple weighting schemas [22]. However, it is not the same for OHSUMED. As far as we know, the results here presented are the best ever achieved, leading us to the conclusion that for some type of data such as MEDLINE documents, we should try to increment the number of features and the amount of information they contain, as confirmed in table 1.

6   Results

Next tables show the results obtained in previous works followed by the results achieved by applying the new proposed algorithm for feature selection and weighting over the same training and test sets. The best values are in boldface.

Table 2 shows the micro-averaged BEP performance calculated over the 23 diseases categories. Noun phrases gets a global 3.8% improvement, also outperforming almost all categories individually.

Table 3 contains both the micro and macro averaged F1 performance over the 102 categories of the heart diseases sub-tree. For this corpus we have achieved more than 10% improvements (10.6% for micro and 10.3% for macro measures respectively).

Table 2. Break even point on 5 most frequent categories and micro-averaged performance over all 23 diseases categories [20]

	SVM (words)	SVM (noun phrases)
Pathology	**58.1**	52.7
Cardiovascular	77.6	**80.9**
Immunologic	73.5	**77.1**
Neoplasms	70.7	**81.5**
Digestive system	73.8	**77.5**
Micro avg (23 cat.)	66.1	**68.6**

Table 3. Averaged F1 performance over 102 heart diseases categories [10]

	SVM (words)	SVM (noun phrases)
Micro avg	63.2	**69.9**
Macro avg	50.3	**55.5**

7 Conclusions

Using a proper feature selection and weighting schema is known to be decisive. This work proposes a particular way to choose, extract and weight special n-grams from documents in plain text format so that we get a high performing representation. Moreover, the new algorithm is fast, easy to implement and it contains some necessary adjustments to automatically fit both existing and incoming MEDLINE documents.

References

1. Sebastiani, F.: A tutorial on automated text categorisation. In: Amandi, A., Zunino, R. (eds.) Proceedings of ASAI 1999, 1st Argentinian Symposium on Artificial Intelligence, Buenos Aires, AR, pp. 7–35 (1999)
2. Aas, K., Eikvil, L.: Text categorisation: A survey. Technical report, Norwegian Computer Center (June 1999)
3. Yang, Y., Liu, X.: A re-examination of text categorization methods. In: Hearst, M.A., Gey, F., Tong, R. (eds.) Proceedings of SIGIR 1999, 22nd ACM International Conference on Research and Development in Information Retrieval, Berkeley, US, pp. 42–49. ACM Press, New York (1999)
4. Scott, S., Matwin, S.: Feature engineering for text classification. In: Bratko, I., Dzeroski, S. (eds.) Proceedings of ICML 1999, 16th International Conference on Machine Learning, Bled, SL, pp. 379–388. Morgan Kaufmann Publishers, San Francisco (1999)
5. Tan, C.M., Wang, Y.F., Lee, C.D.: The use of bigrams to enhance text categorization. Information Processing and Management 38(4), 529–546 (2002)
6. Tesar, R., Strnad, V., Jezek, K., Poesio, M.: Extending the single words-based document model: a comparison of bigrams and 2-itemsets. In: DocEng 2006: Proceedings of the 2006 ACM symposium on Document engineering, pp. 138–146. ACM Press, New York (2006)
7. Antonie, M., Zaane, O.: Text document categorization by term association. In: IEEE International Conference on Data Mining (ICDM), pp. 19–26 (2002)
8. Zhang, Y., Zhang, L., Yan, J., Li, Z.: Using association features to enhance the performance of naive bayes text classifier. In: Fifth International Conference on Computational Intelligence and Multimedia Applications, ICCIMA 2003, pp. 336–341 (2003)
9. Basili, R., Moschitti, A., Pazienza, M.T.: Language-sensitive text classification. In: Proceeding of RIAO 2000, 6th International Conference Recherche d'Information Assistee par Ordinateur, Paris, FR, pp. 331–343 (2000)
10. Granitzer, M.: Hierarchical text classification using methods from machine learning. Master's thesis, Graz University of Technology (2003)
11. Moschitti, A., Basili, R.: Complex linguistic features for text classification: A comprehensive study. In: McDonald, S., Tait, J.I. (eds.) ECIR 2004. LNCS, vol. 2997, pp. 181–196. Springer, Heidelberg (2004)

12. Buckley, C.: The importance of proper weighting methods. In: Bates, M. (ed.) Human Language Technology. Morgan Kaufman, San Francisco (1993)
13. Singhal, A., Buckley, C., Mitra, M.: Pivoted document length normalization. Department of Computer Science, Cornell University, Ithaca, NY 14853 (1996)
14. Ruiz-Rico, F., Vicedo, J.L., Rubio-Sánchez, M.C.: Newpar: an automatic feature selection and weighting schema for category ranking. In: Proceedings of DocEng 2006, 6th ACM symposium on Document engineering, pp. 128–137 (2006)
15. Màrquez, L., Giménez, J.: A general pos tagger generator based on support vector machines. Journal of Machine Learning Research (2004), www.lsi.upc.edu/~nlp/SVMTool
16. Kongovi, M., Guzman, J.C., Dasigi, V.: Text categorization: An experiment using phrases. In: Crestani, F., Girolami, M., van Rijsbergen, C.J.K. (eds.) ECIR 2002. LNCS, vol. 2291, pp. 213–228. Springer, Heidelberg (2002)
17. Joachims, T.: Making large-Scale SVM Learning Practical. Advances in Kernel Methods - Support Vector Learning (1999), http://svmlight.joachims.org/
18. Joachims, T.: Support Vector and Kernel Methods. In: SIGIR 2003 Tutorial (2003)
19. Zu, G., Ohyama, W., Wakabayashi, T., Kimura, F.: Accuracy improvement of automatic text classification based on feature transformation. In: Proceedings of DOCENG 2003, ACM Symposium on Document engineering, Grenoble, FR, pp. 118–120. ACM Press, New York (2003)
20. Joachims, T.: Text categorization with support vector machines: learning with many relevant features. In: Nédellec, C., Rouveirol, C. (eds.) ECML 1998. LNCS, vol. 1398, pp. 137–142. Springer, Heidelberg (1998)
21. Joachims, T.: Estimating the generalization performance of a svm efficiently. In: Langley, P. (ed.) Proceedings of ICML 2000, 17th International Conference on Machine Learning, Stanford, US, pp. 431–438. Morgan Kaufmann Publishers, San Francisco (2000)
22. Dumais, S.T., Platt, J., Heckerman, D., Sahami, M.: Inductive learning algorithms and representations for text categorization. In: Gardarin, G., French, J.C., Pissinou, N., Makki, K., Bouganim, L. (eds.) Proceedings of CIKM 1998, 7th ACM International Conference on Information and Knowledge Management, Bethesda, US, pp. 148–155. ACM Press, New York (1998)

Representing and Reasoning with Temporal Constraints in Clinical Trials Using Semantic Technologies

Ravi D. Shankar[1], Susana B. Martins[1], Martin J. O'Connor[1], David B. Parrish[2], and Amar K. Das[1]

[1] Stanford Medical Informatics, Stanford University
251 Campus Drive, MSOB X215, Stanford, California, USA
{ravi.shankar,smartins,moconnor,amar.das}@stanford.edu
[2] Immune Tolerance Network, Pittsburgh, Pennsylvania, USA
dparrish@immunetolerance.org

Abstract. Clinical trial protocols include schedule of clinical trial activities such as clinical tests, procedures, and medications. The schedule specifies temporal constraints on the sequence of these activities, on their start times and duration, and on their potential repetitions. There is an enormous requirement to conform to the constraints found in the protocols during the conduct of the clinical trials. In this paper, we present our approach to formally represent temporal constraints found in clinical trials, and to facilitate reasoning with the constraints. We have identified a representative set of temporal constraints found in clinical trials in the immune tolerance area, and have developed a temporal constraint ontology that allows us to formulate the temporal constraints to the extent required to support clinical trials management. We use the ontology to specify temporal annotation on clinical activities in an encoded clinical trial protocol. We have developed a temporal model to encapsulate time-stamped data, and to facilitate interval-based temporal operations on the data. Using semantic web technologies, we are building a knowledge-based framework that integrates the temporal constraint ontology with the temporal model to support queries on clinical trial data. Using our approach, we can formally specify temporal constraints, and reason with the temporal knowledge to support management of clinical trials.

Keywords: Ontology, temporal reasoning, clinical trials, biomedical informatics, Semantic Web, OWL.

1 Introduction

Clinical trials are formal studies on participants to systematically evaluate the safety and efficacy of new or unproven approaches in the prevention and treatment of medical conditions in humans. A clinical trial protocol is a document that includes study objectives, study design, participant eligibility criteria, enrollment schedule, and study plan. It specifies a temporal schedule of clinical trial activities such as clinical tests, procedures, and medications. The schedule includes temporal constraints on the sequence of these activities, on their duration, and on potential cycles. A temporal constraint is defined as an interval-based temporal annotation on a domain entity in

A. Fred, J. Filipe, and H. Gamboa (Eds.): BIOSTEC 2008, CCIS 25, pp. 520–530, 2008.

relationship with other entities. Temporal constraints are fundamental to the descriptions of protocol entities, such as the following specifications: *Participants will be enrolled at least two days apart; Participant is ineligible if he/she had vaccination with a live virus within the last 6 weeks before enrollment; The first dose will be infused over a minimum of 12 hours; Visit 10 for the participant occurs 3 weeks ± 2 days from the day of transplant.* There is an enormous requirement on the execution of a clinical trial to conform to the temporal constraints found in the protocol. Studies need to be tracked for the purposes of general planning, gauging progression, monitoring patient safety, and managing personnel and clinical resources. The tracking effort is compounded by the fact that a trial often is carried out at multiple sites, geographically distributed, sometimes across the world. The validity of the findings of the clinical trial depends on the clinical trial personnel and the participants performing clinical trial activities as planned in the protocol. More importantly, the treatment and assessment schedules should be strictly followed to ensure the safety of participants.

We have developed an ontological framework that we call Epoch [1,2], to support the management of clinical trials at the Immune Tolerance Network, or ITN [3,4]. As part of this effort, we have developed a suite of ontologies that, along with semantic inferences and rules, provide a formal protocol definition for clinical trial applications. We use the OWL Web Ontology language [5], which is a W3C standard language for use in Semantic Web where machines can provide enhanced services by reasoning with facts and definitions expressed in OWL. Central to our ontological effort is the modeling of temporal constraints that we identified in clinical trial protocols. We have created the *temporal constraint ontology* to formally represent temporal constraints. The ontological representation can then be used to construct rules that can be used in turn, for reasoning with temporal constraints. Thus, at protocol specification phase, a domain expert can capture the essence of temporal constraints using higher-level ontological constructs. At a later time, a software developer can fully encode the constraints by creating rules in terms of temporal patterns and other protocol entities in the ontologies. We are using SWRL, the Semantic Web Rule Language [6] to write the rules. At execution time of the protocol, the rule elements use the protocol knowledge specified in the Epoch ontologies, and the clinical trial data collected in the clinical trial databases to reason with the temporal constraints. In this paper, we discuss our work in identifying temporal constraints found in ITN's clinical trial protocols. We then discuss our temporal constraint ontology using some patterns that we found in the temporal constraints. We then show how we use the temporal constraint ontology along with other Epoch ontologies to create rules that can be executed at runtime to support clinical trial management.

2 Temporal Constraints in Clinical Trials

A clinical trial protocol defines a protocol schema that divides the temporal span of the study into phases such as the treatment phase and follow-up phase, and specifies the temporal sequence of the phases. It also includes a schedule of activities that enumerates a sequence of protocol visits that are planned at each phase, and, for each visit, specifies the time window when the visit should happen and a list of clinical activities (assessments, procedures and tests) that are planned at that visit. Activities such as medication need not be confined to visits and can be planned to occur in a time window within a protocol phase. An activity can have sub activities that can impose additional temporal constraints. For example, an assessment activity can include collection and processing of biological specimens with its own set of temporal constraints.

Here is a representative set of temporal constraints that we found in the ITN protocols that we are encoding:

1. *Visit 17 must occur at least 1 week but no later than 4 weeks after the end of 2003 ragweed season.*
2. *Administer Rapamune 1 week from Visit 0 daily for 84 days.*
3. *Visit 1 should occur 2 weeks ± 3 days after transplant.*
4. *Screening visit evaluations must occur between 30 days prior to Visit -1 and 45 days prior to Visit 0.*
5. *The vital signs of the participant should be obtained at routine time points starting at 10 minutes post infusion, then at 20-minute intervals until the participant is discharged.*
6. *Administer study medication at weekly intervals for 3 months.*
7. *Clinical assessments are required twice a week until Day 28 or discharge from hospital.*
8. *The first and second blood draws are 10 days apart, and the third draw is 11-14 days after the second.*
9. *On days that both IT and omalizumab are administered, omalizumab will be injected 60 minutes after the IT.*
10. *Monitor cyclosporine levels 3 times per week while in-patient, then weekly as out-patient.*

As evident in the constraints, clinical activities —we are using the terms *activity* and *event* interchangeably— are temporally dependent on each other. The temporal annotations in the constraints are specified in relative terms typically with reference to one or more clinical events. At the protocol execution time, the actual times of these events found in the clinical data will be used to reason with the constraints. There can also be fuzziness in the relative start and end times as well as in the duration of the activity. An activity can be repeated at a periodic interval for a specific number of times or until a condition is satisfied. The periodic interval can be a single offset or a set of offsets. The temporal annotation of an activity or the temporal ordering of activities can be conditional on other events.

3 Temporal Representation

We have developed a temporal constraint ontology that can be used to formally specify the temporal constraints found in the clinical trial protocols. We briefly describe the core entities of the underlying temporal representation below:

Anchor defines an unbound time point that can be used to specify temporal relations among activities. It can be used as a reference point to define the start of another event before or after the anchor. In example 1 (of the constraints listed earlier), *end of 2003 ragweed season* is an anchor used to define the start of *Visit 17*. During the execution of the protocol, an anchor is bound to the absolute time of the anchor as recorded in the clinical trial data.

Duration is the difference between two time points. It is used typically to specify how long an activity lasts. In example 2, *84 days* is the duration.

Anchored Duration relates two activities with a temporal offset. In example 2, the activity *administer Rapamune* is offset from the anchor *Visit 0* by *1 week*.

Varying Duration is defined as duration with a high variance and a low variance. In example 3, *2 weeks ± 3 days* specifies a varying offset between *transplant* time and *Visit 1* start time.

Start and End Expression constrains the start and the end of an activity and is expressed as offsets before or after one or more reference events. In example 4, the start of the activity *Screening visit evaluations* is 30 days before the anchor *Visit -1* and the end is 45 days before another anchor *Visit 0*.

Cyclical Plan Expression formulates events that are repeated at periodic intervals. The repetition ends typically when a specific number of cycles is reached or until a specific condition is satisfied. There are two types of cyclical plans with subtle differences. The first type has a single anchor point with potentially multiple intervals. In example 5, the *vital signs* assessments are planned at 10, 30, 60, 90, 120, and 180 minutes after *infusion*. If the participant gets off schedule because the assessment is made at minute 35 instead of minute 30, then the participant gets back on schedule with the next assessment at minute 60. This type of cyclical plan is used generally with assessments and tests where evaluations need to be made at specific intervals after a clinical intervention. The second type of cyclical plan can potentially have multiple anchors with a single offset. In example 6, the plan is to administer medication at weekly intervals for 3 months. The initial anchor is the event of administering the first dose. According to the schedule, the second dose will be 1 week later, and the third 1 week later from the second dose. If the participant gets off schedule because the drug was administered 5 days after first dose and not 7 days, then the participant gets back on schedule with the next dose at 7 days from the last dose. This type of cyclical plan is used typically with drug administration where fixed intervals between dosages need to be maintained for safety and efficacy purposes.

Conditional Expression allows associating different temporal annotations with a single activity based on a condition. There are three patterns of conditional expressions – *if-then, if-then-else* and *until-then* patterns. Example 9 illustrates the *if-then* pattern – the temporal constraint between the administrations of two drugs is dependent on the condition that the two drugs are administered on the same day. Example 10 illustrates the *until-then* pattern – the monitoring activity is performed 3 times a week until the participant is in in-patient status, and when the status changes to out-patient then the activity is performed weekly.

4   Epoch Ontologies

In order to support clinical trial management activities, the Epoch knowledge-based approach provides three methods: 1. knowledge acquisition methods that allow users to encode protocols, 2. ontology-database mapping methods that integrate the protocol and biomedical knowledge with clinical trial data including clinical results and operational data stored in the ITN data repository, and 3. concept-driven querying methods that support integrated data management, and that can be used to create high-level abstractions of clinical data during analysis of clinical results. At the center of all these methods is the suite of Epoch ontologies that provide a common nomenclature and semantics of clinical trial protocol elements:

- The *clinical trial ontology* is the overarching ontology that encapsulates the knowledge structure of a clinical trial protocol. It simplifies the complexity inherent in the full structure of the protocol by focusing only on concepts required to support clinical trial management. Other concepts are either ignored or partially represented.
- The *constraint expression ontology* models the class of temporal constraints (see Section 3) and logical constraints found in clinical trial protocols.
- The *virtual trial data ontology* encapsulates the study data that is being collected, such as participant clinical record, specimen workflow logs, and site related data. A mapping component can then map clinical trial data (found in a relational database) to these virtual data records using a mapping ontology. The data model concept is similar to the Virtual Medical Record [7] specification promoted in the clinical guideline modeling efforts.
- The *organization ontology* provides a structure to specify study sites, clinical and core laboratories, and bio-repositories that participate in the implementation of a specific protocol.
- The *assay ontology* models characteristics of mechanistic studies relevant to immune disorders. An assay specification includes the clinical specimen that can be analyzed using that assay, and the workflow of the specimen processing at the core laboratories.
- The *labware ontology* models a laboratory catalog that mainly lists specimen containers used in the clinical trials.
- The *measurement ontology* has concepts of physical measurements such as volume and duration, and units of measurement such as milliliter and month

4.1 Temporal Model in the Virtual Trial Data Ontology

The *virtual trial data ontology* uses a valid-time model to represent the temporal component of clinical trial data. In this model, all facts have temporal extent and are associated with instants or intervals denoting the times that they are held to be true. The core concept in the model is the *extended proposition* class that represents information that extends over time. There are two types of extended propositions in the model: 1. *extended primitive propositions* that represent data derived directly from secondary storage, and 2. *extended abstract propositions* that are abstracted from other propositions. These extended propositions can be used to consistently represent temporal information in ontologies. For example, a set of participant visits in a clinical trial data can be represented by defining a class called *VisitRecord* that inherits the *valid time* property from *extended proposition* class. The *valid time* property will then hold a visit's actual occurrence time. Similarly, an extended primitive proposition can be used to represent a drug regimen, with a value of type string to hold the drug name and a set of periods in the valid time property to hold drug delivery times. A more detailed discussion of the temporal model can be found elsewhere in the literature [8].

5   Implementation Using Semantic Technologies

We have developed the Epoch ontologies in OWL where *classes* encapsulate the protocol concepts, and *properties* relate the classes to each other. Clinical trial data can be represented as instances of classes —referred to as *individuals*. OWL is a powerful constraint language for precisely defining how concepts in ontology should be interpreted. The Semantic Web Rule Language (SWRL) allows users to write Horn-like rules that can be expressed in terms of OWL concepts, and that can reason about OWL individuals. SWRL provides deductive reasoning capabilities that can infer new knowledge from an existing OWL knowledge base. We use SWRL to specify temporal constraints. Once all temporal information is represented consistently using the temporal model, SWRL rules can be written in terms of this model and the temporal constraint ontology. However, the core SWRL language has limited temporal reasoning capabilities. A few temporal predicates called *built-ins* are included in the set of standard predicates, but they have limited expressive power. SWRL provides an extension mechanism to add user-defined predicates. We used this mechanism to define a set of temporal predicates to operate on temporal values. These predicates support the standard Allen temporal operators [9]. Using these built-in operators in conjunction with the temporal model, we can express complex temporal rules.

Here is an example SWRL rule to check if clinical trial participants conform to a visit schedule (Figure 1) specified in the protocol:

```
Participant(?p) ^
hasVisitRecord(?p, ?vr) ^
hasVisitId(?vr, ?vid1) ^
hasValidTime(?vr, ?vt) ^
Visit(?v) ^
hasVisitId(?vr, ?vid2) ^
hasPlannedTiming(?v, ?pt) ^
hasRelativeStartTime(?pt, ?st) ^
swrlb:getStartInterval(?se, ?p, ?st) ^
swrlb:equal(?vid1, ?vid2) ^
temporal:inside(?vt, ?se) ^
-> ConformingParticipant(?p)
```

The rule uses concepts such as *Participant* and *Visit* from the *clinical trial ontology* and the concept of *Relative Start Time* as Anchored Duration in the *temporal constraint ontology*. The actual visits undertaken by a participant is encapsulated as the *VisitRecord* in the *virtual trial data ontology*, and is an extended proposition in the

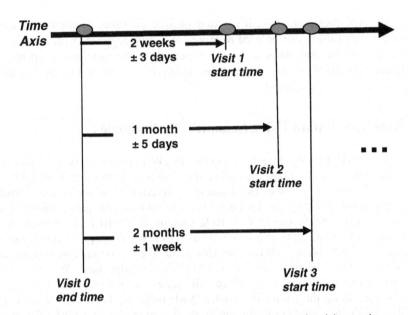

Fig. 1. A sample visit schedule specifies temporal constraints on the visit start times

temporal model. The rule uses two built-ins – *equal*, that checks if two strings are equal, and *inside*, which is a built-in that we developed to check if an absolute time is within a time interval (see Section 3). Protégé [10,11] is a software tool that supports the specification and maintenance of terminologies, ontologies and knowledge-bases in OWL. It has a plug-in called SWRL Tab [12], an editor for SWRL rules. We used Protégé to create the ontologies in OWL and SWRL. We then encoded specific protocols using Protégé's knowledge-acquisition facilities. The data generated from the implementation and execution of clinical trials is stored in a relational database. The types of data include participant enrollment data, specimen shipping and receiving logs, participant visits and activities, and clinical results. We have implemented a dynamic OWL-to-relational mapping method and have used SWRL to provide a high-level query language that uses this mapping methodology. A *schema ontology* describes the schema of an arbitrary relational database. A *mapping ontology* describes the mapping of data stored in tables in a relational database to entities in an OWL ontology. A *mapping software* uses the data source and the mapping ontologies to dynamically map trial data to entities in the *virtual trial data ontology* (Figure 2). A detailed description of the mapping techniques can be found elsewhere in the literature [13]. We are currently using JESS [14], a production rule-engine, to selectively execute the SWRL rules based on the context. For example, the rule that specifies the constraint on a visit time window will alone need to be executed when checking if a specific participant's visit satisfied the constraint. Thus, a temporal constraint is defined first using the temporal constraint ontology, then is formulated as a rule, finally, is reasoned with real clinical data using dynamic mappings between ontological concepts and relational database elements.

6 Related Work

Over the years, many expressive models have been developed to represent temporal constraints [15-18]. Shahar's approach [19] identifies temporal abstractions of data and properties using interpolation-based techniques and knowledge-based reasoning. In recent years, there have been a number of initiatives to create clinical trial protocol models that encapsulate clinical trial activities and associated temporal constraints found in a protocol. These ontologies are then used to automate different clinical trial management activities such as eligibility determination, participant tracking, and site management. The ontologies can also be used when subsequently analyzing the clinical trial data. Our Epoch framework employs a task-based paradigm that combines an explicit representation of the clinical trial domain with rules that capture the logical conditions and temporal constraints found in the trial management process. There have been a number of proposals on task-based clinical guideline representation formats – EON [20], PRO*forma* [21], GLIF [22], etc. that deal with temporal constraints on patient data and on activities found in clinical guidelines.

In the area of clinical trials, several modeling efforts have addressed different requirements of trial management activities. An ontology to represent temporal information and cyclical event patterns in clinical trial protocols has been proposed by Weng et al. [23]. The Trial Bank Project [24] is a trial registry that uses a protocol ontology to capture information such as intervention, outcomes, and eligibility criteria on randomized clinical trials. The underlying knowledge base can support systematic reviewing and evidence-based practice.

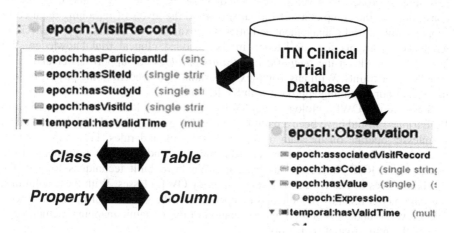

Fig. 2. The OWL classes and properties of the *virtual trial data ontology* is mapped to tables and columns of a clinical trial database using the model-database mapper

There is an ongoing effort by CDISC [25], an industry-lead, multidisciplinary organization, to develop and support the electronic acquisition, exchange, submission and archiving of clinical trials data. As part of this effort, CDISC is developing the Structured Protocol Representation that identifies standard elements of a clinical trial protocol that can be codified to facilitate the data interchange among systems and stakeholders including regulatory authorities, biopharmaceutical industry, statisticians, project managers, etc. A parallel effort is the BRIDG [26,27] project, a partnership of several organizations including CDISC, the HL7 standards body [28], the National Cancer Institute and the Federal Drug Administration, that consumes the Trial Design Model work to build a comprehensive domain analysis model representing protocol-driven biomedical/clinical research. The BRIDG model is a work in progress to elaborately define functions and behaviors throughout clinical trials, and uses the Unified Modeling Language (UML) for representation. The model, in its current state, lacks formalization of and reasoning with temporal constraints, and thus, cannot fully support the requirements of ITN's clinical trial management.

7 Discussion

The increasing complexity of clinical trials has generated an enormous requirement for knowledge and information management at all stages of the trials – planning, specification, implementation, and analysis. Our focus is currently on two application areas: 1. tracking participants of the trial as they advance through the studies, and 2. tracking clinical specimens as they are processed at the trial laboratories. The core of the Epoch framework is a suite of ontologies that encodes knowledge about the clinical trial domain that is relevant to trial management activities. Our focus on supporting trial management activities is also reflected in our approach to temporal constraint reasoning. Thus, in developing the temporal constraint ontology and in our reasoning approach with rules, we limited ourselves to the types and of temporal constraints, to the complexity of formalism and to the levels of reasoning required to support clinical trial management activities. For example, we do not support checking temporal constraints for consistency in the encoded clinical trial knowledge. We continue to work on the *temporal constraints ontology* to support newer and more complex constraints. With any complex constraint, one concern is the power, or lack thereof, of our reasoning approach with SWRL rules.

Since we use OWL ontologies and SWRL rules, native RDF Store (storing data as RDF triples) would have been a natural solution for storing clinical trial data, and then seamlessly operate on the data using our ontologies and rules. ITN uses a legacy relational database system to store clinical trial data, and therefore, prevents us from using native RDF Stores as our backend. We have built techniques to map the database tables to our *virtual trial data ontology* OWL classes. With these solutions, our data model remains flexible and independent of the structure of the data sources. We are yet to undertake a thorough evaluation of our dynamic mapping methodology especially in the area of scalability.

An often over-looked aspect of knowledge-based reasoning approaches is the task of knowledge-acquisition. Currently, we use the Protégé-OWL editor to build the Epoch models. Based on the class and property definitions, Protégé automatically generates graphical user interface (GUI) forms that can be used to create instances of these classes (OWL *individuals*). Thus, domain specialists can use to enter a specification of a protocol, say for a transplant clinical trial, using these Protégé generated forms. Unfortunately, domain specialists find it cumbersome and non-intuitive to use the generic user interfaces as they are exposed to the complexities of the Epoch ontologies, the OWL expressions and the SWRL rules. We are building custom graphical user interfaces that hide the complexities of the knowledge models, and that facilitate guided knowledge-acquisition. Providing a friendly user interface to enter SWRL rules can be challenging though.

The knowledge requirements borne out of the need for managing clinical trials align well with the touted strengths of semantic web technologies – uniform domain-specific semantics, flexible information models, and inference technology. Using these technologies, we have built a knowledge-based framework for temporal constraints reasoning that is, above all, practical.

Acknowledgements. This work was supported in part by the Immune Tolerance Network, which is funded by the National Institutes of Health under Grant NO1-AI-15416.

References

1. Shankar, R.D., Martins, S.B., O'Connor, M.J., Parrish, D.B., Das, A.K.: Epoch: an ontological framework to support clinical trials management. In: Proceedings of the International Workshop on Health Information and Knowledge Management, pp. 25–32 (2006)
2. Shankar, R.D., Martins, S.B., O'Connor, M.J., Parrish, D.B., Das, A.K.: Towards Semantic Interoperability in a Clinical Trials Management System. In: Proceedings of the Fifth International Semantic Web Conference, pp. 901–912 (2006)
3. Immune Tolerance network, http://www.immunetolerance.org
4. Rotrosen, D., Matthews, J.B., Bluestone, J.A.: The Immune Tolerance Network: a New Paradigm for Developing Tolerance-Inducing Therapies. Journal of Allergy and Clinical Immunology 110(1), 17–23 (2002)
5. OWL, http://www.w3.org/2004/OWL
6. SWRL, http://www.w3.org/Submission/SWRL
7. Johnson, P.D., Tu, S.W., Musen, M.A., Purves, I.: A Virtual Medical Record for Guideline-Based Decision Support. In: Proceedings of the 2001 AMIA Annual Symposium, pp. 294–298 (2001)
8. Snodgrass, R.T.(ed.): The TSQL2 temporal query language. Kluwer Academic Publishers, Boston (1995)
9. Allen, J.F.: Maintaining knowledge about temporal intervals. Communications of the ACM 26(11), 832–843 (1993)
10. Protégé, http://protege.stanford.edu
11. Knublauch, H., Fergerson, R.W., Noy, N.F., Musen, M.A.: The Protégé OWL Plugin: An Open Development Environment for Semantic Web applications. In: Proceedings of the Third International Semantic Web Conference, pp. 229–243 (2004)
12. O'Connor, M.J., Knublauch, H., Tu, S.W., Grossof, B., Dean, M., Grosso, W.E., Musen, M.A.: Supporting Rule System Interoperability on the Semantic Web with SWRL. In: Proceedings of the Fourth International Semantic Web Conference, pp. 974–986 (2005)
13. O'Connor, M.J., Shankar, R.D., Tu, S.W., Nyulas, C., Musen, M.A., Das, A.K.: Using Semantic Web Techonologies for Knowledge-Driven Queries in Clinical Trials. In: Proceedings of the 11th Conference on Artificial Intelligence in Medicine (2007)
14. JESS, http://www.jessrules.com

15. Bettini, C., Jajodia, S., Wang, X.: Solving multi-granularity constraint networks. Artificial Intelligence 140(1-2), 107–152 (2002)
16. Combi, C., Franceschet, M., Peron, A.: Representing and Reasoning about Temporal Granularities. Journal of Logic and Computation 14(1), 51–77 (2004)
17. Terenziani, P.: Toward a Unifying Ontology Dealing with Both User-Defined Periodicity and Temporal Constraints About Repeated Events. Computational Intelligence 18(3), 336–385 (2002)
18. Duftschmid, G., Miksch, S., Gall, W.: Verification of temporal scheduling constraints in clinical practice guidelines. Artificial Intelligence in Medicine 25(2), 93–121 (2002)
19. Shahar, Y., Musen, M.A.: Knowledge-Based Temporal Abstraction in Clinical Domains. Artificial Intelligence in Medicine 8, 267–298 (1996)
20. Musen, M.A., Tu, S.W., Das, A.K., Shahar, Y.: EON: A component-based approach to automation of protocol-directed therapy. Journal of the American Medical Informatics Association 3(6), 367–388 (1996)
21. Fox, J., Johns, N., Rahmanzadeh, A., Thomson, R.: PROforma: A method and language for specifying clinical guidelines and protocols. In: Proceedings of Medical Informatics Europe (1996)
22. Boxwala, A.A., Peleg, M., Tu, S.W., Ogunyemi, O., Zeng, Q.T., Wang, D., Patel, V.L., Greenes, R.A., Shortliffe, E.H.: GLIF3: A Representation Format for Sharable Computer-Interpretable Clinical Practice. Journal of Biomedical Informatics 37(3), 147–161 (2004)
23. Weng, C., Kahn, M., Gennari, J.H.: Temporal Knowledge Representation for Scheduling Tasks in Clinical Trial Protocols. In: Proceedings of the American Medical Informatics Association Fall Symposium, pp. 879–883 (2002)
24. Sim, I., Olasov, B., Carini, S.: The Trial Bank system: capturing randomized trials for evidence-based medicine. In: Proceedings of the American Medical Informatics Association Fall Symposium, p. 1076 (2003)
25. CDISC, http://www.cdisc.org
26. BRIDG, http://www.bridgproject.org
27. Fridsma, D.B., Evans, J., Hastak, S., Mead, C.N.: The BRIDG project: a technical report. Journal of American Medical Informatics Association 15(2), 130–137 (2007)
28. HL7, http://www.hl7.org

Gesture Therapy: A Vision-Based System for Arm Rehabilitation after Stroke

L. Enrique Sucar[1], Gildardo Azcárate[1], Ron S. Leder[2], David Reinkensmeyer[3], Jorge Hernández[4], Israel Sanchez[4], and Pedro Saucedo[5]

[1] Departamento de Computación, INAOE, Tonantzintla, Puebla, Mexico
esucar@inaoep.mx, yayo_tec@hotmail.com
[2] División de Ingeniería Eléctrica, UNAM, Mexico D.F., Mexico
rleder@ieee.org
[3] Department of Mechanical and Aerospace Engineering, UC Irvine, US
dreinken@uci.edu
[4] Unidad de Rehabilitación, INNN, Mexico D.F., Mexico
jhfranco@medicapolanco.com, drisavi_mx@yahoo.com.mx
[5] Universidad Anáhuac del Sur, Mexico D.F., Mexico
saucedouribe@gmail.com

Abstract. Each year millions of people in the world survive a stroke, in the U.S. alone the figure is over 600,000 people per year. Movement impairments after stroke are typically treated with intensive, hands-on physical and occupational therapy for several weeks after the initial injury. However, due to economic pressures, stroke patients are receiving less therapy and going home sooner, so the potential benefit of the therapy is not completely realized. Thus, it is important to develop rehabilitation technology that allows individuals who had suffered a stroke to practice intensive movement training without the expense of an always-present therapist. Current solutions are too expensive, as they require a robotic system for rehabilitation. We have developed a low-cost, computer vision system that allows individuals with stroke to practice arm movement exercises at home or at the clinic, with periodic interactions with a therapist. The system integrates a web based virtual environment for facilitating repetitive movement training, with state-of-the art computer vision algorithms that track the hand of a patient and obtain its 3-D coordinates, using two inexpensive cameras and a conventional personal computer. An initial prototype of the system has been evaluated in a pilot clinical study with promising results.

Keywords: Rehabilitation, stroke, therapeutic technology, computer vision.

1 Introduction

Each year millions of people in the world survive a stroke, in the U.S. alone the figure is over 600,000 people per year (ASA 2004). Approximately 80% of acute stroke survivors lose arm and hand movement skills. Movement impairments after stroke are typically treated with intensive, hands-on physical and occupational therapy for several weeks after the initial injury. Unfortunately, due to economic pressures on health care providers, stroke patients are receiving less therapy and going home sooner. The

A. Fred, J. Filipe, and H. Gamboa (Eds.): BIOSTEC 2008, CCIS 25, pp. 531–540, 2008.
© Springer-Verlag Berlin Heidelberg 2008

ensuing home rehabilitation is often self directed with little professional or quantitative feedback. Even as formal therapy declines, a growing body of evidence suggests
that both acute and chronic stroke survivors can improve movement ability with intensive, supervised training. Thus, an important goal for rehabilitation engineering is
to develop technology that allows individuals with stroke to practice intensive movement training without the expense of an always-present therapist.

Fig. 1. Screen shot of T-WREX. Arm and hand movements are focused as a mouse pointer to
activate an object in the simulation. In this case a hand interacts with a basketball. The upper
left insert shows the camera views, frontal and side, of the patient's hand tracked by the system.

We have developed a prototype of a low-cost, computer vision system that allows
individuals with stroke to practice arm movement exercises at home or at the clinic,
with periodic interactions with a therapist. The system makes use of our previous
work on a low-cost, highly accessible, web-based system for facilitating repetitive
movement training, called "Java Therapy", which has evolved into T-WREX (Fig. 1)
(Reinkensmeyer 2000 and Sanchez 2006). T-WREX provides simulation activities relevant to daily life. The initial version of Java Therapy allowed users to log into a
Web site, perform a customized program of therapeutic activities using a mouse or a
joystick, and receive quantitative feedback of their progress. In preliminary studies of
the system, we found that stroke subjects responded enthusiastically to the quantitative feedback provided by the system. The use of a standard mouse or joystick as the
input device also limited the functional relevance of the system. We have developed
an improved input device that consists of an instrumented, anti-gravity orthosis that
allows assisted arm movement across a large workspace. However, this orthosis costs
about $4000 to manufacture, limiting its accessibility. Using computer vision this

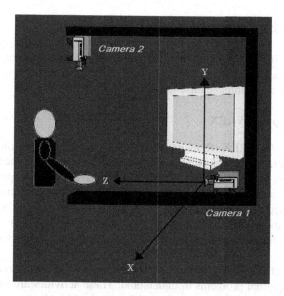

Fig. 2. Set up for the Gesture Therapy system. The patient is seated in front of a table that serves as a support for the impaired arm, and its movements are followed by two cameras. The patient watches in a monitor the simulated environment and his/her control of the simulated actuator.

system becomes extremely attractive because it can be implemented with low cost components, such as inexpensive web cameras and a conventional computer.

For "Gesture Therapy" (GT) we combine T-WREX with state-of-the art computer vision algorithms that track the hand of a patient and obtain its 3-D coordinates, using two inexpensive cameras (web cams) and a conventional personal computer (Fig. 2). The vision algorithms locate and track the hand of the patient using color and motion information, and the views obtained from the two cameras are combined to estimate the position of the hand in 3-D space. The coordinates of the hand (X, Y, Z) are sent to T-WREX so that the patient interacts with a virtual environment by moving his/her impaired arm, performing different tasks designed to mimic real life situations and thus oriented for rehabilitation. In this way we have a low-cost system which increases the motivation of stroke subjects to follow their rehabilitation program, and with which they can continue their arm exercises at home.

A prototype of this system, called "Gesture Therapy", has been installed at the rehabilitation unit at the National Institute of Neurology and Neurosurgery (INNN) in Mexico City. A pilot study was conducted with 11 stroke patients, 5 interacted with Gesture Therapy and 6 received conventional therapy. The results based on the therapist and patients' opinions are positive, although there is not a significant difference between both groups in terms of a clinical evaluation based on the Fugl-Meyer scale (Fugl-Meyer, 1975). A more extensive controlled clinical trial is required to evaluate the impact of the system in stroke rehabilitation.

The rest of the paper is structured as follows. Section 2 describes the virtual environment for rehabilitation called Java Therapy, and Section 3 the visual gesture tracking system. The pilot study is presented in Section 4. We conclude with a summary and directions for future work.

2 Simulated Environment for Rehabilitation

Gesture Therapy integrates a simulated environment for rehabilitation (Java Therapy) with a gesture tracking software in a low-cost system for rehabilitation after stroke. Next we describe the simulated environment.

2.1 Java Therapy/T-WREX

The Java Therapy/T-WREX (Reinkensmeyer 2000) web-based user interface has three key elements: therapy activities that guide movement exercise and measure movement recovery, progress charts that inform users of their rehabilitation progress, and a therapist page that allows rehabilitation programs to be prescribed and monitored.

The therapy activities are presented in the software simulation like games and the system configuration allows therapists to customize the software to enhance the therapeutic benefits for each patient, by selecting a specific therapy activity among others in the system.

The therapy activities were designed to be intuitive even for patients with minimal cognitive or perceptual problems to understand. These activities are for repetitive daily task-specific practice and were selected by its functional relevance and inherent motivation like grocery shopping, car driving, playing basketball, self feeding, etc.

Additionally, the system gives objective visual feedback of patient task performance, and patient progress can be illustrated easily by the therapist by a simple statistical chart. The visual feedback has the effect of enhancing motivation and endurance along the rehabilitation process by a patient's awareness of his/her progress.

There are several simulation/games in the current prototype, including: Basket-ball (see Fig 1), Car Racing, Wall Painting, Supermarket Shopping, and Cooking Eggs, among others.

3 Gesture Tracking

Based on a stereo system (two web cameras) and a computer, the hand of the user is detected and tracked in a sequence of images to obtain its 3-D coordinates in each frame, which are sent to the T-WREX environment. This process involves the following steps:

1. Calibration,
2. Segmentation,
3. Tracking,
4. 3-D reconstruction.

Next we describe each stage.

3.1 Calibration

To have a precise estimation of the 3-D position in space of the hand, the camera system has to be calibrated. The calibration consists in obtaining the intrinsic (focal length, pixel size) and extrinsic (position and orientation) parameters of the cameras.

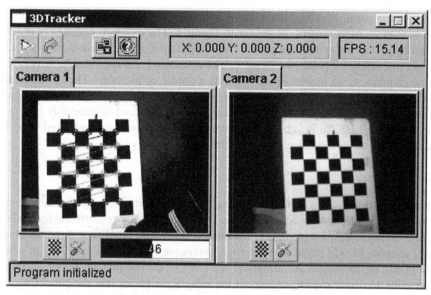

Fig. 3. Calibration procedure: reference pattern used for obtaining the intrinsic parameters of each camera (left: camera 1, right: camera 2). The crossing points (left image) of the checker board pattern are used to estimate the intrinsic parameters of the camera.

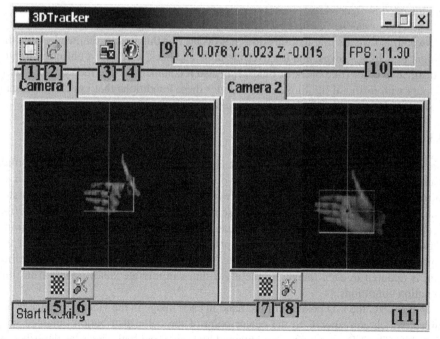

Fig. 4. Hand detection and segmentation in both images. The approximate hand region is shown as a rectangle, in which the center point is highlighted, used later for finding the 3-D co-ordinates of the hand.

The intrinsic parameters are obtained via a reference pattern (checker board) that is put in front of each camera, as shown in figure 3. The extrinsic parameters are obtained by giving the system the position and orientation of each camera in space with respect to a reference point, see figure 2. The reference point could be the lens of one of the cameras, or an external point such as a corner of the table. The colors on the checker board pattern and the status bar shown in figure 3 above indicate the progress of the calibration process. Note that the calibration procedure is done only once and stored in the system, so in subsequent sessions this procedure does not need to be repeated, unless the cameras are moved or changed for other models.

3.2 Segmentation

The hand of the patient is localized and segmented in the initial image combining color and motion information. Skin color is a good clue to point potential regions where there is a hand/face of a person. We trained a Bayesian classifier with many (thousand) samples of skin pixels in HSV (hue, saturation, value), which is used to detect skin pixels in the image. Additionally, we use motion information based on image subtraction to detect moving objects in the images, assuming that the patient will be moving his impaired arm. Regions that satisfy both criteria, skin color and motion, are extracted by an intersection operation; this region corresponds to the hand of the person. This segment is used as the initial position of the hand for tracking it in the image sequence, as described in the next section. This procedure is applied to both images, as illustrated in figure 4.

The system can be confused with objects that have a similar color as human skin (i.e wood), so we assume that this does not occur. For this it is recommended that the patient uses long sleeves, and to cover the table and back wall with a uniform cloth in a distinctive color (like black or blue). It is also recommended that the system is used indoors with artificial lighting (white). Under these conditions that system can localize and track the hand quite robustly in real time.

3.3 Tracking

Hand tracking is based on the Camshift algorithm (Bradski, 1998). This algorithm uses only color information to track an object in an image sequence. Based on an initial object window, obtained in the previous stage, Camshift builds a color histogram of the object of interest, in this case the hand. Using a search window (define heuristically according to the size of the initial hand region) and the histogram, Camshift obtains a probability of each pixel in the search region to be part of the object, and the center of the region is the "mean" of this distribution. The distribution is updated in each image, so the algorithm can tolerate small variation in illumination conditions.

In this way, the 2-D position of the hand in each image in the video sequence is obtained, which corresponds to the center point of the color distribution obtained with Camshift. The 3-D coordinates are obtained by combining both views, as described in the next section.

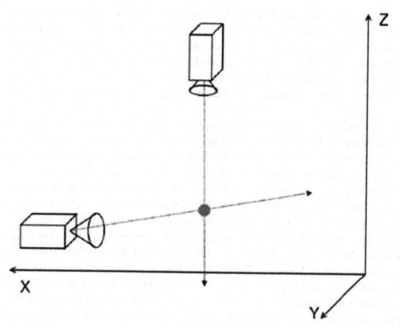

Fig. 5. Estimation of the 3-D position of the hand by intersecting the projection lines of the center points of the hand regions, obtained from both images

3.4   Three Dimensional Reconstruction

Based on the 2-D coordinates of the center point of the image region in each image, the 3-D coordinates are obtained in the following way (Gonzalez 2004). For each image, a line in 3-D space is constructed by connecting the center of the hand region and the center of the camera lens, based on the camera parameters. This is depicted in figure 5. Once the two projection lines are obtained, their intersection provides an estimation of the coordinates in 3-D (X, Y, Z) of the hand of the patient.

Thus, we have the 3-D position of the hand for each processed image pair (about 15 frames per second in a standard PC), which are sent to T-WREX so that the patient can interact with the virtual environments.

4   Pilot Studies

We performed two pilot studies with *Gesture Therapy* at the National Institute for Neurology and Neurosurgery (INNN) in Mexico City. The purpose of these pilot studies was to improve the protocol for a larger clinical trial with Gesture Therapy, to anticipate problems and to gain experience using the technology in a clinical setting.

In the first study, one patient interacted with *Gesture Therapy*. The patient was diagnosed with ischemic stroke, left hemi paresis, with a time of evolution of 4 years. An evaluation with the Fugl-Meyer (Fugl-Meyer, 1975) scale was performed at the start and end of the study.

The patient used Gesture Therapy for 6 sessions, between 20 and 45 minutes each session. The main objective of the exercises was the control of the distal portion of the upper extremity and hand. The patient performed pre exercises for stretching, relaxation, and contraction of the fingers and wrist flexors and extensors. The patient performed several of the simulated exercises in the virtual environment, increasing in difficulty as the sessions progressed (clean stove, clean windows, basketball, paint room, car race).

After the 6 sessions the patient increased his capacity to voluntarily extend and flex the wrist through relaxation of the extensor muscles. He also tried to do bimanual activities (such as take and throw a basket ball) even if he maintained the affected left hand closed; he increased use of the affected extremity to close doors.

In the therapist's opinion: "The GT system favors the movement of the upper extremity by the patient. It makes the patient maintain the control of his extremity even if he does not perceive it. GT maintains the motivation of the patient as he tries to perform the activity better each time (more control in positioning the extremity, more speed to do the task, more precision). This particular patient gained some degree of range of movement of his wrist. There are still many problems with the fingers flexor synergy, but he feels well and motivated with his achievements. It is also important to note the motivation effect the system has on patient endurance to complete the treatment until the last day by increasing the enthusiasm of the patient in executing the variety of rehabilitation exercises."

In the patient's opinion: "At the beginning I felt that my arm was too *heavy*, and at the shoulder I felt as if there was something cutting me, now I feel it less heavy and the cutting sensation has also been reduced."

An "analog visual scale" in the range 1-10 (very bad to excellent) was applied, asking the patient about the treatment based on GT, he gave it a 10. Asked about if he will like to continue using GT, his answer was "YES".

The second pilot study was comparative with a total of 11 patients in two groups; 5 patients used Gesture Therapy in a similar fashion to that of the first pilot study just described above (test group), and 6 patients received traditional occupational therapy (control group). Patients in both groups were evaluated before and after the treatment with the *Fugl-Meyer scale*, which concentrates on the function of the affected arm; the *Motricity Index*, which evaluated motor recovery; and the *Motor Activity Log*, which evaluated upper limb function. In the Motor Activity Log, two scores are given for each activity, one for the amount of use (AOU) and one for the quality of movement (QOM) of the paretic arm.

The two groups showed a significant improvements en all scales ($p<0.005$). However, when comparing the results between them, no significant therapeutic advantage was observed with the use of the Gesture Therapy vs. traditional occupational therapy. There are several hypotheses for this. On the one hand, the size of the study group was small, so it is difficult to reach significant conclusions; it may also be the scales are not sensitive enough to detect a difference. On the other hand, the scales we used were not designed to evaluate motivation or attachment to the computer simulated reality treatment. As in the first study, the patients in the second study that used

Gesture Therapy were motivated by the system and were keen on continue using if for their therapy; more so than patients using conventional therapy. Since motivation can play a significant role in the long term outcome (Woods et al. 2003) we suggest using a scale to estimate motivation for the therapy as part of the clinical evaluation and in comparisons between therapies.

The initial conclusion of this study is that GT is at least as effective as occupational therapy; but more motivating to patients. A more extensive clinical study is required to obtain more significant conclusions.

5 Conclusions and Future Work

These pilot studies show the importance of quantifying patient motivation in rehabilitation. Involving the patient in simulated daily activities helps the psychological rehabilitation component as well. The potential ease of use, motivation promoting characteristics, and objective quantitative potential are obvious advantages to this Gesture Therapy system. The patient can work independently with reduced therapist interaction. With current technology the system can be adapted to a portable low-cost device for the home including communications for remote interaction with a therapist and medical team.

It is possible to extend the system to full arm tracking, including wrist, hand and fingers for more accurate movement measurement. Movement trajectories can be compared and used to add a new metric of patient progress. To make the system easier to use a GUI tool is planned for system parameters configuration, including the camera. Future work includes more games to increase the variety of therapy solutions and adaptability to patient abilities, so that a therapist or patient can match the amount of challenge necessary to keep the rehabilitation advancing.

In the current low-cost, vision-based system the table top serves as an arm support for 2D movement until the patients are strong enough to lift their arms into 3D. Extending the system to wrist, hand, and finger movement is planned to make a full superior extremity rehabilitation system.

Wrist accelerometers can be used to increase the objectivity of clinical studies by adding long term, continuous, quantitative movement data to subjective reports of patients and caregivers; especially when the patient spends less time in the clinic. (Uswatte 2006). fMRI of patients' brains, pre and post training, are planned for increasing our understanding of the biological basis for rehabilitation (Johansen-Berg 2002).

We are currently developing a single camera, 3D monocular version of the system for home use that would not require prior calibration.

Acknowledgements. This work was supported in part by a grant from UC-MEXUS/CONACYT # CN-15-179, and by a grant from the U.S. Department of Education This research is supported by the U.S. Department of Education, National Institute on Disability and Rehabilitation Research (NIDRR) Grant Number H133E070013. (This content does not necessarily represent the policy of the U.S. Department of Education, and one should not assume endorsement by the U.S. Federal Government).

References

1. American Stroke Association (2004) (Retrieved July 10, 2007),
   http://www.strokeassociation.org
2. Bradski, G.R.: Computer vision face tracking as a component of a perceptual user interface. In: Workshop on Applications of Computer Vision, pp. 214–219 (1998)
3. Fugl-Meyer, A.R., Jaasko, L., Leyman, I., Olsson, S., Steglind, S.: The post-stroke hemiplegic patient: a method for evaluation of physical performance. Scand. J. Rehabil. Med. 7, 13–31 (1975)
4. González, P., Cañas, J.: Seguimiento tridimensional usando dos cámaras. Technical Report, Universidad Rey Juan Carlos, Spain (2004) (in Spanish)
5. Johansen-Berg, H., Dawes, H., Guy, C., Smith, S.M., Wade, D.T., Matthews, P.M.: Correlation between motor improvements and altered fMRI activity after rehabilitative therapy. Brain Journal 125, 2731–2742 (2002)
6. Reinkensmeyer, D., Pang, C., Nessler, J., Painter, C.: Web-based telerehabilitation for the upper extremity after stroke. IEEE Trans. Neural Sci. Rehabil. Eng. 10, 1–7 (2000)
7. Reinkensmeyer, D., Housman, S., Le Vu Rahman, T., Sanchez, R.: Arm-Training with T-WREX After Chronic Stroke: Preliminary Results of a Randomized Controlled Trial. In: ICORR 2007, 10th International Conference on Rehabilitation Robotics, Noordwijk (2007)
8. Reinkensmeyer, D., Housman, S.: If I can't do it once, why do it a hundred times?: Connecting volition to movement success in a virtual environment motivates people to exercise the arm after stroke. In: IWVR (2007)
9. Sanchez, R.J., Liu, J., Rao, S., Shah, P., Smith, T., Rahman, T., Cramer, S.C., Bobrow, J.E., Reinkensmeyer, D.: Automating arm movement training following severe stroke: Functional exercise with quantitative feedback in a gravity-reduced environment. IEEE Trans. Neural Sci. Rehabil. Eng. 14(3), 378–389 (2006)
10. Woods, et al.: Motivating, Game-Based Stroke Rehabilitation: A Brief Report. Top. Stroke Rehabil. 10(2), 134–140 (2003)
11. Uswatte, G., Giuliani, C., Winstein, C., Zeringue, A., Hobbs, L., Wolf, S.L.: Validity of accelerometry for monitoring real-world arm activity in patients with sub acute stroke: evidence from the extremity constraint-induced therapy evaluation trial. Arch. Med. Rehabil. 86, 1340–1345 (2006)

Author Index

Communications
in Computer and Information Science

For information about Vols. 1–7

please contact your bookseller or Springer